United States Nuclear Regulatory Commission

Protecting People and the Environment

NUREG-0980
Vol. 1, No. 10

Nuclear Regulatory Legislation

112ᵗʰ Congress; 2ⁿᵈ Session

Office of the General Counsel

AVAILABILITY OF REFERENCE MATERIALS
IN NRC PUBLICATIONS

NRC Reference Material

As of November 1999, you may electronically access NUREG-series publications and other NRC records at NRC's Public Electronic Reading Room at http://www.nrc.gov/reading-rm.html. Publicly released records include, to name a few, NUREG-series publications; *Federal Register* notices; applicant, licensee, and vendor documents and correspondence; NRC correspondence and internal memoranda; bulletins and information notices; inspection and investigative reports; licensee event reports; and Commission papers and their attachments.

NRC publications in the NUREG series, NRC regulations, and Title 10, "Energy," in the *Code of Federal Regulations* may also be purchased from one of these two sources.
1. The Superintendent of Documents
   U.S. Government Printing Office
   Mail Stop SSOP
   Washington, DC 20402–0001
   Internet: bookstore.gpo.gov
   Telephone: 202-512-1800
   Fax: 202-512-2250
2. The National Technical Information Service
   Springfield, VA 22161–0002
   www.ntis.gov
   1–800–553–6847 or, locally, 703–605–6000

A single copy of each NRC draft report for comment is available free, to the extent of supply, upon written request as follows:
Address: U.S. Nuclear Regulatory Commission
         Office of Administration
         Publications Branch
         Washington, DC 20555-0001
E-mail: DISTRIBUTION.RESOURCE@NRC.GOV
Facsimile: 301–415–2289

Some publications in the NUREG series that are posted at NRC's Web site address http://www.nrc.gov/reading-rm/doc-collections/nuregs are updated periodically and may differ from the last printed version. Although references to material found on a Web site bear the date the material was accessed, the material available on the date cited may subsequently be removed from the site.

Non-NRC Reference Material

Documents available from public and special technical libraries include all open literature items, such as books, journal articles, transactions, *Federal Register* notices, Federal and State legislation, and congressional reports. Such documents as theses, dissertations, foreign reports and translations, and non-NRC conference proceedings may be purchased from their sponsoring organization.

Copies of industry codes and standards used in a substantive manner in the NRC regulatory process are maintained at—
   The NRC Technical Library
   Two White Flint North
   11545 Rockville Pike
   Rockville, MD 20852–2738

These standards are available in the library for reference use by the public. Codes and standards are usually copyrighted and may be purchased from the originating organization or, if they are American National Standards, from—
   American National Standards Institute
   11 West 42nd Street
   New York, NY 10036–8002
   www.ansi.org
   212–642–4900

United States Nuclear Regulatory Commission

Protecting People and the Environment

NUREG-0980
Vol. 1, No. 10

Nuclear Regulatory Legislation

112th Congress; 2nd Session

Prepared by:

Office of the General Counsel
U.S. Nuclear Regulatory Commission
Washington, DC 20555-0001

Date Published: September 2013

FOREWORD

This compilation of statutes and materials pertaining to nuclear regulatory legislation through the 112th Congress, 2nd Session, has been prepared by the Office of the General Counsel, U.S. Nuclear Regulatory Commission, with the assistance of staff, for use as an internal resource document. The compilation is not to be used as an authoritative citation in lieu of the primary legislative sources. Furthermore, while every effort has been made to ensure the completeness and accuracy of this material, neither the United States Government, the Nuclear Regulatory Commission, nor any of their employees makes any expressed or implied warranty or assumes liability for the accuracy or completeness of the material presented in this compilation.

If you have any questions concerning this compilation, please contact:

Anthony de Jesús
Legislative Specialist
Office of the General Counsel
U.S. Nuclear Regulatory Commission
Washington, DC 20555-0001
Anthony.deJesus@nrc.gov

Contents

1. Commissioner Tenure

1. Commissioner Tenure

A. ATOMIC ENERGY COMMISSION, 1946–1975

	From	To	Remarks
David E. Lilienthal, Chairman	Nov. 1, 1946	Feb. 15, 1950	Resigned
Robert F. Bacher	Nov. 1, 1946	May 10, 1949	Resigned
Sumner T. Pike	Oct. 31, 1947	Dec. 15, 1951	Resigned
William W. Waymack	Nov. 5, 1946	Dec. 21, 1948	Resigned
Lewis L. Strauss	Nov. 12, 1946	Apr. 15, 1950	Resigned
Chairman	July 2, 1953	June 30, 1958	Term expired
Henry De Wolf Smyth	May 30, 1949	Sept. 30, 1954	Resigned
Gordon Dean	May 24, 1949	June 30, 1953	Term expired
Chairman	July 11, 1950	June 30, 1953	
Thomas E. Murray	May 9, 1950	June 30, 1957	Term expired
Thomas Keith Glennan	Oct. 2, 1950	Nov. 1, 1952	Resigned
Eugene M. Zuckert	Feb. 25, 1952	June 30, 1954	Term expired
Joseph Campbell	July 27, 1953	Nov. 30, 1954	Resigned
Willard F. Libby	Oct. 5, 1954	June 30, 1959	Resigned
John Von Neumann	Mar. 15, 1955	Feb. 8, 1957	Deceased
Harold S. Vance	Oct. 31, 1955	Aug. 31, 1959	Deceased
John S. Graham	Sept. 12, 1957	June 30, 1962	Resigned
John Forrest Floberg	Oct. 1, 1957	June 23, 1960	Resigned
John A. McCone, Chairman	July 14, 1958	Jan. 20, 1961	Resigned
John H. Williams	Aug. 13, 1959	June 30, 1960	Resigned
Robert E. Wilson	Mar. 22, 1960	Jan. 31, 1964	Resigned
Loren K. Olson	June 23, 1960	June 30, 1962	Term expired
Glenn T. Seaborg, Chairman	Mar. 1, 1961	Aug. 16, 1971	Resigned
Leland J. Haworth	Apr. 17, 1961	June 30, 1963	Resigned
John G. Palfrey	Aug. 31, 1962	June 30, 1966	Resigned
James T. Ramey	Aug. 31, 1962	June 30, 1973	Term expired
Gerald F. Tape	July 15, 1963	Apr. 30, 1969	Resigned
Mary I. Bunting	June 29, 1964	June 30, 1965	Term expired
Wilfrid E. Johnson	Aug. 1, 1966	June 30, 1972	Term expired
Samuel M. Nabrit	Aug. 1, 1966	Aug. 1, 1967	Resigned
Francesco Costagliola	Oct. 1, 1968	June 30, 1969	Term expired
Theos J. Thompson	June 12, 1969	Nov. 25, 1970	Deceased
Clarence E. Larson	Sept. 2, 1969	June 30, 1974	Term expired
James R. Schlesinger, Chairman	Aug. 17, 1971	Feb. 5, 1973	Resigned
William O. Doub	Aug. 17, 1971	Aug. 17, 1974	Resigned
Dixy Lee Ray[1]	Aug. 8, 1972	Jan. 19, 1975	AEC abolished
Chairman	Feb. 6, 1973	Jan. 19, 1975	
William E. Kriegsman	June 12, 1973	Dec. 31, 1974	Resigned
William A. Anders	Aug. 6, 1973	Jan. 19, 1975	AEC abolished

[1] Designated Chairman February 6, 1973.

B. U.S. NUCLEAR REGULATORY COMMISSION, 1975–PRESENT

From	From	To	Remarks
William A. Anders, Chairman	Jan. 19, 1975	Apr. 20, 1976	Resigned
Marcus A. Rowden	Jan. 19, 1975	Apr. 20, 1976	
Chairman	Apr. 21, 1976	June 30, 1977	Term Expired
Edward A. Mason	Jan. 19, 1975	Jan. 15, 1977	Resigned
Victor Gilinksy	Jan. 19, 1975	June 30, 1984	Term Expired[2]
Richard T. Kennedy	Jan. 19, 1975	June 30, 1980	Term Expired
Joseph M. Hendrie, Chairman	Aug. 9, 1977	Dec. 7, 1979[3]	
Commissioner	Dec. 8, 1979	Mar. 2, 1981	
Chairman	Mar. 3, 1981[4]	June 30, 1981	Term Expired
Peter A. Bradford	Aug. 15, 1977	Mar. 12, 1982	Resigned
John F. Ahearne	July 31, 1978	Dec. 7, 1979	
Chairman	Dec. 7, 1979[5]	Mar. 2, 1981	
Commissioner	Mar. 3, 1981[6]	June 30, 1983	Term Expired
Nunzio J. Palladino, Chairman	July 1, 1981	June 30, 1986	Term Expired
Thomas M. Roberts	Aug. 3, 1981	June 30, 1990[7]	Term Expired
James K. Asselstine	May 17, 1982	June 30, 1987[8]	Term Expired
Frederick M. Bernthal	Aug. 4, 1983	June 30, 1988	Term Expired
Lando W. Zech, Jr.	July 5, 1984	June 30, 1986[9]	
Chairman	July 1, 1986	June 30, 1989	Term Expired
Kenneth M. Carr	Aug. 14, 1986	June 30, 1989	
Chairman	July 1, 1989	June 30, 1991	Term Expired

---

[2] Victor Gilinsky served two terms. The Senate reconfirmed his nomination for a term of 5 years on June 27, 1979.

[3] On December 7, 1979, Joseph M. Hendrie vacated the chairmanship but remained as a Commissioner.

[4] On March 3, 1981, Joseph M. Hendrie resumed the chairmanship.

[5] On December 7, 1979, John F. Ahearne assumed the chairmanship.

[6] On March 3, 1981, John F. Ahearne vacated the chairmanship but remained as a Commissioner.

[7] Thomas M. Roberts' first term expired on June 30, 1985; he took the oath of office for a second term on July 12, 1985.

[8] James K. Asselstine completed Peter A. Bradford's term and was appointed to a full 5–year term.

[9] On June 28, 1984, Lando W. Zech, Jr., was nominated by the President. He received a recess appointment on July 3, 1984, and took office on July 5, 1984. On January 3, 1985, the President resubmitted the nomination to the 99th Congress for a full 5–year appointment. The Senate subsequently confirmed the nomination, and he took office for the full 5–year term on March 6, 1985. On July 1, 1986, Mr. Zech assumed the chairmanship.

B. U.S. NUCLEAR REGULATORY COMMISSION, 1975–PRESENT

From	From	To	Remarks
Kenneth C. Rogers	Aug. 7, 1987	June 30, 1997[10]	Term Expired
James R. Curtiss	Oct. 20, 1988	June 30, 1993	Term Expired
Forrest J. Remick	Dec. 1, 1989	June 30, 1994	Term Expired
Ivan Selin, Chairman	July 1, 1991	June 30, 1995[11]	Resigned
E. Gail de Planque	Dec. 16, 1991	June 30, 1995	Term Expired
Shirley A. Jackson	May 2, 1995	June 30, 1995	
Chairman	July 1, 1995	June 30, 1999	Term Expired
Greta J. Dicus	February 15, 1996	June 30, 1998	Term Expired
	October 27, 1998	June 30, 1999[12]	
Chairman	July 1, 1999	October 29, 1999[13]	
Commissioner	October 29, 1999[14]	June 30, 2003	Term Expired
Nils J. Diaz	August 23, 1996	June 30, 2001	Term Expired
Nils J. Diaz	October 4, 2001	March 31, 2003	Term Expired
Chairman	April 1, 2003[15]	June 30, 2006	Term Expired
Edward McGaffigan, Jr.	August 28, 1996	June 30, 2000	Term Expired
Edward McGaffigan, Jr.	July 1, 2000	June 30, 2005	Term Expired
Edward McGaffigan, Jr.	October 12, 2005[16]	September 2, 2007	Deceased
Jeffrey S. Merrifield	October 23, 1998	June 30, 2002	Term Expired
Jeffrey S. Merrifield	August 5, 2002	June 30, 2007	Term Expired
Richard A. Meserve, Chairman	October 29, 1999	March 31, 2003[17]	Resigned

---

[10] Kenneth C. Rogers served as Commissioner from August 7, 1987, to June 30, 1992, and was reappointed as Commissioner from July 1, 1992, to June 30, 1997.

[11] Ivan Selin resigned on June 30, 1995.

[12] Greta J. Dicus was renominated to a new 5–year term as Commissioner on May 22, 1998. Her nomination was confirmed by the Senate on October 21, 1998.

[13] Greta J. Dicus assumed the chairmanship on July 1, 1999.

[14] On October 29, 1999, Greta J. Dicus vacated the chairmanship but remained as a Commissioner.

[15] On April 1, 2003, Nils J. Diaz assumed the chairmanship.

[16] The office was vacant from July 1, 2005, through October 11, 2005. Commissioner Edward McGaffigan, Jr., was nominated on July 28, 2005, and confirmed by the Senate on October 7, 2005.

[17] On March 31, 2003, Richard A. Meserve resigned.

B. U.S. NUCLEAR REGULATORY COMMISSION, 1975–PRESENT

From	From	To	Remarks
Gregory B. Jaczko	January 21, 2005	Recess Appointment[18]	
Gregory B. Jaczko	May 31, 2006	June 30, 2008	Term Expired
Gregory B. Jaczko	July 1, 2008	May 13, 2009	
Chairman	May 13, 2009	June 29, 2012[19]	Resigned
Peter B. Lyons	January 25, 2005	Recess Appointment[20]	
Peter B. Lyons	May 31, 2006	June 30, 2009	Term Expired
Dale E. Klein, Chairman	July 1, 2006	May 13, 2009	
Commissioner	May 13, 2009[21]	March 29, 2010	Resigned
Kristine L. Svinicki	March 28, 2008	June 30, 2012	Term Expired
Kristine L. Svinicki	July 1, 2012	June 30, 2017	Term Expires
William D. Magwood	April 1, 2010	June 30, 2015	Term Expires
William C. Ostendorff	April 1, 2010	June 30, 2011	Term Expired
William C. Ostendorff	July 1, 2011	June 30, 2016	Term Expires
George Apostolakis	April 23, 2010	June 30, 2014	Term Expires
Allison M. Macfarlane, Chairman	July 9, 2012	June 30, 2013	Term Expires

---

[18] On January 19, 2005, President Bush appointed Gregory B. Jaczko during the Senate recess (effective to the end of the Senate's session in 2006).

[19] Gregory B. Jaczko assumed the chairmanship on May 13, 2009. Chairman Jaczko's resignation was effective upon Senate confirmation of Allison M. Macfarlane as his successor.

[20] On January 19, 2005, President Bush appointed Peter B. Lyons during the Senate recess (effective to the end of Senate's session in 2006).

[21] On May 13, 2009, Dale E. Klein vacated the chairmanship but remained as a Commissioner. On March 29, 2010, Dale E. Klein resigned his position as Commissioner.

2. Atomic Energy Act of 1954, as Amended

2. Atomic Energy Act of 1954, as Amended
Contents

A. THE ATOMIC ENERGY ACT OF 1954, AS AMENDED

Public Law 83–703   **68 Stat. 919**

August 30, 1954

Title I–Atomic Energy

Chapter 1–Declaration, Findings, and Purpose

42 USC 2011.
Declaration.

Sec. 1. Declaration

Atomic energy is capable of application for peaceful as well as military purposes. It is therefore declared to be the policy of the United States that[1]–

a. the development, use, and control of atomic energy shall be directed so as to make the maximum contribution to the general welfare, subject at all times to the paramount objective of making the maximum contribution to the common defense and security; and

b. the development, use, and control of atomic energy shall be directed so as to promote world peace, improve the general welfare, increase the standard of living, and strengthen free competition in private enterprise.

42 USC 2012.
Findings.

Sec. 2. Findings

The Congress of the United States hereby makes the following findings concerning the development, use and control of atomic energy:[2]

a. The development, utilization, and control of atomic energy for military and for all other purposes are vital to the common defense and security. –

c.[3] The processing and utilization of source, byproduct, and special nuclear material affect interstate and foreign commerce and must be regulated in the national interest.

d. The processing and utilization of source, byproduct, and special nuclear material must be regulated in the national interest and in order to provide for the common defense and security and to protect the health and safety of the public.

e. Source and special nuclear material, production facilities, and utilization facilities are affected with the public interest, and regulation by the United States of the production and utilization of atomic energy and of the facilities used in connection therewith is necessary in the

---

[1]   Added by P.L. 102–486, 106 Stat. 2943 (1992).

[2]   PL. 88–489, § 20, 78 Stat. 602 (1964), the Private Ownership of Special Nuclear
       Materials Act, reads as follows:
       Nothing in this Act shall be deemed to diminish existing authority of the
       United States, or of the Atomic Energy Commission under the Atomic Energy
       Act of 1954, as amended to regulate source, byproduct, and special nuclear
       material and production and utilization facilities or to control such materials
       and facilities exported from the United States by imposition of governmental
       guarantees and security safeguards with respect thereto, in order to assure the
       common defense and security and to protect the health and safety of the
       public, or to reduce the responsibility of the Atomic Energy Commission to
       achieve such objectives.

[3]   Amended by P.L. 88–489, § 1, 78 Stat. 602 (1964), which deleted subsection 2b.
       Subsection 2b read as follows:
       b. In permitting the property of the United States to be used by others such use
       must be regulated in the national interest and in order to provide for the
       common defense and security and to protect the health and safety of the
       public.

national interest to assure the common defense and security and to protect the health and safety of the public.

f. The necessity for protection against possible interstate damage occurring from the operation of facilities for production or utilization of source or special nuclear material places the operation of those facilities in interstate commerce for the purposes of this Act.

g. Funds of the United States may be provided for the development and use of atomic energy under conditions which will provide for the common defense and security and promote the general welfare.

i.[4] In order to protect the public and to encourage the development of the atomic energy industry, in the interest of the general welfare and of the common defense and security, the United States may make funds available for a portion of the damages suffered by the public from nuclear incidents, and may limit the liability of those persons liable for such losses.[5]

42 USC 2013.
Purpose.

Sec. 3. Purpose

It is the purpose of this Act to effectuate the policies set forth above by providing for–

a. A program of conducting, assisting, and fostering research and development in order to encourage maximum scientific and industrial progress;

b. A program for the dissemination of unclassified scientific and technical information and for the control, dissemination, and declassification of Restricted Data, subject to appropriate safeguards, so as to encourage scientific and industrial progress;

c. A program for Government control of the possession, use, and production of atomic energy and special nuclear material, whether owned by the Government or others, so directed as to make the maximum contribution to the common defense and security and the national welfare, and to provide continued assurance of the Government's ability to enter into and enforce agreements with nations or groups of nations for the control of special nuclear materials and atomic weapons.[6]

d. A program to encourage widespread participation in the development and utilization of atomic energy for peaceful purposes to the maximum extent consistent with the common defense and security and with the health and safety of the public;

e. A program of international cooperation to promote the common defense and security and to make available to cooperating nations the benefits of peaceful applications of atomic energy as widely as expanding technology and considerations of the common defense and security will permit; and

f. A program of administration which will be consistent with the foregoing policies and programs, with international arrangements, and with agreements for cooperation, which will enable the Congress to be

---

[4]   Amended by P.L. 88–489, § 2, 78 Stat. 602 (1964), which deleted subsection 2h. Subsection 2h read as follows:

   h. It is essential to the common defense and security that title to all special nuclear material be in the United States while such special nuclear material is within the United States.

[5]   Added by P.L. 85–256, § 1, 71 Stat. 576 (1957).

[6]   Amended by P.L. 88–489, § 3, 78 Stat. 602 (1964). Before amendment, it read as follows:

   c. A program for Government control of the possession, use, and production of atomic energy and special nuclear material so directed as to make the maximum contribution to the common defense and security and the national welfare;

currently informed so as to take further legislative action as may be appropriate.

Chapter 2–Definitions

42 USC 2014.
Definitions.

Sec. 11. Definitions

The intent of Congress in the definitions as given in this section should be construed from the words or phrases used in the definitions. As used in this Act:

Agency of the U.S.

a. The term "agency of the United States" means the executive branch of the United States, or any Government agency, or the legislative branch of the United States, or any agency, committee, commission, office, or other establishment in the legislative branch, or the judicial branch of the United States, or any office, agency, committee, commission, or other establishment in the judicial branch.

Agreement for cooperation.

b. The term "agreement for cooperation" means any agreement with another nation or regional defense organization authorized or permitted by sections 54, 57, 64, 82, 91c., 103, 104, or 144, and made pursuant to section 123.[7]

Atomic energy.

c. The term "atomic energy" means all forms of energy released in the course of nuclear fission or nuclear transformation.

Atomic weapon.

d. The term "atomic weapon" means any device utilizing atomic energy, exclusive of the means for transporting or propelling the device (where such means is a separable and divisible part of the device), the principal purpose of which is for use as, or for development of, a weapon, a weapon prototype, or a weapon test device.

Byproduct material.

e. The term "byproduct material" means–

(1) any radioactive material (except special nuclear material) yielded in or made radioactive by exposure to the radiation incident to the process of producing or utilizing special nuclear material;

(2) the tailings or wastes produced by the extraction or concentration of uranium or thorium from any ore processed primarily for its source material content;

(3)(A) any discrete source of radium–226 that is produced, extracted, or converted after extraction, before, on, or after August 8, 2005, for use for a commercial, medical, or research activity; or

(B) any material that–

(i) has been made radioactive by use of a particle accelerator; and

(ii) is produced, extracted, or converted after extraction, before, on, or after the date of enactment of this paragraph for use for a commercial, medical, or research activity; and

(4) any discrete source of naturally occurring radioactive material, other than source material, that –

(A) the Commission, in consultation with the Administrator of the Environmental Protection Agency, the Secretary of Energy, the Secretary of Homeland Security, and the head of any other appropriate Federal agency, determines would pose a threat similar to the threat posed by a discrete source of radium–226 to the public health and safety or the common defense and security; and

---

[7]   Amended by P.L. 87–206, § 2, 75 Stat. 475 (1961).

(B) before, on, or after August 8, 2005 is extracted or converted after extraction for use in a commercial, medical, or research activity.[8]

---

[8]   Amended by P.L. 109–58, § 651(e)(1), 119 Stat. 806 (2005), which revised subsection (e) of § 11 ("Definitions") of the Atomic Energy Act of 1954 by adding new paragraphs (3) and (4) to the definition of byproduct material. The former subsection (e) read as follows:

(e) The term 'byproduct material' means (1) any radioactive material (except special nuclear material) yielded in or made radioactive by exposure to the radiation incident to the process of producing or utilizing special nuclear material, and (2) the tailings or wastes produced by the extraction or concentration of uranium or thorium from any ore processed primarily for its source material content.

NOTE: P.L. 109–58 (119 Stat. 808), section 651(e)(4) and (5), added the following relating to the material added to the Atomic Energy Act by paragraphs (3) and (4) of section 11e. of the Act:

(4) FINAL REGULATIONS.–
    (A) REGULATIONS.–
Deadline.   (i) IN GENERAL.–Not later than 18 months after the date of enactment of this Act, the Commission, after consultation with States and other stakeholders, shall issue final regulations establishing such requirements as the Commission determines to be necessary to carry out this section and the amendments made by this section.
    (ii) INCLUSIONS.–The regulations shall include a definition of the term "discrete source " for purposes of paragraphs (3) and (4) of section 11e. of the Atomic Energy Act of 1954 (42 U.S.C. 2014(e)) (as amended by paragraph (1)).
    (B) COOPERATION.–In promulgating regulations under paragraph (1), the Commission shall, to the maximum extent practicable–
    (i) cooperate with States; and
    (ii) use model State standards in existence on the date of enactment of this Act.
    (C) TRANSITION PLAN.–
    (i) DEFINITION OF BYPRODUCT MATERIAL.–In this paragraph, the term 'byproduct material' has the meaning given the term in paragraphs (3) and (4) of section 11e. of the Atomic Energy Act of 1954 (42 U.S.C. 2014(e)) (as amended by paragraph (1)).
    (ii) PREPARATION AND PUBLICATION.–To facilitate an orderly transition of regulatory authority with respect to byproduct material, the Commission, in issuing regulations under subparagraph (A), shall prepare and publish a transition plan for–
    (I) States that have not, before the date on which the plan is published, entered into an agreement with the Commission under section 274b. of the Atomic Energy Act of 1954 (42 U.S.C. 2021(b)); and
    (II) States that have entered into an agreement with the Commission under that section before the date on which the plan is published.
    (iii) INCLUSIONS.—The transition plan under clause (ii) shall include–
    (I) a description of the conditions under which a State may exercise authority over byproduct material; and
    (II) a statement of the Commission that any agreement covering byproduct material, as defined in paragraph (1) or (2) of section 11e. of the Atomic Energy Act of 1954 (42 U.S.C. 2014(e)), entered into between the Commission and a State under section 274b. of that Act (42 U.S.C. 2021(b)) before the date of publication of the transition plan shall be considered to include byproduct material, as defined in paragraph (3) or (4) of section 11e. of that Act (42 U.S.C. 2014(e)) (as amended by paragraph (1)), if the Governor of the State certifies to the Commission on the date of publication of the transition plan that–
    (aa) the State has a program for licensing byproduct material, as defined in paragraph (3) or (4) of section 11e. of the Atomic Energy Act of 1954, that is adequate to protect the public health and safety, as determined by the Commission; and
    (bb) the State intends to continue to implement the regulatory responsibility of the State with respect to the byproduct material.

f. The term "Commission" means the Atomic Energy Commission.

g. The term "common defense and security" means the common defense and security of the United States.

h. The term "defense information" means any information in any category determined by any Government agency authorized to classify information, as being information respecting, relating to, or affecting the national defense.

i. The term "design" means (1) specifications, plans drawings, blueprints, and other items of like nature; (2) the information contained therein; or (3) the research and development data pertinent to the information contained therein.

j. The term "extraordinary nuclear occurrence" means any event causing a discharge or dispersal of source, special nuclear, or byproduct material from its intended place of confinement in amounts off-site, or causing radiation levels off-site, which the Nuclear Regulatory Commission or the Secretary of Energy, as appropriate, determines to be substantial, and which the Nuclear Regulatory Commission or the Secretary of Energy, as appropriate determines has resulted or will probably result in substantial damages to persons off-site or property off-site. Any determination by the Nuclear Regulatory Commission or the Secretary of Energy, as appropriate, that such an event has, or has not, occurred shall be final and conclusive, and no other official or any court shall have power or jurisdiction to review any such determination. The Nuclear Regulatory Commission or the Secretary of Energy, as appropriate, shall establish criteria in writing setting forth the basis upon which such determination shall be made. As used in this subsection, "off-

---

(D) AVAILABILITY OF RADIOPHARMACEUTICALS.–In promulgating regulations under subparagraph (A), the Commission shall consider the impact on the availability of radiopharmaceuticals to–
(i) physicians; and
(ii) patients the medical treatment of which relies on radiopharmaceuticals.
(5) WAIVERS.–
(A) IN GENERAL.–Except as provided in subparagraph (B), the Commission may grant a waiver to any entity of any requirement under this section or an amendment made by this section with respect to a matter relating to byproduct material (as defined in paragraphs (3) and (4) of section 11e. of the Atomic Energy Act of 1954 (42 U.S.C. 2014(e)) (as amended by paragraph (1)) if the Commission determines that the waiver is in accordance with the protection of the public health and safety and the promotion of the common defense and security.–
(B) EXCEPTIONS.–
(i) IN GENERAL.–The Commission may not grant a waiver under subparagraph (A) with respect to–
(I) any requirement under the amendments made by subsection (c)(1);
(II) a matter relating to an importation into, or exportation from, the United States for a period ending after the date that is 1 year after the date of enactment of this Act; or
(III) any other matter for a period ending after the date that is 4 years after the date of enactment of this Act.
(ii) WAIVERS TO STATES.–The Commission shall terminate any waiver granted to a State under subparagraph (A) if the Commission determines that–
(I) the State has entered into an agreement with the Commission under section 274b. of the Atomic Energy Act of 1954 (42 U.S.C. 2021(b));
(II) the agreement described in subclause (I) covers byproduct material (as described in paragraph (3) or (4) of section 11e. of the Atomic Energy Act of 1954 (42 U.S.C. 2014(e)) (as amended by paragraph (1)); and
(III) the program of the State for licensing such byproduct material is adequate to protect the public health and safety.
(C) PUBLICATION–The Commission shall publish in the *Federal Register* a notice of any waiver granted under this subsection.

site" means away from "the location" or "the contract location" as defined in the applicable Nuclear Regulatory Commission or the Secretary of Energy, as appropriate, indemnity agreement, entered into pursuant to section 170.[9]

Financial protection.

k. The term "financial protection" means the ability to respond in damages for public liability and to meet the costs of investigating and defending claims and settling suits for such damages.[10]

Government agency.

l. The term "Government agency" means any executive department, commission, independent establishment, corporation, wholly or partly owned by the United States of America which is an instrumentality of the United States, or any board, bureau, division, service, office, officer, authority, administration, or other establishment in the executive branch of the Government.

Indemnitor.

m. The term "indemnitor" means (1) any insurer with respect to his obligations under a policy of insurance furnished as proof of financial protection; (2) any licensee, contractor or other person who is obligated under any other form of financial protection, with respect to such obligations; and (3) the Nuclear Regulatory Commission or the Secretary of Energy, as appropriate, with respect to any obligation undertaken by it in an indemnity agreement entered into pursuant to section 170.[11]

International arrangement.

n. The term "international arrangement" means any international agreement hereafter approved by the Congress or any treaty during the time such agreement or treaty is in full force and effect, but does not include any agreement for cooperation.

Congressional Committee.

o. Energy Committees' means the Committee on Energy and Natural Resources of the Senate and the Committee on Energy and Commerce of the House of Representatives.[12]

Licensed activity.
Nuclear incident.

p. The term "licensed activity" means an activity licensed pursuant to this Act and covered by the provisions of section 170a.[13]

q. The term "nuclear incident" means any occurrence, including an extraordinary nuclear occurrence,[14] within the United States causing, within or outside the United States, bodily injury, sickness, disease, or death, or loss of or damage to property, or loss of use of property, arising out of or resulting from the radioactive, toxic, explosive, or other hazardous properties of source, special nuclear, or byproduct material: *Provided, however,* That as the term is used in section 170 l., it shall include any such occurrence outside of the United States: *and provided further,* That as the term is used in section 170d., it shall include any such occurrence outside the United States if such occurrence involves source, special nuclear, or byproduct material owned by, and used by or under contract with, the United States: *and provided further,* That as the term is used in section 170c., it shall include any such occurrence outside both the United States and any other nation if such occurrence arises out of or results from the radioactive, toxic, explosive, or other hazardous properties of source, special nuclear, or byproduct material licensed pursuant to Chapters 6, 7, 8, and 10 of this Act, which is used in connection with the operation of a licensed stationary production utilization facility or which moves outside the territorial limits of the

42 USC 2091.
42 USC 2111.
42 USC 2121.
42 USC 2151.

[9]   Added by P.L. 89–645, § 1, 80 Stat. 891 (1966).
[10]   Added by P.L. 85–256, § 3, 71 Stat. 576 (1957).
[11]   Added by P.L. 89–645, § 1, 80 Stat. 891 (1966).
[12]   Added by P.L. 103–437, § 15(f)(1), 108 Stat. 4581 (1994) to reflect the current names of congressional committees.
[13]   Added by P.L. 85–256, § 3, 71 Stat. 576 (1957).
[14]   Amended by P.L. 89–645, § 1, 80 Stat. 891 (1966).

United States in transit from one person licensed by the Nuclear Regulatory Commission to another person licensed by the Nuclear Regulatory Commission.[15]

Operator.

r. The term "operator" means any individual who manipulates the controls of a utilization or production facility.

Person.

s. The term "person" means (1) any individual, corporation, partnership, firm, association, trust, estate, public or private institution, group, Government agency other than the Commission, any State or any political subdivision of, or any political entity within a State, any foreign government or nation or any political subdivision of any such government or nation, or other entity; and (2) any legal successor, representative, agent, or agency of the foregoing.

Person indemnified.

t. The term "person indemnified" means (1) with respect to a nuclear incident occurring within the United States or outside the United States as the term is used in section 170c., and with respect to any nuclear incident in connection with the design, development, construction, operation, repair, maintenance, or use of the nuclear ship Savannah, the person with whom an indemnity agreement is executed or who is required to maintain financial protection, and any other person who may be liable for public liability or (2) with respect to any other nuclear incident occurring outside the United States, the person with whom an indemnity agreement is executed and any other person who may be liable for public liability by reason of his activities under any contract with the Secretary of Energy or any project to which indemnification under the provisions of section 170d. has been extended or under any subcontract, purchase order or other agreement, of any tier, under any such contract or project.[16]

u. The term "produce", when used in relation to special nuclear material, means (1) to manufacture, make, produce, or refine special nuclear material; (2) to separate special nuclear material from other substances in which such material may be contained; or (3) to make or to produce new special nuclear material.

---

[15]   Added by P.L. 85–256, § 3, 71 Stat. 576 (1957). Before amendment by P.L. 89–645, 80 Stat. 891 (1966) (see footnote 9 above), the subsection was amended by P.L. 87–615, § 4, 76 Stat. 409 (1962). Before amendment, it read as follows:

> o. The term "nuclear incident" means any occurrence within the United States causing bodily injury, sickness, disease, or death, or loss of or damage to property, or for loss of use of property, arising out of or resulting from the radioactive, toxic, explosive, or other hazardous properties of source, special nuclear, or byproduct material; *Provided however*, That as the term is used in subsection 170.1., it shall mean any such occurrence outside of the United States rather than within the United States.

P.L. 84–197 (89 Stat. 1111) (1975), section 1, amended the second proviso in subsection 11q. Before amendment, the proviso read as follows:

> And provided further, That as the term is used in section 170d., it shall include any such occurrence outside of the United States if such occurrence involves a facility or device owned by, and used by or under contract with, the United States.

[16]   Added by P.L. 85–256, § 3, 71 Stat. 576 (1957). P.L. 87–615, § 5, 76 Stat. 409 (1962), amended the subsection. Before amendment, it read as follows:

> r. The term "person indemnified" means the person with whom an indemnity agreement is executed and any other person who may be liable for public liability.

P.L. 94–197 (89 Stat. 1111) (1975), section 1, amended subsection 11t. as follows:

> By adding the phrases "or outside the United States as the term is used in subsection 170c." and "or who is required to maintain financial protection." to the definition of the term person "indemnified."

Production
facility.

v.[17] The term "production facility" means (1) any equipment or device determined by rule of the Commission to be capable of the production of special nuclear material in such quantity as to be of significance to the common defense and security, or in such manner as to affect the health and safety of the public; or (2) any important component part especially designed for such equipment or device as determined by the Commission. Except with respect to the export of a uranium enrichment production facility,[18] [19] such term as used in Chapters 10 and 16 shall not include any equipment or device (or important component part especially designed for such equipment or device) capable of separating the isotopes of uranium or enriching uranium in the isotope 235.[20]

Public liability.

w. The term "public liability"[21] means any legal liability arising out of or resulting from a nuclear incident or precautionary evacuation (including all reasonable additional costs incurred by a State, or political subdivision of a State, in the course of responding to a nuclear incident or precautionary evacuation) except: (i) claims under State or Federal workmen's compensation acts of employees or persons indemnified who are employed at the site of and in connection with the activity where the nuclear incident occurs; (ii) claims arising out of an act of war; and (iii) whenever used in subsections a., c., and k., of section 170, claims for loss of, or damage to property which is located at the site of and used in connection with licensed activity where the nuclear incident occurs. "Public liability" also includes damage to property of persons indemnified: *Provided,* That such property is covered under the terms of the financial protection required, except property which is located at the site of and used in connection with the activity where the nuclear incident occurs.

Research and
development.

x. The term "research and development" means (1) theoretical analysis, exploration, or experimentation; or (2) the extension of investigative findings and theories of a scientific or technical nature into practical application for experimental and demonstration purposes, including the experimental production and testing of models, devices, equipment, materials, and processes.

Restricted
Data.

y. The term "restricted data" means all data concerning (1) design, manufacture, or utilization of atomic weapons; (2) the production of special nuclear material; or (3) the use of special nuclear material in the

Source
material.

---

[17]   Amended by P.L. 101–575, § 5(a), 104 Stat. 2834 (1990).
[18]   Amended by P.L. 102–486, 106 Stat. 2955 (1992). Before amendment, the last sentence read as follows:
Except with respect to the export or a uranium enrichment production facility, such term as used in Chapters 10 and 16 shall not include any equipment or device (or important component part especially designed for such equipment or device) capable of separating the isotopes of uranium or enriching uranium in the isotope 235.
[19]   Amended by P.L. 104–134, 110 Stat. 1321–349 (1996).
[20]   Amended by P.L. 104–134, Title III, § 3116(b)(1), 110 Stat. 1321–349 (1996).
[21]   Added by P.L. 85–256, § 3, 71 Stat. 576 (1957). P.L. 87–206, § 3, 75 Stat. 475 (1961), amended the subsection. Before amendment, it read as follows:
u. The term "public liability" means any legal liability arising out of or resulting from a nuclear incident, except claims under State or Federal Workmen's Compensation Acts of employees or persons indemnified who are employed at the site of and in connection with the activity where the nuclear incident occurs, and except for claims arising out of an act of war. "Public liability" also included damage to property of persons indemnified: *Provided,* That such property is covered under the terms of the financial protection required, except property which is located at the site of and used in connection with the activity where the nuclear incident occurs.

production of energy, but shall not include data declassified or removed from the Restricted Data category pursuant to section 142.

z. The term "source material" means (1) uranium, thorium, or any other material which is determined by the Commission pursuant to the provisions of section 61 to be source material; or (2) ores containing one or more of the foregoing materials, in such concentration as the Commission may by regulation determine from time to time.

Special nuclear material.

aa. The term "special nuclear material" means (1) plutonium, uranium enriched in the isotope 233 or in the isotope 235, and any other material which the Commission, pursuant to the provisions of section 51, determines to be special nuclear material, but does not include source material; or (2) any material artificially enriched by any of the foregoing, but does not include source material.

United States.

bb. The term "United States" when used in a geographical sense includes all territories and possessions of the United States, the Canal Zone and Puerto Rico.[22]

Utilization facility.

cc. The term "utilization facility" means (1) any equipment or device, except an atomic weapon, determined by rule of the Commission to be capable of making use of special nuclear material in such quantity as to be of significance to the common defense and security, or in such manner as to affect the health and safety of the public, or peculiarly adapted for making use of atomic energy in such quantity as to be of significance to the common defense and security, or in such manner as to affect the health and safety of the public; or (2) any important component part especially designed for such equipment or device as determined by the Commission.

dd.[23] The terms "high-level radioactive waste" and "spent nuclear fuel" have the meanings given such terms in section 2 of the Nuclear Waste Policy Act of 1982 (42 USC 10101).

ee. The term "transuranic waste" means material contaminated with elements that have an atomic number greater than 92, including neptunium, plutonium, americium, and curium, and that are in concentrations greater than 10 nano-curies per gram, or in such other concentrations as the Nuclear Regulatory Commission may prescribe to protect the public health and safety.

ff. The term "nuclear waste activities", as used in section 170, means activities subject to an agreement of indemnification under subsection d. of such section, that the Secretary of Energy is authorized to undertake, under this Act or any other law, involving the storage, handling, transportation, treatment, or disposal of, or research and development on, spent nuclear fuel, high-level radioactive waste, or transuranic waste, including (but not limited to) activities authorized to be carried out under the Waste Isolation Pilot Project under section 213 of Public Law 96-164 (93 Stat. 1265).

gg. The term "precautionary evacuation" means an evacuation of the public within a specified area near a nuclear facility, or the transportation route in the case of an accident involving transportation of source material, special nuclear material, byproduct material, high-level

---

[22]   Amended by P.L. 84–1006, § 1, 70 Stat. 1069 (1956). Before amendment, it read as follows:
"u. The term "United States" when used in a geographical sense, includes all territories and possessions of the United States, and the Canal Zone."

[23]   Amended by P.L. 100–408, 102 Stat. 1066 (1988), which added subsections dd-jj.

radioactive waste, spent nuclear fuel, or transuranic waste to or from a production or utilization facility, if the evacuation is–

(1) the result of any event that is not classified as a nuclear incident but that poses imminent danger of bodily injury or property damage from the radiological properties of source material, special nuclear material, byproduct material, high-level radioactive waste, spent nuclear fuel, or transuranic waste, and causes an evacuation; and

(2) initiated by an official of a State or a political subdivision of a State, who is authorized by State law to initiate such an evacuation and who reasonably determined that such an evacuation was necessary to protect the public health and safety.

hh. The term "public liability action", as used in section 170, means any suit asserting public liability. A public liability action shall be deemed to be an action arising under section 170, and the substantive rules for decision in such action shall be derived from the law of the State in which the nuclear incident involved occurs, unless such law is inconsistent with the provisions of such section.

jj. Legal Costs.–As used in section 170, the term "legal costs" means the costs incurred by a plaintiff or a defendant in initiating, prosecuting, investigating, settling, or defending claims or suits for damages arising under such section.

Chapter 3–Organization

42 USC 2033.
Office.

Sec. 23. Office

The principal office of the Commission shall be in or near the District of Columbia, but the Commission or any duly authorized representative may exercise any or all of its powers in any place; however, the Commission shall maintain an office for the service of processing papers within the District of Columbia.[24]

---

[24]   Amended by P.L. 93–438, § 104(a), 88 Stat. 1233 (1974), repealed section 21 and section 22. Before repeal, section 21 read as follows:
Sec. 21. Atomic Energy Commission.–There is hereby established an Atomic Energy Commission, which shall be composed of five members, each of whom shall be a citizen of the United States. The President shall designate one member of the Commission as Chairman thereof to serve as such during the pleasure of the President. The Chairman may from time to time designate any other member of the Commission as Acting Chairman to act in the place and stead of the Chairman during his absence. The Chairman (or the Acting Chairman in the absence of the Chairman) shall preside at all meetings of the Commission and a quorum for the transaction of business shall consist of at least three members present. Each member of the Commission, including the Chairman, shall have equal responsibility and authority in all decisions and actions of the Commission, shall have full access to all information relating to the performance of his duties or responsibilities, and shall have one vote.
    Action of the Commission shall be determined by a majority vote of the members present. The Chairman (or Acting Chairman in the absence of the Chairman) shall be the official spokesman of the Commission in its relations with the Congress. Government agencies, persons or the public, and on behalf of the Commission, shall see to the faithful execution of the policies and decisions of the Commission, and shall report thereon to the Commission from time to time or as the Commission may direct. The Commission shall have an official seal which shall be judicially noticed.
P.L. 84–337, § 3, 69 Stat. 630 (1955), had previously amended the fifth sentence of § 21. Before amendment, this sentence read as follows:
    Each member of the Commission, including the Chairman, shall have equal responsibility and authority in all decisions and actions of the Commission and shall have one vote.

42 USC 2034.
General
Manager,
Deputy and
Assistant
General
Managers.

Sec. 24. General Manager, Deputy and Assistant General Managers

There is hereby established within the Commission:[25]

a. A General Manager, who shall be the chief executive officer of the Commission, and who shall discharge such of the administrative and executive functions of the Commission as the Commission may direct. The General Manager shall be appointed by the Commission, shall serve at the pleasure of the Commission, and shall be removable by the Commission.[26]

b. A Deputy General Manager, who shall act in the stead of the General Manager during his absence when so directed by the General Manager, and who shall perform such other administrative and executive functions as the General Manager shall direct. The Deputy General Manager shall be appointed by the General Manager with the approval of the Commission, shall serve at the pleasure of the General Manager, and shall be removable by the General Manager.[27]

c. Assistant General Managers, or their equivalents (not to exceed a total of three positions), who shall perform such administrative and executive functions as the General Manager shall direct. They shall be appointed by the General Manager with the approval of the Commission, shall serve at the pleasure of the General Manager, and shall be removable by the General Manager.[28]

Prior to repeal, § 22 read as follows:
        Sec. 22. Members.—
    a. Members of the Commission shall be appointed by the President, by and
    with the advice and consent of the Senate. In submitting any nomination to the
    Senate, the President shall set forth the experience and qualifications of the
    nominee. The term of office of each member of the Commission taking office
    after June 30, 1950, shall be five years, except that (1) the terms of office of
    the members first taking office after June 30, 1950, shall expire, as designated
    by the President at the time of the appointment, one at the end of one year, one
    at the end of two years, one at the end of three years, one at the end of four
    years, and one at the end of five years, after June 30, 1950: and (2) any
    member appointed to fill a vacancy occurring prior to the expiration of the
    term for which his predecessor was appointed, shall be appointed for the
    remainder of such term. Any member of the Commission may be removed by
    the President for inefficiency, neglect of duty, or malfeasance in office.
    b. No member of the Commission shall engage in any business, vocation, or
    employment other than that of serving as a member of the Commission.
P.L. 88–426 (78 Stat. 400) (1964), section 305(10)(A) previously amended section 22a. by
repealing the last sentence, which read:
        Each member, except the Chairman, shall receive compensation at the rate of
        $22,000 per annum; and the member designated as Chairman shall receive
        compensation at the rate of $22,500 per annum.
P.L. 85–287 (71 Stat. 612) (1957), section 1, had amended that sentence by substituting
$22,000 for $18,000, and by substituting $22,500 for $20,000.
[25]   Amended by P.L. 85–287, § 2, 71 Stat. 612 (1957). Before amendment,
        section 24 read as follows:
    Sec. 24. General Manager.–There is hereby established within the
    Commission a General Manager, who shall discharge such of the
    administrative and executive functions of the Commission as the Commission
    may direct. The General Manager shall be appointed by the Commission, shall
    serve at the pleasure of the Commission, shall be removable by the
    Commission, and shall receive compensation at a rate determined by the
    Commission, but not in excess of $20,000 per annum.
[26]   Amended by P.L. 88–426, § 306(f), 78 Stat. 400 (1964).
[27]   Amended by P.L. 88–426, § 306(f), 78 Stat. 400 (1964).
[28]   Amended by P.L. 88–426, § 306(f), 78 Stat. 400 (1964).

42 USC 2035.
Assistant
General
Manager for
Military
Application.
Divisions and
offices.

Program
divisions.

General
Counsel.

Inspection
Division.

Sec. 25. Divisions, Offices, and Positions

There is hereby established within the Commission:[29]

a. A Division of Military Application and such other program divisions (not to exceed ten in number) as the Commission may determine to be necessary to the discharge of its responsibilities, including a division or divisions the primary responsibilities of which include the development and application of civilian uses of atomic energy. The Division of Military Application shall be under the direction of an Assistant General Manager for Military Application, who shall be appointed by the Commission and shall be an active commissioned officer of the Armed Forces serving in general or flag officer rank or grade, as appropriate. Each other program division shall be under the direction of a Director who shall be appointed by the Commission. The Commission shall require each such division to exercise such of the Commission's administrative and executive powers as the Commission may determine;[30]

b. An Office of the General Counsel under the direction of the General Counsel who shall be appointed by the Commission;[31] and

c. An Inspection Division under the direction of a Director who shall be appointed by the Commission.[32] The Inspection Division shall be responsible for gathering information to show whether or not the contractors, licensees, and officers and employees of the Commission are complying with the provisions of this Act (except those provisions for which the Federal Bureau of Investigation is responsible) and the appropriate rules and regulations of the Commission.

d. such other executive management positions (not to exceed six in number) as the Commission may determine to be necessary to the discharge of its responsibilities. Such positions shall be established by the General Manager with the approval of the Commission. They shall be appointed by the General Manager with the approval of the Commission,

---

[29]   Amended by P.L. 85–287, § 3, 71 Stat. 612 (1957).

[30]   Amended by P.L. 90–190, section 5, 81 Stat. 575 (1967). Before amendment, section 25a read as follows:

a. A Division of Military Application and such other program divisions (not to exceed ten in number) as the Commission may determine to be necessary to the discharge of its responsibilities, including a division or divisions the primary responsibilities of which include the development and application of civilian uses of atomic energy. Each such division shall be under the direction of a Director who shall be appointed by the Commission. The Director of the Division of Military Application shall be an active member of the Armed Forces. The Commission shall require each such division to exercise such of the Commission's administrative and executive powers as the Commission may determine.

P.L. 88–426 (78 Stat. 400) (1964), section 306(f), earlier had amended the second sentence of section 25a. by deleting the last part, which read: "and shall receive compensation at a rate determined by the Commission, but not in excess of $19,000 per annum." P.L. 85–287 (71 Stat. 612) (1957), section 3, had amended that sentence by substituting $19,000 for $16,000.

[31]   Amended by P.L. 88–426, § 306(f), 78 Stat. 400 (1964), which amended section 25b by deleting the last part, which read as follows: "and shall receive compensation at a rate determined by the Commission, but not in excess of $19,500 per annum." P.L. 85–287, section 3, 71 Stat. 612 (1957), had amended section 25b by substituting $19,500 for $16,000.

[32]   Amended by P.L. 88–426, § 306(f), 78 Stat. 400 (1964), which amended the first sentence of section 25c by deleting the last part, which read as follows: "and shall receive compensation at a rate determined by the Commission, but not in excess of $19,000 per annum." P.L. 85–287, section 3, 71 Stat. 612 (1957), had amended that sentence by substituting $19,000 for $16,000.

shall serve at the pleasure of the General Manager, and shall be removable by the General Manager.[33]

Sec. 26. General Advisory Committee
(Repealed[34])

Sec. 27. Military Liaison Committee
(Repealed[35])

42 USC 2038.
Appointment of
Army, Navy, or
Air Force
Officers.

Sec. 28. Appointment of Army, Navy, or Air Force Officers
Notwithstanding the provisions of any other law, the officer of the Army, Navy, or Air Force serving as Assistant General Manager for Military Application shall serve without prejudice to his commissioned status as such officer. Any such officer serving as Assistant General Manager for Military Application shall receive in addition to his pay and allowances, including special and incentive pays, for which pay and allowances the Commission shall reimburse his service, an amount equal to the difference between such pay and allowances, including special and incentive pays, and the compensation established for this position.[36] Notwithstanding the provisions of any other law, any active or retired officer of the Army, Navy, or Air Force may serve as Chairman of the Military Liaison Committee without prejudice to his active or retired status as such officer. Any such active officer serving as Chairman of the Military Liaison Committee shall receive, in addition to his pay and allowances, including special and incentive pays, an amount equal to the difference between such pay and allowances, including special and incentive pays, and the compensation fixed for such Chairman. Any such retired officer serving as Chairman of the Military Liaison Committee shall receive the compensation fixed for such Chairman and his retired pay.[37]

Chairman,
Military
Liaison
Committee.

42 USC 2039.
Committee on
Reactor
Safeguards.

Sec. 29. Advisory Committee on Reactor Safeguards
There is hereby established an Advisory Committee on Reactor Safeguards consisting of a maximum of fifteen members appointed by the Commission for terms of four years each. The Committee shall review safety studies and facility license applications referred to it and shall make reports thereon, shall advise the Commission with regard to the hazards of proposed or existing reactor facilities and the adequacy of proposed reactor safety standards, and shall perform such other duties as the Commission may request. One member shall be designated by the

---

[33]   Added by P.L. 85–287, § 3, 71 Stat. 612 (1957). P.L. 88–426, § 306(f), 78 Stat. 400 (1964), amended the last sentence of this subsection by inserting "and" immediately before "shall be removable by the General Manager," and by deleting the last part of the sentence, which read as follows: "and shall receive compensation at a rate determined by the General Manager, but not in excess of $19,000 per annum."

[34]   Repealed by P.L. 95–91, § 709(c)(1), 91 Stat. 608 (1977).

[35]   Repealed by P.L. 99–661, Div. C, Title I, Part C, section 3137(c), 100 Stat. 4066 (1986).

[36]   Amended by P.L. 90–190, § 6, 81 Stat. 575 (1967). Prior to this amendment, these sentences read as follows:
  Notwithstanding the provisions of any other law, any active officer of the Army, Navy, or Air Force may serve as Director of the Division of Military Application without prejudice to his commissioned status as such officer. Any such officer serving as Director of the Division of Military Application shall receive in addition to his pay and allowances, including special and incentive pays, an amount equal to the difference between such pay and allowances, including special and incentive pays, and the compensation established for this position pursuant to section 303, or section 309 of the Federal Executive Salary Act of 1964.

[37]   Amended by P.L. 107–107, Division A, Title X, Subtitle E, section 1048(i)(11), 115 Stat. 1230 (2001).

Committee as its Chairman. The members of the Committee shall receive a per diem compensation for each day spent in meetings or conferences, or other work of the Committee, and all members shall receive their necessary traveling or other expenses while engaged in the work of the Committee. The provisions of section 163 shall be applicable to the Committee.[38, 39]

Chapter 4–Research

Sec. 31. Research Assistance

a. The Commission is directed to exercise its powers in such manner as to insure the continued conduct of research and development and training[40] activities in the fields specified below, by private or public institutions or persons, and to assist in the acquisition of an ever-expanding fund of theoretical and practical knowledge in such fields. To this end the Commission is authorized and directed to make arrangements (including contracts, agreements, and loans) for the conduct of research and development activities relating to–

(1) nuclear processes;

(2) the theory and production of atomic energy, including processes, materials, and devices related to such production;

(3) utilization of special nuclear material and radioactive material for medical, biological, agricultural, health, or military purposes;

(4) utilization of special nuclear material, atomic energy, and radioactive material and processes entailed in the utilization or production of atomic energy or such material for all other purposes, including industrial or commercial uses, the generation of usable energy, and the demonstration of advances in the commercial or industrial application of atomic energy;[41]

(5) the protection of health and the promotion of safety during research and production activities; and

(6) the preservation and enhancement of a viable environment by developing more efficient methods to meet the Nation's energy needs.[42]

b. GRANTS AND CONTRIBUTIONS.–The Commission is authorized–

(1) to make grants and contributions to the cost of construction and operation of reactors and other facilities and other equipment to colleges, universities, hospitals, and eleemosynary or charitable

---

[38]   Added by P.L. 85–256, § 5, 71 Stat. 576 (1957).

[39]   Amended by P.L. 105–362, 112 Stat. 3292 (1998), which struck the following two sentences, which had previously been added by P.L. 99–209, § 5, 91 Stat. 1483 (1977):

In addition to its other duties under this section, the committee, making use of all available sources, shall undertake a study of reactor safety research and prepare and submit annually to the Congress a report containing the results of such study. The first such report shall be submitted to the Congress no later than December 31, 1977.

[40]   Amended by P.L.84–1006, § 2, 70 Stat. 1069 (1956), added the words "and training."

[41]   Amended by P.L. 91–560, § 1, 84 Stat. 1472 (1970). Previously paragraph 31a(4), read as follows:

Utilization of special nuclear material, atomic energy, and radioactive material and processes entailed in the utilization or production of atomic energy or such material for all other purposes, including industrial use, the generation of usable energy, and the demonstration of the practical value of utilization or production facilities for industrial or commercial purposes; and.

[42]   Added by P.L. 92–84, § 201(a), 85 Stat. 304 (1971).

institutions for the conduct of educational and training activities relating to the fields in subsection (a) of this section; and

(2) to provide grants, loans, cooperative agreements, contracts, and equipment to institutions of higher education (as defined in section 102 of the Higher Education Act of 1965 (20 U.S.C. 1002)) to support courses, studies, training, curricula, and disciplines pertaining to nuclear safety, security, or environmental protection, or any other field that the Commission determines to be critical to the regulatory mission of the Commission.[43]

c. The Commission may (1) make arrangements pursuant to this section, without regard to the provisions of section 3709 of the Revised Statutes, as amended, upon certification by the Commission that such action is necessary in the interest of the common defense and security, or upon a showing by the Commission that advertising is not reasonably practicable; (2) make partial and advance payments under such arrangements; and (3) make available for use in connection therewith such of its equipment and facilities as it may deem desirable.

d. The arrangements made pursuant to this section shall contain such provisions (1) to protect health, (2) to minimize danger to life or property, and (3) to require the reporting and to permit the inspection of work performed thereunder, as the Commission may determine. No such arrangement shall contain any provisions or conditions which prevent the dissemination of scientific or technical information, except to the extent such dissemination is prohibited by law.

Sec. 32. Research by the Commission

The Commission is authorized and directed to conduct, through its own facilities, activities and studies of the types specified in section 31.

Sec. 33. Research for Others

Where the Commission finds private facilities or laboratories are inadequate for the purpose, it is authorized to conduct for other persons, through its own facilities, such of those activities and studies of the types specified in section 31 as it deems appropriate to the development of energy.[44] To the extent the Commission determines that private facilities or laboratories are inadequate to the purpose, and that the Commission's facilities, or scientific or technical resources have the potential of lending significant assistance to other persons in the fields of protection of public health and safety, the Commission may also assist other persons in these fields by conducting for such persons, through the Commission's own facilities, research and development or training activities and studies. The Commission is authorized to determine and make such charges as in its discretion may be desirable for the conduct of the activities and studies referred to in this section.[45]

Margin notes:

41 USC 252(c) (See 41 USC 260(b)).

42 USC 2052. Research by the Commission.

42 USC 2053. Research for others.

---

[43]   Amended by P.L. 109–58, Title VI, § 651(c)(1), 119 Stat. 801 (2005).

[44]   Amended by P.L. 92–84, section 201(b), 85 Stat. 304 (1971). Before amendment, it read as follows:

Where the Commission finds private facilities or laboratories are inadequate to the purpose, it is authorized to conduct for other persons, through its own facilities, such of those activities and studies of the types specified in section 31 as it deems appropriate to the development of atomic energy.

[45]   Amended by P.L. 90–190, section 7, 81 Stat. 575 (1967). Before amendment, the section read as follows:

Sec. 33. RESEARCH FOR OTHERS.–Where the Commission finds private facilities or laboratories are inadequate to the purpose, it is authorized to conduct for other persons, through its own facilities, such of those activities and studies of the types specified in section 31 as it deems appropriate to the development of atomic energy. The Commission is authorized to determine

Chapter 5–Production of Special Nuclear Material

Sec. 41. Ownership and Operation of Production Facilities

a. Ownership of Production Facilities.–The Commission, as agent of and on behalf of the United States, shall be the exclusive owner of all production facilities other than facilities which (1) are useful in the conduct of research and development activities in the fields specified in section 31, and do not, in the opinion of the Commission, have a potential production rate adequate to enable the user of such facilities to produce within a reasonable period of time a sufficient quantity of special nuclear material to produce an atomic weapon; (2) are licensed by the Commission under this title; or (3) are owned by the United States Enrichment Corporation.[46]

b. Operation of the Commission's Production Facilities.–The Commission is authorized and directed to produce or to provide for the production of special nuclear material in its own production facilities. To the extent deemed necessary, the Commission is authorized to make, or to continue in effect, contracts with persons obligating them to produce special nuclear material in facilities owned by the Commission. The Commission is also authorized to enter into research and development contracts authorizing the contractor to produce special nuclear material in facilities owned by the Commission to the extent that the production of such special nuclear material may be incident to the conduct of research and development activities under such contracts. Any contract entered into under this section shall contain provisions (1) prohibiting the contractor from subcontracting any part of the work he is obligated to perform under the contract, except as authorized by the Commission; and (2) obligating the contractor (A) to make such reports pertaining to activities under the contract to the Commission as the Commission may require (B) to submit to inspection by employees of the Commission of all such activities, and (C) to comply with all safety and security regulations which may be prescribed by the Commission. Any contract made under the provisions of this paragraph may be made without regard to the provisions of section 3079 of the Revised Statutes, as amended, upon certification by the Commission that such action is necessary in the interest of the common defense and security, or upon a showing by the Commission that advertising is not reasonable practicable. Partial and advance payments may be made under such contracts.[47]

c. Operation of Other Production Facilities.–Special nuclear material may be produced in the facilities which under this section are not required to be owned by the Commission.

---

and make such charges as in its discretion may be desirable for the conduct of such activities and studies.

[46]   Amended by P.L. 101–575, § 5(c), 104 Stat. 2835 (1990), and P.L. 102–486, 106 Stat. 2943 (1992), added new section (3).

[47]   Amended by P.L. 90–190, section 8, 81 Stat. 575 (1967), which deleted the last sentence of section 41b, which read as follows:

The President shall determine in writing at least once each year the quantities of special nuclear material to be produced under this section and shall specify in such determination the quantities of special nuclear material to be available for distribution by the Commission pursuant to section 53 or section 54.

42 USC 2062.
Irradiation of
materials.

Sec. 42. Irradiation of Materials

The Commission and persons lawfully producing or utilizing special nuclear material are authorized to expose materials of any kind to the radiation incident to the processes of producing or utilizing special nuclear material.

42 USC 2063.
Acquisition of
production
facilities.

44 USC 252(c)
(See 41 USC
260(b)).

Sec. 43. Acquisition of Production Facilities

The Commission is authorized to purchase any interest in facilities for the production of special nuclear materials, or in real property on which such facilities are located, without regard to the provisions of section 3709 of the Revised Statutes, as amended, upon certification by the Commission that such action is necessary in the interest of the common defense and security, or upon a showing by the Commission that advertising is not reasonably practicable. Partial and advance payments may be made under contracts for such purposes. The Commission is further authorized to requisition, condemn, or otherwise acquire any interest in such production facilities, or to condemn or otherwise acquire such real property, and just compensation shall be made therefor.

42 USC 2064.
Disposition of
energy.

Sec. 44. Disposition of Energy

If energy is produced at production facilities of the Commission or is produced in experimental utilization facilities of the Commission, such energy may be used by the Commission, or transferred to other Government agencies, or sold to publicly, cooperatively, or privately owned utilities or users at reasonable and nondiscriminatory prices. If the energy produced is electric energy, the price shall be subject to regulation by the appropriate agency having jurisdiction. In contracting for the disposal of such energy, the Commission shall give preference and priority to public bodies and cooperatives or to privately owned utilities providing electric utility services to high cost areas not being served by public bodies or cooperatives. Nothing in this Act shall be construed to authorize the Commission to engage in the sale or distribution of energy for commercial use except such energy as may be produced by the Commission incident to the operation of research and development facilities of the Commission, or of production facilities of the Commission.

Chapter 6–Special Nuclear Material

42 USC 2071.
Special nuclear
material.

Sec. 51. Special Nuclear Material

The Commission may determine from time to time that other material is special nuclear material in addition to that specified in the definition as special nuclear material. Before making any such determination, the Commission must find that such material is capable of releasing substantial quantities of atomic energy and must find that the determination that such material is special nuclear material is in the interest of the common defense and security, and the President must have expressly assented in writing to the determination. The Commission's determination, together with the assent of the President, shall be submitted to the Energy[48] Committee[49] and a period of thirty days shall elapse while Congress is in session (in computing such thirty days, there shall be excluded the days on which either House is not in session because of an adjournment for more than three days) before the

[48] Amended by P.L. 103–437, § 15(f))(2), 108 Stat. 4592 (1994).
[49] See P.L. 95–110, § 301b.

determination of the Commission may become effective: *Provided, however*, That the Energy[50] Committee, after having received such determination, may by resolution in writing, waive the conditions of or all or any portion of such thirty-day period.

42 USC 2073.
Nuclear material licenses.

Sec. 53. Domestic Distribution of Special Nuclear Material

a.[51] The Commission is authorized (i) to issue licenses to transfer or receive in interstate commerce, transfer, deliver, acquire, possess, own, receive possession of or title to, import, or export under the terms of an agreement for cooperation arranged pursuant to section 123, special nuclear material, (ii) to make special nuclear material available for the period of the license, and, (iii) to distribute special nuclear material within the United States to qualified applicants requesting such material–[52]

(1) for the conduct of research and development activities of the types specified in section 31;

(2) for use in the conduct of research and development activities or in medical therapy under a license issued pursuant to section 104;

(3) for use under a license issued pursuant to section 103;

(4) for such other uses as the Commission determines to be appropriate to carry out the purposes of this Act.[53]

b. The Commission shall establish, by rule, minimum criteria for the issuance of specific or general licenses for the distribution of special nuclear material depending upon the degree of importance to the common defense and security or to the health and safety of the public of–

(1) the physical characteristics of the special nuclear material to be distributed;

(2) the quantities of special nuclear material to be distributed; and

(3) the intended use of the special nuclear material to be distributed.

---

[50]   Amended by P.L. 103–437, § 15(f))(2), 108 Stat. 4592 (1994).

[51]   P.L. 88–489, § 4, 78 Stat. 602 (1964), repealed Sec. 52. P.L. 88–489 read as follows:
Section 52 of the Atomic Energy Act of 1954, as amended, is repealed. All rights, title, and interest in and to any special nuclear material vested in the United States solely by virtue of the provisions of the first sentence of such section 52, and not by any other transaction authorized by the Atomic Energy Act of 1954, as amended, or other applicable law, are hereby extinguished.

Section 52 read as follows:
Sec. 52. Government Ownership of All Special Nuclear Material.–All rights, title, and interest in or to any special nuclear material within or under the jurisdiction of the United States, now or hereafter produced, shall be the property of the United States and shall be administered and controlled by the Commission as agent of and on behalf of the United States by virtue of this Act. Any person owning any interest in any special nuclear material at the time when such material is hereafter determined to be a special nuclear material shall be paid just compensation therefor. Any person who lawfully produces any special nuclear material, except pursuant to a contract with the Commission under the provisions of section 31 or section 41, shall be paid a fair price, determined pursuant to section 56, for producing such material.

[52]   Amended by P.L. 88–489, § 5, 78 Stat. 602 (1964). Before amendment, this subsection read as follows:
a. The Commission is authorized to issue licenses for the possession of, to make available for the period of the license, and to distribute special nuclear material within the United States to qualified applicants requesting such material–

[53]   Added by P.L. 85–681, § 1, 72 Stat. 632 (1958).

Distribution.

c. (1) The Commission may distribute special nuclear material licensed under this section by sale, lease, lease with option to buy, or grant.[54] *Provided however*, That unless otherwise authorized by law, the Commission shall not after December 31, 1970, distribute special nuclear material except by sale[55] to any person who possesses or operates a utilization facility under a license pursuant to section 103 or 104b. for use in the course of activities under such license; nor shall the Commission permit any such person after June 30, 1973, to continue leasing for use in the course of such activities special nuclear material previously leased to such person by the Commission.

(2) The Commission shall establish reasonable sales prices for the special nuclear material licensed and distributed by sale under this section. Such sales prices shall be established on a nondiscriminatory basis which, in the opinion of the Commission, will provide reasonable compensation to the Government for such special nuclear material.

Agreements.

(3) The Commission is authorized to enter into agreements with licensees for such period of time as the Commission may deem necessary or desirable to distribute to such licensees such quantities of special nuclear material as may be necessary for the conduct of the licensed activity. In such agreements, the Commission may agree to repurchase any special nuclear material licensed and distributed by sale which is not consumed in the course of the licensed activity, or any uranium remaining after irradiation of such special nuclear material, at a repurchase price not to exceed the Commission's sale price for comparable special nuclear material or uranium in effect at the time of delivery of such material to the Commission.

Charges.

(4) The Commission may make a reasonable charge, determined pursuant to this section, for the use of special nuclear material licensed and distributed by lease under subsection 53a.(1), (2) or (4)[56] and shall make a reasonable charge determined pursuant to this section for the use of special nuclear material licensed and distributed by lease under subsection 53a.(3). The Commission shall establish criteria in writing for the determination of whether special nuclear material will be distributed by grant and for the determination of whether a charge will be made for the use of special nuclear material licensed and distributed by lease under subsection 53a.(1), (2) or (4), considering, among other things, whether the licensee is a nonprofit or eleemosynary institution and the purposes for which the special nuclear material will be used.[57]

---

[54]   Amended by P.L. 90–190, § 10, 81 Stat. 575 (1967).
[55]   Amended by P.L. 102–486, 106 Stat. 2943 (1992).
[56]   Amended by P.L. 85–681, § 2, 72 Stat. 632 (1958).
[57]   Amended by P.L. 88–489, § 6, 78 Stat. 602 (1964). Before amendment, this subsection read as follows:

   c. The Commission may make a reasonable charge, determined pursuant to this section, for the use of special nuclear material licensed and distributed under subsection 53a.(1), (2) or (4) and shall make a reasonable charge determined pursuant to this section for the use of special nuclear material licensed and distributed under subsection 53a.(3). The Commission shall establish criteria in writing for the determination of whether a charge will be made for the use of special nuclear material licensed and distributed under subsection 53a.(1), (2) or (4) considering, among other things, whether the licensee is a nonprofit or eleemosynary institution and the purposes for which the special nuclear material will be used.

d. In determining the reasonable charge to be made by the Commission for the use of special nuclear material distributed by lease[58] to licensees of utilization or production facilities licensed pursuant to section 103 or section 104, in addition to consideration of the cost thereof, the Commission shall take into consideration–

(1) the use to be made of the special nuclear material;

(2) the extent to which the use of the special nuclear material will advance the development of the peaceful uses of atomic energy;

(3) the energy value of the special nuclear material in the particular use for which the license is issued;

(4) whether the special nuclear material is to be used in facilities licensed pursuant to section 103 or section 104. In this respect, the Commission shall, insofar as practicable, make uniform, nondiscriminatory charges for the use of special nuclear material distributed to facilities licensed pursuant to section 103; and

(5) with respect to special nuclear material consumed in a facility licensed pursuant to section 103, the Commission shall make a further charge equivalent to the sale price for similar special nuclear material established by the Commission in accordance with subsection 53c.(2), and the Commission may make such a charge with respect to such material consumed in a facility licensed pursuant to section 104.[59]

License conditions.

e. Each license issued pursuant to this section shall contain and be subject to the following conditions–

(2)[60] no right to the special nuclear material shall be conferred by the license except as defined by the license;

(3) neither the license nor any right under the license shall be assigned or otherwise transferred in violation of the provisions of this Act;

(4) all special nuclear material shall be subject to the right of recapture or control reserved by section 108 and to all other provisions of this Act;

(5) no special nuclear material may be used in any utilization or production facility except in accordance with the provisions of this Act;

(6) special nuclear material shall be distributed only on terms, as may be established by rule of the Commission, such that no user will be permitted to construct an atomic weapon;

(7) special nuclear material shall be distributed only pursuant to such safety standards as may be established by rule of the Commission to protect health and to minimize danger to life or property; and

(8) except to the extent that the indemnification and limitation of liability provisions of section 170 apply, the licensee will hold the United States and the Commission harmless from any damages

[58] Amended by P.L. 88–489, § 7, 78 Stat. 602 (1964).
[59] Amended by P.L. 88–489, § 7, 78 Stat. 602 (1964). Before amendment, this paragraph read as follows:
(5) with respect to special nuclear material consumed in a facility licensed pursuant to section 103, the Commission shall make a further charge based on the cost to the Commission, as estimated by the Commission, or the average fair price paid for the production of such special nuclear material as determined by section 56, whichever is lower.
[60] Amended by P.L. 88–489, § 8, 78 Stat. 602 (1964), which deleted subsection 53e(1). Subsection 53e(1) read as follows: "   (1) title to all special nuclear material shall at all times be in the United States:"

resulting from the use or possession of special nuclear material by the licensee.[61]

Distribution for
independent
research, etc.

f. The Commission is directed to distribute within the United States sufficient special nuclear material to permit the conduct of widespread independent research and development activities to the maximum extent practicable.[62] In the event that applications for special nuclear material exceed the amount available for distribution, preference shall be given to those activities which are most likely, in the opinion of the Commission, to contribute to basic research, to the development of peacetime uses of atomic energy, or to the economic and military strength of the Nation.

42 USC 2074.

Sec. 54. Foreign Distribution of Special Nuclear Material

Foreign
distribution of
special nuclear
material.

a. The Commission is authorized to cooperate with any nation or group of nations by distributing special nuclear material and to distribute such special nuclear material, pursuant to the terms of an agreement for cooperation to which such nation or group of nations is a party and which is made in accordance with section 123. Unless hereafter otherwise authorized by law the Commission shall be compensated for special nuclear material so distributed at not less than the Commission's published charges applicable to the domestic distribution of such material, except that the Commission to assist and encourage research on peaceful uses or for medical therapy may so distribute without charge during any calendar year only a quantity of such material which at the time of transfer does not exceed in value $10,000 in the case of one nation or $50,000 in the case of any group of nations. The Commission may distribute to the International Atomic Energy Agency, or to any group of nations, only such amounts of special nuclear materials and for such period of time as are authorized by Congress: *Provided, however,* That, (i) notwithstanding this provision, the Commission is hereby authorized, subject to the provisions of section 123, to distribute to the Agency, five thousand kilograms of contained uranium-235, five hundred grams of uranium-233, and three kilograms of plutonium, together with the amounts of special nuclear material which will match in amount the sum of all quantities of special nuclear materials made available by all other members of the Agency to June 1, 1960; and (ii) notwithstanding the foregoing provisions of this subsection, the Commission may distribute to the International Atomic Energy Agency, or to any group of nations, such other amounts of special nuclear materials and for such other periods of time as are established in writing by the Commission: *Provided, however,* That before they are established by the Commission pursuant to this subdivision (ii), such proposed amounts and periods shall be submitted to the Congress and referred to the Energy Committees[63] and a period of sixty days shall elapse while Congress is in session (in computing such sixty days, there shall be excluded the days on which either House is not in session because of adjournment of more than three days): *and provided further,* That any such proposed amounts and periods shall not become effective if during such sixty-day period the Congress passes a concurrent resolution stating in substance that it does not favor the proposed action: *and provided further,* That prior to the elapse of the

---

61   Amended by P.L. 85–256, section 2, 71 Stat. 576. Before amendment, this subsection
      read as follows: "(8) the licensee will hold the United States and the Commission
      harmless from any damages resulting from the use or possession of special nuclear
      material by the licensee."
62   Amended by P.L. 90–190, section 9, 81 Stat. 575 (1967).
63   Amended by See P.L.95–110, § 301b; P.L. 103–437, § 15(f)(3)(A), 108 Stat. 4581
      (1994).

Purchase of special nuclear material.

first thirty days of any such sixty-day period the Energy Committees[64] shall submit to their respective houses reports of their views and recommendations respecting the proposed amounts and periods and an accompanying proposed concurrent resolution stating in substance that the Congress favors, or does not favor, as the case may be, the proposed amounts or periods. The Commission may agree to repurchase any special nuclear material distributed under a sale arrangement pursuant to this subsection which is not consumed in the course of activities conducted in accordance with the agreement for cooperation, or any uranium remaining after irradiation of such special nuclear material, at a repurchase price not to exceed the Commission's sale price for comparable special nuclear material or uranium in effect at the time of delivery of such material to the Commission. The Commission may also agree to purchase, consistent with and within the period of the agreement for cooperation, special nuclear material produced in a nuclear reactor located outside the United States through the use of special nuclear material which was leased or sold pursuant to this subsection. Under any such agreement the Commission shall purchase only such material as is delivered to the Commission during any period when there is in effect a guaranteed purchase price for the same material produced in a nuclear reactor by a person licensed under section 104, established by the Commission pursuant to section 56, and the price to be paid shall be the price so established by the Commission and in effect for the same material delivered to the Commission.

Foreign distribution of certain materials.

b. Notwithstanding the provisions of sections 123, 124, and 125, the Commission is authorized to distribute to any person outside the United States (1) plutonium containing 80 percent centrum or more by weight of plutonium-238, and (2) other special nuclear material when it has, in accordance with subsection 57d., exempted certain classes or quantities of such other special nuclear material or kinds of uses or users thereof from the requirements for a license set forth in this chapter. Unless hereafter otherwise authorized by law, the Commission shall be compensated for special nuclear material so distributed at not less than the Commission's published charges applicable to the domestic distribution of such material. The Commission shall not distribute any plutonium containing 80 per centrum or more by weight of plutonium-238 to any person under this subsection if, in its opinion, such distribution would be inimical to the common defense and security. The Commission may require such reports regarding the use of material distributed pursuant to the provisions of this subsection as it deems necessary.

c. The Commission is authorized to license or otherwise permit others to distribute special nuclear material to any person outside the United States under the same conditions, except as to charges, as would be applicable if the material were distributed by the Commission.[65]

---

[64]   Amended by P.L. 103–437, § 15(f)(3)(B), 108 Stat. 4581 (1994).
[65]   Amended by P.L. 93–377, 88 Stat. 473 (1974). Previously, section 54 read as follows:
    Sec. 54. Foreign Distribution of Special Nuclear Material.–The Commission is authorized to cooperate with any nation by distributing special nuclear material and to distribute such special nuclear material, pursuant to the terms of an agreement for cooperation to which such nation is a party and which is made in accordance with section 123. Unless hereafter otherwise authorized by law the Commission shall be compensated for special nuclear material so distributed at not less than the Commission's published charges applicable to the domestic distribution of such material, except that the Commission to

d. The authority to distribute special nuclear material under this section other than under an export license granted by the Nuclear Regulatory Commission shall extend only to the following small quantities of special nuclear material (in no event more than five hundred grams per year of the uranium isotope 233, the uranium isotope 235, or plutonium contained in special nuclear material to any recipient):

(1) which are contained in laboratory samples, medical devices, or monitoring or other instruments; or

(2) the distribution of which is needed to deal with an emergency situation in which time is of the essence.

e. The authority in this section to commit United States funds for any activities pursuant to any subsequent arrangement under section 131a.(2)(E) shall be subject to the requirements of section 131.[66]

Sec. 55. Acquisition

42 USC 2075.
Acquisition.

The Commission is authorized, to the extent it deems necessary to effectuate the provisions of this Act, to purchase without regard to the limitations in section 54 or any guaranteed purchase prices established pursuant to section 56, and to take, requisition, condemn, or otherwise acquire any special nuclear material or any interest therein. Any contract of purchase made under this section may be made without regard to the provisions of section 3709 of the Revised Statutes, as amended, upon certification by the Commission that such action is necessary in the interest of the common defense and security, or upon a showing by the Commission that advertising is not reasonably practical. Partial and advance payments may be made under contracts for such purposes. Just

41 USC 252(c)
(See 41 USC 260(b)).

assist and encourage research on peaceful uses or for medical therapy may so distribute without charge during any calendar year only a quantity of such material which at the time of transfer does not exceed in value $10,000 in the case of one nation or $50,000 in the case of any group of nations. The Commission may distribute to the International Atomic Energy Agency, or to any group of nations, only such amounts of special nuclear materials and for such periods of time as are authorized by Congress: *Provided, however,* That notwithstanding this provision, the Commission is hereby authorized subject to the provisions of section 123, to distribute to the Agency five thousand kilograms of contained uranium–235, five hundred grams of uranium 233 and three kilograms of plutonium together with the amounts of special nuclear material which will match in amount the sum of all quantities of special nuclear materials made available by all other members of the Agency to July 1, 1960. The Commission may agree to repurchase any special nuclear material distributed under a sale arrangement pursuant to this section which is not consumed in the course of the activities conducted in accordance with the agreement for cooperation, or any uranium remaining after irradiation of such special nuclear material, at a repurchase price not to exceed the Commission's sale price for comparable special nuclear material or uranium in effect at the time of delivery of such material to the Commission. The Commission may also agree to purchase, consistent with and within the period of the agreement for cooperation, special nuclear material produced in a nuclear reactor located outside the United States through the use of special nuclear material which was leased or sold pursuant to this section. Under any such agreement, the Commission shall purchase only such material as is delivered to the Commission during any period when there is in effect a guaranteed purchase price for the same material produced in a nuclear reactor by a person licensed under section 104, established by the Commission pursuant to section 56, and the price to be paid shall be the price so established by the Commission and in effect for the same material delivered to the Commission.

P.L. 88–487, 78 Stat. 602 (1964) added the last three sentences to Section 54. P.L. 87–206, 75 Stat. 475 (1961) §4, added the words "five hundred grams of uranium 233 and three kilograms of plutonium" to the proviso in this section. P.L. 85–177, 71 Stat. 453 (1957), 7, had added the second and third sentences, including the proviso, to Sec. 54.
[66] Amended by P.L. 95–242, 92 Stat. 125 (1978), § 301(a) and § 303(b)(1), added subsection 54(d) and subsection 54(e), respectively.

compensation shall be made for any right, property, or interest in property taken, requisitioned, or condemned under this section.[67] *Providing,* That the authority in this section to commit United States funds for any activities pursuant to any subsequent arrangement under section 131a.(2)(E) shall be subject to the requirements of section 131.[68]

42 USC 2076.
Guaranteed
purchase
prices.

Sec. 56. Guaranteed Purchase Prices

The Commission shall establish guaranteed purchase prices for plutonium produced in a nuclear reactor by a person licensed under section 104 and delivered to the Commission before January 1, 1971. The Commission shall also establish for such periods of time as it may deem necessary but not to exceed ten years as to any such period, guaranteed purchase prices for uranium enriched in the isotope 233 produced in a nuclear reactor by a person licensed under section 103 or section 104 and delivered to the Commission within the period of the guarantee.[69] Guaranteed purchase prices established under the authority of this section shall not exceed the Commission's determination of the estimated value of plutonium or uranium enriched in the isotope 233 as fuel in nuclear reactors, and such prices shall be established on a non-discriminatory basis: *Provided,* That the Commission is authorized to establish such guaranteed purchase prices only for such plutonium or uranium enriched in the isotope 233 as the Commission shall determine is produced through the use of special nuclear material which was leased or sold by the Commission pursuant to section 53.[70]

42 USC 2077.
Unauthorized
handling.

Sec. 57. Prohibition

a. Unless authorized by a general or specific license issued by the Commission, which the Commission is authorized to issue pursuant to section 53, no person may transfer or receive in interstate commerce, transfer, deliver, acquire, own, possess, receive possession of or title to, or import into or export from the United States any special nuclear material.

---

[67] Amended by P.L. 88–489, § 10, 78 Stat. 602 (1964), amended Sec. 55 by substituting a completely new Sec. 55. Before amendment, Sec. 55 read as follows:
   Sec. 55 Acquisition.–The Commission is authorized to purchase or otherwise acquire any special nuclear material or any interest therein outside the United States without regard to the provisions of section 3709 of the Revised Statutes, as amended, upon certification by the Commission that such action is necessary in the interest of the common defense and security, or upon a showing by the Commission that advertising is not reasonably practicable. Partial and advance payments may be made under contracts for such purposes.

[68] Amended by P.L. 95–242, § 303(b)(2), 92 Stat. 131 (1978), added the proviso at the end of section 55.

[69] Amended by P.L. 91–560, § 2, 84 Stat. 1472 (1970).

[70] Amended by P.L. 88–489, § 11, 78 Stat. 602 (1964). Before amendment, section 56 read as follows:
   Sec. 56. Fair Price.–In determining the fair price to be paid by the Commission pursuant to section 52 for the production of any special nuclear material, the Commission shall take into consideration the value of the special nuclear material for its intended use by the United States and may give such weight to the actual cost of producing that material as the Commission finds to be equitable. The fair price, as may be determined by the Commission, shall apply to all licensed producers of the same material: *Provided, however,* That the Commission may establish guaranteed fair prices for all special nuclear material delivered to the Commission for such period of time as it may deem necessary but not to exceed seven years.

42 USC 2077.
Special nuclear
material
production.
Technology
transfers.

b. It shall be unlawful for any person to directly or indirectly engage or participate in the development or production of any special nuclear material[71] outside of the United States except (1) as specifically authorized under an agreement for cooperation made pursuant to section 123, including a specific authorization in a subsequent arrangement under section 131 of this Act, or (2) upon authorization by the Secretary of Energy after a determination that such activity will not be inimical to the interest of the United States: *Provided,* That any such determination by the Secretary of Energy shall be made only with the concurrence of the Department of State and after consultation with[72] the Nuclear Regulatory Commission, the Department of Commerce, and the Department of Defense. The Secretary of Energy shall, within ninety days after the enactment of the Nuclear Non-Proliferation Act of 1978, establish orderly and expeditious procedures, including provision for necessary administrative actions and inter-agency memoranda of understanding,

Authorization
requests,
procedures.

which are mutually agreeable to the Secretaries of State, Defense, Commerce,[73] and the Nuclear Regulatory Commission for the consideration of requests for authorization under this subsection. Such procedures shall include, at a minimum, explicit direction on the handling of such requests, express deadlines for the solicitation and collection of the views of the consulted agencies (with identified officials responsible for meeting such deadlines), an inter-agency coordinating authority to monitor the processing of such requests, predetermined procedures for the expeditious handling of intra-agency and inter-agency disagreements and appeals to higher authorities, frequent meetings of inter-agency administrative coordinators to review the status of all pending requests, and similar administrative mechanisms. To the extent practicable, an applicant should be advised of all the information required of the applicant for the entire process for every agency's needs at the beginning of the process. Potentially controversial requests should be identified as quickly as possible so that any required policy decisions or diplomatic consultations can be initiated in a timely manner. An immediate effort should be undertaken to establish quickly any necessary standards and

Standards and
criteria.

criteria, including the nature of only required assurances or evidentiary showings, for the decision required under this subsection. The processing of any requests proposed and filed as of the date of enactment of the Nuclear Non-Proliferation Act of 1978 shall not be delayed pending the development and establishment of procedures to implement the

Trade secrets,
protection.

requirements of this subsection. Any trade secrets or proprietary information submitted by any person seeking an authorization under this subsection shall be afforded the maximum degree of protection allowable

42 USC 2014.
42 USC 2074.
42 USC 2094.
42 USC 7172.

by law: *Provided further,* That the export of component parts as defined in subsection 11v.(2) or 11cc.(2), or shall be governed by sections 109 and 126 of this Act: *Provided further,* That notwithstanding subsection 402(d) of the Department of Energy Organization Act (Public Law 95-91), the Secretary of Energy and not the Federal Energy Regulatory Commission, shall have sole jurisdiction within the Department of Energy over any matter arising from any function of the Secretary of Energy in this section, section 54d., section 64, or section 111b.[74]

---

[71]   Amended by P.L. 108–458, 118 Stat. 3768 (2004).
[72]   Amended by P.L. 105–277, 112 Stat. 2681–774 (1998).
[73]   Amended by P.L. 105–277, 112 Stat. 2681–774 (1998).
[74]   Amended by P.L. 95–242, § 302, 92 Stat. 126 (1978). Before amendment, subsection 57 b. read as follows:

c. The Commission shall not–

(1) distribute any special nuclear material to any person for a use which is not under the jurisdiction of the United States except pursuant to the provisions of section 54; or

(2) distribute any special nuclear material or issue a license pursuant to section 53 to any person within the United States if the Commission finds that the distribution of such special nuclear material or the issuance of such license would be inimical to the common defense and security or would constitute an unreasonable risk to the health and safety of the public.

Certain exemptions.

d. The Commission is authorized to establish classes of special nuclear material and to exempt certain classes or quantities of special nuclear material or kinds of uses or users from the requirements for a license set forth in this section when it makes a finding that the exemption of such classes or quantities of special nuclear material or such kinds of uses or users would not be inimical to the common defense and security and would not constitute unreasonable risk to the health and safety of the public.[75]

e. Special nuclear material, as defined in section 11, produced in facilities licensed under section 103 or section 104 may not be transferred, reprocessed, used, or otherwise made available by any instrumentality of the United States or any other person for nuclear explosive purposes.[76]

42 USC 2078.
Review.

Sec. 58. Review

Before the Commission establishes any guaranteed purchase price or guaranteed purchase price period in accordance with the provisions of section 56, or establishes any criteria for the waiver of any charge for the use of special nuclear material licensed and distributed under section 53, the proposed guaranteed purchase price, guaranteed purchase price

---

b. It shall be unlawful for any person to directly or indirectly engage in the production of any special nuclear material outside of the United States except
(1) under an agreement for cooperation made pursuant to section 123, or
(2) upon authorization by the Commission after a determination that such activity will not be inimical to the interest of the United States.

[75]   Added by P.L. 93–377, 88 Stat. 475 (1974). Previously, P.L. 88–489, § 12, 78 Stat. 602 (1964), amended section 57 by substituting a completely new section 57. Before amendment, section 57 read as follows:
Sec. 57. Prohibition.–
a. It shall be unlawful for any person to–
(1) possess or transfer any special nuclear material which is the property of the United States except as authorized by the Commission pursuant to subsection
53 a.;
(2) transfer or receive any special nuclear material in interstate commerce except as authorized by the Commission pursuant to subsection 53a., or export from or import into the United States any special nuclear material; and
(3) directly or indirectly engage in the production of any special nuclear material outside of the United States except (A) under an agreement for cooperation made pursuant to section 123, or (B) upon authorization by the Commission after a determination that such activity will not be inimical to the interest of the United States.
b.
The Commission shall not distribute any special nuclear material—
(1) to any person for a use which is not under the jurisdiction of the United States except pursuant to the provisions of section 54; or
(2) to any person within the United States, if the Commission finds that the distribution of such special nuclear material to such person would be inimical to the common defense and security.

[76]   Added by P.L. 97–415, 96 Stat. 2067 (1983).

period, or criteria for the waiver of such charge shall be submitted to the Energy Committees and a period of forty-five days shall elapse while Congress is in session (in computing such forty-five days there shall be excluded the days in which either House is not in session because of adjournment for more than three days): *Provided, however*, That the Energy Committees, after having received the proposed guaranteed purchase price, guaranteed purchase price period, or criteria for the waiver of such charge, may by resolution in writing waive the conditions of, or all or any portion of, such forty-five day period.[77] [78]

Chapter 7–Source Material

42 USC 2091.
Source
material.

Sec. 61. Source Material

The Commission may determine from time to time that other material is source material in addition to those specified in the definition of source material. Before making such determination, the Commission must find that such material is essential to the production of special nuclear material and must find that the determination that such material is source material is in the interest of the common defense and security, and the President must have expressly assented in writing to the determination. The Commission's determination, together with the assent of the President, shall be submitted to the Energy Committees[79] and a period of thirty days shall elapse while Congress is in session (in computing such thirty days, there shall be excluded the days on which either House is not in session because of an adjournment of more than three days) before the determination of the Commission may become effective: Provided, however, That the Energy Committees, after having received such determination, may by resolution in writing waive the conditions of or all or any portion of such thirty-day period.

Submittal of
determination
to Energy
Committees.

42 USC 2092.
License for
transfers
required.

Sec. 62. License for Transfers Required

Unless authorized by a general or specific license issued by the Commission, which the Commission is authorized to issue, no person may transfer or receive in interstate commerce, transfer, deliver, receive possession of or title to, or import into or export from the United States any source material after removal from its place of deposit in nature, except that licenses shall not be required for quantities of source material which, in the opinion of the Commission, are unimportant.

---

[77] Added byP.L. 85–79, 71 Stat. 274 (1957).

[78] Amended by P.L. 88–489, § 13, 78 Stat. 602 (1964). Before amendment, section 58 read as follows:

Sec. 58. Review.–Before the Commission establishes any fair price or guaranteed fair price period in accordance with the provisions of section 56, or establishes any criteria for the waiver of any charge for the use of special nuclear material licensed or distributed under section 53, the proposed fair price, guaranteed fair price period, or criteria for the waiver of such charge shall be submitted to the Joint Committee, and a period of forty–five days shall elapse while Congress is in session (in computing such forty–five days there shall be excluded the days in which either House is not in session because of adjournment for more than three days): *Provided, however,* That the Joint Committee, after having received the proposed fair price, guaranteed fair prices period, or criteria for the waiver of such charge, may by resolution waive the conditions of, or all or any portion of, such forty–five day period.

Also, amended by P.L. 103–437, §15 (f)(4), 108 Stat. 4581 (1994), struck "Joint Committee" and substituted "Energy Committees" throughout the section.

[79] See P.L. 95–110, § 301b, 91 Stat. 884 (1977); P.L. 103 437, § 15(f)(4), 108 Stat. 4581 (1994), struck "Joint Committee" and substituted "Energy Committees" throughout the section.

42 USC 2093.
Domestic
distribution of
source material.

Charges.

42 USC 2094.
Foreign
distribution of
material.

42 USC 2095.
Reporting.

Sec. 63. Domestic Distribution of Source Material

a. The Commission is authorized to issue licenses for and to distribute source material within the United States to qualified applicants requesting such material–

(1) for the conduct of research and development activities of the types specified in section 31;

(2) for use in the conduct of research and development activities or in medical therapy under a license issued pursuant to section 104;

(3) for use under a license issued pursuant to section 103; or

(4) for any other use approved by the Commission as an aid to science or industry.

b. The Commission shall establish, by rule, minimum criteria for the issuance of specific or general licenses for the distribution of source material depending upon the degree of importance to the common defense and security or to the health and safety of the public of–

(1) the physical characteristics of the source material to be distributed;

(2) the quantities of source material to be distributed; and

(3) the intended use of the source material to be distributed.

c. The Commission may make a reasonable charge determined pursuant to subsection 161m. for the source material licensed and distributed under subsection 63a.(1), subsection 63a.(2), or subsection 63a.(4), and shall make a reasonable charge determined pursuant to subsection 161m., for the source material licensed and distributed under subsection 63a.(3). The Commission shall establish criteria in writing for the determination of whether a charge will be made for the source material licensed and distributed under subsection 63a.(1), subsection 63a.(2), or subsection 63a.(4), considering, among other things, whether the licensee is a nonprofit or eleemosynary institution and the purposes for which the source material will be used.

Sec. 64. Foreign Distribution of Source Material

The Commission is authorized to cooperate with any nation by distributing source material and to distribute source material pursuant to the terms of an agreement for cooperation to which such nation is a party and which is made in accordance with section 123. The Commission is also authorized to distribute source material outside of the United States upon a determination by the Commission that such activity will not be inimical to the interests of the United States. The authority to distribute source material under this section other than under an export license granted by the Nuclear Regulatory Commission shall in no case extend to quantities of source material in excess of three metric tons per year per recipient.[80]

Sec. 65. Reporting

The Commission is authorized to issue such rules, regulations, or orders requiring reports of ownership, possession, extraction, refining, shipment, or other handling of source material as it may deem necessary, except that such reports shall not be required with respect to (a) any source material prior to removal from its place of deposit in nature, or (b) quantities of source material which in the opinion of the Commission are unimportant or the reporting of which will discourage independent prospecting for new deposits.

---

[80]   Amended by P.L. 95–242, § 301.(b), 92 Stat. 126 (1978).

42 USC 2096.
Acquisitions.

Sec. 66. Acquisition

The Commission is authorized and directed, to the extent it deems necessary to effectuate the provisions of this Act–

a. to purchase, take, requisition, condemn, or otherwise acquire supplies of source material;

b. to purchase, condemn, or otherwise acquire any interest in real property containing deposits of source material; and

c. to purchase, condemn, or otherwise acquire rights to enter upon any real property deemed by the Commission to have possibilities of containing deposits of source material in order to conduct prospecting and exploratory operations for such deposits.

41 USC 252(c)
(See 41 USC
260(b)).

Any purchase made under this section may be made without regard to the provisions of section 3709 of the Revised Statutes, as amended, upon certification by the Commission that such action is necessary in the interest of the common defense and security, or upon a showing by the Commission that advertising is not reasonably practicable. Partial and advanced payments may be made under contracts for such purposes. The Commission may establish guaranteed prices for all source material delivered to it within a specified time. Just compensation shall be made for any right, property, or interest in property taken, requisitioned, condemned, or otherwise acquired under this section.

42 USC 2097.

Operations on
lands belonging
to the United
States.

Sec. 67. Operations on Lands Belonging to the United States

The Commission is authorized, to the extent it deems necessary to effectuate the provisions of this Act, to issue leases or permits for prospecting for, exploration for, mining of, or removal of deposits of source material in lands belonging to the United States: *Provided, however,* That notwithstanding any other provisions of law, such leases or permits may be issued for lands administered for national park, monument, and wildlife purposes only when the President by Executive Order declares that the requirements of the common defense and security make such action necessary.

42 USC 2098.
Public and
acquired lands.

Sec. 68. Public and Acquired Lands

a.[81] No individual, corporation, partnership, or association, which had any part, directly or indirectly, in the development of the atomic energy program, may benefit by any location, entry, or settlement upon the public domain made after such individual, corporation, partnership, or association took part in such project, if such individual, corporation, partnership, or association, by reason of having had such part in the development of the atomic energy program, acquired confidential official information as to the existence of deposits of such uranium, thorium, or other materials in the specific lands upon which such location, entry, or settlement is made, and subsequent to the date of the enactment of this Act made such location, entry, or settlement or caused the same to be made for his, or its, or their benefit.

Release of
reservation.

b. Any reservation of radioactive mineral substances, fissionable materials, or source material, together with the right to enter upon the land and prospect for, mine, and remove the same, inserted pursuant to Executive Order 9613 of September 13, 1945, Executive Order 9701 of March 4, 1946, the Atomic Energy Act of 1946, or Executive Order 9908 of December 5, 1947, in any patent, conveyance, lease, permit, or other authorization or instrument disposing of any interest in public or acquired lands of the United States, is hereby released, remised, and quitclaimed to

---

[81]   Amended by P.L. 85–681, § 3, 72 Stat. 623 (1958).

the person or persons entitled upon the date of this Act under the grant from the United States or successive grants to the ownership, occupancy, or use of the land under applicable Federal or State laws: *Provided, however*, That in cases where any such reservation on acquired lands of the United States has been heretofore released, remised, or quitclaimed subsequent to August 12, 1954, in reliance upon authority deemed to have been contained in the Atomic Energy Act of 1946, as amended, or the Atomic Energy Act of 1954, as heretofore amended, the same shall be valid and effective in all respects to the same extent as if public lands and not acquired lands had been involved. The foregoing release shall be subject to any rights which may have been granted by the United States pursuant to any such reservation, but the releases shall be subrogated to the rights of the United States.[82]

30 USC 501–505.

c. Notwithstanding the provisions of the Atomic Energy Act of 1946, as amended, and particularly section 5(b)(7) thereof,[83] or the provisions of the Act of August 12, 1953 (67 Stat. 539), and particularly section 3 thereof, any mining claim, heretofore located under the mining laws of the United States, for or based upon a discovery of a mineral deposit which is a source material and which, except for the possible contrary construction of said Atomic Energy Act, would have been locatable under such mining laws, shall, insofar as adversely affected by such possible contrary construction, be valid and effective, in all respects to the same extent as if said mineral deposit were a locatable mineral deposit other than a source material.

42 USC 2099.
Prohibition.

Sec. 69. Prohibition

The Commission shall not license any person to transfer or deliver, receive possession of or title to, or import into or export from the United States any source material if, in the opinion of the Commission, the issuance of a license to such person for such purpose would be inimical to the common defense and security or the health and safety of the public.

---

[82]   Amended by P.L. 85–681, § 3, 72 Stat. 632 (1958). Before amendment, subsection b. read as follows:

> b. In cases where any patent, conveyance, lease, permit, or other authorization has been issued, which reserved to the United States source materials and the right to enter upon the land and prospect for, mine, and remove the same, the head of the Government agency which issued the patent, conveyance, lease, permit, or other authorization shall, on application of the holder thereof, issue a new or supplemental patent, conveyance, lease, permit, or other authorization without such reservation. If any rights have been granted by the United States pursuant to any such reservation then such patent shall be made subject to those rights, but the patentee shall be subrogated to the rights of the United States.

[83]   See Atomic Energy Act of 1946, Appendix B, § 5(b)(7).

Chapter 8–Byproduct Material

42 USC 2111.
Domestic
distribution.

Sec. 81. Domestic Distribution

a. IN GENERAL–No person may[84] transfer or receive in interstate commerce, manufacture, produce, transfer, acquire, own, possess, import, or export any byproduct material, except to the extent authorized by this section, section 82 or section 84.[85] The Commission is authorized to issue general or specific licenses to applicants seeking to use byproduct material for research or development purposes, for medical therapy, industrial uses, agricultural uses, or such other useful applications as may be developed. The Commission may distribute, sell, loan, or lease such byproduct material as it owns to qualified applicants[86] with or without charge: *Provided, however,* That, for byproduct material to be distributed by the Commission for a charge, the Commission shall establish prices on such equitable basis as, in the opinion of the Commission, (a) will provide reasonable compensation to the Government for such material, (b) will not discourage the use of such material or the development of sources of supply of such material independent of the Commission, and (c) will encourage research and development. In distributing such material, the Commission shall give preference to applicants proposing to use such material either in the conduct of research and development or in medical therapy. The Commission shall not permit the distribution of any byproduct material to any licensee, and shall recall or order the recall of any distributed material from any licensee, who is not equipped to observe or who fails to observe such safety standards to protect health as may be established by the Commission or who uses such material in violation of law or regulation of the Commission or in a manner other than as disclosed in the application therefor or approved by the Commission. The Commission is authorized to establish classes of byproduct material and to exempt certain classes or quantities of material or kinds of uses or users from the requirements for a license set forth in this section when it makes a finding that the exemption of such classes or quantities of such material or such kinds of uses or users will not constitute an unreasonable risk to the common defense and security and to the health and safety of the public.

b. REQUIREMENTS–

(1) IN GENERAL–Except as provided in paragraph (2), byproduct material, as defined in paragraphs (3) and (4) of section 11e., may only be transferred to and disposed of in a disposal facility that–

(A) is adequate to protect public health and safety; and

(B)(i) is licensed by the Commission; or

(ii) is licensed by a State that has entered into an agreement with the Commission under section 274b., if the licensing requirements of the State are compatible with the licensing requirements of the Commission.

(2) EFFECT OF SUBSECTION–Nothing in this subsection affects the authority of any entity to dispose of byproduct material, as

---

[84]   Amended by P.L. 109–58, § 651(e)(3)(A)(i), 119 Stat. 807 (2005).

[85]   Amended by P.L. 95–604, § 205(b), 92 Stat. 3039 (1978). Before amendment, it read as follows:

   No person may transfer or receive in interstate commerce, manufacture, produce, transfer, acquire, own, possess, import, or export any byproduct material, except to the extent authorized by this section or by section 82.

[86]   Amended by P.L. 93–377, §4, 88 Stat. 475 (1974).

defined in paragraphs (3) and (4) of section 11e., at a disposal facility in accordance with any Federal or State solid or hazardous waste law, including the Solid Waste Disposal Act (42 U.S.C. 6901 *et seq.*)

c. TREATMENT AS LOW-LEVEL RADIOACTIVE WASTE–Byproduct material, as defined in paragraphs (3) and (4) of section 11e., disposed of under this section shall not be considered to be low-level radioactive waste for the purposes of–

(1) section 2 of the Low-Level Radioactive Waste Policy Act (42 U.S.C. 2021b); or

(2) carrying out a compact that is–

(A) entered into in accordance with that Act (42 U.S.C. 2021b *et seq.*); and

(B) Approved by Congress.[87]

42 USC 2112.
Foreign distribution of byproduct material.

Sec. 82. Foreign Distribution of Byproduct Material

a. The Commission is authorized to cooperate with any nation by distributing byproduct material, and to distribute byproduct material, pursuant to the terms of an agreement for cooperation to which such nation is party and which is made in accordance with section 123.

b. The Commission is also authorized to distribute byproduct material to any person outside the United States upon application therefor by such person and demand such charge for such material as would be charged for the material if it were distributed within the United States: *Provided, however,* That the Commission shall not distribute any such material to any person under this section if, in its opinion, such distribution would be inimical to the common defense and security: *and provided further,* That the Commission may require such reports regarding the use of material distributed pursuant to the provisions of this section as it deems necessary.

c. The Commission is authorized to license others to distribute byproduct material to any person outside the United States under the same conditions, except as to charges, as would be applicable if the material were distributed by the Commission.

42 USC 2113.

Sec. 83. Ownership and Custody of Certain Byproduct Material and Disposable Sites

a. Any license issued or renewed after the effective date of this section under section 62 or section 81 for any activity which results in the production of any byproduct material, as defined in section 11e.(2), shall contain terms and conditions as the Commission determines to be necessary to assure that, prior to termination of such license–

42 USC 2014.
42 USC 2111.

(1) the licensee will comply with decontamination, decommissioning, and reclamation standards prescribed by the Commission for sites (A) at which ores were processed primarily for their source material content and (B) at which such byproduct material is deposited, and

42 USC 2014.

(2) ownership of any byproduct material, as defined in section 11e.(2), which resulted from such licensed activity shall be transferred to (A) the United States or (B) in the State in which such activity occurred if such State exercises the option under subsection b.(1) to acquire land used for the disposal of byproduct material.

---

[87]   Added by P.L. 109–58, § 651(e)(3), 119 Stat. 807, 808 (2005).
Note: P.L. 109–58, Title VI, § 651(e)(3), 119 Stat. 808 (2005), redefined low–level radioactive waste in section 2(9) of the Low Level Radioactive Waste Policy Amendments Act to take into account the addition of paragraphs (3) and (4) of section 11e of the Atomic Energy Act.

Rule,
regulation or
order.

Any license which is in effect on the effective date of this section and which is subsequently terminated without renewal shall comply with paragraphs (1) and (2) upon termination.[88]

b. (1)(A) The Commission shall require by rule, regulation, or order that prior to the termination of any license which is issued after the effective date of this section, title to the land, including any interest therein (other than land owned by the United States or by a State) which is used for the disposal of any byproduct material, as defined by section 11e.(2), pursuant to such license shall be transferred to:

(i) the United States or–

(ii) the State in which such land is located, at the option of such State

unless[89] the Commission determines prior to such termination that transfer of title to such land and such byproduct material is not necessary or desirable to protect the public health, safety, or welfare or to minimize or eliminate danger to life or property. Such determination shall be made in accordance with section 181 of this Act. Notwithstanding any other provision of law or any such determination, such property and materials shall be maintained pursuant to a license issued by the Commission pursuant to section 81 of this Act[90] in such manner as will protect the public health, safety, and the environment.

(B) If the Commission determines by order that use of the surface or subsurface estates, or both, of the land transferred to the United States or to a State under sub-paragraph (A) would not endanger the public health, safety, welfare, or environment, the Commission, pursuant to such regulations as it may prescribe, shall permit the use of the surface or subsurface estates, or both, of such land in a manner consistent with the provisions of this section. If the Commission permits such use of such land, it shall provide the person who transferred such land with the right of first refusal with respect to such use of such land.

(2) If transfer to the United States of title to such byproduct material and such land is required under this section, the Secretary of Energy or any Federal agency designated by the President shall, following the Commission's determination of compliance under subsection c., assume title and custody of such byproduct material and land transferred as provided in this subsection. Such Secretary or Federal agency shall maintain such material and land in such manner as will protect the public health and safety and the environment. Such custody may be transferred to another officer or instrumentality of the United States only upon approval of the President.

(3) If transfer to a State of title to such byproduct material is required in accordance with this subsection, such State shall, following the Commission's determination of compliance under subsection d., assume title and custody of such byproduct material and land transferred as provided in this subsection. Such State

---

[88]   Amended by P.L. 96–106, § 22(c), 93 Stat. 800 (1979). Before amendment, this sentence read as follows:

Any license in effect on the date of the enactment of this section shall either contain such terms and conditions on renewal thereof after the effective date of this section, or comply with paragraphs (1) and (2) upon the termination of such license, whichever first occurs.

[89]   Amended by P.L. 96–106, § 22(e)(1), 93 Stat. 800 (1979).

[90]   Amended by P.L. 96–106, § 22(e)(2), 93 Stat. 800 (1979).

42 USC 2092.

42 USC 2014.

shall maintain such material and land in such manner as will protect the public health, safety, and the environment.

(4) In the case of any such license under section 62, which was in effect on the effective date of this section, the Commission may require, before the termination of such license, such transfer of land and interest therein (as described in paragraph (1) of this subsection) to the United States or a State in which land is located, at the option of such State, as may be necessary to protect the public health, welfare, and the environment from any effects associated with such byproduct material. In exercising the authority of this paragraph, the Commission shall take into consideration the status of the ownership of such land and interest therein and the ability of the licensee to transfer title and custody thereof to the United States or a State.

(5) The Commission may, pursuant to a license, or by rule or order, require the Secretary or other Federal agency or State having custody of such property and materials to undertake such monitoring, maintenance, and emergency measures as are necessary to protect the public health and safety and such other actions as the Commission deems necessary to comply with the standards promulgated pursuant to section 84 of this Act. The Secretary or such other Federal agency is authorized to carry out maintenance, monitoring, and emergency measures, but shall take no other action pursuant to such license, rule or order, with respect to such property and materials unless expressly authorized by Congress after the date of enactment of this Act.

(6) The transfer of title to land or byproduct materials, as defined in section 11e.(2), to a State or the United States pursuant to this subsection shall not relieve any licensee of liability for any fraudulent or negligent acts done prior to such transfer.

(7) Material and land transferred to the United States or a State in accordance with this subsection shall be transferred without cost to the United States or a State (other than administrative and legal costs incurred in carrying out such transfer). Subject to the provisions of paragraph (1)(B) of this subsection, the United States or a State shall not transfer title to material or property acquired under this subsection to any person, unless such transfer is in the same manner as provided under section 104(h) of the Uranium Mill Tailings Radiation Control Act of 1978.

(8) The provisions of this subsection respecting transfer of title and custody to land shall not apply in the case of lands held in trust by the United States for any Indian tribe or lands owned by such Indian tribe subject to a restriction against alienation imposed by the United States. In the case of such lands which are used for the disposal of byproduct material, as defined in section 11e.(2), the licensee shall be required to enter into such arrangements with the Commission as may be appropriate to assure the long-term maintenance and monitoring of such lands by the United States.

c. Upon termination on any license to which this section applies, the Commission shall determine whether or not the licensee has complied with all applicable standards and requirements under such license.[91]

---

[91]   Added by P.L. 95–604, § 202(a), 92 Stat. 3033 (1978).

42 USC 2114.

Sec. 84. Authorities of Commission Respecting Certain Byproduct Material[92]

a. The Commission shall insure that the management of any byproduct material, as defined in section 11e.(2), is carried out in such manner as–

(1) the Commission deems appropriate to protect the public health and safety and the environment from radiological and non-radiological hazards associated with the processing and with the possession and transfer of such material taking into account the risk to the public health, safety, and the environment, with due consideration of the economic costs and such other factors as the Commission determines to be appropriate,[93]

(2) conforms with applicable general standards promulgated by the Administration of the Environmental Protection Agency under section 275, and

(3) conforms to general requirements established by the Commission, with the concurrence of the Administrator, which are, to the maximum extent practicable, at lease comparable to requirements applicable to the possession, transfer, and disposal of similar hazardous material regulated by the Administrator under the Solid Waste Disposal Act, as amended.

42 USC 6901 note.
42 USC 2112.
Rule, regulation of order.

b. In carrying out its authority under this section, the Commission is authorized to–

(1) by rule, regulation, or order require persons, officers, or instrumentalities, exempted from licensing under section 81 of this Act to conduct monitoring, perform remedial work, and to comply with such other measures as it may deem necessary or desirable to protect health or to minimize danger to life or property, and in connection with the disposal or storage of such byproduct material; and

(2) make such studies and inspections and to conduct such monitoring as may be necessary.

Civil penalty.

Any violation by any person other than the United States or any officer or employee of the United States or a State of any rule, regulation, or order or licensing provision, of the Commission established under this section or section 83 shall be subject to a civil penalty in the same manner and in the same amount as violations subject to a civil penalty under section 234. Nothing in this section affects any authority of the Commission under any other provisions of this Act.

42 USC 2282.
42 USC 2014.
42 USC 2114.

c. In the case of sites at which ores are processed primarily for their source material content or which are used for the disposal of byproduct material as defined in section 11e.(2), a licensee may propose alternatives to specific requirements adopted and enforced by the Commission under this Act. Such alternative proposals may take into account local or regional conditions, including geology, topography, hydrology and meteorology. The Commission may treat such alternatives as satisfying Commission requirements if the Commission determines that such alternatives will achieve a level of stabilization and containment of the sites concerned, and a level of protection for public health, safety, and the environment from radiological and non-radiological hazards associated

---

[92] Added by P.L. 95–604, § 205(a), 92 Stat. 3039 (1978).
[93] Amended by P.L. 97–415, § 22, 96 Stat. 2067 (1983).

42 USC 2022.

with such sites, which is equivalent to, to the extent practicable, or more stringent than the level which would be achieved by standards and requirements adopted and enforced by the Commission for the same purpose and any final standards promulgated by the Administrator of the Environmental Protection Agency in accordance with section 275.[94]

Chapter 9–Military Application of Atomic Energy

42 USC 2121.
Authority.

Sec. 91. Authority

a. The Commission is authorized to–

(1) conduct experiments and do research and development work in the military application of atomic energy;

(2) engage in the production of atomic weapons, or atomic weapon parts, except that such activities shall be carried on only to the extent that the express consent and direction of the President of the United States has been obtained, which consent and direction shall be obtained at least once each year;

(3) provide for safe storage, processing, transportation, and disposal of hazardous waste (including radioactive waste) resulting from nuclear materials production, weapons production and surveillance programs, and naval nuclear propulsion programs;

(4) carry out research on and development of technologies needed for the effective negotiation and verification of international agreements on control of special nuclear materials and nuclear weapons; and

(5) under applicable law (other than this paragraph) and consistent with other missions of the Department of Energy, make transfers of federally owned or originated technology to State and local governments, private industry, and universities or nonprofit organizations so that the prospects for commercialization of such technology are enhanced.

b. The President from time to time may direct the Commission (1) to deliver such quantities of special nuclear material or atomic weapons to the Department of Defense for such use as he deems necessary in the interest of national defense, or (2) to authorize the Department of Defense to manufacture, produce, or acquire any atomic weapon or utilization facility for military purposes: *Provided, however,* That such authorization shall not extend to the production of special nuclear material other than that incidental to the operation of such utilization facilities.

c. The President may authorize the Commission or the Department of Defense, with the assistance of the other, to cooperate with another nation and, notwithstanding the provisions of section 57, 62, or 81, to transfer by sale, lease, or loan to that nation, in accordance with terms and conditions of a program approved by the President–

(1) nonnuclear parts of atomic weapons provided that such nation has made substantial progress in the development of atomic weapons, and other nonnuclear parts of atomic weapons systems involving Restricted Data provided that such transfer will not contribute significantly to that nation's atomic weapon design, development or fabrication capability; for the purpose of improving that nation's state of training and operational readiness;

(2) utilization facilities for military applications; and

---

[94]   Added by P.L. 97–415, § 20, 96 Stat. 2067 (1983).

(3) source, byproduct, or special nuclear material for research on, development of, production of, or use in utilization facilities for military applications; and

(4) source, byproduct, or special nuclear material for research on, development of, or use in atomic weapons: *Provided, however,* That the transfer of such material to that nation is necessary to improve its atomic weapon design, development, or fabrication capability: *and provided further,* That such nation has made substantial progress in the development of atomic weapons,

whenever the President determines that the proposed cooperation and each proposed transfer arrangement for the nonnuclear parts of atomic weapons and atomic weapons systems, utilization facilities or source, byproduct, or special nuclear material will promote and will not constitute an unreasonable risk to the common defense and security, while such other nation is participating with the United States pursuant to an international arrangement by substantial and material contributions to the mutual defense and security: *Provided, however,* That the cooperation is undertaken pursuant to an agreement entered into in accordance with section 123: *and provided further,* That if an agreement for cooperation arranged pursuant to this subsection provides for transfer of utilization facilities for military applications the Commission, or the Department of Defense with respect to cooperation it has been authorized to undertake, may authorize any person to transfer such utilization facilities for military applications in accordance with the terms and conditions of this subsection and of the agreement for cooperation[95].

42 USC 2122.
Prohibition.

Sec. 92. Prohibition

a. It shall be unlawful, except as provided in section 91, for any person inside or outside of the United States to knowingly participate in the development of, manufacture, produce, transfer, acquire, receive, possess, import, export, or use, or possess and threaten to use, any atomic weapon. Nothing in this section shall be deemed to modify the provisions of subsection 31a. or section 101.[96]

"b. Conduct prohibited by subsection a. is within the jurisdiction of the United States if –

"(1) the offense occurs in or affects interstate or foreign commerce; the offense occurs outside of the United States and is committed by a national of the United States;

"(2) the offense is committed against a national of the United States while the national is outside the United States;

"(3) the offense is committed against any property that is owned, leased, or used by the United States or by any department or agency of the United States, whether the property is within or outside the United States; or

"(4) an offender aids or abets any person over whom jurisdiction exists under this subsection in committing an offense under this section or conspires with any person over whom jurisdiction exists under this subsection to commit an offense under this section."[97]

---

[95] Amended by P.L. 83–703, Title I, Ch. 9, section 91, 68 Stat. 936, (1954); P.L. 85–479, § I, 72 Stat. 276, (1958); P.L. 101–189, Div. C, Title XXXI, Part E, § 3157, 103 Stat. 1684, (1989); P.L. 102–486, Title IX, § 902(a)(8), 106 Stat. 2944 (1992), renumbered Title I.
[96] Added by P.L. 108–458, 118 Stat. 3771 (2004).
[97] Added by P.L. 108–458, 118 Stat. 3771 (2004).

Chapter 10–Atomic Energy Licenses

42 USC 2131.
License
required.

Sec. 101. License Required

It shall be unlawful, except as provided in section 91, for any person within the United States to transfer or receive in interstate commerce, manufacture, produce, transfer, acquire, possess, use,[98] import, or export any utilization or production facility except under and in accordance with a license issued by the Commission pursuant to section 103 or section 104.

42 USC 2132.

Sec. 102. Utilization and Production Facilities for Industrial or Commercial Purposes

a. Except as provided in subsections b. and c., or otherwise specifically authorized by law, any license hereafter issued for a utilization or production facility for industrial or commercial purposes shall be issued pursuant to section 103.

b. Any license hereafter issued for a utilization or production facility for industrial or commercial purposes, the construction or operation of which was licensed pursuant to subsection 104b. prior to enactment into law of this subsection, shall be issued under subsection 104b.

c. Any license for a utilization or production facility for industrial or commercial purposes constructed or operated under an arrangement with the Commission entered into under the Cooperative Power Reactor Demonstration Program shall, except as otherwise specifically required by applicable law, be issued under subsection 104b.[99]

42 USC 2133.
Commercial
licenses.

Sec. 103. Commercial Licenses

a. The Commission is authorized to issue licenses to persons applying therefor to transfer or receive in interstate commerce, manufacture, produce, transfer, acquire, possess, use[100] import, or export under the terms of an agreement for cooperation arranged pursuant to section 123, utilization or production facilities for industrial or commercial purposes.[101] Such licenses shall be issued in accordance with the provisions of chapter 16 and subject to such conditions as the Commission may by rule or regulation establish to effectuate the purposes and provisions of this Act.

b. The Commission shall issue such licenses on a nonexclusive basis to persons applying therefor (1) whose proposed activities will serve a useful purpose proportionate to the quantities of special nuclear material or source material to be utilized; (2) who are equipped to observe and who agree to observe such safety standards to protect health and to minimize danger to life or property as the Commission may by rule

---

[98]   Amended by P.L. 84–1006, § 11, 70 Stat. 1069 (1956).
[99]   Amended by P.L. 91–560, § 3, 84 Stat. 1472 (1970). Prior to amendment, it read as follows:

> Sec. 102. Finding of Practical Value–Whenever the Commission has made a finding in writing that any type of utilization or production facility has been sufficiently developed to be of practical value for industrial or commercial purposes, the Commission may thereafter issue licenses for such type of facility pursuant to section 103.

[100]   Amended by P.L. 84–1006, § 12, 70 Stat. 1069 (1956).
[101]   Amended by P.L. 91–560, § 4, 84 Stat. 1472 (1970). Before amendment, it read as follows:

> Subsequent to a finding by the Commission as required in section 102, the Commission may issue licenses to transfer or receive in interstate commerce, manufacture, produce, transfer, acquire, possess, use, import, or export under the terms of an agreement for cooperation arranged pursuant to section 123, such type of utilization or production facility.

establish; and (3) who agree to make available to the Commission such technical information and data concerning activities under such licenses as the Commission may determine necessary to promote the common defense and security and to protect the health and safety of the public. All such information may be used by the Commission only for the purposes of the common defense and security and to protect the health and safety of the public.

c. Each such license shall be issued for a specified period, as determined by the Commission, depending on the type of activity to be licensed, but not exceeding forty years from the authorization to commence operations[102] and may be renewed upon the expiration of such period.

d. No license under this section may be given to any person for activities which are not under or within the jurisdiction of the United States, except for the export of production or utilization facilities under terms of an agreement for cooperation arranged pursuant to section 123, or except under the provisions of section 109. No license may be issued to an alien or any[103] corporation or other entity if the Commission knows or has reason to believe it is owned, controlled, or dominated by an alien, a foreign corporation, or a foreign government. In any event, no license may be issued to any person within the United States if, in the opinion of the Commission, the issuance of a license to such person would be inimical to the common defense and security or to the health and safety of the public.

f. Each license issued for a utilization facility under this section or section 104b. shall require as a condition thereof that in case of any accident which could result in an unplanned release of quantities of fission products in excess of allowable limits for normal operation established by the Commission, the licensee shall immediately so notify the Commission. Violation of the condition prescribed by this subsection may, in the Commission's discretion, constitute grounds for license revocation. In accordance with section 187 of this Act, the Commission shall promptly amend each license for a utilization facility issued under this section or section 104b. which is in effect on the date of enactment of this subsection to include the provisions required under this subsection.[104]

Sec. 104. Medical Therapy and Research and Development

a. The Commission is authorized to issue licenses to persons applying therefore for utilization facilities for use in medical therapy. In issuing such licenses the Commission is directed to permit the widest amount of effective medical therapy possible with the amount of special nuclear material available for such purposes and to impose the minimum amount of regulation consistent with its obligations under this Act to promote the common defense and security and to protect the health and safety of the public.

b. As provided for in subsection 102b., or 102c., or where specifically authorized by law, the Commission is authorized to issue licenses under this subsection to persons applying therefor for utilization and production facilities for industrial and commercial purposes. In issuing licenses under this subsection, the Commission shall impose the minimum amount

42 USC 2133.

42 USC 2237.

42 USC 2134.

Medical therapy and research and development.

---

[102]   Amended by P.L. 109–58, section 621, 119 Stat. 782, (2005).
[103]   Amended by P.L. 84–1006, §13, 70 Stat. 1069 (1956).
[104]   Amended by P.L. 96–295, §201, 94 Stat. 786 (1980).

of such regulations and terms of license as will permit the Commission to fulfill its obligations under this Act.[105]

c. The Commission is authorized to issue licenses to persons applying therefor for utilization and production facilities useful in the conduct of research and development activities of the types specified in section 31 and which are not facilities of the type specified in subsection 104b. The Commission is directed to impose only such minimum amount of regulation of the licensee as the Commission finds will permit the Commission to fulfill its obligations under this Act to promote the common defense and security and to protect the health and safety of the public and will permit the conduct of widespread and diverse research and development.

d. No license under this section may be given to any person for activities which are not under or within the jurisdiction of the United States, except for the export of production or utilization facilities under terms of an agreement for cooperation arranged pursuant to section 123 or except under the provisions of section 109. No license may be issued to any corporation or other entity if the Commission knows or has reason to believe it is owned, controlled, or dominated by an alien, a foreign corporation, or a foreign government. In any event, no license may be issued to any person within the United States if, in the opinion of the Commission, the issuance of a license to such person would be inimical to the common defense and security or to the health and safety of the public.

42 USC 2135.
Antitrust
provisions.

Sec. 105. Antitrust Provisions

a. Nothing contained in this Act[106] shall relieve any person from the operation of the following Acts, as amended, An Act to protect trade and commerce against unlawful restraints and monopolies, approved July second, eighteen hundred and ninety: sections seventy-three to seventy-six[107] inclusive, of an Act entitled 'An Act to reduce taxation, to provide revenue for the Government, and for other purposes approved August twenty-seven, eighteen hundred and ninety-four; 'An Act to supplement existing laws against unlawful restraints and monopolies, and for other purposes, approved October fifteen, nineteen hundred and fourteen; and 'An Act to create a Federal Trade Commission, to define its powers and duties, and for other purposes, approved September twenty-six, nineteen hundred and fourteen. In the event a licensee is found by a court of competent jurisdiction, either in an original action in that court or in a proceeding to enforce or review the findings or orders of any

---

[105]   Amended by P.L. 91–560, § 5, 84 Stat. 1472 (1970). Before amendment, it read as follows:

b. The Commission is authorized to issue licenses to persons applying therefor for utilization and production facilities involved in the conduct of research and development activities leading to the demonstration of the practical value of such facilities for industrial or commercial purposes. In issuing licenses under this subsection, the Commission shall impose the minimum amount of such regulations and terms of license as will permit the Commission to fulfill its obligations under this Act to promote the common defense and security and to protect the health and safety of the public and will be compatible with the regulations and terms of license which would apply in the event that a commercial license were later to be issued pursuant to section 103 for that type of facility. In issuing such licenses, priority shall be given to those activities which will, in the opinion of the Commission, lead to major advances in the application of atomic energy for industrial or commercial purposes.

[106]   Amended by P.L. 88–489, § 14, 78 Stat. 602 (1964).
[107]   Amended by P.L. 107–273, Div. C, Title IV, § 14102(c)(2)(D), 116 Stat. 1921 (2002).

Government agency having jurisdiction under the laws cited above, to have violated any of the provisions of such laws in the conduct of the licensed activity, the Commission may suspend, revoke, or take such other action as it may deem necessary with respect to any license issued by the Commission under the provisions of this Act.

b. The Commission shall report promptly to the Attorney General any information it may have with respect to any utilization of special nuclear material or atomic energy which appears to violate or to tend toward the violation of any of the foregoing Acts, or to restrict free competition in private enterprise.

c. (1) The Commission shall promptly transmit to the Attorney General a copy of any license application provided for in paragraph (2) of this subsection, and a copy of any written request provided for in paragraph (3) of this subsection; and the Attorney General shall, within a reasonable time, but in no event to exceed 180 days after receiving a copy of such application or written request, render such advice to the Commission as he determines to be appropriate in regard to the finding to be made by the Commission pursuant to paragraph (5) of this subsection. Such advice shall include an explanatory statement as to the reasons or basis therefor.

(2) Paragraph (1) of this subsection shall apply to an application for a license to construct or operate a utilization or production facility under section 103: *Provided, however,* That paragraph (1) shall not apply to an application for a license to operate a utilization or production facility for which a construction permit was issued under section 103 unless the Commission determines such review is advisable on the ground that significant changes in the licensee's activities or proposed activities have occurred subsequent to the previous review by the Attorney General and the Commission under this subsection in connection with the construction permit for the facility.

(3) With respect to any Commission permit for the construction of a utilization or production facility issued pursuant to subsection 104b. prior to the enactment into law of this subsection, any person who intervened or who sought by timely written notice to the Commission to intervene in the construction permit proceeding for the facility to obtain a determination of antitrust considerations or to advance a jurisdiction basis for such determination shall have the right, upon a written request to the Commission, to obtain an antitrust review under this section of the application for an operating license. Such written request shall be made within 25 days after the date of initial Commission publication in the Federal Register of notice of the filing of an application for an operating license for the facility or the date of enactment into law of this subsection, whichever is later.

(4) Upon the request of the Attorney General, the Commission shall furnish or cause to be furnished such information as the Attorney General determines to be appropriate for the advice called for in paragraph (1) of this subsection.

(5) Promptly upon receipt of the Attorney General's advice, the Commission shall publish the advice in the Federal Register. Where the Attorney General advises that there may be adverse antitrust aspects and recommends that there be a hearing, the Attorney General or his designee may participate as a party in the proceedings thereafter held by the Commission on such licensing matter in connection with the subject matter of his advice. The Commission shall give due consideration to the advice received from the Attorney General and to such evidence as may be provided during the proceedings in connection with such subject

matter, and shall make a finding as to whether the activities under the license would create or maintain a situation inconsistent with the antitrust laws as specified in subsection 105a.

(6) In the event the Commission's finding under paragraph (5) is in the affirmative, the Commission shall also consider, in determining whether the license should be issued or continued, such other factors, including the need for power in the affected area, as the Commission in its judgment deems necessary to protect the public interest. On the basis of its findings, the Commission shall have the authority to issue or continue a license as applied for, to refuse to issue a license, to rescind a license or amend it, and to issue a license with such conditions as it deems appropriate.

(7) The Commission, with the approval of the Attorney General, may except from any of the requirements of this subsection such classes or types of licenses as the Commission may determine would not significantly affect the applicant's activities under the antitrust laws as specified in subsection 105a.

(8) With respect to any application for a construction permit on file at the time of enactment into law of this subsection, which permit would be for issuance under section 103, and with respect to any application for an operating license in connection with which a written request for an antitrust review is made as provided for in paragraph (3), the Commission, after consultation with the Attorney General, may, upon determination that such action is necessary in the public interest to avoid unnecessary delay, establish by rule or order periods for Commission notification and receipt of advice differing from those set forth above and may issue a construction permit or operating license in advance of consideration of and findings with respect to the matters covered in this subsection: *Provided*, That any construction permit or operating license so issued shall contain such conditions as the Commission deems appropriate to assure that any subsequent findings and orders of the Commission with respect to such matters will be given full force and effect.[108]

(9) APPLICABILITY–This subsection does not apply to an application for a license to construct or operate a utilization facility or production facility under section 103 or 104b. that is filed on or after the date of enactment of this paragraph.[109]

---

[108]   Amended by P.L. 91–560, section 6, 84 Stat. 1472 (1970). Before amendment, it read as follows:

c. Whenever the Commission proposes to issue any license to any persons under section 103, it shall notify the Attorney General of the proposed license and the proposed terms and conditions thereof, except such classes or type of licenses, as the Commission, with the approval of the Attorney General, may determine would not significantly affect the licensee's activities under the antitrust laws as specified in subsection 150a. Within a reasonable time, in no event to exceed 90 days after receiving such notification, the Attorney General shall advise the Commission whether, insofar as he can determine, the proposed license would tend to create or maintain a situation inconsistent with the antitrust laws, and such advice shall be published in the Federal Register. Upon the request of the Attorney General, the Commission shall furnish or cause to be furnished such information as the Attorney General determines to be appropriate or necessary to enable him to give the advice called for by this section.

[109]   Amended by P.L. 109–58, § 625, 119 Stat. 784 (2005).

42 USC 2136.
Classes of
facilities.

Sec. 106. Classes of Facilities

The Commission may–

a. group the facilities licensed either under section 103 or under section 104 into classes which may include either production or utilization facilities or both, upon the basis of the similarity of operating and technical characteristics of the facilities;

b. define the various activities to be carried on at each such class of facility; and

c. designate the amounts of special nuclear material available for use by each such facility.

42 USC 2137.
Operators'
licenses.

Sec. 107. Operators' Licenses

The Commission shall–

a. prescribe uniform conditions for licensing individuals as operators of any of the various classes of production and utilization facilities licensed in this Act;

b. determine the qualifications of such individuals;

c. issue licenses to such individuals in such form as the Commission may prescribe; and

d. suspend such licenses for violations of any provision of this Act or any rule or regulation issued thereunder whenever the Commission deems such action desirable.

42 USC 2138.
War or national
emergency.

Sec. 108. War or National Emergency

Whenever the Congress declares that a state of war or national emergency exists, the Commission is authorized to suspend any licenses granted under this Act if in its judgment such action is necessary to the common defense and security. The Commission is authorized during such period, if the Commission finds it necessary to the common defense and security, to order the recapture of any special nuclear material[110] or to order the operation of any facility licensed under section 103 or section 104, and is authorized to order the entry into any plant or facility in order to recapture such material, or to operate such facility. Just compensation shall be paid for any damages caused by the recapture of any special nuclear material or by the operation of any such facility.

42 USC 2139.
Domestic
activities
licenses,
issuance,
authorization.

Export licenses.

Sec. 109. Component and Other Parts of Facilities

a. With respect to those utilization and production facilities which are so determined by the Commission pursuant to subsection 11v.(2) or 11cc.(2) the Commission may issue general licenses for domestic activities required to be licensed under section 101, if the Commission determines in writing that such general licensing will not constitute an unreasonable risk to the common defense and security.

b. After consulting with the Secretaries of State, Energy, and Commerce,[111] the Commission is authorized and directed to determine which component parts as defined in subsection 11v.(2) or 11cc.(2) and which other items or substances are especially relevant from the standpoint of export control because of their significance for nuclear explosive purposes. Except as provided in section 126 b.(2), no such component, substance, or item which is so determined by the Commission shall be exported unless the Commission issues a general or specific license for its export after finding, based on a reasonable judgment of the assurances provided and other information available to the Federal Government, including the Commission, that the following

---

[110]   Amended by P.L. 86–373, § 2, 73 Stat. 688 (1959).
[111]   Amended by P.L. 105–277, 112 Stat. 2681–774 (1998).

criteria or their equivalent are met: (1) IAEA safeguards as required by Article III(2) of the Treaty will be applied with respect to such component, substance, or item; (2) no such component, substance, or item will be used for any nuclear explosive device or for research on or development of any nuclear explosive device; and (3) no such component, substance, or item will be re-transferred to the jurisdiction of any nation or group of nations unless the prior consent of the United States is obtained for such retransfer; and after determining in writing that the issuance of each such general or specific license or category of licenses will not be inimical to the common defense and security; *Provided,* That a specific license shall not be required for an export pursuant to this section if the component, item or substance is covered by a facility license issued pursuant to section 126 of this Act.

c. The Commission shall not issue an export license under the authority of subsection b. if it is advised by the executive branch, in accordance with the procedures established under subsection 126 a., that the export would be inimical to the common defense and security of the United States.[112]

42 USC 2140.
Exclusions.

Sec. 110. Exclusions

Nothing in this subchapter shall be deemed

a. to require a license for (1) the processing, fabricating, or refining of special nuclear material, or the separation of special nuclear material, or the separation of special nuclear material from other substances, under contract with and for the account of the Commission; or (2) the construction or operation of facilities under contract with and for the account of the Commission; or

b. to require a license for the manufacture, production, or acquisition by the Department of Defense of any utilization facility authorized pursuant to section 91, or for the use of such facility by the Department of Defense or a contractor thereof.

42 USC 2141.

Sec. 111. Distribution by the Department of Energy[113]

a. The Nuclear Regulatory Commission is authorized to license the distribution of special nuclear material, source material, and byproduct material by the Department of Energy, pursuant to sections 54, 64, and 82 of this Act, respectively, in accordance with the same procedures established by law for the export licensing of such material by any person: *Provided,* That nothing in this section shall require the licensing of the distribution of byproduct material by the Department of Energy under section 82 of this Act.

42 USC 2112.

b. The Department of Energy shall not distribute any special nuclear material or source material under section 54 or section 64 of this Act other than under an export license issued by the Nuclear Regulatory Commission until (1) the Department has obtained the concurrence of the

---

[112]   Amended by P.L. 95–242, § 309(a), 92 Stat. 141 (1978). Before amendment, section 109 read as follows:

Sec. 109. Component Parts of Facilities–With respect to those utilization and production facilities which are so determined by this Commission pursuant to subsection 11v.(2) or 11cc.(2) the Commission may (a) issue general licenses for activities required to be licensed under section 101, if the Commission determines in writing that such general licensing will not constitute an unreasonable risk to the common defense and security, and (b) issue licenses for the export of such facilities, if the Commission determines in writing that each export will not constitute an unreasonable risk to the common defense and security.

Amended by P.L. 89–645, § 1, 80 Stat. 891, (1966); P.L. 87–615, § 9, 76 Stat. 409 (1962).

[113]   Added by P.L. 95–242, § 301(c), 92 Stat. 125 (1978).

Department of State and has consulted with[114] the Nuclear Regulatory Commission, and the Department of Defense under mutually agreed procedures which shall be established within not more than ninety days after the date of enactment of this provision and (2) the Department finds based on a reasonable judgment of the assurances provided and the information available to the United States Government, that the criteria in section 127 of this Act or their equivalent and any applicable criteria in subsection 128 are met, and that the proposed distribution would not be inimical to the common defense and security.

Chapter 11–International Activities

42 USC 2151.
Effect of
international
arrangements.

Sec. 121. Effect of International Arrangements

Any provision of this Act or any action of the Commission to the extent and during the time that it conflicts with the provisions of any international arrangements made after the date of enactment of this Act shall be deemed to be of no force or effect.

42 USC 2152.

Policies
contained
in international

Sec. 122. Policies Contained in International Arrangements

In the performance of its functions under this Act, the Commission shall give maximum effect to the policies contained in any international arrangement made after the date of enactment of this Act.

42 USC 2153.

Sec. 123. Cooperation with Other Nations

42 USC 2073,
2074, 2077,
2094, 2112,
2121, 2133,
2134, 2164.
Cooperative
agreements,
submitted to
President.
Contents.

No cooperation with any nation, group of nations or regional defense organization pursuant to section 53, 54a., 57, 64, 82, 91, 103, 104, or 144 shall be undertaken until–

a. the proposed agreement for cooperation has been submitted to the President, which proposed agreement shall include the terms, conditions, duration, nature, and scope of the cooperation; and shall include the following requirements:

(1) a guaranty by the cooperating party that safeguards as set forth in the agreement for cooperation will be maintained with respect to all nuclear materials and equipment transferred pursuant thereto, and with respect to all special nuclear material used in or produced through the
use of such nuclear materials and equipment, so long as the material or equipment remains under the jurisdiction or control of the cooperating party, irrespective of the duration of other provisions in the agreement or whether the agreement is terminated or suspended for any reason;

(2) in the case of non-nuclear-weapon states, a requirement, as a condition of continued United States nuclear supply under the agreement for cooperation, that IAEA safeguards be maintained with respect to all nuclear materials in all peaceful nuclear activities within the territory of such state, under its jurisdiction, or carried out under its control anywhere;

(3) except in the case of those agreements for cooperation arranged pursuant to subsection 91c., a guaranty by the cooperating party that no nuclear materials and equipment or sensitive nuclear technology to be transferred pursuant to such agreement, and no special nuclear material produced through the use of any nuclear materials and equipment or sensitive nuclear technology transferred

---

[114]   Amended by P.L. 105–277, 112 Stat. 2681–774 (1998).

pursuant to such agreement, will be used for any nuclear explosive device, or for research on or development of any nuclear explosive device, or for any other military purpose;

(4) except in the case of those agreements for cooperation arranged pursuant to subsection 91c. and agreements for cooperation with nuclear-weapon states, a stipulation that the United States shall have the right to require the return of any nuclear materials and equipment transferred pursuant thereto and any special nuclear material produced through the use thereof if the cooperating party detonates a nuclear   explosive device or terminates or abrogates an agreement providing for IAEA safeguards;

42 USC 2121.
42 USC 2164.

(5) a guaranty by the cooperating party that any material or any Restricted Data transferred pursuant to the agreement for cooperation and, except in the case of agreements arranged pursuant to subsection 91c., 144b., 144c., or 144d.,[115] any production or utilization facility transferred pursuant to the agreement for cooperation or any special nuclear material produced through the use of any such facility or through the use of any material transferred pursuant to the agreement, will not be transferred to unauthorized persons or beyond the jurisdiction or control of the cooperating party without the consent of the United States;

(6) a guaranty by the cooperating party that adequate physical security will be maintained with respect to any nuclear material transferred pursuant to such agreement and with respect to any special nuclear material used in or produced through the use of any material, production facility, or utilization facility transferred pursuant to such agreement;

(7) except in the case of agreements for cooperation arranged pursuant to subsection 91c., 144b., 144c., or 144d.,[116] a guaranty by the cooperating party that no material transferred pursuant to the agreement for cooperation and no material used in or produced through the use of any material, production facility, or utilization facility transferred pursuant to the agreement for cooperation will be reprocessed, enriched or (in the case of plutonium, uranium 233, or uranium enriched to greater than twenty percent in the isotope 235, or other nuclear materials which have been irradiated) otherwise altered in form or content without the prior approval of the United States;

(8) except in the case of agreements for cooperation arranged pursuant to subsection 91c., 144b., 144c., or 144d.,[117] a guaranty by the cooperating party that no plutonium, no uranium 233, and no uranium enriched to greater than twenty percent in the isotope 235, transferred pursuant to the agreement for cooperation, or recovered from any source or special nuclear material so transferred or from any source or special nuclear material used in any production facility or utilization facility transferred pursuant to the agreement for cooperation, will be stored in any facility that has not been approved in advance by the United States; and

(9) except in the case of agreements for cooperation arranged pursuant to subsection 91c., 144b., 144c., or 144d.,[118] a guaranty by the cooperating party that any special nuclear material, production

[115]   Amended by P.L. 103–337, 108 Stat. 2663 (1994).
[116]   Amended by P.L. 103–337, 108 Stat. 2663 (1994).
[117]   Amended by P.L. 103–337, 108 Stat. 2663 (1994).
[118]   Amended by P.L. 103–337, 108 Stat. 2663 (1994).

facility, or utilization facility produced or constructed under the jurisdiction of the cooperating party by or through the use of any sensitive nuclear technology transferred pursuant to such agreement for cooperation will be subject to all the requirements specified in this subsection.

42 USC 2121.
42 USC 2164.
Agreement
requirements
Presidential
exemptions.
Nuclear
Proliferation
Assessment
Statement,
submitted to
President.
Proposed
cooperation
agreements
submittal to
President.

The President may exempt a proposed agreement for cooperation (except an agreement arranged pursuant to subsection 91c., 144b., 144c., or 144d.[119]) from any of the requirements of the foregoing sentence if he determines that inclusion of any such requirement would be seriously prejudicial to the achievement of United States non-proliferation objectives or otherwise jeopardize the common defense and security. Except in the case of those agreements for cooperation arranged pursuant to subsection 91c., 144b., 144c., or 144d.[120] any proposed agreement for cooperation shall be negotiated by the Secretary of State, with the technical assistance and concurrence of the Secretary of Energy and[121] after consultation with the Commission shall be submitted to the President jointly by the Secretary of State and the Secretary of Energy accompanied by the views and recommendations of the Secretary of State, the Secretary of Energy and the Nuclear Regulatory Commission. Each Nuclear Proliferation Assessment Statement prepared pursuant to this Act shall be accompanied by a classified annex, prepared in consultation with the Director of Central Intelligence, summarizing relevant classified information. The Secretary of State shall also provide to the president an unclassified Nuclear Proliferation Assessment Statement (A) which shall analyze the consistency of the text of the proposed agreement for cooperation with all the requirements of this Act, with specific attention to whether the proposed agreement is consistent with each of the criteria set forth in this subsection, and (B)[122] regarding the adequacy of the safeguards and other control mechanisms and the peaceful use assurances contained in the agreement for cooperation to ensure that any assistance furnished thereunder will not be used to further any military or nuclear explosive purpose. In the case of those agreements for cooperation arranged pursuant to subsection 91c., 144b., 144c., or 144d.,[123] any proposed agreement for cooperation shall be submitted to the President by the Secretary of Energy or, in the case of those agreements for cooperation arranged pursuant to subsection 91c., or 144b., which are to be implemented by the Department of Defense, by the Secretary of Defense;

b. the President has submitted text of the proposed agreement for cooperation, except an agreement arranged pursuant to section 91c., 144b., 144c., or 144d. of section 144,[124] together with the accompanying unclassified Nuclear Proliferation Assessment Statement, to the Committee on Foreign Relations of the Senate and the Committee on Foreign Affairs of the House of Representatives, the President has consulted with such Committees for a period of not less than thirty days of continuous session (as defined in section 130g. of this Act) concerning

---

[119]   Amended by P.L. 103–337, 108 Stat. 3092 (1994).
[120]   Amended by P.L. 103–337, 108 Stat. 3092 (1994).
[121]   Amended by P.L. 105–277, 112 Stat. 2681–774 (1998).
[122]   Amended by P.L. 99–64, § 301(a)(1).
[123]   Amended by P.L. 103–337, 108 Stat. 3092 (1994).
[124]   Amended by P.L. 103–337, 108 Stat. 3092 (1994).

the consistency of the terms of the proposed agreement with all the requirements of this Act, and[125] the President has approved and authorized the execution of the proposed agreement for cooperation and has made a determination in writing that the performance of the proposed agreement will promote and will not constitute an unreasonable risk to, the common defense and security;

c. the proposed agreement for cooperation (if not an agreement subject to subsection d.), together with the approval and determination of the President, has been submitted to the Committee on Foreign Affairs[126] of the House of Representatives and the Committee on Foreign Relations of the Senate for a period of thirty days of continuous session (as defined in subsection 130g.): *Provided, however,* That these committees, after having received such agreement for cooperation, may by resolution in writing waive the conditions of all or any portion of such thirty-day period; and

d. the proposed agreement for cooperation (if arranged pursuant to subsection 91c., 144b., 144c., or 144d., or if entailing implementation of sections 53, 54a., 103, or 104 in relation to a reactor that may be capable of producing more than five thermal megawatts or special nuclear material for use in connection therewith) has been submitted to the Congress, together with the approval and determination of the President, for a period of sixty days of continuous session (as defined in subsection 130g. of this Act) and referred to the Committee on Foreign Affairs of the House of Representatives and the Committee on Foreign Relations of the Senate, and in addition, in the case of a proposed agreement for cooperation arranged pursuant to subsection 91c., 144b., 144c., or 144d., the Committee on Armed Services of the House of Representatives and the Committee on Armed Services of the Senate, but such proposed agreement for cooperation shall not become effective if during such sixty-day period the Congress adopts, and there is enacted, a joint resolution[127] stating in substance that the Congress does not favor the proposed agreement for cooperation: *Provided,* That the sixty-day period shall not begin until a Nuclear Proliferation Assessment Statement prepared by the Secretary of State, and any annexes thereto,[128] when required by subsection 123a., have been submitted to the Congress: *Provided further,* That an agreement for cooperation exempted by the President pursuant to subsection a. from any requirement contained in that subsection, or an agreement exempted pursuant to section 104(a)(1) of the Henry J. Hyde United States-India Peaceful Atomic Energy Cooperation Act of 2006,[129] shall not become effective unless the Congress adopts, and there is enacted, a joint resolution stating that the Congress does favor such agreement.[130] During the sixty-day period the Committee on Foreign Affairs of the House of Representatives and the Committee on Foreign Relations of the Senate shall each hold hearings on the proposed agreement for cooperation and submit a report to their respective bodies recommending whether it should be approved or

Submittal to congressional committees.

42 USC 2073.
42 USC 2074.
42 USC 2133.
42 USC 2134.

---

[125] Amended by P.L. 99–64, § 301(a)(2).
[126] Amended by P.L. 103–437, § 15(f)(5), 108 Stat. 4581 (1994).
[127] Amended by P.L. 99–64, § 301(a)(1).
[128] Amended by P.L. 105–277, 112 Stat. 774 (1998).
[129] Amended by P.L. 109–401, 120 Stat. 2734 (2006) to include the exemption clause from the Henry J. Hyde United States–India Peaceful Atomic Energy Cooperation Act of 2006.
[130] Amended by P.L. 99–64, § 301(b)(2).

disapproved.[131] Any such proposed agreement for cooperation shall be considered pursuant to the procedures set forth in section 130i. of this Act.[132]

Following submission of a proposed agreement for cooperation (except an agreement for cooperation arranged pursuant to subsection 91c., 144b., 144c., or 144d.) to the Committee on Foreign Affairs of the House of Representatives and the Committee on Foreign Relations of the Senate, the Nuclear Regulatory Commission, the Department of State, the Department of Energy, and the Department of Defense shall, upon the request of either of those committees, promptly furnish to those committees their views as to whether the safeguards and other controls contained therein provide an adequate framework to ensure that any export as contemplated by such agreement will not be inimical to or constitute an unreasonable risk to the common defense and security.

If, after the date of enactment of the Nuclear Non-Proliferation Act of 1978, the Congress fails to disapprove a proposed agreement for cooperation which exempts the recipient nation from the requirement set forth in subsection 123a.(2), such failure to act shall constitute a failure to adopt a resolution of disapproval pursuant to subsection 128b.(3) for purposes of the Commission's consideration of applications and requests under section 126a.(2) and there shall be no congressional review pursuant to section 128 of any subsequent license or authorization with respect to that state until the first such license or authorization which is issued after twelve months from the elapse of the sixty-day period in which the agreement for cooperation in question is reviewed by the Congress.[133]

e. The President shall keep the Committee on Foreign Affairs of the House of Representatives and the Committee on Foreign Relations of the Senate fully and currently informed of any initiative or negotiations relating to a new or amended agreement for peaceful nuclear cooperation pursuant to this section (except an agreement arranged pursuant to section 91 c., 144 b., 144 c., or 144 d., or an amendment thereto).[134]

Sec. 124. International Atomic Pool

The President is authorized to enter into an international arrangement with a group of nations providing for international cooperation in the nonmilitary applications of atomic energy and he may thereafter cooperate with that group of nations pursuant to section 54a, 57, 64, 82, 103, 104, or 144a.: *Provided, however,* That the cooperation is undertaken pursuant to an agreement for cooperation entered into in accordance with section 123.

Sec. 125. Cooperation with Berlin[135]

The President may authorize the Commission to enter into agreements for cooperation with the Federal Republic of Germany in accordance with section 123, on behalf of Berlin, which for the purposes of this Act comprises those areas over which the Berlin Senate exercises jurisdiction (the United States, and French sectors) and the Commission may thereafter cooperate with Berlin pursuant to section 54a,[136] 57, 64, 82,

---

[131]   Amended by P.L. 99–64, § 301(a)(3).
[132]   Amended by P.L. 99–64, § 301(b)(3).
[133]   Amended by P.L. 95–242, § 401, 92 Stat. 142 (1978); P.L. 88–489, § 15, 78 Stat. 602 (1964); P.L. 93–377, § 5, 88 Stat. 475 (1974); P.L. 85–479, § 3, § 4, 72 Stat. 276 (1958); P.L. 85–681, § 4, 72 Stat. 632 (1958), P.L. 93–485, 88 Stat. 1460 (1974).
[134]   Added by P.L. 110–369, 122 Stat. 4028 (2008).
[135]   Added by P.L. 85–14, 71 Stat. 11 (1957).
[136]   Amended by P.L. 93–377, § 5, 88 Stat. 475 (1974).

103, or 104; *Provided,* That the guaranties required by section 123 shall be made by Berlin with the approval of the allied commandants.

42 USC 2155.

42 USC 2112.
Executive
branch
judgment
notice to
commission.
Exemption.

Contents.
Procedure.

Standards and
criteria.

Sec. 126. Export Licensing Procedures

a. No license may be issued by the Nuclear Regulatory Commission (the "Commission") for the export of any production or utilization facility, or any source material or special nuclear material, including distributions of any material by the Department of Energy under section 54, 64, or 82, for which a license is required or requested, no exemption from any requirement for such an export license may be granted by the Commission, as the case may be, until–

(1) the Commission has been notified by the Secretary of State that it is the judgment of the executive branch that the proposed export or exemption will not be inimical to the common defense and security, or that any export in the category to which the proposed export belongs would not be inimical to the common defense and security because it lacks significance for nuclear explosive purposes. Secretary of State shall, within ninety days after the enactment of this section, establish orderly and expeditious procedures, including provision for necessary administrative actions and inter-agency memoranda of understanding, which are mutually agreeable to the Secretaries of Energy, Defense, and Commerce,[137] and the Nuclear Regulatory Commission for the preparation of the executive branch judgment on export applications under this section. Such procedures shall include, at a minimum, explicit direction on the handling of such applications, express deadlines for the solicitation and collection of the views of the consulted agencies (with identified officials responsible for meeting such deadlines), an inter-agency coordinating authority to monitor the processing of such applications, predetermined procedures for the expeditious handling of intra-agency and inter-agency disagreements and appeals to higher authorities, frequent meetings of inter-agency administrative coordinators to review the status of all pending applications, and similar administrative mechanisms. To the extent practicable, an applicant should be advised of all the information required of the applicant for the entire process for every agency's needs at the beginning of the process. Potentially controversial applications should be identified as quickly as possible so that any required policy decisions or diplomatic consultations can be initiated in a timely manner. An immediate effort should be undertaken to establish quickly any necessary standards and criteria, including the nature of any required assurances or evidentiary showing, for the decisions required under this section. The processing of any export application proposed and filed as of the date of enactment of this section shall not be delayed pending the development and establishment of procedures to implement the requirements of this section. The executive branch judgment shall be completed in not more than sixty days from receipt of the application or request, unless the Secretary of State in his discretion specifically authorizes additional time for consideration of the application or request because it is in the national interest to allow such additional time. The Secretary shall notify the Committee on Foreign Relations of the Senate and the Committee on Foreign Affairs[138] of the House of

---

[137]   Amended by P.L. 105–277, 112 Stat. 774 (1998).
[138]   Amended by P.L. 103–437, § 15(f)(5), 108 Stat. 4581 (1994).

Notice to
congressional
committees.

Representatives of any such authorization. In submitting any such judgment, the Secretary of State shall specifically address the extent to which the export criteria then in effect are met and the extent to which the cooperating party has adhered to the provisions of the applicable agreement for cooperation. In the event he considers it warranted, the Secretary may also address the following additional factors, among others:

(A) whether issuing the license or granting the exemption will materially advance the non-proliferation policy of the United States by encouraging the recipient nation to adhere to the Treaty, or to participate in the undertakings contemplated by section 403 or 404(a) of the Nuclear Non-Proliferation Act of 1978;

(B) whether failure to issue the license or grant the exemption would otherwise be seriously prejudicial to the non-proliferation objectives of the United States; and

(C) whether the recipient nation or group of nations has agreed that conditions substantially identical to the export criteria set forth in section 127 of this Act will be applied by another nuclear supplier nation or group of nations to the proposed United States export, and whether in the Secretary's judgment those conditions will be implemented in a manner acceptable to the United States.

Data and
recommendation
~

The Secretary of State shall provide appropriate data and recommendations, subject to requests for additional data and recommendations, as required by the Commission or the Secretary of Energy, as the case may be; and

Data and
recommendations.

(2) the Commission finds, based on a reasonable judgment of the assurances provided and other information available to the Federal Government, including the Commission, that the criteria in section 127 of this Act or their equivalent, and any other applicable statutory requirements, are met: *Provided,* That continued cooperation under

42 USC 2154.
Findings.

an agreement for cooperation as authorized in accordance with section 124 of this Act shall not be prevented by failure to meet the provisions of paragraph (4) or (5) of section 127 for a period of thirty days after enactment of this section, and for a period of twenty-three months thereafter if the Secretary of State notifies the commission that the nation or group of nations bound by the relevant agreement has agreed to negotiations as called for in section 404(a) of the Nuclear Non-Proliferation act of 1978; however, nothing in this subsection shall be

Findings.

deemed to relinquish any rights which the United States may have under agreements for cooperation in force on the date of enactment of

Extension,
notice
to Congress.

this section: *Provided further,* That if, upon the expiration of such twenty four month period, the President determines that failure to continue cooperation with any group of nations which has been exempted pursuant to the above proviso from the provisions of paragraph (4) or (5) of section 127 of this Act, but which has not yet agreed to comply with those provisions would be seriously prejudicial to the achievement of United States non-proliferation objectives or otherwise jeopardize the common defense and security, he may, after notifying the Congress of his determination, extend by Executive order the duration of the above proviso for a period of twelve months, and may further extend the duration of such proviso by one year increments annually thereafter if he again makes such determination and so notifies the Congress. In the event that the Committee on Foreign Affairs of the House of Representatives or the Committee on Foreign Relations of the Senate reports a joint resolution to take any action with respect to any such extension, such joint resolution will

Findings.

be considered in the House or Senate, as the case may be, under procedures identical to those provided for the consideration of resolutions pursuant to section 120 of this Act: *and additionally provided,* That the Commission is authorized to (A) make a single finding under this subsection for more than a single application or request, where the applications or requests involve exports to the same country, in the same general time frame, of similar significance for nuclear explosive purposes and under reasonably similar circumstances and (B) make a finding under this subsection that there is no material changed circumstance associated with a new application or request from those existing at the time of the last application or request for an export to the same country, where the prior application or request was approved by the Commission using all applicable procedures of this section, and such finding of no material changed circumstance shall be deemed to satisfy the requirement of this paragraph for findings of the Commission. The decision not to make any such finding in lieu of the findings which would otherwise be required to be made under this paragraph shall not be subject to judicial review: *and provided further,* That nothing contained in this section is intended to require the Commission independently to conduct or prohibit the Commission from independently conducting country or site specific visitations in the Commission's consideration of the application of IAEA safeguards.

Judicial review. Exception.

b. (1) Timely consideration shall be given by the Commission to requests for export license and exemptions and such requests shall be granted upon a determination that all applicable statutory requirements have been met.

(2) If, after receiving the executive branch judgment that the issuance of a proposed export license will not be inimical to the common defense and security, the Commission does not issue the proposed license on a timely basis because it is unable to make the statutory determinations required under this Act, the Commission shall publicly issue its decision to that effect, and shall submit the license application to the President. The Commission's decision shall include an explanation of the basis for the decision and any dissenting or separate views. If, after receiving the proposed license application and reviewing the Commission's decision, the President determines that withholding the proposed export would be seriously prejudicial to the achievement of United States non-proliferation objectives, or would otherwise jeopardize the common defense and security, the proposed export may be authorized by Executive order: *Provided,* That prior to any such export, the President shall submit the Executive order, together with his explanation of why, in light of the Commission's decision, the export should nonetheless be made, to the Congress for a period of sixty days of continuous session (as defined in subsection 130g.) and shall be referred to the Committee on Foreign Affairs of the House of Representatives and the Committee on Foreign Relations of the Senate, but any such proposed export shall not occur if during such sixty-day period the Congress adopt a concurrent resolution stating in substance that it does not favor the proposed export. Any such Executive order shall be considered pursuant to the procedures set forth in section 130 of this Act for the consideration of Presidential submissions. *And provided further,* That the procedures established pursuant to subsection (b) of section 304 of the Nuclear Non-Proliferation Act of 1978 shall provide that the Commission shall immediately initiate review of any application for a

Presidential review.

Report to Congress and congressional Committee.

Review.

license under this section and to the maximum extent feasible shall expeditiously process the application concurrently with the executive branch review while awaiting the final executive branch judgment. In initiating its review the Commission may identify a set of concerns and requests for information associated with the projected issuance of such license and shall transmit such concerns and requests to the executive branch which shall address such concerns and requests in its written communications with the Commission. Such procedures shall also provide that if the Commission has not completed action on the application within sixty days after the receipt of an executive branch judgment that the proposed export or exemption is not inimical to the common defense and security or that any export in the category to which the proposed export belongs would not be inimical to the common defense and security because it lacks significance for nuclear explosive purposes, the Commission shall inform the applicant in writing of the reason for delay and provide follow-up reports as appropriate. If the Commission has not completed action by the end of an additional sixty days (a total of one hundred and twenty days from receipt of the executive branch judgment), the President may authorize the proposed export by Executive order, upon a finding that further delay would be excessive and upon making the findings required for such Presidential authorizations under this subsection, and subject to the Congressional review procedures set forth herein. However, if the Commission has commenced procedures for public participation regarding the proposed export under regulations promulgated pursuant to subsection (b) of section 304 of the Nuclear Non-Proliferation Act of 1978, or–within sixty days after receipt of the executive branch judgment on the proposed export–the Commission has identified and transmitted to the executive branch a set of additional concerns or requests for information, the President may not authorize the proposed export until sixty days after public proceedings are completed or sixty days after a full executive branch response to the Commission's additional concerns or requests has been made consistent with subsection a.(1) of this section: *Provided further,* That nothing in this section shall affect the right of the Commission to obtain data and recommendations from the Secretary of State at any time as provided in subsection a.(1) of this section.

c. In the event that the House of Representatives or the Senate passes a joint resolution which would adopt one or more additional export criteria, or would modify any existing export criteria under this Act, any such joint resolution shall be referred in the other House to the Committee on Foreign Relations of the Senate or the Committee on Foreign Affairs of the House of Representatives, as the case may be, and shall be considered by the other House under applicable procedures provided for the consideration of resolutions pursuant to section 130 of this Act.[139]

Sec. 127. Criteria Governing United States Nuclear Exports

The United States adopts the following criteria which, in addition to other requirements of law, will govern exports for peaceful nuclear uses from the United States of source material, special nuclear material, production or utilization facilities, and any sensitive nuclear technology:

Margin notes:

Concerns and request, transmittal to executive branch.

Referral to congressional committees.

42 USC 2156.

---

[139]   Amended by P.L. 95–242, § 304(a), 92 Stat. 131 (1978).

(1) IAEA safeguards as required by Article III(2) of the Treaty will be applied with respect to any such material or facilities proposed to be exported, to any such material or facilities previously exported and subject to the applicable agreement for cooperation, and to any special nuclear material used in or produced through the use thereof.

(2) No such material, facilities, or sensitive nuclear technology proposed to be exported or previously exported and subject to the applicable agreement for cooperation, and no special nuclear material produced through the use of such materials, facilities, or sensitive nuclear technology, will be used for any nuclear explosive device or for research on or development of any nuclear explosive device.

(3) Adequate physical security measures will be maintained with respect to such material or facilities proposed to be exported and to any special nuclear material used in or produced through the use thereof. Following the effective date of any regulations promulgated by the Commission pursuant to section 304(d) of the Nuclear Non-Proliferation Act of 1978, physical security measures shall be deemed adequate if such measures provide a level of protection equivalent to that required by the applicable regulations.

(4) No such materials, facilities, or sensitive nuclear technology proposed to be exported, and no special nuclear material produced through the use of such material, will be re-transferred to the jurisdiction of any other nation or group of nations unless the prior approval of the United States is obtained for such re-transfer. In addition to other requirements of law, the United States may approve such retransfer only if the nation or group of nations designated to receive such retransfer agrees that it shall be subject to the conditions required by this section.

(5) No such material proposed to be exported and no special nuclear material produced through the use of such material will be reprocessed, and no irradiated fuel elements containing such material removed from a reactor shall be altered in form or content, unless the prior approval of the United states is obtained for such reprocessing or alteration.

(6) No such sensitive nuclear technology shall be exported unless the foregoing conditions shall be applied to any nuclear material or equipment which is produced or constructed under the jurisdiction of the recipient nation or group of nations by or through the use of any exported sensitive nuclear technology.[140]

42 USC 2157.

Sec. 128. Additional Export Criterion and Procedures

a. (1) As a condition of continued United States export of source material, special nuclear material, production or utilization facilities, and any sensitive nuclear technology to non-nuclear-weapon states, no such export shall be made unless IAEA safeguards are maintained with respect to all peaceful nuclear activities in, under the jurisdiction of, or carried out under the control of such state at the time of the export.

(2) The President shall seek to achieve adherence to the foregoing criterion by recipient non-nuclear weapon states.

---

[140]   Amended by P.L. 95–242, § 305, 92 Stat. 136 (1978).

b. The criterion set forth in subsection a. shall be applied as an export criterion with respect to any application for the export of materials, facilities, or technology specified in subsection a. which is filed after eighteen months from the date of enactment of this section, or for any such application under which the first export would occur at least twenty-four months after the date of enactment of this section, except as provided in the following paragraphs:

(1) If the Commission or the Department of Energy, as the case may be, is notified that the President has determined that failure to approve an export to which this subsection applies because such criterion has not yet been met would be seriously prejudicial to the achievement of United States non-proliferation objectives or otherwise jeopardize the common defense and security, the license or authorization may be issued subject to other applicable requirements of law: *Provided,* That no such export of any production or utilization facility or of any source or special nuclear material (intended for use as fuel in any production or utilization facility) which has been licensed or authorized pursuant to this subsection shall be made to any non-nuclear-weapon state which has failed to meet such criterion until the first such license or authorization with respect to such state is submitted to the Congress (together with a detailed assessment of the reasons underlying the President's determination, the judgement of the executive branch required under section 126 of this Act, and any Commission opinion and views) for a period of sixty days of continuous session (as defined in subsection 130g. of this Act) and referred to the Committee on Foreign Affairs[141] of the House of Representatives and the Committee on Foreign Relations of the Senate, but such export shall not occur if during such sixty-day period the Congress adopts a concurrent resolution stating in substance that the Congress does not favor the proposed export. Any such license or authorization shall be considered pursuant to the procedures set forth in section 130 of this Act for the consideration of Presidential submissions.

(2) If the Congress adopts a resolution of disapproval pursuant to paragraph (1), no further export of materials, facilities, or technology specified in subsection a. shall be permitted for the remainder of that Congress, unless such state meets the criterion or the President notifies the Congress that he has determined that significant progress has been made in achieving adherence to such criterion by such state or that United States foreign policy interests dictate reconsideration and the Congress, pursuant to the procedure of paragraph (1), does not adopt a concurrent resolution stating in substance that it disagrees with the President's determination.

(3) If the Congress does not adopt a resolution of disapproval with respect to a license or authorization submitted pursuant to paragraph (1), the criterion set forth in subsection a. shall not be applied as an export criterion with respect to exports of materials, facilities and technology specified in subsection a. to that state: *Provided,* That the first license or authorization with respect to that state which is issued pursuant to this paragraph after twelve months from the elapse of the sixty-day period specified in paragraph (1), and the first such license or authorization which is issued after each twelve-month period

---

[141]   Amended by P.L. 103–437, § 15(f)(5), 108 Stat. 4581 (1994).

thereafter, shall be submitted to the Congress for review pursuant to the procedures specified in paragraph (1): *Provided further,* That if the Congress adopts a resolution of disapproval during any review period provided for by this paragraph, the provisions of paragraph (2) shall apply with respect to further exports to such state.[142]

42 USC 2158.

Export terminations, criterion.

Sec. 129. Conduct Resulting in Termination of Nuclear Exports

a. No nuclear materials and equipment or sensitive nuclear technology shall be exported to–[143]

(1) any non-nuclear-weapon state that is found by the President to have, at any time after the effective date of this section,

(A) detonated a nuclear explosive device; or

(B) terminated or abrogated IAEA safeguards; or

(C) materially violated an IAEA safeguards agreement; or

(D) engaged in activities involving source or special nuclear material and having direct significance for the manufacture or acquisition of nuclear explosive devices, and has failed to take steps which, in the President's judgment, represent sufficient progress toward terminating such activities; or

(2) any nation or group of nations that is found by the President to have, at any time after the effective date of this section,

(A) materially violated an agreement for cooperation with the United States, or, with respect to material or equipment not supplied under an agreement for cooperation, materially violated the terms under which such material or equipment was supplied or the terms of any commitments obtained with respect thereto pursuant to section 402(a) of the Nuclear Non-Proliferation Act of 1978; or

(B) assisted, encouraged, or induced any non-nuclear-weapon state to engage in activities involving source or special nuclear material and having direct significance for the manufacture or acquisition of nuclear explosive devices, and has failed to take steps which, in the President's judgment, represent sufficient progress toward terminating such assistance, encouragement, or inducement; or

(C) entered into an agreement after the date of enactment of this section for the transfer of reprocessing equipment, materials, or technology to the sovereign control of a non-nuclear-weapon state except in connection with an international fuel cycle evaluation in which the United States is a participant or pursuant to a subsequent international agreement or understanding to which the United States subscribes;

unless the President determines that cessation of such exports would be seriously prejudicial to the achievement of United States non-proliferation objectives or otherwise jeopardize the common defense and security: *Provided,* That prior to the effective date of any such determination, the President's determination, together with a report containing the reasons for his determination, shall be submitted to the Congress and referred to the Committee on Foreign Affairs[144] of the House of Representatives and the Committee on Foreign Relations of the Senate for a period of sixty days of continuous session (as defined in subsection 130g. of this act), but any such determination shall not

Report to Congress.

---

[142]   Amended by P.L. 95–242, § 306, 92 Stat. 137 (1978).
[143]   Amended by P.L. 109–58, § 632(a)(1), 119 Stat. 788 (2005).
[144]   Amended by P.L. 103–437, § 15(f)(5), 108 Stat. 4581 (1994).

become effective if during such sixty-day period the Congress adopts, and there is enacted, a joint resolution[145] stating in substance that it does not favor the determination. Any such determination shall be considered pursuant to the procedures set forth in section 130 of this Act for the consideration of Presidential submissions.[146]

b.(1) Notwithstanding any other provision of law, including specifically section 121 of this Act, and except as provided in para-graphs (2) and (3), no nuclear materials and equipment or sensitive nuclear technology, including items and assistance authorized by section 57b. of this Act and regulated under part 810 of title 10, Code of Federal Regulations, and nuclear-related items on the Commerce Control List maintained under part 774 of title 15 of the Code of Federal Regulations, shall be exported or reexported, or transferred or retransferred whether directly or indirectly, and no Federal agency shall issue any license, approval, or authorization for the export or reexport, or transfer, or retransfer, whether directly or indirectly, of these items or assistance (as defined in this paragraph) to any country whose government has been identified by the Secretary of State as engaged in state sponsorship of terrorist activities (specifically including any country the government of which has been determined by the Secretary of State under section 620A(a) of the Foreign Assistance Act of 1961 (22 U.S.C. 2371(a)), section 6(j)(1) of the Export Administration Act of 1979 (50 U.S.C. App. 2405(j)(1), or section 40(d) of the Arms Export Control Act (22 U.S.C. 2780(d)) to have repeatedly provided support for acts of international terrorism).

(2) This subsection shall not apply to exports, reexports, transfers, or retransfers of radiation monitoring technologies, surveillance equipment, seals, cameras, tamper-indication devices, nuclear detectors, monitoring systems, or equipment necessary to safely store, transport, or remove hazardous materials, whether such items, services, or information are regulated by the Department of Energy, the Department of Commerce, or the Commission, except to the extent that such technologies, equipment, seals, cameras, devices, detectors, or systems are available for use in the design or construction of nuclear reactors or nuclear weapons.

(3) The President may waive the application of paragraph (1) to a country if the President determines and certifies to Congress that the waiver will not result in any increased risk that the country receiving the waiver will acquire nuclear weapons, nuclear reactors, or any materials or components of nuclear weapons and–

(A) the government of such country has not within the preceding 12-month period willfully aided or abetted the international proliferation of nuclear explosive devices to individuals or groups or willfully aided and abetted an individual or groups in acquiring unsafeguarded nuclear materials;

(B) in the judgment of the President, the government of such country has provided adequate, verifiable assurances that it will cease its support for acts of international terrorism;

(C) the waiver of that paragraph is in the vital national security interest of the United States; or

---

[145]   Amended by P.L. 110–369, § 203, 122 Stat. 4028 (2008).
[146]   Amended by P.L. 95–242, § 307, 92 Stat. 138 (1978).

(D) such a waiver is essential to prevent or respond to a serious radiological hazard in the country receiving the waiver that may or does threaten public health and safety.

42 USC 2158 note.

(b) APPLICABILITY TO EXPORTS APPROVED FOR TRANSFER BUT NOT TRANSFERRED Subsection b. of section 129 of Atomic Energy Act of 1954, as added by subsection (a) of this section, shall apply with respect to exports that have been approved for transfer as of the date of the enactment of this Act but have not yet been transferred as of that date.[147]

42 USC 2159.

Sec. 130. Congressional Review Procedures

a. Not later than forty-five days of continuous session of Congress after the date of transmittal to the Congress of any submission of the

42 USC 2121.
42 USC 2164.
Congressional committee reports.

President required by subsection[148] 126a.(2), 126b.(2), 128b., 129, 131a.(3), or 131f.(1)(A) of this Act, the Committee on Foreign Relations of the Senate and the Committee on Foreign Affairs of the House of Representatives[149] shall each submit a report to its respective House on its views and recommendations respecting such Presidential submission together with a resolution, as defined in subsection f., stating in substance that the Congress approves or disapproves such submission, as the case may be: *Provided,* That if any such committee has not reported such a resolution at the end of such forty-five day period, such committee shall be deemed to be discharged from further consideration of such submission.[150] If no such resolution has been reported at the end of such period, the first resolution, as defined in subsection f., which is introduced within five days thereafter within such House shall be placed on the appropriate calendar of such House.

b. When the relevant committee or committees have reported such a resolution (or have been discharged from further consideration of such a resolution pursuant to subsection a.) or when a resolution has been introduced and placed on the appropriate calendar pursuant to subsection a., as the case may be, it is at any time thereafter in order (even though a previous motion to the same effect has been disagreed to) for any Member of the respective House to move to proceed to the consideration of the resolution. The motion is highly privileged and is not debatable. The motion shall not be subject to amendment, or to a motion to postpone, or to a motion to proceed to the consideration of other business. A motion to reconsider the vote by which the motion is agreed to or disagreed to shall not be in order. If a motion to proceed to the consideration of the resolution is agreed to, the resolution shall remain the unfinished business of the respective House until disposed of.

c. Debate on the resolution, and on all debatable motions and appeals in connection therewith, shall be limited to not more than ten hours, which shall be divided equally between individuals favoring and individuals opposing the resolution. A motion further to limit debate is in order and not debatable. An amendment to a motion to postpone, or a motion to recommit the resolution, or a motion to proceed to the consideration of other business is not in order. A motion to reconsider the vote by which the resolution is agreed to or disagreed to shall not be in order. No amendment to any concurrent resolution pursuant to the procedures of this section is in order except as provided in subsection d.

---

[147]   Amended by P.L. 109–58, § 632(a)(1), 119 Stat. 789 (2005).
[148]   Amended by P.L. 99–64, § 301(c)(1)(A)(i), (1985).
[149]   Amended by P.L. 99–64, § 301(c)(1)(A)(ii), (1985).
[150]   Amended by P.L. 99–64, § 301(c)(1)(B), (1985).

d. Immediately following (1) the conclusion of the debate on such concurrent resolution, (2) a single quorum call at the conclusion of debate if requested in accordance with the rules of the appropriate House, and (3) the consideration of an amendment introduced by the Majority Leader or his designee to insert the phrase, "does not" in lieu of the word "does" if the resolution under consideration is a concurrent resolution of approval, the vote on final approval of the resolution shall occur.

e. Appeals from the decisions of the Chair relating to the application of the rules of the Senate or the House of Representatives, as the case may be, to the procedure relating to such a resolution shall be decided without debate.

Resolution.

f. For the purposes of subsections a. through e. of this section, the term "resolution" means a concurrent resolution of the Congress, the matter after the resolving clause of which is as follows: That the Congress (does or does not) favor the transmitted to the Congress by the President on _____, the blank spaces therein to be appropriately filled, and the affirmative or negative phrase within the parenthetical to be appropriately selected.

Continuous
sessions of
Congress.

g. (1) Except as provided in paragraph (2), for the purposes of this section–

Computation.

(A) continuity of session is broken only by an adjournment of Congress sine die; and

(B) the days on which either House is not in session because of an adjournment of more than three days to a day certain are excluded in the computation of any period of time in which Congress is in continuous session.

(2) For purposes of this section insofar as it applies to section 123–

(A) continuity of session is broken only by an adjournment of congress sine die at the end of a Congress; and

(B) the days on which either House is not in session because of an adjournment of more than three days are excluded in the computation of any period of time in which Congress is in continuous session.

h. This section is enacted by Congress–

(1) as an exercise of the rulemaking power of the Senate and the House of Representatives, respectively, and as such they are deemed a part of the rules of each House, respectively, but applicable only with respect to the procedure to be followed in that House in the case of resolutions described by subsection f. of this section; and they supersede other rules only to the extent that they are inconsistent therewith; and

(2) With full recognition of the constitutional right of either House to change the rules (so far as relating to the procedure of that House) at any time, in the same manner and to the same extent as in the case of any other rule of that House.[151]

i. (1) For the purposes of this subsection, the term "joint resolution" means–

(A) for an agreement for cooperation pursuant to section 123 of this Act, a joint resolution, the matter after the resolving H. R. 7081–7 clause of which is as follows: "That the Congress (does

---

[151]   Amended by P.L. 95–242, § 308, 92 Stat. 138 (1978).

or does not) favor the proposed agreement for cooperation transmitted to the Congress by the President on _____.",

(B) for a determination under section 129 of this Act, a joint resolution, the matter after the resolving clause of which is as follows: "That the Congress does not favor the determination transmitted to the Congress by the President on _____.", or

(C) for a subsequent arrangement under section 201 of the United States-India Nuclear Cooperation Approval and Nonproliferation Enhancement Act, a joint resolution, the matter after the resolving clause of which is as follows: "That the Congress does not favor the subsequent arrangement to the Agreement for Cooperation Between the Government of the United States of America and the Government of India Concerning Peaceful Uses of Nuclear Energy that was transmitted to Congress by the President on September 10, 2008.", with the date[152] of the transmission of the proposed agreement for cooperation inserted in the blank, and the affirmative or negative phrase within the parenthetical appropriately selected.

(2) On the day on which a proposed agreement for cooperation is submitted to the House of Representatives and the Senate under section 123d., a joint resolution with respect to such agreement for cooperation shall be introduced (by request) in the House by the chairman of the Committee on Foreign Affairs,[153] for himself and the ranking minority member of the Committee, or by Members of the House designated by the chairman and ranking minority member; and shall be introduced (by request) in the Senate by the majority leader of the Senate, for himself and the minority leader of the Senate, or by Members of the Senate designated by the majority leader and minority leader of the Senate. If either House is not in session on the day on which such an agreement for cooperation is submitted, the joint resolution shall be introduced in that House, as provided in the preceding sentence, on the first day thereafter on which that House is in session.

(3) All joint resolutions introduced in the House of Representatives shall be referred to the appropriate committee or committees, and all joint resolutions introduced in the Senate shall be referred to the Committee on Foreign Relations and in addition, in the case of a proposed agreement for cooperation arranged pursuant to section 91c., 144b., or 144c., the Committee on Armed Services.

(4) If the committee of either House to which a joint resolution has been referred has not reported it at the end of 45 days after its introduction (or in the case of a joint resolution related to a subsequent arrangement under section 201 of the United States-India Nuclear Cooperation Approval and Nonproliferation Enhancement Act, 15 days after its introduction),[154] the committee shall be discharged from further consideration of the joint resolution or of any other joint resolution introduced with respect to the same matter; except that, in the case of a joint resolution which has been referred to more than one committee, if before the end of that 45-day period (or in the case of a joint resolution related to a subsequent arrangement

---

[152]   Amended by P.L. 110–369, section 205(1), 122 Stat. 4028 (2008).
[153]   Amended by P.L. 103–437, section 15(f)(5), 108 Stat. 4581 (1994).
[154]   Amended by P.L. 110–369, § 205(2)(A), 122 Stat. 4028 (2008).

under section 201 of the United States-India Nuclear Cooperation Approval and Nonproliferation Enhancement Act, 15-day period)[155] one such committee has reported the joint resolution, any committee to which the joint resolution was referred shall be discharged from further consideration of the joint resolution or of any other joint resolution introduced with respect to the same matter.

(5) A joint resolution under this subsection shall be considered in the Senate in accordance with the provisions of section 601(b)(4) of the International Security Assistance and Arms Export Control Act of 1976. For the purpose of expediting the consideration and passage of joint resolutions reported or discharged pursuant to the provisions of this subsection, it shall be in order for the committee on Rules of the House of Representatives to present for consideration a resolution of the House of Representatives providing procedures for the immediate consideration of a joint resolution under this subsection which may be similar, if applicable, to the procedures set forth in section 601(b)(4) of the International Security Assistance and Arms Export Control Act of 1976.

(6) In the case of a joint resolution described in paragraph (1), if prior to the passage by one House of a joint resolution of that House, that House receives a joint resolution with respect to the same matter from the other House, then–

(A) the procedure in that House shall be the same as if no joint resolution had been received from the other House; but

(B) the vote on final passage shall be on the joint resolution of the other House.

42 USC 2160.

Sec. 131. Subsequent Arrangements[156]

a.(1) Prior to entering into any proposed subsequent arrangement under an agreement for cooperation (other than an agreement for cooperation arranged pursuant to subsection 91c., 144b., or 144c. of this Act), the Secretary of Energy shall obtain the concurrence of the Secretary of State and shall consult with the Commission, and the Secretary of Defense: *Provided,* That the Secretary of State shall have the leading role in any negotiations of a policy nature pertaining to any proposed subsequent arrangement regarding arrangements for the storage or disposition of irradiated fuel elements or approvals for the transfer, for which prior approval is required under an agreement for cooperation, by a recipient of source or special nuclear material, production or utilization facilities, or nuclear technology. Notice of any proposed subsequent arrangement shall be published in the Federal Register, together with the written determination of the Secretary of Energy that such arrangement will not be inimical to the common defense and security, and such proposed subsequent arrangement shall not take effect before fifteen days after publication. Whenever the Secretary of State is required[157] to prepare a Nuclear Proliferation Assessment Statement pursuant to paragraph (2) of this subsection, notice of the proposed subsequent arrangement which is the subject of the requirement to prepare a Nuclear Proliferation Assessment Statement[158] shall not be published until after the receipt by the Secretary of Energy of such Statement or the expiration of the

42 USC 2121.
42 USC 2164.
Consultation.

Notice
publication
in the *Federal
Register.*

[155]   Amended by P.L. 110–369, § 205(2)(A), 122 Stat. 4028 (2008).
[156]   Added by P.L. 95–242, § 303(a), 92 Stat. 127 (1978).
[157]   Amended by P.L. 105–277, 112 Stat. 2681–774 (1998).
[158]   Amended by P.L. 105–277, 112 Stat. 2681–774 (1998).

time authorized by subsection c. for the preparation of such Statement, whichever occurs first.

(2) If in the view of the Secretary of State, Secretary of Energy, Secretary of Defense, or the Commission, a proposed subsequent arrangement might significantly contribute to proliferation, the Secretary of State, in consultation with such Secretary or the Commission shall prepare an unclassified Nuclear Proliferation Assessment Statement with regard to such proposed subsequent arrangement regarding the adequacy of the safeguards and other control mechanisms and the application of the peaceful use assurances of the relevant agreement to ensure that assistance to be furnished pursuant to the subsequent arrangement will not be used to further any military or nuclear explosive purpose. For the purposes of this section,

the term "subsequent arrangements" means arrangements entered into by any agency or department of the United States Government with respect to cooperation with any nation or group of nations (but not purely private or domestic arrangements) involving–

(A) contracts for the furnishing of nuclear materials and equipment;

(B) approvals for the transfer, for which prior approval is required under an agreement for cooperation, by a recipient of any source or special nuclear material, production or utilization facility, or nuclear technology;

(C) authorization for the distribution of nuclear materials and equipment pursuant to this Act which is not subject to the procedures set forth in section 111b., section 126, or section 109b.;

(D) arrangements for physical security;

(E) arrangements for the storage or disposition of irradiated fuel elements;

(F) arrangements for the application of safeguards with respect to nuclear materials and equipment; or

(G) any other arrangement which the President finds to be important from the standpoint of preventing proliferation.

(3) The United States will give timely consideration to all requests for prior approval, when required by this Act, for the reprocessing of material proposed to be exported, previously exported and subject to the applicable agreement for cooperation, or special nuclear material produced through the use of such material or a production or utilization facility transferred pursuant to such agreement for cooperation, or to the altering of irradiated fuel elements containing such material, and additionally, to the maximum extent feasible, will attempt to expedite such consideration when the terms and conditions for such actions set forth in such agreement for cooperation or in some other international agreement executed by the United States and subject to congressional review procedures comparable to those set forth in section 123 of this Act.

(4) All other statutory requirements under other sections of this Act for the approval or conduct of any arrangement subject to this subsection shall continue to apply and any other such requirements for prior approval or conditions for entering such arrangements shall also be satisfied before the arrangement takes effect pursuant to subsection a.(1).

b. With regard to any special nuclear material exported by the United States or produced through the use of any nuclear materials and equipment or sensitive nuclear technology exported by the United States–

Report to
congressional
committees.

(1) the Secretary of Energy may not enter into any subsequent arrangement for the retransfer of any such material to a third country for reprocessing, for the reprocessing of any such material, or for the subsequent retransfer of any plutonium in quantities greater than 500 grams resulting from the reprocessing of any such material, until he has provided the Committee on Foreign Affairs[159] of the House of Representatives and the Committee on Foreign Relations of the Senate with a report containing his reasons for entering into such arrangement and a period of 15 days of continuous session (as defined in subsection 130g. of this Act) has elapsed: *Provided, however,* That if in the view of the President an emergency exists due to unforeseen circumstances requiring immediate entry into a subsequent arrangement, such period shall consist of fifteen calendar days;

(2) the Secretary of Energy may not enter into any subsequent arrangement for the reprocessing of any such material in a facility which has not processed power reactor fuel assemblies or been the subject of a subsequent arrangement therefor prior to the date of enactment of the Nuclear Non-Proliferation Act of 1978 or for subsequent retransfer to a non-nuclear-weapon state of any plutonium in quantities greater than 500 grams resulting from such reprocessing, unless in his judgment, and that of the Secretary of State, such reprocessing or retransfer will not result in a significant increase of the risk of proliferation beyond that which exists at the time that approval is requested. Among all the factors in making this judgment, foremost consideration will be given to whether or not the reprocessing or retransfer will take place under conditions that will ensure timely warning to the United States of any diversion well in advance of the time at which the non-nuclear-weapon state could transform the diverted material into a nuclear explosive device; and

(3) the Secretary of Energy shall attempt to ensure, in entering into any subsequent arrangement for the reprocessing of any such material in any facility that has processed power reactor fuel assemblies or been the subject of a subsequent arrangement therefor prior to the date of enactment of the Nuclear Non-Proliferation Act of 1978, or for the subsequent retransfer to any non-nuclear-weapon state of any plutonium in quantities greater than 500 grams resulting from such reprocessing, that such reprocessing or retransfer shall take place under conditions comparable to those which in his view, and that of the Secretary of State, satisfy the standards set forth in paragraph (2).

Nuclear
materials,
reprocessing or
transfer
procedures.

c. The Secretary of Energy shall, within ninety days after the enactment of this section, establish orderly and expeditious procedures, including provision for necessary administrative actions and inter-agency memoranda of understanding, which are mutually agreeable to the Secretaries of State, Defense, and Commerce, and the Nuclear Regulatory Commission for the consideration of requests for subsequent arrangements under this section. Such procedures shall include, at a minimum, explicit direction on the handling of such requests, express deadlines for the solicitation and collection of the views of the consulted agencies (with identified officials responsible for meeting such deadlines), an inter-agency coordinating authority to monitor the processing of such requests, predetermined procedures for the

---

[159]   Amended by P.L. 103–437, § 15(f)(6)(A), 108 Stat. 4581 (1994).

expeditious handling of intra-agency and inter-agency disagreements and appeals to higher authorities, frequent meetings of inter-agency administrative coordinators to review the status of all pending requests, and similar administrative mechanisms. To the extent practicable, an applicant should be advised of all the information required of the applicant for the entire process for every agency's needs at the beginning

of the process. Potentially controversial request should be identified as quickly as possible so that any required policy decisions or diplomatic consultations can be initiated in a timely manner. An immediate effort should be undertaken to establish quickly any necessary standards and criteria, including the nature of any required assurance or evidentiary showings, for the decisions required under this section. Further, such procedures shall specify that if he intends to prepare a Nuclear Proliferation Assessment Statement, the Secretary of State shall so

declare in his response to the Department of Energy. If the Secretary of State declares that he intends to prepare such a Statement, he shall do so within sixty days of his receipt of a copy of the proposed subsequent arrangement (during which time the Secretary of Energy may not enter into the subsequent arrangement), unless pursuant to the Secretary of State's request, the President waives the sixty-day requirement and notifies the Committee on Foreign Affairs of the House of Representatives and the Committee on Foreign Relations of the Senate of such waiver and the justification therefor.[160] The processing of any subsequent arrangement proposed and filed as of the date of enactment of this section shall not be delayed pending the development and establishment of procedures to implement the requirements of this section.

d. Nothing in this section is intended to prohibit, permanently or unconditionally, the reprocessing of spent fuel owned by a foreign nation which fuel has been supplied by the United States, to preclude the United States from full participation in the International Nuclear Fuel cycle Evaluation provided for in section 105 of the Nuclear Non-Proliferation Act of 1978; to in any way limit the presentation or consideration in that evaluation of any nuclear fuel cycle by the United States or any other participation; nor to prejudice open and objective consideration of the results of the evaluation.

e. Notwithstanding subsection 402(d) of the Department of Energy Organization Act (Public Law 95-91), the Secretary of Energy, and not the Federal Energy Regulatory Commission, shall have sole jurisdiction within the Department of Energy over any matter arising from any function of the Secretary of energy in this section.

f.(1) With regard to any subsequent arrangement under subsection a.(2)(E) (for the storage or disposition of irradiated fuel elements), where such arrangement involves a direct or indirect commitment of the United States for the storage or other disposition, interim or permanent, of any foreign spent nuclear fuel in the United States, the Secretary of Energy may not enter into any such subsequent arrangement, unless:

(A)(i) Such commitment of the United States has been submitted to the Congress for a period of sixty days of continuous session (as defined in subsection 130g. of this act) and has been referred to the Committee on Foreign Affairs of the House of representatives and the Committee on Foreign Relations of the

---

[160]   Amended by P.L. 105–277, 112 Stat. 2681–775 (1998).

Senate, but any such commitment shall not become effective if during such sixty-day period the Congress adopts a concurrent resolution stating in substance that it does not favor the commitment, any such commitment to be considered pursuant to the procedures set forth in section 130 of this act for the consideration of Presidential submission; or (ii) if the President has submitted a detailed generic plan for such disposition or storage in the United States to the Congress for a period of sixty days of continuous session (as defined in subsection 130g. of this Act), which plan has been referred to the Committee on Foreign Affairs of the House of Representatives and the Committee on Foreign Relations of the Senate and has not been adoption of a concurrent resolution stating in substance that Congress does not favor the plan; and the commitment is subject to the terms of an effective plan. Any such plan shall be considered pursuant to the procedures set forth in section 130 of this act for the consideration of Presidential submissions:

(B) The Secretary of Energy has complied with subsection a.; and

(C) The Secretary of Energy has complied, or in the arrangement will comply with all other statutory requirements of this Act, under sections 54 and 55 and any other applicable sections, and any other requirements of law.

Notice to congressional committees.

(2) Subsection (1) shall not apply to the storage or other disposition in the United States of limited quantities of foreign spent nuclear fuel if the President determines that (A) a commitment under section 54 or 55 of this Act of the United States for storage or other disposition of such limited quantities in the United States is required by an emergency situation, (B) it is in the national interest to take such immediate action, and (C) he notifies the Committees on Foreign Affairs and Science, Space, and Technology[161] of the House of Representatives and the Committees on Foreign Relations and Energy and Natural Resources of the Senate of the determination and action, with a detailed explanation and justification thereof, as soon as possible.

Plan, contents.

(3) Any plan submitted by the President under subsection f.(1) shall include a detailed discussion, with detailed information, and any supporting documentation thereof, relating to policy objectives, technical description, geographic information, cost data and justifications, legal and regulatory considerations, environmental impact information and any related international agreements, arrangements for understandings.

Foreign spent nuclear fuel.

(4) For the purposes of this subsection, the term "foreign spent nuclear fuel" shall include any nuclear fuel irradiated in any nuclear power reactor located outside of the United States and operated by any foreign legal entity, government or non-government, regardless of the legal ownership or other control of the fuel or the reactor and regardless of the origin or licensing of the fuel or reactor, but not including fuel irradiated in a research reactor.

---

[161]   Amended by P.L. 103–437, § 15 (f)(6)(B), 108 Stat. 4581 (1994).

42 USC 2160b.

President of
U.S.

Sec. 132. Authority to Suspend Nuclear Cooperation with Nations Which Have Not Ratified the Convention on the Physical Security of Nuclear Material[162]

The President may suspend nuclear cooperation under this Act with any nation or group of nations which has not ratified the Convention on the Physical Security of Nuclear Material.

42 USC 2160c.

Sec. 133. Consultation with the Department of Defense Concerning Certain Exports and Subsequent Arrangements[163]

a. In addition to other applicable requirements–

(1) a license may be issued by the Nuclear Regulatory Commission under this Act for the export of special nuclear material described in subsection b.; and

42 USC 2160.

(2) approval may be granted by the Secretary of Energy under section 131 of this Act for the transfer of special nuclear material described in subsection b.; only after the Secretary of Defense has been consulted on whether the physical protection of that material during the export or transfer will be adequate to deter theft, sabotage, and other acts of international terrorism which would result in the diversion of that material. If, in the view of the Secretary of Defense based on all available intelligence information, the export or transfer might be subject to a genuine terrorist threat, the Secretary shall provide to the Nuclear Regulatory commission or the Secretary of Energy, as appropriate, his written assessment of the risk and a description of the actions the Secretary of Defense considers necessary to upgrade physical protection measures.

b. Subsection a. applies to the export or transfer of more than 2 kilograms of plutonium or more than 5[164] kilograms of uranium enriched to more than 20 percent in the isotope 233 or the isotope 235.

42 USC 2160d.

Sec. 134. Further Restrictions on Exports[165]

a. IN GENERAL[166]–Except as provided in subsection b., the Commission[167] may issue a license for the export of highly enriched uranium to be used as a fuel or target in a nuclear research or test reactor only if, in addition to any other requirement of this Act, the Commission determines that–

(1) there is no alternative nuclear reactor fuel or target enriched in the isotope 235 to a lesser percent that the proposed export, that can be used in that reactor;

(2) the proposed recipient of that uranium has provided assurances that, whenever an alternative nuclear reactor fuel or target can be used in that reactor, it will use that alternative in lieu of highly enriched uranium; and

(3) the United States Government is actively developing an alternative nuclear reactor fuel or target than can be used in that reactor.

b. MEDICAL ISOTOPE PRODUCTION.–

(1) DEFINITIONS–In this subsection:

---

[162]   Added by P.L. 99–399, § 602, 100 Stat 853 (1986).
[163]   Added by P.L. 99–399, § 602, 100 Stat 853 (1986).
[164]   Amended by P.L. 103–236, 108 Stat. 521 (1994).
[165]   Added by P.L. 102–486, 106 Stat 2945 (1992).
[166]   Amended by P.L. 109–58, § 630(1), 119 Stat. 785 (2005).
[167]   Amended by P.L. 109–58, § 630(1), 119 Stat. 785 (2005).

(A) HIGHLY ENRICHED URANIUM–The term 'highly enriched uranium' means uranium enriched to include concentration of U–235 above 20 percent.

(B) MEDICAL ISOTOPE–The term 'medical isotope' includes Molybdenum 99, Iodine 131, Xenon 133, and other radioactive materials used to produce a radiopharmaceutical for diagnostic, therapeutic procedures or for research and development.

(C) RADIOPHARMACEUTICAL–The term 'radiopharmaceutical' means a radioactive isotope that--

(i) contains byproduct material combined with chemical or biological material; and

(ii) is designed to accumulate temporarily in a part of the body for therapeutic purposes or for enabling the production of a useful image for use in a diagnosis of a medical condition.

(D) RECIPIENT COUNTRY–The term 'recipient country' means Canada, Belgium, France, Germany, and the Netherlands.

(2) LICENSES–The Commission may issue a license authorizing the export (including shipment to and use at intermediate and ultimate consignees specified in the license) to a recipient country of highly enriched uranium for medical isotope production if, in addition to any other requirements of this Act (except subsection a.), the Commission determines that–

(A) a recipient country that supplies an assurance letter to the United States Government in connection with the consideration by the Commission of the export license application has informed the United States Government that any intermediate consignees and the ultimate consignee specified in the application are required to use the highly enriched uranium solely to produce medical isotopes; and

(B) the highly enriched uranium for medical isotope production will be irradiated only in a reactor in a recipient country that–

(i) uses an alternative nuclear reactor fuel; or

(ii) is the subject of an agreement with the United States Government to convert to an alternative nuclear reactor fuel when alternative nuclear reactor fuel can be used in the reactor.

(3) REVIEW OF PHYSICAL PROTECTION REQUIREMENTS–

(A) IN GENERAL–The Commission shall review the adequacy of physical protection requirements that, as of the date of an application under paragraph (2), are applicable to the transportation and storage of highly enriched uranium for medical isotope production or control of residual material after irradiation and extraction of medical isotopes.

(B) IMPOSITION OF ADDITIONAL REQUIREMENTS–If the Commission determines that additional physical protection requirements are necessary (including a limit on the quantity of highly enriched uranium that may be contained in a single shipment), the Commission shall impose such requirements as license conditions or through other appropriate means.

(4) FIRST REPORT TO CONGRESS–

Contracts.

(A) NAS STUDY–The Secretary shall enter into an arrangement with the National Academy of Sciences to conduct a study to determine–

(i) the feasibility of procuring supplies of medical isotopes from commercial sources that do not use highly enriched uranium;

(ii) the current and projected demand and availability of medical isotopes in regular current domestic use;

(iii) the progress that is being made by the Department of Energy and others to eliminate all use of highly enriched uranium in reactor fuel, reactor targets, and medical isotope production facilities; and

(iv) the potential cost differential in medical isotope production in the reactors and target processing facilities if the products were derived from production systems that do not involve fuels and targets with highly enriched uranium.

(B) FEASIBILITY–For the purpose of this subsection, the use of low enriched uranium to produce medical isotopes shall be determined to be feasible if–

(i) low enriched uranium targets have been developed and demonstrated for use in the reactors and target processing facilities that produce significant quantities of medical isotopes to serve United States needs for such isotopes;

(ii) sufficient quantities of medical isotopes are available from low enriched uranium targets and fuel to meet United States domestic needs; and

(iii) the average anticipated total cost increase from production of medical isotopes in such facilities without use of highly enriched uranium is less than 10 percent.

(C) REPORT BY THE SECRETARY–Not later than 5 years after the date of enactment of the Energy Policy Act of 2005, the Secretary shall submit to Congress a report that–

(i) contains the findings of the National Academy of Sciences made in the study under subparagraph (A); and

(ii) discloses the existence of any commitments from commercial producers to provide domestic requirements for medical isotopes without use of highly enriched uranium consistent with the feasibility criteria described in subparagraph (B) not later than the date that is 4 years after the date of submission of the report.

(5) SECOND REPORT TO CONGRESS–If the study of the National Academy of Sciences determines under paragraph (4)(A)(i) that the procurement of supplies of medical isotopes from commercial sources that do not use highly enriched uranium is feasible, but the Secretary is unable to report the existence of commitments under paragraph (4)(C)(ii), not later than the date that is 6 years after the date of enactment of the Energy Policy Act of 2005, the Secretary shall submit to Congress a report that describes options for developing domestic supplies of medical isotopes in quantities that are adequate to meet domestic demand without the use of highly enriched uranium consistent with the cost increase described in paragraph (4)(B)(iii).

(6) CERTIFICATION–At such time as commercial facilities that do not use highly enriched uranium are capable of meeting domestic requirements for medical isotopes, within the cost increase described in paragraph (4)(B)(iii) and without impairing the reliable supply of medical isotopes for domestic utilization, the Secretary shall submit to Congress a certification to that effect.

(7) SUNSET PROVISION–After the Secretary submits a certification under paragraph (6), the Commission shall, by rule, terminate its review of export license applications under this subsection.[168]

c. As used in this section[169]–

(1) the term "alternative nuclear reactor fuel or target" means a nuclear reactor fuel or target which is enriched to less than 20 percent in the isotope U-235;

(2) the term "highly enriched uranium" means uranium enriched to 20 percent or more in the isotope U-235; and

(3) a fuel or target "can be used" in a nuclear research or test reactor if–

(A) the fuel or target has been qualified by the Reduced Enrichment Research and Test Reactor Program of the Department of Energy, and

(B) use of the fuel or target will permit the large majority of ongoing and planned experiments and isotope production to be conducted in the reactor without a large percentage increase in the total cost of operating the reactor.

Chapter 12–Control of Information

42 USC 2161.
Policy.

Sec. 141. Policy

It shall be the policy of the Commission to control the dissemination and declassification of Restricted Data in such a manner as to assure the common defense and security. Consistent with such policy, the Commission shall be guided by the following principles:

a. Until effective and enforceable international safeguards against the use of atomic energy for destructive purposes have been established by an international arrangement, there shall be no exchange of Restricted Data with other nations except as authorized by section 144; and

b. The dissemination of scientific and technical information relating to atomic energy should be permitted and encouraged so as to provide that free interchange of ideas and criticism which is essential to scientific and industrial progress and public understanding and to enlarge the fund of technical information.

42 USC 2162.

Classification and declassification of restricted data.

Sec. 142. Classification and Declassification of Restricted Data

a. The Commission shall from time to time determine the data, within the definition of Restricted Data, which can be published without undue risk to the common defense and security and shall thereupon cause such data to be declassified and removed from the category of Restricted Data.

b. The Commission shall maintain a continuous review of Restricted Data and of any Classification guides issued for the guidance of those in the atomic energy program with respect to the areas of Restricted Data which have been declassified in order to determine which information may be declassified and removed from the category of Restricted Data without undue risk to the common defense and security.

c. In the case of Restricted Data which the Commission and the Department of Defense jointly determine to relate primarily to the military utilization of atomic weapons, the determination that such data may be published without constituting an unreasonable risk to the

---

[168]   Amended by P.L. 109–58, §§ 630(2), (3), 119 Stat. 785 (2005).
[169]   Amended by P.L. 109–58, § 630(2), 119 Stat. 785 (2005).

common defense and security shall be made by the Commission and the Department of Defense jointly, and if the Commission and the Department of Defense do not agree, the determination shall be made by the President.

d. The Commission shall remove from the Restricted Data category such data as the Commission and the Department of Defense jointly determine relates primarily to the military utilization of atomic weapons and which the Commission and Department of Defense jointly determine can be adequately safeguarded as defense information: *Provided, however,* That no such data so removed from the Restricted Data category shall be transmitted or otherwise made available to any nation or regional defense organization, while such data remains defense information, except pursuant to an agreement for cooperation entered into in accordance with subsection b. or d. of section 144.[170]

e. The Commission shall remove from the Restricted Data category such information concerning the atomic energy programs of other nations as the Commission and the Director of Central Intelligence jointly determine to be necessary to carry out the provisions of section 102(d) of the National Security Act of 1947, as amended, and can be adequately safeguarded as defense information.

Sec. 143. Department of Defense Participation

The Commission may authorize any of its employees, or employees of any contractor, prospective contractor, licensee or prospective licensee of the Commission or any other person authorized access to Restricted Data by the Commission under subsections 145b.[171] and 145c.[172] to permit any employee of an agency of the Department of Defense or of its contractors, or any member of the Armed Forces to have access to Restricted Data required in the performance of his duties and so certified by the head of the appropriate agency of the Department of Defense or his designee: *Provided, however,* That the head of the appropriate agency of the Department of Defense or his designee has determined, in accordance with the established personnel security procedures and standards of such agency, that permitting the member or employee to have access to such Restricted Data will not endanger the common defense and security: *and provided further,* That the Secretary of Defense finds that the established personnel and other security procedures and standards of such agency are adequate and in reasonable conformity to the standards established by the Commission under section 145.

Sec. 144. International Cooperation

a. The President may authorize the Commission to cooperate with another nation and to communicate to that nation Restricted Data on–

(1) refining, purification, and subsequent treatment of source material;

(2) civilian reactor development;

(3) production of special nuclear material;

(4) health and safety;

(5) industrial and other applications of atomic energy for peaceful purposes; and

---

[170]   Amended by P.L. 103–337, 108 Stat. 3092 (1994).
[171]   Amended by P.L. 84–1006, § 14, 70 Stat. 1069 (1956).
[172]   Amended by P.L. 87–206, § 5, 75 Stat. 475 (1961).

50 USC 406(d).

42 USC 2163.
Department of Defense participation.

42 USC 2164.
International cooperation.

(6) research and development relating to the foregoing: *Provided, however,* That no such cooperation shall involve the communication of Restricted Data relating to the design or fabrication of atomic weapons: *and provided further,* That the cooperation is undertaken pursuant to an agreement for cooperation entered into in accordance with section 123, or is undertaken pursuant to an agreement existing on the effective date of this Act.[173]

Cooperation by
Defense
Department.

b. The President may authorize the Department of Defense, with the assistance of the Commission, to cooperate with another nation or with a regional defense organization to which the United States is a party, and to communicate to that nation or organization such Restricted Data (including design information) as is necessary to–

(1) the development of defense plans;

(2) the training of personnel in the employment of and defense against atomic weapons and other military applications of atomic energy;

(3) the evaluation of the capabilities of potential enemies in the employment of atomic weapons and other military applications of atomic energy; and

(4) the development of compatible delivery systems for atomic weapons;

whenever the President determines that the proposed cooperation and the proposed communication of the Restricted Data will promote and will not constitute an unreasonable risk to the common defense and security, while such other nation or organization is participating with the United States pursuant to an international arrangement by substantial and material contributions to the mutual defense and security: *Provided, however,* That the cooperation is undertaken pursuant to an agreement entered into accordance with section 123.[174]

c. In addition to the cooperation authorized in subsections 144a. and 144b., the President may authorize the Commission, with the assistance of the Department of Defense, to cooperate with another nation and–

---

[173]   Amended by P.L. 85–479, § 5, 72 Stat. 276 (1958).

[174]   Amended by P.L. 85–479, § 6, 72 Stat. 276 (1958), by substituting a new subsection b. Before amendment, subsection b. read as follows:

b. The President may authorize the Department of Defense, with the assistance of the Commission, to cooperate with another nation or with a regional defense organization to which the United States is a party, and to communicate to that nation or organization such Restricted Data as is necessary to—

(1) the development of defense plans;

(2) the training of personnel in the employment of and defense against atomic weapons; and

(3) the evaluation of the capabilities of potential enemies in the employment of atomic weapons.

while such other nation or organization is participating with the United States pursuant to an international arrangement by substantial and material contributions to the mutual defense and security: *Provided, however,* That no such cooperation shall involve communication of Restricted Data relating to the design or fabrication of atomic weapons except with regard to external characteristics, including size, weight, and shape, yields and effects, and systems employed in the delivery or use thereof but not including any data in these categories unless in the joint judgment of the Commission and the Department of Defense such data will not reveal important information concerning the design or fabrication of the nuclear components of an atomic weapon: *And provided further,* That the cooperation is undertaken pursuant to an agreement entered into in accordance with section 123.

(1) to exchange with that nation Restricted Data concerning atomic weapons: *Provided,* That communication of such Restricted Data to that nation is necessary to improve its atomic weapon design, development, or fabrication capability and provided that nation has made substantial progress in the development of atomic weapons; and

(2) to communicate or exchange with that nation Restricted Data concerning research, development, or design, of military reactors, whenever the President determines that the proposed cooperation and the communication of the proposed Restricted Data will promote and will not constitute an unreasonable risk to the common defense and security, while such other nation is participating with the United States pursuant to an international arrangement by substantial and material contributions to the mutual defense and security: *Provided, however,* That the cooperation is undertaken pursuant to an agreement entered into in accordance with section 123.

d.[175] (1) In addition to the cooperation authorized in subsections a., b., and c., the President may, upon making a determination described in paragraph (2), authorize the Department of Energy, with the assistance of the Department of Defense, to cooperate with another nation to communicate to that nation such Restricted Data, and the President may, upon making such determination, authorize the Department of Defense, with the assistance of the Department of Energy, to cooperate with another nation to communicate to that nation such data removed from the Restricted Data category under section 142, as is necessary for–

(A) the support of a program for the control of and accounting for fissile material and other weapons material;

(B) the support of the control of and accounting for atomic weapons;

(C) the verification of a treaty; and

(D) the establishment of international standards for the classification of data on atomic weapons, data on fissile material, and related data.

(2) A determination referred to in paragraph (1) is a determination that the proposed cooperation and proposed communication referred to in that paragraph–

(A) will promote the common defense and security interests of the United States and the nation concerned; and

(B) will not constitute an unreasonable risk to such common defense and security interests.

(3) Cooperation under this subsection shall be undertaken pursuant to an agreement for cooperation entered into in accordance with section 123.

e. The President may authorize any agency of the United States to communicate in accordance with the terms and conditions of an agreement for cooperation arranged pursuant to subsection 144a., b., c., or d., such Restricted Data as is determined to be transmissible under the agreement for cooperation involved.[176]

Sec. 145. Restrictions

a. No arrangement shall be made under section 31, no contract shall be made or continued in effect under section 41, and no license shall be issued under section 103 or 104, unless the person with whom such

42 USC 2165.
Restrictions.

---

[175]   Added by P.L. 103–337, 108 Stat. 3091 (1994).
[176]   Amended by P.L.103–337, 108 Stat. 3092 (1994).

Investigations
by CSC.

arrangement is made, the contractor or prospective contractor, or the prospective licensee agrees in writing not to permit any individual to have access to Restricted Data until the Civil Service Commission shall have made an investigation and report to the Commission on the character, associations, and loyalty of such individual, and the Commission shall have determined that permitting such person to have access to Restricted Data will not endanger the common defense and security.

b. Except as authorized by the Commission or the General Manager upon a determination by the Commission or General Manager that such action is clearly consistent with the national interest, no individual shall be employed by the Commission nor shall the Commission permit any individual to have access to Restricted Data until the Civil Service Commission shall have made an investigation and report to the Commission on the character, associations, and loyalty of such individual, and the Commission shall have determined that permitting such person to have access to Restricted Data will not endanger the common defense and security.

c. In lieu of the investigation and report to be made by the Civil Service Commission pursuant to subsection b. of this section, the Commission may accept an investigation and report on the character, associations, and loyalty of an individual made by another Government agency which conducts personnel security investigations, provided that a security clearance has been granted to such individual by another Government agency based on such investigation and report.

Investigations
by FBI.

d. In the event an investigation made pursuant to subsections a. and b. of this section develops any data reflecting that the individual who is the subject of the investigation is of questionable loyalty, the Civil Service Commission shall refer the matter to the Federal Bureau of Investigation for the conduct of a full field investigation, the results of which shall be furnished to the Civil Service commission for its information and appropriate action.

e. (1) If the President deems it to be in the national interest he may from time to time determine that investigations of any group or class which are required by subsections (a), (b), and (c) of this section be made by the Federal Bureau of Investigation.

(2) In the case of an individual employed in a program known as a Special Access Program,[177] any investigation required by subsections a., b., and c. of this section shall be made by the Federal Bureau of Investigation.[178]

f. (1) Notwithstanding the provisions of subsections a., b., and c. of this section, but subject to subsection (e) of this section, a majority of the members of the Commission may direct that an investigation required by such provisions on an individual described in paragraph (2) be carried out by the Federal Bureau of Investigation rather than by the Civil Service Commission.

(2) An individual described in this paragraph is an individual who is employed --

(A) in a program certified by a majority of the members of the Commission to be of a high degree of importance or sensitivity; or

---

[177]   Amended by P.L. 108–136, Div. C, Title XXXI, § 3131, 117 Stat. 1749 (2003).
[178]   Amended by P.L. 106–65, Div. C, Title XXXI, Subtitle D, § 3144(a), 113 Stat. 934 (1999).

(B) in any other specific position certified by a majority of the members of the Commission to be of a high degree of importance or sensitivity.[179]

g. The commission shall establish standards and specifications in writing as to the scope and extent of investigations, the reports of which will be utilized by the Commission in making the determination, pursuant to subsections a., b., and c. of this section, that permitting a person access to restricted data will not endanger the common defense and security. Such standards and specifications shall be based on the location and class or kind of work to be done, and shall, among other considerations, take into account the degree of importance to the common defense and security of the Restricted Data to which access will be permitted.

h. Whenever the Congress declares that a state of war exists, or in the event of a national disaster due to enemy attack, the Commission is authorized during the state of war or period of national disaster due to enemy attack to employ individuals and to permit individuals access to Restricted Data pending the investigation report, and determination required by section 145b., to the extent that and so long as the Commission finds that such action is required to prevent impairment of its activities in furtherance of the common defense and security.[180]

---

[179]   Amended by P.L. 108–136, Div. C, Title XXXI, § 3131, 117 Stat. 1749 (2003).
[180]   Amended by P.L. 87–206, § 6, 75 Stat. 475 (1961), by redesignating subsection c as subsection d and subsection g as subsection h. This amendment also added new subsections "c," "e," "f," and "g." Before amendment, the section read as follows:

  Sec. 145. Restrictions.–

  a. No arrangement shall be made under section 31, no contract shall be made or continued in effect under section 41, and no license shall be issued under section 103 or 104, unless the person with whom such arrangement is made, the contractor or prospective contractor, or the prospective licensee agrees in writing not to permit any individual to have access to Restricted Data until the Civil Service Commission shall have made an investigation and report to the Commission on the character, associations, and loyalty of such individual and the Commission shall have determined that permitting such person to have access to Restricted Data will not endanger the common defense security.

  b. Except as authorized by the Commission or the General Manager upon a determination by the Commission or General Manager that such action is clearly consistent with the national interest, no individual shall be employed by the Commission nor shall the Commission permit any individual to have access to Restricted Data until the Civil Service Commission shall have made an investigation and report to the Commission on the character, associations, and loyalty of such individual, and the Commission shall have determined that permitting such person to have access to Restricted Data will not endanger the common defense and security.

  c. In the event an investigation made pursuant to subsections a. and b. of this section develops any data reflecting that the individual who is the subject of the investigation is of questionable loyalty, the Civil Service Commission shall refer the matter to the Federal Bureau of Investigation of the conduct of a full field investigation, the results of which shall be furnished to the Civil Service Commission for its information and appropriate action.

  d. If the President deems it to be in the national interest, he may from time to time cause investigations of any group or class which are required by subsections a. and b. of this section to be made by the Federal Bureau of Investigation instead of by the Civil Service Commission.

  e. Notwithstanding the provisions of subsections a. and b. of this section, a majority of the members of the Commission shall certify those specific positions which are of a high degree of importance or sensitivity and upon such certification the investigation and reports required by such provisions shall be made by the Federal Bureau of Investigation instead of by the Civil Service Commission.

  f. The Commission shall establish standards and specifications in writing as to the scope and extent of investigations to be made by the Civil Service Commission pursuant to subsections a. and b. of this section. Such standards

42 USC 2166.
General
Provisions.

Sec. 146. General Provisions

a. Sections 141 to 145, inclusive, shall not exclude the applicable provisions of any other laws, except that no Government agency shall take any action under such other laws inconsistent with the provisions of those sections.

b. The Commission shall have no power to control or restrict the dissemination of information other than as granted by this or any other law.

42 USC 2167.
Regulations.

Sec. 147. Safeguards Information[181]

a. In addition to any other authority or requirement regarding protection from disclosure of information, and subject to subsection (b)(3) of section 552 of title 5 of the United States Code, the Commission shall prescribe such regulations, after notice and opportunity for public comment, or issue such orders, as necessary to prohibit the unauthorized disclosure of safeguards information which specifically identifies a licensee's or applicant's detailed–

(1) control and accounting procedures or security measures (including security plans, procedures, and equipment) for the physical protection of special nuclear material, by whomever possessed, whether in transit or at fixed sites, in quantities determined by the Commission to be significant to the public health and safety or the common defense and security;

(2) security measures (including security plans, procedures, and equipment) for the physical protection of source material or byproduct material, by whomever possessed, whether in transit or at fixed sites, in quantities determined by the Commission to be significant to the public health and safety or the common defense and security; or

(3) security measures (including security plans, procedures, and equipment) for the physical protection of and the location of certain plant equipment vital to the safety of production or utilization facilities involving nuclear materials covered by paragraphs (1) and (2).

If the unauthorized disclosure of such information could reasonably be expected to have a significant adverse effect on the health and safety of the public or the common defense and security by significantly increasing the likelihood of theft, diversion, or sabotage of such material or such facility. The Commission shall exercise the authority of this subsection–

(A) so as to apply the minimum restrictions needed to protect the health and safety of the public or the common defense and security, and

---

and specifications shall be based on the location and class or kind of work to be done, and shall, among other considerations, take into account the degree of importance to the common defense and security of the Restricted Data to which access will be permitted.

g. Whenever the Congress declares that a state of war exists, or in the event of a national disaster due to enemy attack, the Commission is authorized during the state of war or period of national disaster due to enemy attack to employ individuals and to permit individual access to Restricted Data pending the investigation report, and determination required by section 145b., to the extent that and so long as the Commission finds that such action is required to prevent impairment of its activities in furtherance of the common defense and security.

[181]   Added by P.L. 96–295, § 207(a)(1), 94 Stat. 788 (1980).

(B) upon a determination that the unauthorized disclosure of such information could reasonably be expected to have a significant adverse effect on the health and safety of the public or the common defense and security by significantly increasing the likelihood of theft, diversion, or sabotage of such material or such facility.

Nothing in this Act shall authorize the Commission to prohibit the public disclosure of information pertaining to the routes and quantities of shipments of source material, by-product material, high level nuclear waste, or irradiated nuclear reactor fuel. Any person, whether or not a licensee of the Commission, who violates any regulations adopted under this section shall be subject to the civil monetary penalties of section 234 of this Act. Nothing in this section shall be construed to authorize the withholding of information from the duly authorized committees of the Congress.

42 USC 2282.

42 USC 2273.

b. For the purpose of section 223 of this Act, any regulations or orders prescribed or issued by the Commission under this section shall also be deemed to be prescribed or issued under section 161b. of this Act.

c. Any determination by the Commission concerning the applicability of this section shall be subject to judicial review pursuant to subsection (a)(4)(B) of section 552 of title 5 of the United States Code.

d. Upon prescribing or issuing any regulation or order under subsection a. of this section, the Commission shall submit to Congress a report that:

(1) specifically identifies the type of information the Commission intends to protect from disclosure under the regulation or order;

(2) specifically states the Commission's justification for determining that unauthorized disclosure of the information to be protected from disclosure under the regulation or order could reasonably be expected to have a significant adverse effect on the health and safety of the public or the common defense and security by significantly increasing the likelihood of theft, diversion, or sabotage of such material or such facility, as specified under subsection (a) of this section; and

(3) provides justification, including proposed alternative regulations or orders, that the regulation or order applies only the minimum restrictions needed to protect the health and safety of the public or the common defense and security.

e. In addition to the reports required under subsection d. of this section, the Commission shall submit to Congress on a quarterly basis a report detailing the Commission's application during that period of every regulation or order prescribed or issued under this section. In particular, the report shall:[182]

(1) identify any information protected from disclosure pursuant to such regulation or order;

(2) specifically state the Commission's justification for determining that unauthorized disclosure of the information protected from disclosure under such regulation or order could reasonably be expected to have a significant adverse effect on the health and safety of the public or the common defense and security by significantly increasing the likelihood of theft, diversion or sabotage of such

---

[182]   As a result of P.L. 104–66, § 3003, 109 Stat. 734 (1995), section 147e ceased to be effective on December 21, 1999.

material or such facility, a specified under subsection a. of this section; and

(3) provide justification that the Commission has applied such regulation or order so as to protect from disclosure only the minimum amount of information necessary to protect the health and safety of the public or the common defense and security.

42 USC 2168.

Sec. 148. Prohibition Against the Dissemination of Certain Unclassified Information[183]

Regulations.

a. (1) In addition to any authority or requirement regarding protection from dissemination of information, and subject to section 552(b)(3) of title 5, United States Code, the Secretary of Energy (hereinafter in this section referred to as the "Secretary" with respect to atomic energy defense programs,[184]) shall prescribe such regulations, after notice and opportunity for public comment thereon, or issue such orders as may be necessary to prohibit the unauthorized dissemination of unclassified information pertaining to—

(A) the design of production facilities or utilization facilities;

(B) security measures (including security plans, procedures, and equipment) for the physical protection of (i) production or utilization facilities, (ii) nuclear material contained in such facilities, or (iii) nuclear material in transit; or

42 USC 2162.

(C) the design, manufacture, or utilization of any atomic weapon or component if the design, manufacture, or utilization of such weapon or component was contained in any information declassified or removed from the Restricted Data category by the Secretary (or the head of the predecessor agency of the Department of Energy) pursuant to section 142.

(2) The Secretary may prescribe regulations or issue orders under paragraph (1) to prohibit the dissemination of any information described in such paragraph only if and to the extent that the Secretary determines that the unauthorized dissemination of such information could reasonably be expected to have a significant adverse effect on the health and safety of the public or the common defense and security by significantly increasing the likelihood of (A) illegal production of nuclear weapons, or (B) theft, diversion, or sabotage of nuclear materials, equipment, or facilities.

(3) In making a determination under paragraph (2), the Secretary may consider what the likelihood of an illegal production, theft, diversion, or sabotage referred to in such paragraph would be if the information proposed to be prohibited from dissemination under this section were at no time available for dissemination.

(4) The Secretary shall exercise his authority under this subsection to prohibit the dissemination of any information described in subsection a.(1)–

(A) so as to apply the minimum restrictions needed to protect the health and safety of the public or the common defense and security; and

(B) upon a determination that the unauthorized dissemination of such information could reasonably be expected to result in a significant adverse effect on the health and safety of the public or the common defense and security by significantly increasing the

---

[183]   Added by P.L. 97–90, § 210(a)(1), 95 Stat. 1163 (1981).
[184]   Amended by P.L. 97–415, § 17, 96 Stat. 2067 (1983).

likelihood of (i) illegal production of nuclear weapons, or (ii) theft, diversion, or sabotage of nuclear materials, equipment, or facilities.

(5) Nothing in this section shall be construed to authorize the Secretary to authorize the withholding of information from the appropriate committees of the Congress.

Penalties.

b. (1) Any person who violates any regulation or order of the Secretary issued under this section with respect to the unauthorized dissemination of information shall be subject to a civil penalty, to be imposed by the Secretary, of not to exceed $100,000 for each such violation. The Secretary may compromise, mitigate, or remit any penalty imposed under this subsection.

42 USC 2282.

(2) The provisions of subsections b. and c. of section 234 of this Act shall be applicable with respect to the imposition of civil penalties by the Secretary under this section in the same manner that such provisions are applicable to the imposition of civil penalties by the Commission under subsection a. of such section.

42 USC 2273.

c. For the purposes of section 223 of this Act, any regulation prescribed or order issued by the Secretary under this section shall also be deemed to be prescribed or issued under section 161b. of this Act.

Judicial review.

d. Any determination by the Secretary concerning the applicability of this section shall be subject to judicial review pursuant to section 552(a)(4)(B) of title 5, United States Code.

Quarterly report.

e. The Secretary shall prepare on a quarterly basis a report to be made available upon the request of any interested person, detailing the Secretary's application during that period of each regulation or order prescribed or issued under this section. In particular, such report shall–

(1) identify any information protected from disclosure pursuant to such regulation or order;

(2) specifically state the Secretary's justification for determining that unauthorized dissemination of the information protected from disclosure under such regulation or order could reasonably be expected to have a significant adverse effect on the health and safety of the public or the common defense and security by significantly increasing the likelihood of illegal production of nuclear weapons, or theft, diversion, or sabotage of nuclear materials, equipment, or facilities, as specified under subsection a.; and

(3) provide justification that the Secretary has applied such regulation or order so as to protect from disclosure only the minimum amount of information necessary to protect the health and safety of the public or the common defense and security.[185]

---

[185]   Amended by P.L. 97–415, § 17, 96 Stat. 2067 (1983), by adding new subsections "d" and "e".

42 USC 2169.

Sec. 149. Fingerprinting for Criminal History Record Checks[186]

a.(1)(A)(i) The Commission shall require each individual or entity described in clause (ii) to fingerprint each individual described in subparagraph (B) before the individual described in subparagraph (B) is permitted access under subparagraph (B).

(ii) The individuals and entities referred to in clause (i) are individuals and entities that, on or before the date on which an individual is permitted access under subparagraph (B)–

(I) are licensed or certified to engage in an activity subject to regulation by the Commission;

(II) have filed an application for a license or certificate to engage in an activity subject to regulation by the Commission; or

Notification.

(III) have notified the Commission in writing of an intent to file an application for licensing, certification, permitting, or approval of a product or activity subject to regulation by the Commission.

(B) The Commission shall require to be fingerprinted any individual who–

(i) is permitted unescorted access to–

(I) a utilization facility; or

(II) radioactive material or other property subject to regulation by the Commission that the Commission determines to be of such significance to the public health and safety or the common defense and security as to warrant fingerprinting and background checks; or

(ii) is permitted access to safeguards information under section 147;

(2) All fingerprints obtained by an individual or entity as required in paragraph (1) shall be submitted to the Attorney General of the United States through the Commission for identification and a criminal history records check.

(3) The costs of an identification or records check under paragraph (2) shall be paid by the individual or entity required to conduct the fingerprinting under paragraph (1)(A).

(4) Notwithstanding any other provision of law–

(A) the Attorney General may provide any result of an identification or records check under paragraph (2) to the Commission; and

---

[186]   Added by P.L. 99–399, § 606, 100 Stat. 853 (1986). P.L. 109–58, 119 Stat. 810 (2005), rewrote subsection (a), which formerly read as follows:

(a) The Nuclear Regulatory Commission (in this section referred to as the 'Commission') shall require each licensee or applicant for a license to operate a utilization facility under section 2133 or 2134(b) of this title to fingerprint each individual who is permitted unescorted access to the facility or is permitted access to safeguards information under section 2167 of this title. All fingerprints obtained by a licensee or applicant as required in the preceding sentence shall be submitted to the Attorney General of the United States through the Commission for identification and a criminal history records check. The costs of any identification and records check conducted pursuant to the preceding sentence shall be paid by the licensee or applicant. Notwithstanding any other provision of law, the Attorney General may provide all the results of the search to the Commission, and, in accordance with regulations prescribed under this section, the Commission may provide such results to the licensee or applicant submitting such fingerprints.

(B) the Commission, in accordance with regulations prescribed under this section, may provide the results to the individual or entity required to conduct the fingerprinting under paragraph (1)(A).

b. The Commission, by rule, may relieve persons from the obligations imposed by this section, under specified terms, conditions, and periods, if the Commission finds that such action is consistent with its obligations to promote the common defense and security and to protect the health and safety of the public.

c. For purposes of administering this section, the Commission shall prescribe, subject to requirements–

(1) to implement procedures for the taking of fingerprints;

(2) to establish the conditions for use of information received from the Attorney General, in order–

(A) to limit the redissemination of such information ;

(B) to ensure that such information is used solely for the purpose of determining whether an individual shall be permitted unescorted access to a utilization facility, radioactive material, or other property described in subsection a.(1)(B) or shall be permitted access to safeguards information under section 147;

(C) to ensure that no final determination may be made solely on the basis of information provided under this section involving–

(i) an arrest more than 1 year old for which there is no information of the disposition of the case; or

(ii) an arrest that resulted in dismissal of the charge or an acquittal; and

(D) to protect individuals subject to fingerprinting under this section from misuse of the criminal history records; and

(3) to provide each individual subject to fingerprinting under this section with the right to complete, correct, and explain information contained in the criminal history records prior to any final adverse determination.

d. The Commission may require a person or individual to conduct fingerprinting under subsection a.(1) by authorizing or requiring the use of any alternative biometric method for identification that has been approved by–

(1) the Attorney General; and

(2) the Commission, by regulation.

e. (1) The Commission may establish and collect fees to process fingerprints and criminal history records under this section.

(2) Notwithstanding section 3302(b) of title 31, United States Code, and to the extent approved in appropriation Acts–

(A) a portion of the amounts collected under this subsection in any fiscal year may be retained and used by the Commission to carry out this section; and

(B) the remaining portion of the amounts collected under this subsection in such fiscal year may be transferred periodically to the Attorney General and used by the Attorney General to carry out this section.

(3) Any amount made available for use under paragraph (2) shall remain available until expended.

Chapter 13–Patents and Inventions

Sec. 151. Inventions Relating to Atomic Weapons, and Filing of Reports

a.[187] No patent shall hereafter be granted for any invention or discovery which is useful solely in the utilization of special nuclear material or atomic energy in an atomic weapon. Any patent granted for any such invention or discovery is hereby revoked, and just compensation shall be made therefor.

b. No patent hereafter granted shall confer any rights with respect to any invention or discovery to the extent that such invention or discovery is used in the utilization of special nuclear material or atomic energy in atomic weapons. Any rights conferred by any patent heretofore granted for any invention or discovery are hereby revoked to the extent that such invention or discovery is so used, and just compensation shall be made therefor.

c. Any person who has made or hereafter makes any invention or discovery useful in the production or utilization of special nuclear material or atomic energy, shall file with the Commission a report containing a complete description thereof unless such invention or discovery is described in an application for a patent filed with the Under Secretary of Commerce for Intellectual Property and Director of the United States Patent and Trademark Office[188] by such person within the time required for the filing of such report. The report covering any such invention or discovery shall be filed on or before the one hundred and eightieth day after such person first discovers or first has reason to believe that such invention or discovery is useful in such production or utilization.[189]

"d. The Under Secretary of Commerce for Intellectual Property and Director of the United States Patent and Trademark Office[190] shall notify the Commission of all applications for patents heretofore or hereafter filed which, in his opinion, disclose inventions or discoveries required to be reported under subsection 151c., and shall provide the Commission access to all such applications.

"e. Reports filed pursuant to subsection c. of this section, and application to which access is provided under subsection d. of this section, shall be kept in confidence by the Commission, and no information concerning the same given without authority of the inventor or owner unless necessary to carry out the provisions of any Act of

---

[187]   Amended by P.L. 87–206, § 7, 75 Stat. 475 (1961).
[188]   Amended by by P.L. 106–113, Division B, § 1000(a)(9), 113 Stat 1536 (1999).
[189]   Amended by P.L. 87–206, § 8, 75 Stat. 475 (1961). Before amendment, section 151c read as follows:

> c. Any person who has made or hereafter makes any invention or discovery useful (1) in the production or utilization of special nuclear material or atomic energy; (2) in the utilization of special nuclear material in an atomic weapon; or (3) in the utilization of atomic energy in an atomic weapon, shall file with the Commission a report containing a complete description thereof unless such invention or discovery is described in an application for a patent filed with the Commission of Patents by such person within the time required for the filing of such report. The report covering any such invention or discovery shall be filed on or before whichever of the following is the later either the ninetieth day after completion of such invention or discovery; or the ninetieth day after such person first discovers or first has reason to believe that such invention or discovery is useful in such production or utilization.

[190]As amended by P.L. 106–113, Division B, § 1000(a)(9), 113 Stat 1536 (1999).

Congress or in such special circumstances as may be determined by the Commission.[191]

42 USC 2182.

Invention
conceived
during
Commission
contracts.

Sec. 152. Inventions Made or Conceived During Commission Contracts

Any invention or discovery, useful in the production or utilization of special nuclear material or atomic energy, made or conceived in the course of or under any contract, subcontract, or arrangement entered into with or for the benefit of the Commission, regardless of whether the contract, subcontract, or arrangement involved the expenditure of funds by the Commission, shall be vested in, and be the property of, the Commission, except that the Commission may waive its claim to any such invention or discovery under such circumstances as the Commission may deem appropriate, consistent with the policy of this section. No patent for any invention or discovery, useful in the production or utilization of special nuclear material or atomic energy, shall be issued unless the applicant files with the application, or within thirty days after request therefor by the Under Secretary of Commerce for Intellectual Property and Director of the United States Patent and Trademark Office[192] (unless the Commission advises the Under Secretary of Commerce for Intellectual Property and Director of the United States Patent and Trademark Office that its rights have been determined and that accordingly no statement is necessary) a statement under oath setting forth the full facts surrounding the making or conception of the invention or discovery described in the application and whether the invention or discovery was made or conceived in the course of or under any contract, subcontract, or arrangement entered into with or for the benefit of the Commission, regardless of whether the contract, subcontract, or arrangement involved the expenditure of funds by the Commission. The Under Secretary of Commerce for Intellectual Property and Director of the United States Patent and Trademark Office shall as soon as the application is otherwise in condition for allowance[193] forward copies of the application and the statement to the Commission.

The Under Secretary of Commerce for Intellectual Property and Director of the United States Patent and Trademark Office may proceed with the application and issue the patent to the applicant (if the invention or discovery is otherwise patentable) unless the Commission, within 90 days after receipt of copies of the application and statement, directs the Under Secretary of Commerce for Intellectual Property and Director of the United States Patent and Trademark Office to issue the patents to the Commission (if the invention or discovery is otherwise patentable) to be held by the Commission as the agent of and on behalf of the United States.

If the Commission files such a direction with the Under Secretary of Commerce for Intellectual Property and Director of the United States Patent and Trademark Office, and if the applicant's statement claims, and the applicant still believes, that the invention or discovery was not made or conceived in the course of or under any contract, subcontract or arrangement entered into with or for the benefit of the Commission entitling the Commission to the title to the applicant or the patent the applicant may, within 30 days after notification of the filing of such a

---

[191]   P.L. 87–206, § 9, 75 Stat. 475 (1961), added subsection e.
[192]   As amended by P.L. 106–113, Division B, § 1000(a)(9), 113 Stat 1536 (1999).
[193]   Amended by P.L. 87–615, § 11, 76 Stat. 409 (1962). Prior to amendment, the word was "allowances."

direction, request a hearing before the Board of Patents Appeals and Interferences. The Board shall have the power to hear and determine whether the Commission was entitled to the direction filed with the Under Secretary of Commerce for Intellectual Property and Director of the United States Patent and Trademark Office. The Board shall follow the rules and procedures established for interference cases and an appeal may be taken by either the applicant or the Commission from the final order of the Board to the United States Court of Appeals for the Federal Circuit in accordance with the procedures governing the appeals from the Board of Patent Appeals an Interferences.

If the statement filed by the applicant should thereafter be found to contain false material statements any notification by the Commission that it has no objections to the issuance of a patent to the applicant shall not be deemed in any respect to constitute a waiver of the provisions of this section or of any applicable civil or criminal statute, and the Commission may have the title to the patent transferred to the Commission on the records of the Under Secretary of Commerce for Intellectual Property and Director of the United States Patent and Trademark Office in accordance with the provisions of this section. A determination of rights by the Commission pursuant to a contractual provision or other arrangement prior to the request of the Under Secretary of Commerce for Intellectual Property and Director of the United States Patent and Trademark Office for the statement, shall be final in the absence of false material statements or nondisclosure of material facts by the applicant.[194]

---

[194]   Amended by P.L. 87–206, § 10, 75 Stat. 475 (1961), amended section 152. Before amendment, this section read as follows:

   Sec. 152. Inventions Made or Conceived During Commission Contracts–Any invention or discovery, useful in the production or utilization of special nuclear material or atomic energy, made or conceived under any contract, subcontract, arrangement, or other relationship with the Commission, regardless of whether the contract or arrangement involved the expenditure of funds by the Commission, shall be deemed to have been made or conceived by the Commission, except the Commission may waive its claim to any such invention or discovery if made or conceived by any person at or in connection with any laboratory under the jurisdiction of the Commission as provided in section 33, or under such other circumstances as the Commission may deem appropriate. No patent for any invention or discovery, useful in the production or utilization of special nuclear material or atomic energy, shall be issued unless the applicant files with the application, or within 30 days after request therefor by the Under Secretary of Commerce for Intellectual Property and Director of the United States Patent and Trademark Office, a statement under oath setting forth the full facts surrounding the making or conception of the invention or discovery described in the application and whether the invention or discovery was made or conceived in the course of, in connection with or under the terms of any contract, subcontract, arrangement, or other relationship with the Commission, regardless of whether the contract or agreement involved the expenditure of funds by the Commission. The Under Secretary of Commerce for Intellectual Property and Director of the United States Patent and Trademark Office shall forthwith forward copies of the application and the statement to the Commission.

   The Under Secretary of Commerce for Intellectual Property and Director of the United States Patent and Trademark Office may proceed with the application and issue the patent to the applicant (if the invention or discovery is otherwise patentable) unless the Commission, within 90 days after receipt of copies of the application and statement, directs the Under Secretary of Commerce for Intellectual Property and Director of the United States Patent and Trademark Office to issue the patent to the Commission (if the invention or discovery is otherwise patentable) to be held by the Commission as the agent of and on behalf of the United States.

   If the Commission files such a direction with the Under Secretary of Commerce for Intellectual Property and Director of the United States Patent

42 USC 2183.
Nonmilitary
utilization.

Sec. 153. Nonmilitary Utilization

a. The Commission may, after giving the patent owner an opportunity for a hearing, declare any patent to be affected with the public interest if (1) the invention or discovery covered by the patent is of primary importance in the production or utilization of special nuclear material or atomic energy; and (2) the licensing of such invention or discovery under this section is of primary importance to effectuate the policies and purposes of this Act.

b. Whenever any patent has been declared affected with the public interest, pursuant to subsection 153a.–

(1) the Commission is hereby licensed to use the invention or discovery covered by such patent in performing any of its powers under this Act;

(2) any person may apply to the Commission for a nonexclusive patent license to use the intervention or discovery covered by such patent, and the Commission shall grant such patent license to the extent that it finds that the use of the invention or discovery is of primary importance to the conduct of an activity by such person authorized under this Act.

c. Any person–

(1) who has made application to the Commission for a license under section 53, 62, 63, 81, 103, or 104, or a permit or lease under section 67;

(2) to whom such license, permit, or lease has been issued by the Commission;

(3) who is authorized to conduct such activities as such applicant is conducting or proposed to conduct under a general license issued by the Commission under section 62 or 81; or

(4) whose activities or proposed activities are authorized under section 31, may at any time make application to the Commission for a patent license for the use of an invention or discovery useful in the production or utilization of special nuclear material or atomic energy covered by a patent. Each such application shall set forth the nature and purpose of the use which the applicant intends to make of the

and Trademark Office, and if the applicant's statement claims, and the applicant still believes, that the invention or discovery was not made or conceived in the course of, in connection with, or under the terms of any contract, subcontract, arrangement, or other relationship with the Commission entitling the Commission to take title to the application or the patent the applicant may, within 30 days after notification of the filing of such a direction, request a hearing before a Board of Patents Interferences. The Boards shall have the power to hear and determine whether the Commission was entitled to the direction filed with the Under Secretary of Commerce for Intellectual Property and Director of the United States Patent and Trademark Office. The Board shall follow the rules and procedures established for interference cases and procedures established an appeal may be taken by either the applicant or the Commission from the final order of the Board to the Court of Customs and Patent Appeals in accordance with the procedures governing the appeals from the Board of Patent Interferences (amended by P.L. 97–164 and P.L. 98–622).

If the statement filed by the applicant should thereafter be found to contain false material statements any notification by the Commission that it has no objections to the issuance of a patent to the applicant shall not be deemed in any respect to constitute a waiver of the provisions of this section or of any applicable civil or criminal statute, and the Commission may have the title to the patent transferred to the Commission on the records of the Under Secretary of Commerce for Intellectual Property and Director of the United States Patent and Trademark Office in accordance with the provisions of this section.

patent license, the steps taken by the applicant to obtain a patent license from the owner of the patent, and a statement of the effects, as estimated by the applicant, on the authorized activities which will result from failure to obtain such patent license and which will result from the granting of such patent license.

d. Whenever any person has made an application to the Commission for a patent license pursuant to subsection 153c.–

(1) the Commission, within 30 days after the filing of such application, shall make available to the owner of the patent all of the information contained in such application, and shall notify the owner of the patent of the time and place at which a hearing will be held by the Commission;

(2) the Commission shall hold a hearing within 60 days after the filing of such application at a time and place designated by the Commission; and

(3) in the event an applicant applies for two or more patent licenses, the Commission may, in it discretion, order the consolidation of such applications, and if the patents are owned by more than one owner, such owners may be made parties to one hearing.

e. If, after any hearing conducted pursuant to subsection 153d, the Commission finds that–

(1) the invention or discovery covered by the patent is of primary importance in the production or utilization of special nuclear material atomic energy;

(2) the licensing of such invention or discovery is of primary importance to the conduct of the activities of the applicant;

(3) the activities to which the patent license are proposed to be applied by such applicant are of primary importance to the furtherance of policies and purposes of this Act; and

(4) such applicant cannot otherwise obtain a patent license from the owner of the patent on terms which the Commission deems to be reasonable for the intended use of the patent to be made by such applicant, the Commission shall license the applicant to use the invention or discovery covered by the patent for the purposes stated in such application on terms deemed equitable by the Commission and generally not less fair than those granted by the patents or by the Commission to similar licensees for comparable use.

f. The Commission shall not grant any patent license pursuant to subsection 153e. for any other purpose than that stated in the application. Nor shall the Commission grant any patent license to any other applicant for a patent license on the same patent without an application being made by such applicant pursuant to subsection 153c., and without separate notification and hearing as provided in subsection 153d., and without a separate finding as provided in subsection 153e.

g. The owner of the patent affected by a declaration or a finding made by the Commission pursuant to subsection 153b. or 153e. shall be entitled to a reasonable royalty fee from the licensee for any use of an invention or discovery licensed by the section. Such royalty fee may be agreed upon by such owner and the patent licensee, or in the absence of such agreement shall be determined for each patent license by the Commission pursuant to subsection 157c.

h. The provisions of this section shall apply to any patent the application for which shall have been filed before September 1, 1979.[195]

42 USC 2184.
Injunctions.

Sec. 154. Injunctions

No court shall have jurisdiction or power to stay, restrain, or otherwise enjoin the use of any invention or discovery by a patent licensee, to the extent that such use is licensed by subsection 153b. or 153e. If, in any action against such patent licensee, the court shall determine that the defendant is exercising such license, the measure of damages shall be the royalty fee determined pursuant to subsection 157c. If any such patent licensee shall fail to pay such royalty fee, the patentee may bring an action in any court of competent jurisdiction for such royalty fee, together with such costs, interest, and reasonable attorney's fees as may be fixed by the court.

42 USC 2185.
Prior art.

Sec. 155. Prior Art

In connection with applications for patents covered by this chapter, the fact that the invention or discovery was known or used before shall be a bar to the patenting of such invention or discovery even though such prior knowledge or use was under secrecy within the atomic energy program of the United States.

42 USC 2186.
Commission
patent licenses.

Sec. 156. Commission Patent Licenses

The Commission shall establish standards specifications upon which it may grant a patent license to use any patent[196] declared to be affected with the public interest pursuant to subsection 153a. Such a patent license shall not waive any of the other provisions of this Act.

42 USC 2187.
Compensation,
awards, and
royalties.

Sec. 157. Compensation, Awards, and Royalties

a. Patent Compensation Board.–The Commission shall designate a patent Compensation Board to consider applications under this section. The members of the Board shall receive a per diem compensation for each day spent in meetings or conferences, and all members shall receive their necessary traveling or other expenses while engaged in the work of the Board. The members of the Board may serve as such without regard to the provisions of section 281, 283, or 284 of title 18[197] of the United States Code, except in so far as such sections may prohibit any such member from receiving compensation in respect of any particular matter which directly involves the Commission or in which the Commission is directly interested.

Eligibility.

b. Eligibility.–

(1) Any owner of a patent licensed under section 158 or subsection 153b. or 153e., or any patent licensed thereunder may make application to the Commission for the determination of a reasonable royalty fee in accordance with such procedures as the Commission by regulation may establish.

(2) Any person seeking to obtain the just compensation provided in section 151 shall make application therefor to the Commission in accordance with such procedures as the Commission may by regulation establish.

(3) Any person making any invention or discovery useful in the production or utilization of special nuclear material or atomic energy, who is not entitled to compensation or a royalty therefor under this Act and who has complied with the provisions of section 151c. hereof

---

[195]   Amended by P.L. 86–50, § 114, 73 Stat. 81 (1959); P.L. 91–161, § 1,83 Stat. 444 (1969); P.L. 93–377, § 6, 88 Stat. 475 (1974).
[196]   Amended by P.L. 96–517, § 7(a), 94 Stat. 3027 (1980).
[197]   These sections have been repealed.

may make application to the Commission for, and the Commission may grant, an award. The Commission may also, after consultation with[198] the General Advisory Committee, and with the approval of the President, grant an award for any especially meritorious contribution to the development, use, or control of atomic energy.

Standards.

c. Standards.–

(1) In determining a reasonable royalty fee as provided for in subsection 153b., or 153e., the Commission shall take into consideration (A) the advice of the Patent Compensation Board; (B) any defense, general or special, that might be pleaded by a defendant in an action for infringement; (C) the extent to which, if any, such patent was developed through federally financed research; and (D) the degree of utility, novelty, and importance of the invention or discovery, and, may consider the cost to the owner of the patent of developing such invention or discovery or acquiring such patent.

(2) In determining what constitutes just compensation as provided for in section 151, or in determining the amount of any award under subsection 157b.(3), the Commission shall take into account the considerations set forth in subsection 157c.(1) and the actual use of such invention or discovery. Such compensation may be paid by the Commission in periodic payments or in a lump sum.

d. Period of Limitations.–Every application under this section shall be barred unless filed within six years after the date on which first accrues the right of such reasonable royalty fee, just compensation, or award for which such application is filed.[199]

42 USC 2188.
Monopolistic use of patents.

Sec. 158. Monopolistic Use of Patents

Whenever the owner of any patent hereafter granter for any invention or discovery or primary use in the utilization or production of special nuclear material or atomic energy is found by a court of competent jurisdiction to have intentionally used such patent in a manner so as to violate any of the antitrust laws specified in subsection 105a., there may be included in the judgement of the court, in its discretion and in addition to any other lawful sanction, a requirement that such owner license such patent to any other licensee of the Commission who demonstrates a need therefor. If the court, at its discretion, deems that such licensee shall pay a reasonable royalty to the owner of the patent, the reasonable royalty shall be determined in accordance with section 157.[200]

42 USC 2189.
Federally financed research.

Sec. 159. Federally Financed Research

Nothing in this Act shall affect the right of the Commission to require the patents granted on inventions made or conceived during the course of federally financed research or operations, be assigned to the United States.

42 USC 2190.
Saving clause.

Sec. 160. Saving Clause

Any patent application on which a patent was denied by the United States Patent Office under section 11(a)(1), 11(a)(2), or 11(b) of the Atomic Energy Act of 1946,[201] and which is not prohibited by section 151 or section 155 of this Act may be reinstated upon application to the Under Secretary of Commerce for Intellectual Property and Director of the United States Patent and Trademark Office within one year after enactment of this Act and shall then be deemed to have been

---

[198]   Amended by P.L. 93–276, § 201, 88 Stat. 115 (1974)..
[199]   Subsection d. added by P.L. 87–206, § 11, 75 Stat. 475 (1961).
[200]   Amended by P.L. 87–206 § 12, 75 Stat. 475 (1961).
[201]   See Atomic Energy Act of 1946, Appendix 4, section 11.

continuously pending since its original filing date: *Provided, however,* That no patent issued upon any patent application so reinstated shall in any way furnish a basis of claim against the Government of the United States.

Chapter 14–General Authority

42 USC 2201.
General
provisions.

Sec. 161. General Provisions

In the performance of its functions the Commission is authorized to–

a. establish advisory boards to advise with and make recommendations to the Commission on legislation, policies, administration, research, and other matters, provided that the Commission issues regulations setting forth the scope, procedure, and limitation of the authority of each such board;

b. establish by rule, regulation, or order, such standards and instructions to govern the possession and use of special nuclear material, source material, and byproduct material as the Commission may deem necessary or desirable to promote the common defense and security or to protect health or to minimize danger to life or property; in addition, the Commission shall prescribe such regulations or orders as may be necessary or desirable to promote the Nation's common defense and security with regard to control, ownership, or possession of any equipment or device, or important component part especially designed for such equipment or device, capable of separating the isotopes of uranium or enriching uranium in the isotope 235;[202]

c. make such studies and investigations, obtain such information, and hold such meetings or hearings as the Commission may deem necessary or proper to assist it in exercising any authority provided in this Act, or in the administration or enforcement of this Act, or any regulations or orders issued thereunder. For such purposes the Commission is authorized to administer oaths and affirmations, and by subpoena to require any person to appear and testify or appear and produce documents, or both, at any designated place. Witnesses subpoenaed under this subsection shall be paid the same fees and mileage as are paid witnesses in the district courts of the United States;[203]

d. Appoint and fix the compensation of such officers and employees as may be necessary to carry out the functions of the Commission. Such officers and employees shall be appointed in accordance with the civil-service laws and their compensation fixed in accordance with the Classification Act of 1949, as amended, except that, to the extent the Commission deems such action necessary to the discharge of its responsibilities, personnel may be employed and their compensation fixed without regard to such laws: *Provided, however,* That no officer or employee (except such officers and employees whose compensation is fixed by law, and scientific and technical personnel up to a limit of the highest rate of grade 18 of the General Schedule of the Classification Act

5 USC 5101.

---

[202]   Amended by P.L. 101–575, 104 Stat. 2835 (1990).
[203]   Amended by P.L. 91–452, § 237, 84 Stat. 922 (1970), which deleted the following sentence from subsection 161c:
    No person shall be excused from complying with any requirements under this paragraph because of his privilege against self–incrimination, but the immunity provisions of the Compulsory Testimony Act of February 11, 1893, shall apply with respect to any individual who specifically claims such privilege.

of 1949, as amended)[204] whose position would be subject to the Classification Act of 1949, as amended, if such Act were applicable to such position, shall be paid a salary at a rate in excess of the rate payable under such Act for positions of equivalent difficulty or responsibility. Such rates of compensation may be adopted by the Commission as may be authorized by the Classification Act of 1949, as amended, as of the same date such rates are authorized for positions subject to such Act.[205] The Commission shall make adequate provision for administrative review of any determination to dismiss any employee;

e. Acquire such material, property, equipment, and facilities, establish or construct such buildings and facilities, and modify such buildings and facilities from time to time, as it may deem necessary, and construct, acquire, provide, or arrange for such facilities and services (at project sites where such facilities and services are not available) for the housing, health, safety, welfare, and recreation of personnel employed by the Commission as it may deem necessary, subject to the provisions of section 174: *Provided, however,* That in the communities owned by the Commission, the Commission is authorized to grant privileges, leases and permits upon adjusted terms which (at the time of the initial grant of any privilege, grant, lease, or permit, or renewal thereof, or in order to avoid inequities or undue hardship prior to the sale by the United States of property affected by such grant)[206] are fair and reasonable to responsible persons to operate commercial businesses without advertising and without advertising (sic) and without securing competitive bids, but taking into consideration, in addition to the price, and among other things (1) the quality and type of services required by the residents of the community, (2) the experience of each concession applicant in the community and its surrounding area, (3) the ability of the concession applicant to meet the needs of the community, and (4) the contribution the concession applicant has made or will make to the other activities and general welfare of the community;[207]

f. with the consent of the agency concerned, utilize or employ the services or personnel of any Government agency or any State or local government, or voluntary or uncompensated personnel, to perform such functions on its behalf as may appear desirable;

g. Acquire, purchase, lease, and hold real and personal property, including patents, as agent of and on behalf of the United States,[208] subject to the provisions of section 174, and to sell, lease, grant, and dispose of such real and personal property as provided in this Act;

h. consider in a single application one or more of the activities for which a license is required by this Act, combine in a single license one or more of such activities, and permit the applicant or licensee to

---

[204] Amended by P.L. 87–793, § 1001(g), 76 Stat. 832 (1962), which added the words "up to a limit of the highest rate of grade 18 of the General Schedule of the Classification Act of 1949, as amended." Prior to this amendment, a limitation of $19,000 had been imposed by P.L. 85–287, § 4, 71 Stat. 612 (1957).

[205] Amended by P.L. 85–681, 72 Stat. 633 (1958).

[206] Amended by P.L. 85–162 § 201, 71 Stat. 403 (1957), which added the clause: (at the time of the initial grant of any privilege grant, lease, or permit, or renewal thereof, or in order to avoid inequalities or undue hardship prior to the sale by the United States of property affected by such grant.)

[207] Amended by P.L. 84–722, 70 Stat. 553 (1956).

[208] The text of Executive Order 9816, providing for the transfer of properties and personnel of the Manhattan Engineer District to the Atomic Energy Commission on January 1, 1947, will be found in Appendix 8.

incorporate by reference pertinent information already filed with the Commission;

    i. prescribe such regulations or order as it may deem necessary

    (1) to protect Restricted Data received by any person in connection with any activity authorized pursuant to this Act,

    (2) to guard against the loss or diversion of any special nuclear material acquired by any person pursuant to section 53 or produced by any person in connection with any activity authorized pursuant to this Act, to prevent any use or disposition thereof which the Commission may determine to be inimical to the common defense and security, including regulations or orders designating activities, involving quantities of special nuclear material which in the opinion of the Commission are important to the common defense and security, that may be conducted only by persons whose character, associations, and loyalty shall have been investigated under standards and specifications established by the Commission and as to whom the Commission shall have determined that permitting each such person to conduct the activity will not be inimical to the common defense and security,[209]

42 USC 2201(K).

    (3) to govern any activity authorized pursuant to this chapter, including standards and restrictions governing the design, location, and operation of facilities used in the conduct of such activity, in order to protect health and to minimize danger to life or property; and

    (4) to ensure that sufficient funds will be available for the decommissioning of any production or utilization facility licensed under section 103 or 104b., including standards and restrictions governing the control, maintenance, use, and disbursement by any former licensee under this Act that has control over any fund for the decommissioning of the facility,[210]

40 USC 471.
40 USC 488
note.

    j. without regard to the provisions of the Federal Property and Administrative Services Act of 1949, as amended, except section 207 of that Act, or any other law, make such disposition as it may deem desirable of (1) radioactive materials, and (2) any other property, the special disposition of which is, in the opinion of the Commission, in the interest of the national security: *Provided, however,* That the property furnished to licensees in accordance with the provisions of subsection 161m. shall not be deemed to be properly disposed of by the commission pursuant to this subsection;

    k. Authorize such of its members, officers, and employees as it deems necessary in the interest of the common defense and security to carry firearms while in the discharge of their official duties. The Commission may also authorize such of those employees of its contractors and subcontractors (at any tier) engaged in the protection of property under the jurisdiction of the United States located at facilities owned by or contracted to the United States or being transported to or from such facilities as it deems necessary in the interests of the common defense and security to carry firearms while in the discharge of their official duties. A person authorized to carry firearms under this subsection may, while in the performance of, and in connection with, official duties, make arrests without warrant for any offense against the United States committed in that person's presence or for any felony cognizable under

---

[209]   Amended by P.L. 93–377, §7, 88 Stat. 475 (1974).
[210]   Amended by P.L. 109–58, § 626, 119 Stat. 784 (2005).

the laws of the United States if that person has reasonable grounds to believe that the individual to be arrested has committed or is committing such felony. An employee of a contractor or subcontractor authorized to carry firearms under this subsection may make such arrests only when the individual to be arrested is within, or in direct flight from, the area of such offense. A person granted authority to make arrests by this subsection may exercise that authority only in the enforcement of (1) laws regarding the property of the United States in the custody of the Department of Energy, the Nuclear Regulatory Commission, or a contractor of the Department of Energy or Nuclear Regulatory Commission, or (2) any provision of this Act that may subject an offender to a fine, imprisonment, or both. The arrest authority conferred by this subsection is in addition to any arrest authority under other laws. The Secretary, with the approval of the Attorney General, shall issue guidelines to implement this subsection;[211]

m.[212] enter into agreements with persons licensed under section 103, 104, 53a.(4), or 63a.(4)[213] for such periods of time as the Commission may deem necessary or desirable (1) to provide for the processing, fabricating, separating, or refining in facilities owned by the Commission of source, byproduct, or other material or special nuclear material owned by or made available to such licensees and which is utilized or produced in the conduct of the licensed activity, and (2) to sell, lease, or otherwise make available to such licensees such quantities of source or byproduct material, and other material not defined as special nuclear material pursuant to this Act, as may be necessary for the conduct of the licensed activities; *Provided, however,* That any such agreement may be canceled by the licensee at any time upon payment of such reasonable cancellation charges as may be agreed upon by the licensee and the Commission: *and provided, further,* That the Commission shall establish prices to be paid by licensees for material or services to be furnished by the Commission pursuant to this subsection, which prices shall be established on such a nondiscriminatory basis as, in the opinion of the Commission, will provide reasonable compensation to the Government for such material or services and will not discourage the development of sources of supply independent of the Commission;

n.[214] delegate to the General Manager or other officers of the Commission any of those functions assigned to it under this Act except

---

[211]   Amended by P.L. 99–661, 100 Stat. 4064 (1986); P.L. 97–90, § 211, 95 Stat. 1163 (1981).

[212]   Amended by P.L. 87–456, § 303(c), 76 Stat. 72 (1962), the Tariff Classification Act of 1962, repealed § 161 l, effective on the 10th day following a Presidential proclamation concerning tariff schedules, import restrictions, and related matters. This proclamation was issued on August 21, 1963 (3 CFR, Proclamation 3548). Section 161 l. read as follows:

> l. Secure the admittance free of duty into the United States of purchases made abroad of source materials, upon certification to the Security of the Treasury that such entry is necessary in the interest of the common defense and security.

[213]   Amended by P.L. 86–300, § 1, 73 Stat. 574 (1959).

[214]   Amended by P.L. 85–507, § 21, 72 Stat. 327 (1958), which repealed former subsection 161n and relettered subsequent subsections accordingly.
Previously, subsection 161 n. read as follows:

> n. Assign scientific, technical, professional, and administrative employees for instruction, education, or training by public or private agencies, institutions of learning, laboratories, or industrial or commercial organizations and to pay the whole or any part of the salaries of such employees, costs of their transportation and per diem in lieu of subsistence in accordance with applicable laws and regulation, and training charges incident to their assignments (including tuition and other related fees): *Provided, however,*

those specified in sections 51, 57b.,[215] 61,[216] 108, 123, 145b. (with respect to the determination of those persons to whom the Commission may reveal Restricted Data in the national interest), 145f.,[217] and 161a.;

o. require by rule, regulation, or order, such reports, and the keeping of such records with respect to, and to provide for such inspections of, activities and studies of types specified in section 31 and of activities under licenses issued pursuant to sections 53, 63, 81, 103, and 104, as may be necessary to effectuate the purposes of this Act, including section 105; and

p. make, promulgate, issue, rescind, and amend such rules and regulations as may be necessary to carry out the purposes of this Act.

Easements for rights-of-way.

q. The Commission is empowered, under such terms and conditions as are deemed advisable by it, to grant easements for rights-of-way over, across, in, and upon acquired lands under its jurisdiction and control, and public lands permanently withdrawn or reserved for the use of the Commission, to any State, political subdivision thereof, or municipality, or to any individual, partnership, or corporation of any State, Territory, or possession of the United States, for (a) railroad tracks; (b) oil pipe lines; (c) substations for electric power transmission lines, telephone lines, and telegraph lines, and pumping stations for gas, water, sewer, and oil pipe lines; (d) canals; (e) ditches; (f) flumes; (g) tunnels; (h) dams and reservoirs in connection with fish and wildlife programs, fish hatcheries, and other fish-cultural improvements; (i) roads and street; and (j) for any other purpose or purposes deemed advisable by the Commission: Provided, That such rights-of-way shall be granted only upon a finding by the Commission that the same will not be incompatible with the public interest: *Provided further*, That such rights-of-way shall not include any more land than is reasonably necessary for the purpose for which granted: *and provided further,* That all or any part of such right-of-way may be annulled and forfeited by the Commission for failure to comply with the terms and conditions of any grant hereunder or for nonuse for a period of two consecutive years or abandonment of rights granted under authority hereof. Copies of all instruments granting easements over public lands pursuant to this section shall be furnished to the Secretary of the Interior.[218]

r. Under such regulations and for such periods and at such prices the Commission may prescribe, the Commission may sell or contract to sell to purchasers within Commission-owned communities or in the immediate vicinity of the Commission community, as the case may be, any of the following utilities and related services, if it is determined that they are not available from another local source and that the sale is in the interest of the national defense or in the public interest:

(1) Electric power.
(2) Steam.

That (1) not more than one per centum of the eligible employees shall be so assigned during any fiscal year, and (2) any such assignment shall be approved in advance by the Commission or shall be in accordance with a training program previously approved by the Commission: *And provided further*, That appropriations or other funds available to the Commission for salaries or expenses shall be available for the purposes of this subsection.

[215]   Amended by P.L. 90–190, § 11, 81 Stat. 575 (1967).
[216]   Amended by P.L. 91–560, § 7, 84 Stat. 1472 (1970), amended subsection 161n by striking out at this point the following: "102 (with respect to the finding of practical value)."
[217]   Amended by P.L. 87–615, § 12, 76 Stat. 409 (1962).
[218]   Amended by P.L. 84–1006, § 4, 70 Stat. 1069, which added subsection q (originally subsection r).

(3) Compressed air.

(4) Water.

(5) Sewage and garbage disposal.

(6) Natural, manufactured, or mixed gas.

(7) Ice.

(8) Mechanical refrigeration.

(9) Telephone service.

Proceeds of sales under this subsection shall be credited to the appropriation currently available for the supply of that utility or service. To meet local needs the Commission may make minor expansions and extensions of any distributing system or facility within or in the immediate vicinity of a Commission-owned community through which a utility or service is furnished under this subsection.[219]

Succession of authority.

s. establish a plan for a succession of authority which will assure the community of direction of the Commission's operations in the event of a national disaster due to enemy activity. Notwithstanding any other provision of this Act, the person or persons succeeding to command in the event of disaster in accordance with the plan established pursuant to this subsection shall be vested with all of the authority of the Commission: *Provided*, That any such succession to authority, and vesting of authority shall be effective only in the event and as long as a quorum of three or more members of the Commission is unable to convene and exercise direction during the disaster period: *Provided further*, That the disaster period includes the period when attack on the United States is imminent and the post-attack period necessary to reestablish normal lines of command;[220]

Processing contracts.

t. enter into contracts for the processing, fabricating, separating, or refining in facilities owned by the Commission of source, byproduct or other material, or special nuclear material, in accordance with and within the period of an agreement for cooperation while comparable services are available to persons licensed under section 103 or 104: *Provided*, That the prices for services under such contracts shall be no less than the prices currently charged by the Commission pursuant to section 161m.;

Long term contract authority.

u. (1) enter into contracts for such periods of time as the Commission may deem necessary or desirable, but not to exceed five years from the date of execution of the contract, for the purchase or acquisition of reactor services or services related to or required by the operation of reactors;

(2) (A) enter into contracts for such periods of time as the Commission may deem necessary or desirable for the purchase or acquisition of any supplies, equipment, materials, or services required by the Commission whenever the Commission determines that: (i) it is advantageous to the Government to make such purchase or acquisition from commercial sources; (ii) the furnishing of such supplies, equipment, materials, or services will require the construction or acquisition of special facilities by the vendors or supplies thereof; (iii) the amortization chargeable to the Commission constitutes an appreciable portion of the cost of contract performance, excluding cost of materials; and (iv) the contract for such period is

---

[219]   Amended by P.L. 85–162, § 204, 71 Stat. 410 (1957), added subsection r (originally subsection s).

[220]   Amended by P.L. 85–681 § 7, 72 Stat. 632 (1958), amended section 161 by adding new subsections t, u, and v. P.L. 87–206, § 13, 75 Stat. 475 (1961), changed the designation of subsections t, u, and v to subsections s, t, and u, respectively.

more advantageous to the Government than a similar contract not executed under the authority of this subsection. Such contracts shall be entered into for periods not to exceed five years each from the date of initial delivery of such supplies, equipment, materials, or services or ten years from the date of execution of the contracts excluding periods of renewal under option.

(B) In entering into such contracts the Commission shall be guided by the following principles: (i) the percentage of the total cost of special facilities devoted to contract performance and chargeable to the Commission should not exceed the ratio between the period of contract deliveries and the anticipated useful life of such special facilities; (ii) the desirability of obtaining options to renew the contract for reasonable periods at prices not to include charges for special facilities already amortized; and (iii) the desirability of reserving in the Commission the right to take title to the special facilities under appropriate circumstances; and

(3) include in contracts made under this subsection provisions which limit the obligation of funds to estimated annual deliveries and services and the unamortized balance of such amounts due for special facilities as the parties shall agree is chargeable to the performance of the contract. Any appropriation available at the time of termination or thereafter made available to the Commission for operating expenses shall be available for payment of such costs which may arise from termination as the contract may provide. The term "special facilities" as used in this subsection means any land and any depreciable buildings, structures, utilities, machinery, equipment, and fixtures necessary for the production or furnishing of such supplies, equipment, materials, and services and not available to the vendors or suppliers for the performance of the contract.[221]

Contract authority.

v. provide services in support of the United States Enrichment Corporation, except that the Secretary of Energy shall annually collect payments and other charges from the Corporation sufficient to ensure recovery of the costs (excluding depreciation and imputed interest on original plant investments in the Department's gaseous diffusion plants and costs under section 1403(d)) incurred by the Department of Energy after the date of the enactment of the Energy Policy Act of 1992 in performing such services;.[222]

w. prescribe and collect from any other Government agency, which applies to the Commission for, or is issued by the Commission, a license or certificate, any fee, charge, or price which it may require, in accordance with the provisions of section 9701 of Title 31 or any other law.[223]

---

[221]   Amended by P.L. 86–300, § 1, 73 Stat. 574 (1959).

[222]   Added by P.L. 88–489, § 16, 78 Stat. 602 (1964). Amended by P.L. 102–486, 106 Stat. 2944 (1992).

[223]   Added by P.L. 92–314, § 301, 86 Stat. 222 (1972). Amended by P.L. 102–486, 106 Stat. 2944 (1992). P.L. 109–58, § 623, 119 Stat. 783 (2005), rewrote subsection (w), which formerly read as follows:

> w. prescribe and collect from any other Government agency, which applies for or is issued a license for a utilization facility designed to produce electrical or heat energy pursuant to section 103 or 104b., including standards and restrictions governing the control, maintenance, use, and disbursement by any former licensee under this chapter that has control over any fund for the decommissioning of the facility.

42 USC 2231.

42 USC 2014.

42 USC 2201a.

x. Establish by rule, regulation, or order, after public notice, and in accordance with the requirements of section 181 of this Act, such standards and instructions as the Commission may deem necessary or desirable to ensure–

(1) that an adequate bond, surety, or other financial arrangement (as determined by the Commission) will be provided, before termination of any license for byproduct materials as defined in section 11e.(2), by a licensee to permit the completion of all requirements established by the Commission for the decontamination, decommissioning, and reclamation of sites, structures, and equipment used in conjunction with byproduct material as so defined, and

(2) that–

(A) in the case of any such license issued or renewed after the date of the enactment of this subsection, the need for long-term maintenance and monitoring of such sites, structures and equipment after termination of such license will be minimized and, to the maximum extent practicable, eliminated; and

(B) in the case of each license for such material (whether in effect on the date of the enactment of this section or issued or renewed thereafter), if the Commission determines that any such long-term maintenance and monitoring is necessary, the licensee, before termination of any license for byproduct material as defined in section 11e.(2), will make available such bonding, surety, or other financial arrangements as may be necessary to assure such long-term maintenance and monitoring.

Such standards and instructions promulgated by the Commission pursuant to this subsection shall take into account, as determined by the Commission, so as to avoid unnecessary duplication and expense, performance bonds or other financial arrangements which are required by other Federal agencies or State agencies and/or other local governing bodies for such decommissioning, decontamination, and reclamation and long-term maintenance and monitoring except that nothing in this paragraph shall be construed to require that the Commission accept such bonds or arrangements if the Commission determines that such bonds or arrangements are not adequate to carry out subparagraphs (1) and (2) of this subsection.[224]

Sec. 161A. Use of Firearms by Security Personnel[225]

a. DEFINITIONS–In this section, the terms 'handgun', 'rifle', 'shotgun', 'firearm', 'ammunition', 'machine gun', 'short-barreled shotgun', and 'short-barreled rifle' have the meanings given the terms in section 921(a) of title 18, United States Code.

b. AUTHORIZATION–Notwithstanding subsections (a)(4), (a)(5), (b)(2), (b)(4), and (o) of section 922 of title 18, United States Code, section 925(d)(3) of title 18, United States Code, section 5844 of the Internal Revenue Code of 1986, and any law (including regulations) of a State or a political subdivision of a State that prohibits the transfer, receipt, possession, transportation, importation, or use of a handgun, a rifle, a shotgun, a short-barreled shotgun, a short-barreled rifle, a machinegun, a semiautomatic assault weapon, ammunition for any such gun or weapon, or a large capacity ammunition feeding device, in carrying out the duties of the Commission, the Commission may authorize the security personnel of any licensee or certificate holder of

---

[224]   Subsection x. added by P.L. 95–604, § 203, 92 Stat. 3036 (1978).
[225]   Added by P.L. 109–58, Title VI, § 653, 119 Stat. 811 (2005).

the Commission (including an employee of a contractor of such a licensee or certificate holder) to transfer, receive, possess, transport, import, and use 1 or more such guns, weapons, ammunition, or devices, if the Commission determines that–

(1) the authorization is necessary to the discharge of the official duties of the security personnel; and

(2) the security personnel–

(A) are not otherwise prohibited from possessing or receiving a firearm under Federal or State laws relating to possession of firearms by a certain category of persons;

(B) have successfully completed any requirement under this section for training in the use of firearms and tactical maneuvers;

(C) are engaged in the protection of–

(i) a facility owned or operated by a licensee or certificate holder of the Commission that is designated by the Commission; or

(ii) radioactive material or other property owned or possessed by a licensee or certificate holder of the Commission, or that is being transported to or from a facility owned or operated by such a licensee or certificate holder, and that has been determined by the Commission to be of significance to the common defense and security or public health and safety; and

(D) are discharging the official duties of the security personnel in transferring, receiving, possessing, transporting, or importing the weapons, ammunition, or devices.

c. BACKGROUND CHECKS–A person that receives, possesses, transports, imports, or uses a weapon, ammunition, or a device under subsection (b) shall be subject to a background check by the Attorney General, based on fingerprints and including a background check under section 103(b) of the Brady Handgun Violence Prevention Act (Public Law 103–159; 18 U.S.C. 922 note) to determine whether the person is prohibited from possessing or receiving a firearm under Federal or State law.

d. EFFECTIVE DATE–This section takes effect on the date on which guidelines are issued by the Commission, with the approval of the Attorney General, to carry out this section.

42 USC 2202.
Contracts.

Sec. 162. Contracts

The President may, in advance, exempt any specific action of the Commission in a particular matter from the provisions of law relating to contracts whenever he determines that such action is essential in the interest of the common defense and security.

42 USC 2203.
Advisory
committees.

Sec. 163. Advisory Committees

The members of the General Advisory Committee established pursuant to section 26 and the members of advisory boards established pursuant to section 161a. may serve as such without regard to the provisions of section 281, 283, or 284 of title 18[226] of the United States Code, except insofar as such sections may prohibit any such member from receiving compensation from a source other than a nonprofit educational institution[227] in respect of any particular matter which

---

[226]   These sections have been repealed.
[227]   Amended by P.L. 86–300, § 2, 73 Stat. 574 (1959).

directly involves the Commission or in which the Commission is directly interested.[228]

42 USC 2204.
Electric utility
contracts.

Sec. 164. Electric Utility Contracts

The Commission is authorized in connection with the construction or operations of the Oak Ridge, Paducah, and Portsmouth installations of the Commission, without regard to section 3679 of the Revised Statutes, as amended, to enter into new contracts or modify or confirm existing contracts to provide for electric utility serves for periods not exceeding twenty-five years, and such contracts shall be subject to termination by the Commission upon payment of cancellation costs as provided in such contracts, and any appropriation presently or hereafter made available to the Commission shall be available for the payment of such cancellation costs. Any such cancellation payments shall be taken into consideration in determination of the rate to be charged in the event the Commission or any other agency of the Federal Government shall purchase electric utility services from the contractors subsequent to the cancellation and during the life of the original contract. The authority of the Commission under this section to enter into new contracts or modify or confirm existing contracts to provide for electric utility services includes, in case such electric utility services are to be furnished to the Commission by the Tennessee Valley Authority, authority to contract with any person to furnish electric utility services to the Tennessee Valley Authority in replacement thereof. Any contract hereafter entered into by the Commission pursuant to this section shall be submitted to the Energy Committees[229] and a period of thirty days shall elapse while Congress is in session (in computing such thirty days, there shall be excluded the days on which either House is not in session because of adjournment for more than three days) before the contract of the Commission shall become effective: *Provided, however,* That the Energy Committees, after having received the proposed contract, may by resolution in writing, waive the conditions of or all or any portion of such thirty-day period.

42 USC 2205.
Contract
practices.

Sec. 165. Contract Practices

a. In carrying out the purposes of this Act the Commission shall not use the cost-plus-percent age-of-cost system of contracting.

b. No contract entered into under the authority of this Act shall provide, and no contract entered into under the authority of the Atomic Energy Act of 1946, as amended, shall be modified or amended after the date of enactment of this Act to provide, for direct payment or direct reimbursement by the Commission of any Federal income taxes on behalf of any contractor performing such contract for profit.

42 USC 2206.
Comptroller
General audit.

Sec. 166. Comptroller General Audit

No moneys appropriated for the purposes of this Act shall be available for payments under any contract with the Commission, negotiated without advertising, except contracts with any foreign government or any agency thereof and contracts with foreign producers, unless such contract includes a clause to the effect that the Comptroller General of the United States or any of his duly authorized representatives shall, until the expiration of three years after final payment, have access

---

[228] Amended by P.L. 87–849, § 2, 76 Stat. 1119 (1962), which revised the existing conflict of interest laws. All exemptions from the provisions of sections 281, 283, and 284 of Title 18 of the U.S. Code are deemed to be exemptions from the corresponding sections of the new conflict-of-interest law "except to the extent that they affect officers or employees of the executive branch of the United States Government [or] of any independent agency of the United States, * * * as to whom they are no longer applicable."

[229] Amended by P.L. 103–437, § 15(f)(7), 108 Stat. 4581 (1994).

to and the right to examine any directly pertinent books, documents, papers, and records of the contractor or any of his subcontractors engaged in the performance of, and involving transactions related to such contracts or subcontracts: *Provided, however,* That no moneys so appropriated shall be available for payment under such contract which includes any provision precluding an audit by the Government Accountability Office of any transaction under such contract: *and provided further,* That nothing in this section shall preclude the earlier disposal of contractor and subcontractor records in accordance with records disposal schedules agreed upon between the Commission and the General Accounting Office.[230]

42 USC 2207.

Sec. 167. Claims Settlements

The Commission, acting on behalf of the United States, is authorized to consider, ascertain, adjust, determine, settle, and pay, any claim for money damage of $5,000 or less against the United States for bodily injury, death, or damage to or loss of real or personal property resulting from any detonation, explosion, or radiation produced in the conduct of any program undertaken by the Commission involving the detonation of an explosive device, where such claim is presented to the Commission in writing within one year after the accident or incident out of which the claim arises: *Provided, however,* That the damage to or loss of property, or bodily injury or death, shall not have been caused in whole or in part by any negligence or wrongful act on the part of the claimant, his agents, or employees. Any such settlement under the authority of this section shall be final and conclusive for all purposes, notwithstanding any other provision of law to the contrary. If the Commission considers that a claim in excess of $5,000 is meritorious and would otherwise be covered by this section, the Commission may report the facts and circumstances thereof to the Congress for its consideration.[231]

42 USC 2208.
Payments in
lieu of taxes.

Sec. 168. Payments in Lieu of Taxes

In order to render financial assistance to those States and localities in which the activities of the Commission are carried on, and in which the Commission has acquired property previously subject to State and local taxation, the Commission is authorized to make payments to State and local governments in lieu of property taxes. Such payments may be in the amounts, at times, and upon the terms the Commission deems appropriate, but the Commission shall be guided by the policy of not making payments in excess of the taxes which would have been payable for such property in the condition in which it was acquired, except in cases where special burdens have been cast upon the State or local

---

[230]   Amended by P.L. 85–681, §8, 72 Stat. 632 (1958); P.L. 108–271, §8(b), 118 Stat. 814 (2004).
[231]   Amended by P.L. 87–206, §14, 75 Stat. 474 (1961). Prior to amendment, this section read as follows:
> Sec. 167. Claim Settlements–The Commission, acting on behalf of the United States, is authorized to consider, ascertain, adjust, determine, settle, and pay, any claim for money damage of $5,000 or less against the United States for bodily injury, death, or damage to or loss of real or personal property resulting from any detonation, explosion, or radiation produced in the conduct of the Commission's program for testing atomic weapons, where such claim is presented to the Commission in writing within one year after the accident or incident out of which the claim arises: *Provided, however,* That the damage to or loss of property, or bodily injury or death, shall not have been caused in whole or in part by any negligence or wrongful act on the part of the claimant, his agents, or employees. Any such settlement under the authority of this section shall be final and conclusive for all purposes, notwithstanding any other provision of law to the contrary.

government by activities of the Commission, the Manhattan Engineer District or their agents. In any such case, any benefit accruing to the State or local government by reason of such activities shall be considered in determining the amount of the payment.

42 USC 2209.
No subsidy.

Sec. 169. No Subsidy

No funds of the Commission shall be employed in the construction or operation of facilities licensed under section 103 or 104 except under contract or other arrangement entered into pursuant to section 31.

42 USC 2210.

Sec. 170. Indemnification and Limitation of Liability[232]

a. REQUIREMENT OF FINANCIAL PROTECTION FOR LICENSEES[233]–Each license issued under section 103 or 104 and each construction permit issued under section 185 shall, and each license issued under section 53, 63, or 81 may, for the public purposes cited in section 2i., have as a condition of the license a requirement that the licensee have and maintain financial protection of such type and in such amounts as the Nuclear Regulatory Commission (in this section referred to as the "Commission")[234] in the exercise of its licensing and regulatory authority and responsibility shall require in accordance with subsection b. to cover public liability claims. Whenever such financial protection is required, it may be a further condition of the license that the licensee execute and maintain an indemnification agreement in accordance with subsection c. The Commission may require, as a further condition of issuing a license, that an applicant waive any immunity from public liability conferred by Federal or State law.[235]

Indemnification agreement.

Waiver.

Liability insurance.

b. AMOUNT AND TYPE OF FINANCIAL PROTECTION FOR LICENSEES[236]

(1)[237] The amount of primary financial protection[238] required shall be the amount of liability insurance available from private sources, except that the Commission may establish a lesser amount on the basis of criteria set forth in writing, which it may revise from time to time, taking into consideration such factors as the following: (A) the cost and terms of private insurance, (B) the type, size, and location of the licensed activity and other factors pertaining to the hazard, and (C) the nature and purpose of the licensed activity: *Provided,* That for facilities designed for producing substantial amounts of electricity and having a rated capacity of 100,000 electrical kilowatts or more, the amount of primary financial protection[239] required shall be the

---

[232]   Added by P.L. 85–256, § 4, 71 Stat. 576 (1957).
[233]   Amended by P.L. 100–408, §16(e)(1), 102 Stat. 1080 (1988).
[234]   Amended by P.L. 100–408, § 16(a)(2), 102 Stat. 1079 (1988).
[235]   Amended by P.L. 94–197, § 2, 89 Stat. 1111 (1975). Prior to amendment, subsection 170a read as follows:

   a. Each license issued under section 103 or 104 and each construction permit issued under section 185 shall, and each license issued under section 53, 63, or 81 may, have as a condition of the license a requirement that the licensee have and maintain financial protection of such type and in such amounts as the Commission shall require in accordance with subsection 170b. to cover public liability claims. Whenever such financial protection is required, it shall be a further condition of the license that the licensee execute and maintain an indemnification agreement in accordance with subsection 170c. The Commission may require, as a further condition of issuing a license, that an applicant waive any immunity from public liability conferred by Federal or State law.

[236]   Amended by P.L. 100–408, § 16(e)(2), 102 Stat. 1080, (1988).
[237]   Amended by P.L. 100–408, § 2(C)(1), 102 Stat. 1066 (1988).
[238]   Amended by P.L. 100–408, §§ 2(a)(1), (2), (C)(2), 102 Stat. 1066 (1988).
[239]   Amended by P.L. 100–408, § 2(a)(1), 102 Stat. 1066 (1988).

maximum amount available at reasonable cost and on reasonable terms from private sources (excluding the amount of private liability insurance available under the industry retrospective rating plan required in this subsection). Such primary financial protection may include private insurance, private contractual indemnities, self insurance, other proof of financial responsibility, or a combination of such measures and shall be subject to such terms and conditions as the Commission may, be rule, regulation, or order, prescribe. The Commission shall require licensees that are required to have and maintain primary financial protection equal to the maximum amount of liability insurance available from private sources to maintain, in addition to such primary financial protection, private liability insurance available under an industry retrospective rating plan providing for premium charges deferred in whole of major part until public liability from a nuclear incident exceeds or appears likely to exceed the level of the primary financial protection required of the licensee involved in the nuclear incident: *Provided,* That such insurance is available to, and required of, all of the licensees of such facilities without regard to the manner in which they obtain other types or amounts of such primary financial protection: *and provided further,* That the maximum amount of the standard deferred premium that may be charged a licensee following any nuclear incident under such a plan shall not be more than $95,800,000[240] (subject to adjustment for inflation under subsection t.) but not more than $15,000,000 in any 1 year (subject to adjustment for inflation under subsection t.),[241] for each facility for which licensee is required to maintain the maximum amount of primary financial protection: *and provided further,* That the amount which may be charged a licensee following any nuclear incident shall not exceed the licensee's pro rata share of the aggregate public liability claims and costs (excluding legal costs subject to subsection o.(1)(D), payment of which has not been authorized under such subsection) arising out of the nuclear incident. Payment of any State premium taxes which may be applicable to any deferred premium provided for in this Act shall be the responsibility of the licensee and shall not be included in the retrospective premium established by the Commission.[242]

(2)[243](A) The Commission may, on a case by case basis, assess annual deferred premium amounts less than the standard annual deferred premium amount assessed under paragraph (1)–

> (i) for any facility, if more than one nuclear incident occurs in any one calendar year; or

> (ii) for any licensee licensed to operate more than one facility, if the Commission determines that the financial impact of assessing the standard annual deferred premium

---

[240]   Amended by P.L. 109–58, Title VI, § 603(1), 119 Stat. 780 (2005), which struck out "$63,000,000" and inserted "$95,800,000."

[241]   Amended by P.L. 109–58, Title VI, § 603(1), 119 Stat. 780 (2005), which struck out "10,000,000 in any 1 year" and inserted "15,000,000 in any 1 year (subject to adjustment for inflation under subsection (t) of this section)."

[242]   Amended by P.L. 100–408, § 2(a)(1), (3), (b), 102 Stat. 1066 (1988). The third sentence of paragraph (1) read as follows:

> The Commission shall require licensees that are required to have and maintain primary financial protection equal to the maximum amount of liability insurance available from private sources to maintain, in addition to such primary financial protection,

[243]   Added by P.L. 100–408, § 2(C)(4), 102 Stat. 1066 (1988).

amount under paragraph (1) would result in undue financial hardship to such licensee or the ratepayers of such licensee.

(B) In the event that the Commission assesses a lesser annual deferred premium amount under subparagraph (A), the Commission shall require payment of the difference between the standard annual deferred premium assessment under paragraph (1) and any such lesser annual deferred premium assessment within a reasonable period of time, with interest at a rate determined by the Secretary of Treasury on the basis of the current average market yield on outstanding marketable obligations of the United States of comparable maturities during the month preceding the date that the standard annual deferred premium assessment under paragraph (1) would become due.[244]

(3) The Commission shall establish such requirements as are necessary to assure availability of funds to meet any assessment of deferred premiums within a reasonable time when due, and may provide reinsurance or shall otherwise guarantee the payment of such premiums in the event it appears that the amount of such premiums will not be available on a timely basis through the resources of private industry and insurance. Any agreement by the Commission with a licensee or indemnitor to guarantee the payment of deferred premiums may contain such terms as the Commission deems appropriate to carry out the purposes of this section and to assure reimbursement to the Commission for its payments made due to the failure of such licensee or indemnitor to meet any of its obligations arising under or in connection with financial protection required under this subsection including without limitation terms creating liens upon the licensed facility and the revenues derived there from or any other property or revenues of such licensee to secure such reimbursement and consent to the automatic revocation of any license.[245]

Claims.

(4)[246] (A) In the event that the funds available to pay valid claims in any year are insufficient as a result of the limitation on the amount of deferred premiums that may be required of a licensee in any year under paragraph (1) or (2), or the Commission is required to make

[244] Amended by P.L. 100–408, § 2(C)(3), 102 Stat. 1066 (1988), which struck out the fifth and sixth sentences of existing paragraph (1), which had authorized the Commissioner to establish a minimum amount which the aggregate deferred premiums charged for each facility within one calendar year could not exceed and which had authorized the Commissioner to establish amounts less than the standard premium for individual facilities taking into account such factors as the facility's size, location, and other factors pertaining to the hazard. See paragraph (2) for successor provisions.
[245] Amended by P.L. 94–197, § 3, 89 Stat. 1111 (1975). Prior to amendment, subsection 170b read as follows:
    b. The amount of financial protection required shall be in the amount of liability insurance available from private sources, except that the Commission may establish a lesser amount on the basis of criteria set forth in writing, which it may revise from time to time, taking into consideration such factors as the following: (1) the cost and terms of private insurance, (2) the type, size and location of the licensed activity and other factors pertaining to the hazard, and (3) the nature and purpose of the licensed activity: *Provided,* That for facilities designed for producing substantial amounts of electricity and having a rated capacity of 100,000 electrical kilowatts or more, the amount of financial protection required shall be the maximum amount available from private sources. Such financial protection may include private insurance, private contractual indemnities, self insurance, other proof of financial responsibility, or a combination of such measures.
[246] Added by P.L. 100–408, § 2(d)(2), 102 Stat. 1067 (1988).

reinsurance or guaranteed payments under paragraph (3), the Commission shall, in order to advance the necessary funds–

(i) request the Congress to appropriate sufficient funds to satisfy such payments; or

(ii) to the extent approved in appropriation Acts, issue to the Secretary of the Treasury obligations in such forms and denominations, bearing such maturities, and subject to such terms and conditions as may be agreed to by the Commission and the Secretary of the Treasury.

(B) Except for funds appropriated for purposes of making reinsurance or guaranteed payments under paragraph (3), any funds appropriated under subparagraph (A)(i) shall be repaid to the general fund of the United States Treasury from amounts made available by standard deferred premium assessments, with interest at a rate determined by the Secretary of Treasury on the basis of the current average market yield on outstanding marketable obligations of the United States of comparable maturities during the month preceding the date that the funds appropriated under such subparagraph are made available.

Securities.

(C) Except for funds appropriate for purposes of making reinsurance or guaranteed payments under paragraph (3), redemption of obligations issued under subparagraph (A)(ii) shall be made by the Commission from amounts made available by standard deferred premium assessments. Such obligations shall bear interest at a rate determined by the Secretary of Treasury by taking into consideration the average market yield on outstanding marketable obligations to the United States of comparable maturities during the month preceding the issuance of the obligations under this paragraph. The Secretary of the Treasury shall purchase any issued obligations, and for such purpose the Secretary of the Treasury may use as a public debt transaction the proceeds from the sale of any securities issued under chapter 31 of title 31, United States Code, and the purposes for which securities may be issued under such chapter are extended to include any purchase of such obligations. The Secretary of the Treasury may at any time sell any of the obligations acquired by the Secretary of the Treasury under this paragraph. All redemptions, purchases, and sales by the Secretary of the Treasury of obligations under this paragraph shall be treated as public debt transactions of the United States.

(5)[247] (A) For purposes of this section only, the Commission shall consider a combination of facilities described in subparagraph (B) to be a single facility having a rated capacity of 100,000 electrical kilowatts or more.

(B) A combination of facilities referred to in subparagraph (A) is two or more facilities located at a single site, each of which has a rated capacity of 100,000 electrical kilowatts or more but not more than 300,000 electrical kilowatts, with a combined rated capacity of not more than 1,300,000 electrical kilowatts.

c.   INDEMNIFICATION   OF   NUCLEAR   REGULATORY COMMISSION LICENSEES–The Commission shall, with respect to

---

[247]   Added by P.L. 109–58, Title VI, § 608, 119 Stat. 781 (2005).

licenses issued between August 30, 1954, and December 31, 2025,[248] for which it requires financial protection of less than $560,000,000, agree to indemnify and hold harmless the licensee and other persons indemnified, as their interest may appear, from public liability arising from nuclear incidents which is in excess of the level of financial protection required of the licensee. The aggregate indemnity for all persons indemnified in connection with each nuclear incident shall not exceed $500,000,000 excluding costs of investigating and settling claims and defending suits for damage: *Provided, however,* That this amount of indemnity shall be reduced by the amount that the financial protection required shall exceed $60,000,000. Such a contract of indemnification shall cover public liability arising out of or in connection with the licensed activity. With respect to any production or utilization facility for which a construction permit is issued between August 30, 1954, and December 31, 2025, the requirements of this subsection shall apply to any license issued for such facility subsequent to December 31, 2025.[249]

> d.(1)(A) In addition to any other authority the Secretary of Energy (in this section referred to as the "Secretary") may have, the Secretary shall, until December 31, 2025, enter into agreements of indemnification under this subsection with any person who may conduct activities under a contract with the Department of Energy that involve the risk of public liability and that are not subject to financial protection requirements under subsection b. or agreements of indemnification under subsection c. or k.[250]

---

[248]   Amended by P.L. 108–7, Div. O, § 101, 117 Stat. 551 (2003); P.L. 109–58, Title VI, § 602(a), 119 Stat. 779 (2005).

[249]   Amended by P.L. 109–58, Title VI, § 602(b), 119 Stat. 779 (2005); P.L. 91–197 §§ 5(a) and (b), 89 Stat. 1111 (1975); P.L. 89–210, § 1, 79 Stat 855 (1965), had previously amended subsection 170c. Prior to amendment, this subsection read as follows:

> c. The commission shall, with respect to licenses issued between August 30, 1954 and August 1, 1967, for which it requires financial protection, agree to indemnify and hold harmless the licensee and other persons indemnified, as their interest may appear, from public liability arising from nuclear incidents which is in excess of the level of financial protection required of the license. The aggregate indemnity for all persons indemnified in connection with each nuclear incident shall not exceed $500,000,000 including the reasonable costs of investigating and settling claims and defending suits for damage. Such a contract of indemnification shall cover public liability arising out of or in connection with the licensed activity. With respect to any production or utilization facility for which a construction permit is issued between August 30, 1954, and August 1, 1967, the requirements of this subsection shall apply to any license issued for such facility subsequent to August 1, 1967.

P.L. 88–394, § 2, 78 Stat. 376 (1964), had previously amended subsection 170c. by adding the last sentence.

[250] Amended by P.L. 109–58, Title VI, § 602(b), 119 Stat. 779 (2005); P.L. 108–375, Div. C, Title XXXI, § 3141, 118 Stat. 2171 (2004); P.L. 100–408, 102 Stat. 1066 (1988); P.L. 94–197, §§ 5(a) and (b), 89 Stat 1111 (1975); P.L. 89–210, § 2, 79 Stat 855 (1965), had previously amended the first two sentences of subsection 170d. Prior to amendment, these sentences read as follows:

> d. In addition to any other authority the Commission may have, the Commission is authorized until August 1, 1967, to enter into agreements of indemnification with its contractors for the construction or operation of production or utilization facilities or other activities under contracts for the benefit of the United States involving activities under the risk of public liability for a substantial nuclear incident. In such agreements of indemnification the Commission may require its contractor to provide and maintain financial protection of such a type and in such amounts as the Commission shall determine to be appropriate to cover public liability arising out of or in connection with the contractual activity, and shall indemnify the persons indemnified against such claims above the amount of the financial

Effective date.

Claims.

(B)(i)(I) Beginning 60 days after the date of enactment of the Price-Anderson Amendments Act of 1988, agreements of indemnification under subparagraph (A) shall be the exclusive means of indemnification for public liability arising from activities described in such subparagraph, including activities conducted under a contract that contains an indemnification clause under Public Law 85-804 entered into between August 1, 1987, and the date of enactment of the Price-Anderson Amendments Act of 1988.

(II) The Secretary may incorporate in agreements of indemnification under subparagraph (A) the provisions relating to the waiver of any issue or defense as to charitable or governmental immunity authorized in subsection n. (1) to be incorporated in agreements of indemnification. Any such provisions incorporated under this subclause shall apply to any nuclear incident arising out of nuclear waste activities subject to an agreement of indemnification under subparagraph (A).

(ii) Public liability arising out of nuclear waste activities subject to an agreement of indemnification under subparagraph (A) that are funded by the Nuclear Waste Fund established in section 302 of the Nuclear Waste Policy Act of 1982 (42 USC 10222) shall be compensated from the Nuclear Waste Fund in an amount not to exceed the maximum amount of financial protection required of licensees under subsection b.

(2) In an agreements of indemnification entered into under paragraph (1), the Secretary–

(A) may require the contractor to provide and maintain financial protection of such a type and in such amounts as the Secretary shall determine to be appropriate to cover public liability arising out of or in connection with the contractual activity; and

(B) shall indemnify the persons indemnified against such liability above the amount of the financial protection required, in the amount of $10,000,000,000 (subject to adjustment for inflation under subsection t.), in the aggregate, for all persons indemnified in connection with the contract and for each nuclear incident, including such legal costs of the contractor as are approved by the Secretary.[251]

---

protection required, in the amount of $500,000,000 including the reasonable costs of investigating and settling claims and defending suits for damage in the aggregate for all persons indemnified in connection with such contract and for each nuclear incident: *Provided* That in the case of nuclear incidents occurring outside the United States, the amount of the indemnity provided by the Commission shall not exceed $100,000,000.

P.L. 87–615, § 6, 76 Stat. 409 (1962), had previously amended the second sentence of subsection 170d. by adding the proviso providing that in the case of incidents occurring outside the United States, the amount of indemnity provided by the Commission shall not exceed $100 million.

[251]   Amended by P.L. 109–58, Title VI, § 604(a), 119 Stat. 780 (2005), which rewrote paragraph (2). Prior to amendment, the paragraph read as follows:

In agreements of indemnification entered into under paragraph (1), the Secretary may require the contractor to provide and maintain financial protection of such a type and in such amounts as the Secretary shall determine to be appropriate to cover public liability arising out of or in connection with the contractual activity, and shall indemnify the persons indemnified against such claims above the amount of the financial protection required, to the full

(3) All agreements of indemnification under which the Department of Energy (or its predecessor agencies) may be required to indemnify any person under this section shall be deemed to be amended, on the date of enactment of the Price-Anderson Amendments Act of 2005, to reflect the amount of indemnity for public liability and any applicable financial protection required of the contractor under this subsection.[252]

(4) Financial protection under paragraph (2) and indemnification under paragraph (1) shall be the exclusive means of financial protection and indemnification under this section for any Department of Energy demonstration reactor licensed by the Commission under section 202 of the Energy Reorganization Act of 1974 (42 USC 5842).[253]

(5) In the case of nuclear incidents occurring outside the United States, the amount of the indemnity provided by the Commission under this subsection shall not exceed $500,000,000.[254]

(6) The provisions of this subsection may be applicable to lump sum as well as cost type contracts and to contracts and projects financed in whole or in part by the Commission.[255]

(7) A contractor with whom an agreement of indemnification has been executed and who is engaged in activities connected with the underground detonation of a nuclear explosive device shall be liable, to the extent so indemnified under this subsection, for injuries or damage sustained as a result of such detonation in the same manner and to the same extent as would a private person acting as principal, and no immunity or defense founded in the Federal, State, or municipal character of the contractor or of the work to be performed under the contract shall be effective to bar such liability.[256]

e. Limitation On Aggregate Public Liability.–(1)[257] The aggregate public liability for a single nuclear incident of persons indemnified,

---

extent of the aggregate public liability of the persons indemnified for each nuclear incident, including such legal costs of the contractor as are approved by the Secretary.

[252] Amended by P.L. 109–58, Title VI, section 604(b), 119 Stat. 780 (2005), which rewrote paragraph (3). Prior to amendment, the paragraph read as follows:

(3)(A) Notwithstanding paragraph (2), if the maximum amount of financial protection required of licenses under subsection (b) of this section is increased by the Commission, the amount of indemnity, together with any financial protection required of the contractor, shall at all times remain equal to or greater than the maximum amount of financial protection required of licensees under subsection (b) of this section.

(B) The amount of indemnity provided contractors under this subsection shall not, at any time, be reduced in the event that the maximum amount of financial protection required of licensees is reduced.

(C) All agreements of indemnification under which the Department of Energy (or its predecessor agencies) may be required to indemnify any person, shall be deemed to be amended on August 20, 1988, to reflect the amount of indemnity for public liability and any applicable financial protection required of the contractor under this subsection on August 20, 1988.

[253] Paragraphs (3) and (4) added by P.L. 100–408, § 4(a), 102 Stat. 1068 (1988).
[254] Amended by P.L.100–408 § 4(a), 102 Stat. 1068 (1988); P.L. 109–58, Title VI, § 605(a), 119 Stat. 781 (2005) struck out $100,000,000 and inserted $500,000,000.
[255] Amended by P.L. 100–408, § 4(a), 102 Stat. 1068 (1988).
[256] Amended by P.L. 100–408, § 4(a), 102 Stat. 1068 (1988).
[257] Amended by P.L. 100–408, § 2(C)(3), 102 Stat. 1066 (1988), which struck out the fifth and sixth sentences of existing paragraph (1), which had authorized the Commissioner to establish a minimum amount that the aggregate deferred premiums charged for each facility within one calendar year could not exceed and which had authorized the Commissioner to establish amounts less than the standard premium for individual facilities

including such legal costs as are authorized to be paid under subsection o.(1)(D), shall not exceed–

(A) in the case of facilities designed for producing substantial amounts of electricity and having a rated capacity of 100,000 electrical kilowatts or more, the maximum amount of financial protection required of such facilities under subsection b. (plus any surcharge assessed under subsection o.(1)(E));

Contracts.

(B) in the case of contractors with whom the Secretary has entered into an agreement of indemnification under subsection d., the amount of indemnity and financial protection that may be required under paragraph (2) of subsection d.[258], and

(C) in the case of all other licensees of the Commission required to maintain financial protection under this section–

(i) $500,000,000, together with the amount of financial protection required of the licensee; or

(ii) if the amount of financial protection required of the licensee exceeds $60,000,000, $560,000,000 or the amount of financial protection required of the licensee, whichever amount is more.

Claims.

(2) That in the event of a nuclear incident involving damages in excess of the amount of aggregate liability, the Congress will thoroughly review the particular incident and will take whatever action is determined necessary and appropriate to protect the public from the consequences of a disaster of such magnitude.[259]

(3)[260] No provision of paragraph (1) may be construed to preclude the Congress from enacting a revenue measure, applicable to licensees of the Commission required to maintain financial protection pursuant to subsection b., to fund any action undertaken pursuant to paragraph (2).

Contracts.

(4) With respect to any nuclear incident occurring outside of the United States to which an agreement of indemnification entered into under the provisions of subsection d. is applicable, such aggregate public liability shall not exceed the amount of $500,000,000[261], together with the amount of financial protection required of the contractor.

f. Collection of Fees by Nuclear Regulatory Commission[262]–The Commission or the Secretary, as appropriate, is authorized to collect a fee from all persons with whom an indemnification agreement is executed under this section. This fee shall be $30 per year per thousand kilowatts of thermal energy capacity for facilities licensed under section 103:

---

taking into account such factors as the facility's size, location, and other factors pertaining to the hazard. See paragraph (2) for successor provisions.

[258]   Amended by P.L. 109–58, Title VI, § 604(C), 119 Stat. 780 (2005).

[259]   Amended by P.L. 100–408, § 6, 102 Stat. 1070 (1988), which added new language to paragraph (2). Paragraph (2) formerly read:

> In the event of a nuclear incident involving damages in excess of the amount of aggregate public liability under paragraph (1), the Congress will thoroughly review the particular incident in accordance with the procedures set forth in subsection 170i of this section and will in accordance with such procedures, take whatever action is determined to be necessary (including approval of appropriate compensation plans and appropriation of funds) to provide full and prompt compensation to the public for all public liability claims resulting from a disaster of such magnitude.

[260]   Added by P.L. 100–408, § 6, 102 Stat. 1070 (1988).

[261]   Amended by P.L. 109–58, Title VI, § 605(b), 119 Stat. 781 (2005), which struck out "$100,000,000" and inserted "$500,000,000."

[262]   Amended by P.L. 100–408, § 16(e)(4), 102 Stat. 1080 (1988).

Provided, That the Commission or the Secretary, as appropriate, is authorized to reduce the fee for such facilities in reasonable relation to increases in financial protection required above a level of $60,000,000. For facilities licensed under section 104, and for construction permits under section 185, the Commission is authorized to reduce the fee set forth above. The Commission shall establish criteria in writing for determination of the fee for facilities licensed under section 104, taking into consideration such factors as (1) the type, size, and location of facility involved, and other factors pertaining to the hazard, and (2) the nature and purpose of the facility. For other licenses, the Commission shall collect such nominal fees as it deems appropriate. No fee under this subsection shall be less than $100 per year.[263]

Private
insurance
organizations.
Use of services.

41 USC 252(c)
(See 41 USC
260(b)).

g. Use of Services of Private Insurers[264]–In administering the provisions of this section, the Commission or the Secretary, as appropriate, shall use, to the maximum extent practicable, the facilities and services of private insurance organizations, and the Commission or the Secretary, as appropriate, may contract to pay a reasonable compensation for such services. Any contract made under the provisions of this subsection may be made without regard to the provisions of section 3709 of the Revised Statutes (41 USC 5), as amended, upon a showing by the Commission or the Secretary, as appropriate, that advertising is not reasonably practicable and advance payments may be made.

Terms of
settlement.

h. Conditions of Agreements of Indemnification[265]–The agreement of indemnification may contain such terms as the Commission or the Secretary, as appropriate, deems appropriate to carry out the purposes of this section. Such agreement shall provide that, when the Commission or the Secretary, as appropriate, makes a determination that the United States will probably be required to make indemnity payments under this section, the Commission or the Secretary, as appropriate, shall collaborate with any person indemnified and may approve the payment of any claim under the agreement of indemnification, appear through the Attorney General on behalf of the person indemnified, take charge of such action, and settle or defend any such action. The Commission or the Secretary, as appropriate, shall have final authority on behalf of the United States to settle or approve the settlement of any such claim on a fair and reasonable basis with due regard for the purposes of this Act. Such settlement shall not include expenses in connection with the claim incurred by the person indemnified.[266]

i. Compensation Plans.[267]–(1) After any nuclear incident involving damages that are likely to exceed the applicable amount of aggregate public liability under subparagraph (A), (B), or (C) of subsection e. (1), the Secretary or the Commission, as appropriate, shall–

　　　(A) make a survey of the causes and extent of damage; and

---

[263]　Amended by P.L. 100–408, 102 Stat. 1066 (1988); P.L. 94–197, § 7, 89 Stat. 1111 (1975).
[264]　Amended by P.L. 100–408, § 16(e)(5), 102 Stat. 1081 (1988).
[265]　Amended by P.L. 100–408, § 16(e)(6), 102 Stat. 1081 (1988).
[266]　Amended by P.L. 100–408, § 2(C)(3), 102 Stat. 1066 (1988), which struck out the fifth and sixth sentences of existing paragraph (1) that had authorized the Commissioner to establish a minimum amount that the aggregate deferred premiums charged for each facility within one calendar year could not exceed and which had authorized the Commissioner to establish amounts less than the standard premium for individual facilities taking into account such factors as the facility's size, location, and other factors pertaining to the hazard. See paragraph (2) for successor provisions.
[267]　Amended by P.L. 100–408, §7(a), 102 Stat. 1071 (1988).

Reports,
Defense
and national
security.

(B) expeditiously submit a report setting forth the results of such survey to the Congress, to the Representatives of the affected districts, to the Senators of the affected States, and (except for information that will cause serious damage to the national defense of the United States) to the public, to the parties involved, and to the courts.

President of
U.S.

(2) Not later than 90 days after any determination by a court, pursuant to subsection o., that the public liability from a single nuclear incident may exceed the applicable amount of aggregate public liability under subparagraph (A), (B), or (C) of subsection e. (1) the President shall submit to the Congress–

(A) an estimate of the aggregate dollar value of personal injuries and property damage that arises from the nuclear incident and exceeds the amount of aggregate public liability under subsection e. (1);

Claims.

(B) recommendations for additional sources of funds to pay claims exceeding the applicable amount of aggregate public liability under subparagraph (A), (B), or (C) of subsection e.(1), which recommendations shall consider a broad range of possible sources of funds (including possible revenue measures on the sector of the economy, or on any other class, to which such revenue measures might be applied);

Claims.

(C) 1 or more compensation plans, that either individually or collectively shall provide for full and prompt compensation for all valid claims and contain a recommendation or recommendations as to the relief to be provided, including any recommendations that funds be allocated or set aside for the payment of claims that may arise as a result of latent injuries that may not be discovered until a later date; and

(D) any additional legislative authorities necessary to implement such compensation plan or plans.

(3)(A) Any compensation plan transmitted to the Congress pursuant to paragraph (2) shall bear an identification number and shall be transmitted to both Houses of Congress on the same day and to each House while it is in session.

(B) The provisions of paragraphs (4) through (6) shall apply with respect to consideration in the Senate of any compensation plan transmitted to the Senate pursuant to paragraph (2).

(4) No such compensation plan may be considered approved for purposes of subsection 170e.(2) unless between the date of transmittal and the end of the first period of sixty calendar days of continuous session of Congress after the date on which such action is transmitted to the Senate, the Senate passes a resolution described in paragraph 6 of this subsection.

(5) For the purpose of paragraph (4) of this subsection–

(A) continuity of session is broken only by an adjournment of Congress sine die; and

(B) the days on which either House is not in session because of an adjournment of more than three days to a day certain are excluded in the computation of the sixty-day calendar period.

(6)(A) This paragraph is enacted–

(i) as an exercise of the rulemaking power of the Senate and as such it is deemed a part of the rules of the Senate, but applicable only with respect to the procedure to be followed in the Senate in the case of resolutions described by subparagraph (B) and it supersedes other rules only to the extent that it is inconsistent therewith; and

(ii) with full recognition of the constitutional right of the Senate to change the rules at any time, in the same manner and to the same extent as in the case of any other rule of the Senate.

(B) For purposes of this paragraph, the term "resolution" means only a joint resolution of the Congress the matter after the resolving clause of which is as follows: That the _____ approves the compensation plan numbered _____ submitted to the Congress on _____ , 19__, the first blank space therein being filled with the name of the resolving House and the other blank spaces being appropriately filled; but does not include a resolution which specifies more than one compensation plan.

(C) A resolution once introduced with respect to a compensation plan shall immediately be referred to a committee (and all resolutions with respect to the same compensation plan shall be referred to the same committee) by the President of the Senate.

(D)(i) If the committee of the Senate to which a resolution with respect to a compensation plan has been referred has not reported it at the end of twenty calendar days after its referral, it shall be in order to move either to discharge the committee from further consideration of such resolution or to discharge the committee from further consideration with respect to such compensation plan which has been referred to the committee.

(ii) A motion to discharge may be made only by an individual favoring the resolution, shall be highly privileged (except that it may not be made after the committee has reported a resolution with respect to the same compensation plan), and debate thereon shall be limited to not more than one hour, to be divided equally between those favoring and those opposing the resolution. An amendment to the motion shall not be in order, and it shall not be in order to move to reconsider the vote by which the motion was agreed to or disagreed to.

(iii) If the motion to discharge is agreed to or disagreed to the motion may not be renewed, nor may another motion to discharge the committee be made with respect to any other resolution with respect to the same compensation plan.

(E)(i) When the committee has reported, or has been discharged from further consideration of, a resolution, it shall be at any time thereafter in order (even though a previous motion to the same effect has been disagreed to) to move to proceed to the consideration of the resolution. The motion shall be highly privileged and shall not be debatable. An amendment to the motion shall not be in order, and it shall not be in order to move to reconsider the vote by which the motion was agreed to or disagreed to.

(ii) Debate on the resolution referred to in clause (i) of this subparagraph shall be limited to not more than ten hours, which shall be divided equally between those favoring and those opposing such resolution. A motion further to limit debate shall not be debatable. An amendment to, or motion to recommit, the resolution shall not be in order, and it shall not be in order to move to reconsider the vote by which such resolution was agreed to or disagreed to.

(F)(i) Motions to postpone, made with respect to the discharge from committee, or the consideration of a resolution or motions to proceed to the consideration of other business, shall be decided without debate.

(ii) Appeals from the decision of the Chair relating to the application of the rules of the Senate to the procedures relating to a resolution shall be decided without debate.

Contracts in advance of appropriations.

j. Contracts in Advance of Appropriations[268]–In administering the provisions of this section, the Commission or the Secretary, as appropriate, may make contracts in advance of appropriations and incur obligations without regard to sections 1341, 1342, 1349, 1350, and 1351, and subchapter II of chapter 15, of title 31, United States Code

Educational activities.

k.[269] With respect to any license issued pursuant to section 53, 63, 81, 104a., or 104c. for the conduct of educational activities to a person found by the Commission to be a nonprofit educational institution, the Commission shall exempt such licensee from the financial protection requirement of subsection a. With respect to licenses issued between August 30, 1954, and December 31, 2025[270], for which the Commission grants such exemption:

(1) the Commission shall agree to indemnify and hold harmless the licensee and other persons indemnified, as their interests may appear, from public liability in excess of $250,000 arising from nuclear incidents. The aggregate indemnity for all persons indemnified in connection with each nuclear incident shall not exceed $500,000,000, including such legal costs of the licensee as are approved by the Commission;[271]

(2) such contracts of indemnification shall cover public liability arising out of or in connection with the licensed activity; and shall include damage to property of persons indemnified, except property which is located at the site of and used in connection with the activity where the nuclear incident occurs; and

(3) such contracts of indemnification, when entered into with a licensee having immunity from public liability because it is a State agency, shall provide also that the Commission shall make payments under the contract on account of activities of the licensee in the same manner and to the same extent as the Commission would be required to do if the licensee were not such a State agency.

Any licensee may waive an exemption to which it is entitled under this subsection. With respect to any production or utilization facility for which a construction permit is issued between August 30, 1954, and December 31, 2025[272], the requirements of this subsection shall apply to any license issued for such facility subsequent to December 31, 2025.[273]

l. Presidential Commission On Catastrophic Nuclear Accidents.–

(1) Not later than 90 days after the date of the enactment of the Price-Anderson Amendments Act of 1988, the President shall establish a commission (in this subsection referred to as the "study commission") in accordance with the Federal Advisory Committee

---

[268]   Amended by P.L. 100–408, § 16(e)(7), 102 Stat. 1081 (1988).
[269]   Added by P.L. 85–744, 72 Stat. 837 (1958).
[270]   Amended by P.L. 109–58, Title VI, § 602(C), 119 Stat 779 (2005).
[271]   Amended by P.L. 100–408, § 8(a)(2), 102 Stat. 1074 (1988).
[272]   Amended by P.L. 109–58, Title VI, § 602(C), 119 Stat 779 (2005).
[273]   Amended by P.L. 88–394, § 3, 78 Stat. 376 (1964); P.L. 89–210, § 4, 79 Stat. 855 (1965); P.L. 94–197, § 10, 89 Stat. 1111 (1975); P.L. 109–58, Title VI, § 602(C), 119 Stat 779 (2005).

Act (5 USC App.) to study means of fully compensating victims of a catastrophic nuclear accident that exceeds the amount of aggregate public liability under subsection e.(1).

(2)(A) The study commission shall consist of not less than 7 and not more than 11 members, who–

(i) shall be appointed by the President; and

(ii) shall be representative of a broad range of views and interests.

(B) The members of the study commission shall be appointed in a manner that ensures that not more than a mere majority of the members are of the same political party.

(C) Each member of the study commission shall hold office until the termination of the study commission, but may be removed by the President for inefficiency, neglect of duty, or malfeasance in office.

(D) Any vacancy in the study commission shall be filled in the manner in which the original appointment was made.

(E) The President shall designate one of the members of the study commission as chairperson, to serve at the pleasure of the President.

Reports.

(3) The study commission shall conduct a comprehensive study of appropriate means of fully compensating victims of a catastrophic nuclear accident that exceeds the amount of aggregate public liability under subsection e.(1), and shall submit to the Congress a final report setting forth–

(A) recommendations for any changes in the laws and rules governing the liability or civil procedures that are necessary for the equitable, prompt, and efficient resolution and payment of all valid damage claims, including the advisability of adjudicating public liability claims through an administrative agency instead of the judicial system;

(B) recommendations for any standards or procedures that are necessary to establish priorities for the hearing, resolution, and payment of claims when awards are likely to exceed the amount of funds available within a specific time period; and

(C) recommendation for any special standards or procedures necessary to decide and pay claims for latent injuries caused by the nuclear incident.

(4)(A) The chairperson of the study commission may appoint and fix the compensation of a staff of such persons as may be necessary to discharge the responsibilities of the study commission, subject to the applicable provisions of the Federal Advisory Committee Act (5 USC App.) and title 5, United States Code.

(B) To the extent permitted by law and requested by the chairperson of the study commission, the Administrator of General Services shall provide the study commission with necessary administrative services, facilities, and support on a reimbursable basis.

(C) The Attorney General, the Secretary of Health and Human Services, and the Administrator of the Federal Emergency Management Agency shall, to the extent permitted by law and subject to the availability of funds, provide the study commission with such facilities, support, funds and services, including staff, as may be necessary for the effective performance of the functions of the study commission.

(D) The study commission may request any Executive agency to furnish such information, advice, or assistance as it determines to be necessary to carry out its functions. Each such agency is directed, to the extent permitted by law, to furnish such information, advice or assistance upon request by the chairperson of the study commission.

(E) Each member of the study commission may receive compensation at the maximum rate prescribed by the Federal Advisory Committee Act (5 USC App.) for each day such member is engaged in the work of the study commission. Each member may also receive travel expenses, including per diem in lieu of subsistence under sections 5702 and 5703 of title 5, United States Code.

(F) The functions of the President under the Federal Advisory Committee Act (5 USC App.) that are applicable to the study commission, except the function of reporting annually to the Congress, shall be performed by the Administrator of General Services.

Reports.

(5) The final report required in paragraph (3) shall be submitted to the Congress not later than the expiration of the 2-year period beginning on the date of the enactment of the Price-Anderson Amendments Act of 1988.

Termination date.

(6) The study commission shall terminate upon the expiration of the 2-month period beginning on the date on which the final report required in paragraph (3) is submitted.[274]

Emergency Assistance payments.

m. Coordinated Procedures for Prompt Settlement of Claims and Emergency Assistance[275]–The Commission or the Secretary, as appropriate, is authorized to enter into agreements with other indemnitors to establish coordinated procedures for the prompt handling, investigation, and settlement of claims for public liability. The Commission or the Secretary, as appropriate, and other indemnitors may make payments to, or for the aid of, claimants for the purpose of providing immediate assistance following a nuclear incident. Any funds appropriate to the Commission or the Secretary, as appropriate, shall be available for such payments. Such payments may be made without securing releases, shall not constitute an admission of the liability of any person indemnified or of any indemnitor, and shall operate as a satisfaction to the extent thereof of any final settlement or judgment.

Waiver of defenses.

n. Waiver of Defenses and Judicial Procedures[276]–

(1) With respect to any extraordinary nuclear occurrence to which an insurance policy or contract furnished as proof of financial protection or an indemnity agreement applies and which–

(A) arises out of or results from or occurs in the course of the construction, possession, or operation of a production or utilization facility,

(B) arises out of or results from or occurs in the course of transportation of source material, by-product material, or special nuclear material to or from a production of utilization facility,

(C) during the course of the contract activity arises out of or results from the possession, operation, or use by a Department of

---

[274]   Amended by P.L. 100–408, § 4(a), 102 Stat. 1068) (1988).
[275]   Amended by P.L. 100–408, § 16(e)(9), 102 Stat. 1081 (1988).
[276]   Amended by P.L. 100–408, § 16(e)(10), 102 Stat. 1081 (1988).

Energy contractor or subcontractor of a device utilizing special nuclear material or by-product material,

(D) arises out of, results from, or occurs in the course of, the construction, possession, or operation of any facility licensed under section 53, 63, or 81, for which the Commission has imposed as a condition of the license a requirement that the licensee have and maintain financial protection under subsection a.,

(E) arises out of , results from, or occurs in the course of, transportation of source material, byproduct material, or special nuclear material to or from any facility licensed under section 53, 63, or 81, for which the Commission has imposed as a condition of the license a requirement that the licensee have and maintain financial protection under subsection a., or

(F) arises out of, results from, or occurs in the course of nuclear waste activities.

The Commission or the Secretary, as appropriate, may incorporate provisions in indemnity agreements with licensees and contractors under this section, and may require provisions to be incorporated in insurance policies or contracts furnished as proof of financial protection, which waive (i) any issue or defense as to conduct of the claimant or fault of persons indemnified, (ii) any issue or defense as to charitable or governmental immunity, and (iii) any issue or defense based on any statute of limitations if suit is instituted within three years from the date on which the claimant first knew, or reasonable could have know, of his injury or damage and the cause thereof. The waiver of any such issue or defense shall be effective regardless of whether such issue or defense may otherwise be deemed jurisdictional or relating to an element in the cause of action. When so incorporated, such waivers shall be judicially enforceable in accordance with their terms by the claimant against the person indemnified. Such waivers shall not preclude a defense based upon a failure to take reasonable steps to mitigate damages, nor shall such waivers apply to injury or damage to a claimant or to claimants property which is intentionally sustained by the claimant or which results from a nuclear incident intentionally and wrongfully caused by the claimant. The waivers authorized in this subsection shall, as to indemnitors, be effective only with respect to those obligations set forth in the insurance policies or the contracts furnished as proof of financial protection and in the indemnity agreements. Such waivers shall not apply to, or prejudice the prosecution or defense of, any claim or portion of claim which is not within the protection afforded under (i) the terms of insurance policies or contracts furnished as proof of financial protection, or indemnity agreements, and (ii) the limit of liability provisions of subsection e.

42 USC 2210.

(2) With respect to any public liability action arising out of or resulting from a nuclear incident, the United States district court in the district where the nuclear incident takes place, or in the case of a nuclear incident taking place outside the United States, the United States District Court for the District of Columbia, shall have original jurisdiction without regard to the citizenship of any party or the amount in controversy. Upon motion of the defendant or of the Commission or the Secretary, as appropriate, any such action pending in any State court (including any such action pending on the date of the enactment of the Price-Anderson Amendments Act of 1988) or United States district court shall be removed or transferred to the United States district court having venue under this subsection.

Process of such district court shall be effective throughout the United States. In any action that is or becomes removable pursuant to his paragraph, a petition for removal shall be filed within the period provided in section 1446 of title 28, United States Code, or within the 30-day period beginning on the date of the enactment of the Price Anderson Amendments Act of 1988, whichever occurs later.

Courts, U.S.

(3) (A) Following any nuclear incident, the chief judge of the United States district court having jurisdiction under paragraph (2) with respect to public liability actions (or the judicial council of the judicial circuit in which the nuclear incident occurs) may appoint a special caseload management panel (in this paragraph referred to as the 'management panel") to coordinate and assign (but not necessarily hear themselves) cases arising out of the nuclear incident, if–

(i) a court, acting pursuant to subsection o. determines that the aggregate amount of public liability is likely to exceed the amount of primary financial protection available under subsection b. (or an equivalent amount in the case of a contractor indemnified under subsection d.); or

(ii) the chief judge of the United States district court (or the judicial council of the judicial circuit) determines that cases arising out of the nuclear incident will have an unusual impact on the work of the court.

(B)(i) Each management panel shall consist only of members who are United States district judges or circuit judges.

(ii) Members of a management panel may include any United States district judge or circuit judge of another district court or court of appeals, if the chief judge of such other district court or court of appeals consents to such assignment.

(C) It shall be the function of each management panel–

(i) to consolidate related or similar claims for hearing or trial;

(ii) to establish priorities for the handling of different classes of cases;

(iii) to assign cases to a particular judge or special master;

(iv) to appoint special masters to hear particular types of cases, or particular elements or procedural steps of cases;

(v) to promulgate special rules of court, not inconsistent with the Federal Rules of Civil Procedure, to expedite cases or allow more equitable consideration of claims;

(vi) to implement such other measures, consistent with existing law and the Federal Rules of Civil Procedure, as will encourage the equitable, prompt, and efficient resolution of cases arising out of the nuclear incident; and

(vii) to assemble and submit to the President such data, available to the court, as may be useful in estimating the aggregate damages from the nuclear incident.[277]

Allocation of funds.

o.[278] Plan For Distribution of Funds.–(1) Whenever the United States district court in the district where a nuclear incident occurs, or the United States District Court for the District of Columbia in case of a nuclear incident occurring outside the United States, determines upon the petition of any indemnitor or other interested person that public liability from a single nuclear incident may exceed the limit of liability under the

---

[277]   Amended by P.L. 100–408, § 4(a), 102 Stat. 1068 (1988).
[278]   Added by P.L. 89–645, § 3, 80 Stat. 891 (1966).

applicable limit of liability under subparagraph (A), (B), or (C) of subsection e.(1):

(A) Total payments made by or for all indemnitors as a result of such nuclear incident shall not exceed 15 per centum of such limit of liability without the prior approval of such court;

(B) The court shall not authorize payments in excess of 15 per centum of such limit of liability unless the court determines that such payments are or will be in accordance with a plan of distribution which has been approved by the court of such payments are not likely to prejudice the subsequent adoption and implementation by the court or a plan of distribution pursuant to subparagraph (C); and

(C) The Commission or the Secretary, as appropriate, shall, and any other indemnitor or other interested person may, submit to such district court a plan for the disposition of pending claims and for the distribution of remaining funds available. Such a plan shall include an allocation of appropriate amounts for personal injury claims, property damage claims, and possible latent injury claims which may not be discovered until a later time and shall include establishment of priorities between claimants and classes of claims, as necessary to insure the most equitable allocation of available funds.[279] Such court shall have all power necessary to approve, disapprove, or modify plans proposed, or to adopt another plan; and to determine the proportionate share of funds available for each claimant. The Commission or the Secretary, as appropriate, any other indemnitor, and any person indemnified shall be entitled to such orders as may be appropriate to implement and enforce the provisions of this section, including orders limiting the liability of the persons indemnified, orders approving or modifying the plan, orders staying the payment of claims and the execution of court judgments, orders apportioning the payments to be made to claimants, and orders permitting partial payments to be made before final determination of the total claims. The orders of such court shall be effective throughout the United States.

(D)[280] A court may authorize payment of only such legal costs as are permitted under paragraph (2) from the amount of financial protection required by subsection b.

(E) If the sum of public liability claims and legal costs authorized under paragraph (2) arising from any nuclear incident exceeds the maximum amount of financial protection required under subsection b., any licensee required to pay a standard deferred premium under subsection b.(1) shall, in addition to such deferred premium, be charged such an amount as is necessary to pay a pro rata share of such claims and costs, but in no case more than 5 percent of the maximum

---

[279] Amended by P.L. 94–197, § 13, 89 Stat. 1111 (1975) which added "and shall include establishment of priorities between claimants and classes of claims, as necessary to insure the most equitable allocation of available funds."

[280] Subsections (D), (E), and (2) added by P.L. 100–408, § 11(d)(1)(C), 102 Stat. 1077 (1988). P.L. 100–408, § 7(b)(2) struck out paragraph (4) which was inserted by P.L. 94–197, § 13, 89 Stat. 1111 (1975). Previously, paragraph (4) read as follows:

(4) The Commission shall, within ninety days after a court shall have made such determination, deliver to the Joint Committee a supplement to the report prepared in accordance with subsection (i) of this section setting forth the estimated requirements for full compensation and relief of all claimants, and recommendations as to the relief to be provided.

amount of such standard deferred premium described in such subsection.

(2) A court may authorize the payment of legal costs under paragraph (1)(D) only if the person requesting such payment has–

    (A) submitted to the court the amount of such payment requested; and

    (B) demonstrated to the court–

        (i) that such costs are reasonable and equitable; and

        (ii) that such person has–

            (I) litigated in good faith;

            (II) avoided unnecessary duplication of effort with that of other parties similarly situated;

            (III) not made frivolous claims or defenses; and

            (IV) not attempted to unreasonably delay the prompt settlement or adjudication of such claims.

p. Reports To Congress.–The Commission and the Secretary shall submit to the Congress by December 31, 2021[281], detailed reports concerning the need for continuation or modification of the provisions of this section, taking into account the condition of the nuclear industry, availability of private, insurance, and the state of knowledge concerning nuclear safety at that time, among other relevant factors, and shall include recommendations as to the repeal or modification of any of the provisions of this section.[282]

q. Limitation On Awarding of Precautionary Evacuation Costs.–No court may award costs of a precautionary evacuation unless such costs constitute a public liability.

r. Limitation of Liability of Lessors.–No person under a bona fide lease of any utilization or production facility (or part thereof or undivided interest therein) shall be liable by reason of an interest as lessor of such production or utilization facility, for any legal liability arising out of or resulting from a nuclear incidents resulting from such facility, unless such facility is in the actual possession and control of such person at the time of the nuclear incident giving rise to such legal liability.

s. Limitation On Punitive Damages.–No court may award punitive damages in any action with respect to a nuclear incident or precautionary evacuation against a person on behalf of whom the United States is obligated to make payments under an agreement of indemnification covering such incident or evacuation.

t. Inflation Adjustment.–(1) The Commission shall adjust the amount of the maximum total and annual[283] standard deferred premium under subsection b.(1) not less than once during each 5-year period following August 20, 2003[284] in accordance with the aggregate percentage change in the Consumer Price Index since–[285]

---

[281]   Amended by P.L. 109–58, Title VI, § 606, 119 Stat. 781, (2005); Pub.L. 109–295, Title VI, § 612(c), 120 Stat. 1410 (2006).
[282]   Amended by P.L. 100–408, § 4(a), 102 Stat. 1068 (1988).
[283]   Amended by P.L. 109–58, Title VI, §§ 603(2)(A), (B), 119 Stat. 780 (2005).
[284]   Amended by P.L. 108–375, Div. C, Title XXXI, § 3141(D)(1)(A), 118 Stat. 2171 (2004).
[285]   In accordance with the aggregate percentage change in the Consumer Price Index, as of April 2008, the new maximum total deferred premium is $111,900,000, and the maximum annual deferred premium is $17,500,000 (73 Fed. Reg. 56451; September 29, 2008). **NOTE: Next change should be in 2013.**

(A) August 20, 2003,[286] in the case of the first adjustment under this subsection; or

(B) the previous adjustment under this subsection.

(2) The Secretary shall adjust the amount of indemnification provided under an agreement of indemnification under subsection d. not less than once during each 5-year period following July 1, 2003, in accordance with the aggregate percentage change in the Consumer Price Index since–

(A) that date, in the case of the first adjustment under this paragraph; or

(B) the previous adjustment under this paragraph.

(3) For purposes of this subsection, the term "Consumer Price Index" means the Consumer Price Index for all urban consumers published by the Secretary of Labor.[287]

42 USC 2210a.
Conflict of
interest.

Sec. 170A. Conflicts of Interest Relating to Contracts and Other Arrangements[288]

a. The Commission shall, by rule, require any person proposing to enter into a contract, agreement, or other arrangement, whether by competitive bid or negotiation, under this Act or any other law administered by it for the conduct of research, development, evaluation activities, or for technical and management support services, to provide the Commission, prior to entering into any such contract, agreement, or arrangement, with all relevant information, as determined by the Commission, bearing on whether that person has a possible conflict of interest with respect to–

(1) being able to render impartial, technically sound, or objective assistance or advice in light of other activities or relationships with other persons, or

(2) being given an unfair competitive advantage. Such person shall insure, in accordance with regulations prescribed by the Commission, compliance with this section by any subcontractor (other than a supply subcontractor) or such person in the case of any subcontract for more than $10,000.

b. EVALUATION[289]–

(1) IN GENERAL[290]–Except as provided in paragraph (2)[291], the Nuclear Regulatory Commission shall not enter into any such contract agreement or arrangement unless it finds, after evaluating all information provided under subsection a. And any other information otherwise available to the Commission that–

(A) it is unlikely that a conflict of interest would exist, or

(B) such conflict has been avoided after appropriate conditions have been included in such contract, agreement, or arrangement; except that if the Commission determines that such conflict of interest exists and that such conflict of interest cannot be avoided by including appropriate conditions therein, the Commission may enter into such contract, agreement, or arrangement, if the

---

[286]   Amended by P.L. 109–58, Title VI, § 603(2)(C), 119 Stat. 780 (2005).
[287]   Amended by P.L. 109–58, Title VI, § 607, 119 Stat. 781 (2005), which redesignated former paragraph (2) as (3) and inserted a new paragraph (2).
[288]   Added by P.L. 95–601, § 8(a), 92 Stat. 2950 (1978).
[289]   Amended by P.L. 109–58, Title VI, § 639(2), 119 Stat. 794 (2005).
[290]   Amended by P.L. 109–58, Title VI, § 639(2), 119 Stat. 794 (2005), added paragraph (1) by inserting "(1) In general" following the subsection (b) heading, and redesignated former paragraph (1) as subparagraph (A).
[291]   Amended by P.L. 109–58, Title VI, § 639(2), 119 Stat. 794 (2005).

Commission determines that it is in the best interests of the United States to do so and includes appropriate conditions in such contract, agreement, or arrangement to mitigate such conflict.

(2)[292] NUCLEAR REGULATORY COMMISSION.–

Notwithstanding any conflict of interest, the Nuclear Regulatory Commission may enter into a contract, agreement, or arrangement with the Department of Energy or the operator of a Department of Energy facility, if the Nuclear Regulatory Commission determines that–

(A) the conflict of interest cannot be mitigated; and

(B) adequate justification exists to proceed without mitigation of the conflict of interest.[293]

Publication.

c. The Commission shall publish rules for the implementation of this section, in accordance with section 553 of title 5, United States Code (without regard to subsection (a)(2) thereof) as soon as practicable after the date of the enactment of this section, but in no event later than 120 days after such date.

42 USC 2210b.
42 USC 2231.
Report to Congress and President.

Regulations.

Sec. 170B. Uranium Supply[294]

a. The Secretary of Energy shall monitor and for the years 1983 to 1992 report annually to the Congress and to the President a determination of the viability of the domestic uranium mining and milling industry and shall establish by rule, after public notice and in accordance with the requirements of section 181 of this Act, within 9 months of enactment of this section, specific criteria which shall be assessed in the annual reports on the domestic uranium industry's viability. The Secretary of Energy is authorized to issue regulations providing for the collection of such information as the Secretary of Energy deems necessary to carry out the monitoring and reporting requirements of this section.

Proprietary information, disclosure.

b. Upon a satisfactory showing to the Secretary of Energy by any person that any information, or portion thereof obtained under this section, would, if made public, divulge proprietary information of such person, the Secretary shall not disclose such information and disclosure thereof shall be punishable under section 1905 of title 18, United States Code.

Criteria.

c. The criteria referred to in subsection a. shall also include, but not be limited to–

(1) an assessment of whether executed contracts or options for source material or special nuclear material will result in greater than 37½ percent of actual or projected domestic uranium requirements for any two-consecutive-year period being supplied by source material or special nuclear material from foreign sources;

(2) projections of uranium requirements and inventories of domestic utilities for a 10 year period;

(3) present and probable future use of the domestic market by foreign imports;

(4) whether domestic economic reserves can supply all future needs for a future 10 year period;

(5) present and projected domestic uranium exploration expenditures and plans;

[292] Added by P.L. 109–58, Title VI, § 639(3), 119 Stat. 794 (2005).
[293] Amended by P.L. 109–58, Title VI, § 639(1), 119 Stat. 794 (2005), which redesignated former paragraphs (1) and (2) as subparagraphs (A) and (B), respectively.
[294] Added by P.L. 97–415, § 23(b)(1).

(6) present and projected employment and capital investment in the uranium industry;

(7) the level of domestic uranium production capacity sufficient to meet projected domestic nuclear power needs for a 10 year period; and

(8) a projection of domestic uranium production and uranium price levels which will be in effect under various assumptions with respect to imports.

Imported material, impact on domestic industry and national security.

d. The Secretary of Energy, at any time, may determine on the basis of the monitoring and annual reports required under this section that source material or special nuclear material from foreign sources is being imported in such increased quantities as to be a substantial cause of serious injury, or threat thereof, to the United States uranium mining and milling industry. Based on that determination, the United States Trade Representative shall request that the United States International Trade Commission initiate an investigation under section 201 of the Trade Act of 1974 (19 USC 2251).

e. (1) If, during the period 1982 to 1992, the Secretary of Energy determines that executed contracts or options for source material or special nuclear material from foreign sources for use in utilization facilities within or under the jurisdiction of the United States represent greater than 37½ percent of actual or projected domestic uranium requirements for any two-consecutive-year period, or if the Secretary of Energy determines the level of contracts or options involving source material and special nuclear material from foreign sources may threaten to impair the national security, the Secretary of Energy shall request the Secretary of Commerce to initiate under section 232 of the Trade Expansion Act of 1962 (19 USC 1862) an investigation to determine the effects on the national security of imports of source material and special nuclear material. The Secretary of Energy shall cooperate fully with the Secretary of Commerce in carrying out such an investigation and shall make available to the Secretary of Commerce the findings that lead to this request and such other information that will assist the Secretary of Commerce in the conduct of the investigation.

Investigations.

(2) The Secretary of Commerce shall, in the conduct of any investigation requested by the Secretary of Energy pursuant to this section, take into account any information made available by the Secretary of Energy, including information regarding the impact on national security of projected or executed contracts or options for source material or special nuclear material from foreign sources or whether domestic production capacity is sufficient to supply projected national security requirements.

(3) No sooner than 3 years following completion of any investigation by the Secretary of Commerce under paragraph (1), if no recommendation has been made pursuant to such study for trade adjustments to assist or protect domestic uranium production, the Secretary of Energy may initiate a request for another such investigation by the Secretary of Commerce.

42 USC 2210c.

Sec. 170C. Elimination of Pension Offset for Certain Rehired Federal Retirees[295]

a. IN GENERAL.–The Commission may waive the application of section 8344 or 8468 of title 5, United States Code, on a case-by-case basis for employment of an annuitant.–

(1) in a position of the Commission for which there is exceptional difficulty in recruiting or retaining a qualified employee; or

(2) when a temporary emergency hiring need exists.

b. PROCEDURES.–The Commission shall prescribe procedures for the exercise of authority under this section, including–

(1) criteria for any exercise of authority; and

(2) procedures for a delegation of authority.

c. EFFECT OF WAIVER.–An employee as to whom a waiver under this section is in effect shall not be considered an employee for purposes of subchapter II of chapter 83, or chapter 84, of title 5, United States Code.

42 USC 2210d.
Deadline.

Sec. 170D. Security Evaluations[296]

a. SECURITY RESPONSE EVALUATIONS.–Not less often than once every 3 years, the Commission shall conduct security evaluations at each licensed facility that is part of a class of licensed facilities, as the Commission considers to be appropriate, to assess the ability of a private security force of a licensed facility to defend against any applicable design basis threat.

b. FORCE-ON-FORCE EXERCISES.–

(1) The security evaluations shall include force-on-force exercises.

(2) The force-on-force exercises shall, to the maximum extent practicable, simulate security threats in accordance with any design basis threat applicable to a facility.

(3) In conducting a security evaluation, the Commission shall mitigate any potential conflict of interest that could influence the results of a force-on-force exercise, as the Commission determines to be necessary and appropriate.

c. ACTION BY LICENSEES.–The Commission shall ensure that an affected licensee corrects those material defects in performance that adversely affect the ability of a private security force at that facility to defend against any applicable design basis threat.

d. FACILITIES UNDER HEIGHTENED THREAT LEVELS.–The Commission may suspend a security evaluation under this section if the Commission determines that the evaluation would compromise security at a nuclear facility under a heightened threat level.

e. REPORT.–Not less often than once each year, the Commission shall submit to the Committee on Environment and Public Works of the Senate and the Committee on Energy and Commerce of the House of Representatives a report, in classified form and unclassified form, that describes the results of each security response evaluation conducted and any relevant corrective action taken by a licensee during the previous year.

[295] Added by P.L. 109–58 § 624(a), 119 Stat. 783 (2005).
[296] Added by P.L. 109–58, § 651(a)(1), 119 Stat. 799 (2005).

42 USC 2210e.
Deadlines.

Sec. 170E. Design Basis Threat Rulemaking[297]

a. RULEMAKING.–The Commission shall–

(1) not later than 90 days after the date of enactment of this section, initiate a rulemaking proceeding, including notice and opportunity for public comment, to be completed not later than 18 months after that date, to revise the design basis threats of the Commission; or

(2) not later than 18 months after the date of enactment of this section, complete any ongoing rulemaking to revise the design basis threats.

b. FACTORS.–When conducting its rulemaking, the Commission shall consider the following, but not be limited to–

(1) the events of September 11, 2001;

(2) an assessment of physical, cyber, biochemical, and other terrorist threats;

(3) the potential for attack on facilities by multiple coordinated teams of a large number of individuals;

(4) the potential for assistance in an attack from several persons employed at the facility;

(5) the potential for suicide attacks;

(6) the potential for water-based and air-based threats;

(7) the potential use of explosive devices of considerable size and other modern weaponry;

(8) the potential for attacks by persons with a sophisticated knowledge of facility operations;

(9) the potential for fires, especially fires of long duration;

(10) the potential for attacks on spent fuel shipments by multiple coordinated teams of a large number of individuals;

(11) the adequacy of planning to protect the public health and safety at and around nuclear facilities, as appropriate, in the event of a terrorist attack against a nuclear facility; and

---

[297]　Added by P.L. 109–58, § 651(a)(1), 119 Stat. 799 (2005). P.L. 109–58 states the following:

(3) FEDERAL SECURITY COORDINATORS.–

(A) REGIONAL OFFICES.–Not later than 18 months after the date of enactment of this Act, the Nuclear Regulatory Commission (referred to in this section as the "Commission") shall assign a Federal security coordinator, under the employment of the Commission, to each region of the Commission.

(B) RESPONSIBILITIES.–The Federal security coordinator shall be responsible for–

(i) communicating with the Commission and other Federal, State, and local authorities concerning threats, including threats against such classes of facilities as the Commission determines to be appropriate;

(ii) monitoring such classes of facilities as the Commission determines to be appropriate to ensure that they maintain security consistent with the security plan in accordance with the appropriate threat level; and

(iii) assisting in the coordination of security measures among the private security forces at such classes of facilities as the Commission determines to be appropriate and Federal, State, and local authorities, as appropriate.

(b) BACKUP POWER FOR CERTAIN EMERGENCY NOTIFICATION SYSTEMS.–For any licensed nuclear power plants located where there is a permanent population, as determined by the 2000 decennial census, in excess of 15,000,000 within a 50–mile radius of the power plant, not later than 18 months after enactment of this Act, the Commission shall require that backup power to be available for the emergency notification system of the power plant, including the emergency siren warning system, if the alternating current supply within the 10–mile emergency planning zone of the power plant is lost.

(12) the potential for theft and diversion of nuclear materials from such facilities.

42 USC 2210f.   **Sec. 170F. Recruitment Tools**[298]

a. The Commission may purchase promotional items of nominal value for use in the recruitment of individuals for employment.

42 USC 2210g.   **Sec. 170G. Expenses Authorized to be Paid by the Commission**[299]

The Commission may–

(1) pay transportation, lodging, and subsistence expenses of employees who–

(A) assist scientific, professional, administrative, or technical employees of the Commission; and

(B) are students in good standing at an institution of higher education (as defined in section 102 of the Higher Education Act of 1965 (20 U.S.C. 1002)) pursuing courses related to the field in which the students are employed by the Commission; and

(2) pay the costs of health and medical services furnished, pursuant to an agreement between the Commission and the Department of State, to employees of the Commission and dependents of the employees serving in foreign countries.

42 USC 2210h.   **Sec. 170H. Radiation Source Protection**[300]

a. DEFINITIONS.–In this section:

(1) CODE OF CONDUCT.–The term 'Code of Conduct' means the code entitled the 'Code of Conduct on the Safety and Security of Radioactive Sources', approved by the Board of Governors of the International Atomic Energy Agency and dated September 8, 2003.

(2) RADIATION SOURCE.–The term 'radiation source' means–

(A) a Category 1 Source or a Category 2 Source, as defined in the Code of Conduct; and

(B) any other material that poses a threat such that the material is subject to this section, as determined by the Commission, by regulation, other than spent nuclear fuel and special nuclear materials.

Deadline.
Regulations.   b. COMMISSION APPROVAL.–Not later than 180 days after the date of enactment of this section, the Commission shall issue regulations prohibiting a person from–

(1) exporting a radiation source, unless the Commission has specifically determined under section 57 or 82, consistent with the Code of Conduct, with respect to the exportation, that–

(A) the recipient of the radiation source may receive and possess the radiation source under the laws and regulations of the country of the recipient;

(B) the recipient country has the appropriate technical and administrative capability, resources, and regulatory structure to ensure that the radiation source will be managed in a safe and secure manner; and

Deadline.
Notification.   (C) before the date on which the radiation source is shipped–

(i) a notification has been provided to the recipient country; and

---

[298]   Added by P.L. 109–58, § 651(c)(2), 119 Stat. 801 (2005).
[299]   Added by P.L. 109–58, § 651(c)(3), 119 Stat. 801 (2005).
[300]   Added by P.L. 109–58, § 651(d)(1), 119 Stat. 802 (2005).

(ii) a notification has been received from the recipient country;

as the Commission determines to be appropriate;

(2) importing a radiation source, unless the Commission has determined, with respect to the importation, that–

(A) the proposed recipient is authorized by law to receive the radiation source; and

(B) the shipment will be made in accordance with any applicable Federal or State law or regulation; and

(3) selling or otherwise transferring ownership of a radiation source, unless the Commission–

(A) has determined that the licensee has verified that the proposed recipient is authorized under law to receive the radiation source; and

(B) has required that the transfer shall be made in accordance with any applicable Federal or State law or regulation.

Deadline.
Regulations.

c. TRACKING SYSTEM.–

(1)(A) Not later than 1 year after the date of enactment of this section, the Commission shall issue regulations establishing a mandatory tracking system for radiation sources in the United States.

(B) In establishing the tracking system under subparagraph (A), the Commission shall coordinate with the Secretary of Transportation to ensure compatibility, to the maximum extent practicable, between the tracking system and any system established by the Secretary of Transportation to track the shipment of radiation sources.

(2) The tracking system under paragraph (1) shall–

(A) enable the identification of each radiation source by serial number or other unique identifier;

(B) require reporting within 7 days of any change of possession of a radiation source;

(C) require reporting within 24 hours of any loss of control of, or accountability for, a radiation source; and

(D) provide for reporting under subparagraphs (B) and (C) through a secure Internet connection.

d. PENALTY.–A violation of a regulation issued under subsection a. or b. shall be punishable by a civil penalty not to exceed $1,000,000.

e. NATIONAL ACADEMY OF SCIENCES STUDY.–

Deadline.
Contracts.

(1) Not later than 60 days after the date of enactment of this section, the Commission shall enter into an arrangement with the National Academy of Sciences under which the National Academy of Sciences shall conduct a study of industrial, research, and commercial uses for radiation sources.

(2) The study under paragraph (1) shall include a review of uses of radiation sources in existence on the date on which the study is conducted, including an identification of any industrial or other process that–

(A) uses a radiation source that could be replaced with an economically and technically equivalent (or improved) process that does not require the use of a radiation source; or

(B) may be used with a radiation source that would pose a lower risk to public health and safety in the event of an accident or attack involving the radiation source.

(3) Not later than 2 years after the date of enactment of this section, the Commission shall submit to Congress the results of the study under paragraph (1).

f. TASK FORCE ON RADIATION SOURCE PROTECTION AND SECURITY.–

(1) There is established a task force on radiation source protection and security (referred to in this section as the 'task force').

(2)(A) The chairperson of the task force shall be the Chairperson of the Commission (or a designee).

(B) The membership of the task force shall consist of the following:

(i) The Secretary of Homeland Security (or a designee).

(ii) The Secretary of Defense (or a designee).

(iii) The Secretary of Energy (or a designee).

(iv) The Secretary of Transportation (or a designee).

(v) The Attorney General (or a designee).

(vi) The Secretary of State (or a designee).

(vii) The Director of National Intelligence (or a designee).

(viii) The Director of the Central Intelligence Agency (or a designee).

(ix) The Administrator of the Federal Emergency Management Agency (or a designee).[301]

(x) The Director of the Federal Bureau of Investigation (or a designee).

(xi) The Administrator of the Environmental Protection Agency (or a designee).

(3)(A) The task force, in consultation with Federal, State, and local agencies, the Conference of Radiation Control Program Directors, and the Organization of Agreement States, and after public notice and an opportunity for comment, shall evaluate, and provide recommendations relating to, the security of radiation sources in the United States from potential terrorist threats, including acts of sabotage, theft, or use of a radiation source in a radiological dispersal device.

(B) Not later than 1 year after the date of enactment of this section, and not less than once every 4 years thereafter, the task force shall submit to Congress and the President a report, in unclassified form with a classified annex if necessary, providing recommendations, including recommendations for appropriate regulatory and legislative changes, for–

(i) a list of additional radiation sources that should be required to be secured under this Act, based on the potential attractiveness of the sources to terrorists and the extent of the threat to public health and safety of the sources, taking into consideration–

(I) radiation source radioactivity levels;

(II) radioactive half-life of a radiation source;

(III) dispersability;

(IV) chemical and material form;

(V) for radioactive materials with a medical use, the availability of the sources to physicians and patients for medical treatment; and

(VI) any other factor that the Chairperson of the Commission determines to be appropriate;

---

[301]   Amended by P.L. 109–295, Title VI, § 612(c), 120 Stat. 1410 (2006).

(ii) the establishment of, or modifications to, a national system for recovery of lost or stolen radiation sources;

(iii) the storage of radiation sources that are not used in a safe and secure manner as of the date on which the report is submitted;

(iv) modifications to the national tracking system for radiation sources;

(v) the establishment of, or modifications to, a national system (including user fees and other methods) to provide for the proper disposal of radiation sources secured under this Act;

(vi) modifications to export controls on radiation sources to ensure that foreign recipients of radiation sources are able and willing to adequately control radiation sources from the United States;

(vii)(I) any alternative technologies available as of the date on which the report is submitted that may perform some or all of the functions performed by devices or processes that employ radiation sources; and

(II) the establishment of appropriate regulations and incentives for the replacement of the devices and processes described in subclause (I)–

(aa) with alternative technologies in order to reduce the number of radiation sources in the United States; or

(bb) with radiation sources that would pose a lower risk to public health and safety in the event of an accident or attack involving the radiation source; and

(viii) the creation of, or modifications to, procedures for improving the security of use, transportation, and storage of radiation sources, including–

(I) periodic audits or inspections by the Commission to ensure that radiation sources are properly secured and can be fully accounted for;

(II) evaluation of the security measures by the Commission;

(III) increased fines for violations of Commission regulations relating to security and safety measures applicable to licensees that possess radiation sources;

(IV) criminal and security background checks for certain individuals with access to radiation sources (including individuals involved with transporting radiation sources);

(V) requirements for effective and timely exchanges of information relating to the results of criminal and security background checks between the Commission and any State with which the Commission has entered into an agreement under section 274b.;

(VI) assurances of the physical security of facilities that contain radiation sources (including facilities used to temporarily store radiation sources being transported); and

(VII) the screening of shipments to facilities that the Commission determines to be particularly at risk for sabotage of radiation sources to ensure that the shipments do not contain explosives.

Deadline.

g. ACTION BY COMMISSION.–Not later than 60 days after the date of receipt by Congress and the President of a report under subsection f.(3)(B), the Commission, in accordance with the recommendations of the task force, shall–

(1) take any action the Commission determines to be appropriate, including revising the system of the Commission for licensing radiation sources; and

(2) ensure that States that have entered into agreements with the Commission under section 274b. take similar action in a timely manner.

42 USC 2210i.

Sec. 170I. Secure Transfer of Nuclear Materials[302]

Procedures

a.   The Commission shall establish a system to ensure that materials described in subsection b., when transferred or received in the United States by any party pursuant to an import or export license issued pursuant to this Act, are accompanied by a manifest describing the type and amount of materials being transferred or received. Each individual receiving or accompanying the transfer of such materials shall be subject to a security background check conducted by appropriate Federal entities.

b. Except as otherwise provided by the Commission by regulation, the materials referred to in subsection a. are byproduct materials, source materials, special nuclear materials, high-level radioactive waste, spent nuclear fuel, transuranic waste, and low-level radioactive waste (as defined in section 2(16) of the Nuclear Waste Policy Act of 1982 (42 U.S.C. 10101(16)).

Chapter 15–Compensation for Private Property Acquired

42 USC 2221.
Just
compensation.

Sec. 171. Just Compensation

The United States shall make just compensation for any property or interests therein taken or requisitioned pursuant to sections 43,[303] 55,[304] 66, and 108. Except in case of real property or any interest therein, the Commission shall determine and pay such just compensation. If the compensation so determined is unsatisfactory to the person entitled thereto, such person shall be paid 75 per centum of the amount so determined, and shall be entitled to sue the United States Court of Federal Claims,[305] or in any district court of the United States for the district in which such claimant is a resident in the manner provided by section 1346 of title 28 of the United States Code to recover such further sum as added to said 75 per centum will constitute just compensation.

---

[302]   Added by P.L. 109–58, § 656(a), 119 Stat. 813 (2005). P.L. 109–58 states the following:
(b) REGULATIONS.–Not later than 1 year after the date of the enactment of this Act, and from time to time thereafter as it considers necessary, the Nuclear Regulatory Commission shall issue regulations identifying radioactive materials or classes of individuals that, consistent with the protection of public health and safety and the common defense and security, are appropriate exceptions to the requirements of section 170D[302] of the Atomic Energy Act of 1954, as added by subsection (a) of this section.
(c) EFFECTIVE DATE.–The amendment made by subsection (a) shall take effect upon the issuance of regulations under subsection (b), except that the background check requirement shall become effective on a date established by the Commission.
 (d) EFFECT ON OTHER LAW.–Nothing in this section or the amendment made by this section shall waive, modify, or affect the application of chapter 51 of title 49, United States Code, part A of subtitle V of title 49, United States Code, part B of subtitle VI of title 49, United States Code, and title 23, United States Code.
[303]   Amended by P.L. 88–489, § 17, 78 Stat. 602 (1964), deleted the phrase "52 (with respect to the material for which the United States is required to pay just compensation)," after 43.
[304]   Amended by P.L. 88–489, §17, 78 Stat. 602 (1964), which added "55."
[305]   Amended by P.L. 102–572, § 902(b)(1), 106 Stat. 4516 (1992).

42 USC 2222.
Condemnation
of real property.
40 USC 3113,
3116, 3117.

Sec. 172. Condemnation of Real Property

Proceedings for condemnation shall be instituted pursuant to the provisions of the Act approved August 1, 1988, as amended and section 1403 of title 28 of the United States Code. The Act approved February 26, 1931, as amended, shall be applicable to any such proceedings.

42 USC 2223.
Patent
application
disclosures.

Sec. 173. Patent Application Disclosures

In the event that the Commission communicates to any nation any Restricted Data based on any patent application not belonging to the United States, just compensation shall be paid by the United States to the owner of the patent application. The Commission shall determine such compensation. If the compensation so determined is unsatisfactory to the person entitled thereto, such person shall be paid 75 per centum of the amount so determined, and shall be entitled to sue the United States Court of Federal Claims[306] or in any district court of the United States for the district in which such claimant is a resident in a manner provided by section 1346 of title 28 of the United States Code to recover such further sum as added to such 75 per centum will constitute just compensation.

42 USC 2224.
Attorney
General
approval of
title.

40 USC 3111.

Sec. 174. Attorney General Approval of Title

All real property acquired under this Act shall be subject to the provisions of section 355 of the Revised Statutes, as amended: *Provided, however,* That real property acquired by purchase or donation, or other means of transfer may also be occupied, used, and improved for the purposes of this Act prior to approval of title by the Attorney General in those cases where the President determines that such action is required in the interest of the common defense and security.

Chapter 16–Judicial Review and Administrative Procedure

42 USC 2231.
General.

Sec. 181. General

The provisions of the Administrative Procedure Act (Public Law 404, Seventy-ninth Congress, approved June 11, 1946) shall apply to all agency action taken under this Act, and the terms "agency" and "agency action" shall have the meaning specified in the Administrative Procedure Act: *Provided, however,* That in the case of agency proceedings or actions which involve Restricted Data, defense information, safeguards information protected from disclosure under the authority of section 147[307] or information protected from dissemination under authority of section 148[308] the Commission shall provide by regulation for such parallel procedures as will effectively safeguard and prevent disclosure of Restricted Data, defense information, or such safeguards information, or information protected from dissemination under the authority of section 148 to unauthorized persons with minimum impairment of the procedural rights which would be available if Restricted Data, defense information, or such safeguards information, or information protected from dissemination under the authority of section 148 were not involved.

42 USC 2232.
License
applications.

Sec. 182. License Applications

a. Each application for a license hereunder shall be in writing and shall specifically state such information as the Commission, by rule or regulation, may determine to be necessary to decide such of the technical and financial qualifications of the applicant, the character of the applicant, the citizenship of the applicant, or any other qualifications of

---

[306]   Amended by P.L. 102–572, § 902(b)(1), 106 Stat. 4516 (1992).
[307]   Amended by P.L. 96–295, § 207(b)(2), 94 Stat. 789 (1980).
[308]   Amended by P.L. 97–90, § 210(b), 95 Stat. 1163 (1981).

the applicant as the Commission may deem appropriate for the license. In connection with applications for licenses to operate production or utilization facilities, the applicant shall state such technical specifications, including information of the amount, kind, and source of special nuclear material required, the place of the use, the specific characteristics of the facility, and such other information as the Commission may, by rule or regulation, deem necessary in order to enable it to find that the utilization or production of special nuclear material will be in accord with the common defense and security and will provide adequate protection to the health and safety of the public. Such technical specifications shall be a part of any license issued. The Commission may at any time after the filing of the original application, and before the expiration of the license, require further written statements in order to enable the Commission to determine whether the application should be granted or denied or whether a license should be modified or revoked. All applications and statements shall be signed by the applicant or licensee. Applications for, and statements made in connection with, licenses under sections 103 and 104 shall be made under oath or affirmation. The Commission may require any other applications or statements to be made under oath or affirmation.[309]

ACRS Report.

b. The Advisory Committee on Reactor Safeguards shall review each application under section 103 or section 104 b. for a construction permit or an operating license for a facility, any application under section 104c. for a construction permit or an operating license for a testing facility, any application under section 104a. or c. specifically referred to it by the Commission, and any application for an amendment to a construction permit or an amendment to an operating license under section 103 or 104a., b., or c. specifically referred to it by the Commission, and shall submit a report thereon which shall be made part of the record of the application and available to the public except to the extent that security classification prevents disclosure[310]

Commercial power.

c. The Commission shall not issue any license under section 103 for a utilization or production facility for the generation of commercial power until it has given notice in writing to such regulatory agency as may have jurisdiction over the rates and services incident to the proposed activity; until it has published notice of the application in such trade or news publications as the Commission deems appropriate to give reasonable notice to municipalities, private utilities, public bodies, and cooperatives which might have a potential interest in such utilization or production facility; and until it has published notice of such application once each week for four consecutive weeks in the Federal Register, and until four weeks after the last notice.[311]

---

[309]   Amended by P.L. 84–1006, § 5, 70 Stat. 1069 (1956).
[310]   Amended by P.L. 85–256, § 6, 71 Stat. 576 (1957), added subsection b and relettered former subsections b and c as subsections c and d. P.L. 87–615, § 3, 76 Stat. 409 (1962), amended subsection b. Before amendment, it read:

   b. The Advisory Committee on Reactor Safeguards shall review each application under section 103 or 104b. for a license for a facility, any application under section 104c. for a testing facility, any application under section 104a. or c. specifically referred to it by the Commission, and shall submit a report thereon, which shall be made part of the record of the application and available to the public, except to the extent that security classification prevents disclosure.

[311]   Amended by P.L. 91–560, § 9, 84 Stat. 1472 (1970). Before amendment, it read as follows:

d. The Commission, in issuing any license for a utilization or production facility for the generation of commercial power under section 103, shall give preferred consideration to applications for such facilities which will be located in high cost power areas in the United States if there are conflicting applications for a limited opportunity for such license. Where such conflicting applications resulting from limited opportunity for such license include those submitted by public or cooperative bodies such applications shall be given preferred consideration.

42 USC 2233.
Terms of licenses.

Sec. 183. Terms of Licenses

Each license shall be in such form and contain such terms and conditions as the Commission may, by rule or regulation, prescribe to effectuate the provisions of this Act, including the following provisions:

a. (Repealed)[312]

b. No right to the special nuclear material shall be conferred by the license except as defined by the license.

c. Neither the license nor any right under the license shall be assigned or otherwise transferred in violation of the provisions of this Act.

d. Every license issued under this Act shall be subject to the right of recapture or control reserved by section 108, and to all of the other provisions of this Act, now or hereafter in effect and to all valid rules and regulations of the Commission.

42 USC 2234.
Inalienability of licenses.

Sec. 184. Inalienability of Licenses

No license granted hereunder and no right to utilize or produce special nuclear material granted hereby shall be transferred, assigned or in any manner disposed of, either voluntarily or involuntarily, directly or indirectly, through transfer of control of any license to any person, unless the Commission shall, after securing full information, find that the transfer is in accordance with the provisions of this Act, and shall give its consent in writing. The Commission may give such consent to the creation of a mortgage, pledge, or other lien upon any facility or special nuclear material,[313] owned or thereafter acquired by a licensee, or upon any leasehold or other interest to such facility,[314] and the rights of the creditors so secured may thereafter be enforced by any court subject to rules and regulations established by the Commission to protect public health and safety and promote the common defense and security.

42 USC 2235.
Construction permits.

Sec. 185. Construction Permits and Operating Licenses

a. All applicants for licenses to construct or modify production or utilization facilities shall, if the application is otherwise acceptable to the Commission, be initially granted a construction permit. The construction permit shall state the earliest and latest dates for the completion of the construction or modification. Unless the construction or modification of the facility is completed by the completion date, the construction permit

---

c. the Commission shall not issue any license for a utilization or production facility for the generation of commercial power under section 103, until it has given notice in writing to such regulatory agency as may have jurisdiction over the rates and services of the proposed activity, to municipalities, private utilities, public bodies, and cooperatives within transmission distance authorized to engage in the distribution of electric energy and until it has published notice of such application once each week for four consecutive weeks in the Federal Register, and until four weeks after the last notice.

[312] Repealed by P.L. 88–489, § 18, 78 Stat. 602 (1964). Subsection a read as follows: "a. Title to all special nuclear material utilized or produced by facilities pursuant to the license, shall at all times be in the United States."

[313] Amended by P.L. 88–489, § 19, 78 Stat. 602 (1964).

[314] Amended by P.L. 88–489, § 19, 78 Stat. 602 (1964).

shall expire, and all rights thereunder be forfeited, unless upon good cause shown, the Commission extends the completion date. Upon the completion of the construction or modification of the facility, upon the filing of any additional information needed to bring the original application up to date, and upon finding that the facility authorized has been constructed and will operate in conformity with the application as amended and in conformity with the provisions of this Act and of the rules and regulations of the Commission, and in the absence of any good cause being shown to the Commission why the granting of a license would not be in accordance with the provisions of this Act, the Commission shall thereupon issue a license to the applicant. For all other purposes of this Act, a construction permit is deemed to be a "license."

b. After holding a public hearing under section 189a.(1)(A), the Commission shall issue to the applicant a combined construction and operating license if the application contains sufficient information to support the issuance of a combined license and the Commission determines that there is reasonable assurance that the facility will be constructed and will operate in conformity with the license, the provisions of this Act, and the Commission's rules and regulations. The Commission shall identify within the combined license the inspections, tests, and analyses, including those applicable to emergency planning, that the licensee shall perform, and the acceptance criteria that, if met, are necessary and sufficient to provide reasonable assurance that the facility has been constructed and will be operated in conformity with the license, the provisions of this Act, and the Commission's rules and regulations. Following issuance of the combined license, the Commission shall ensure that the prescribed inspections, tests, and analyses are performed and, prior to operation of the facility, shall find that the prescribed acceptance criteria are met. Any finding made under this subsection shall not require a hearing except as provided in section 189a.(1)(B). [315]

42 USC 2236.
Revocation.

Sec. 186. Revocation

a. Any license may be revoked for any material false statement in the application or any statement of fact required under section 182, or because of conditions revealed by such application or statement of fact or any report, record, or inspection or other means which would warrant the Commission to refuse to grant a license on an original application, or for failure to construct or operate a facility in accordance with the terms of the construction permit or license or the technical specifications in the application, or for violation of, or failure to observe any of the terms and provisions of this Act or of any regulation of the Commission.

5 USC 558(b).

b. The Commission shall follow the provisions of section 9(b) of the Administrative Procedure Act in revoking any license.

5 USC 551.

c. Upon revocation of the license, the Commission may immediately retake possession of all special nuclear material held by the licensee. In cases found by the Commission to be of extreme importance to the national defense and security or to the health and safety of the public, the Commission may recapture any special nuclear material held by the licensee or may enter upon and operate the facility prior to any of the

---

[315]   Amended by P.L. 102–486, 106 Stat. 3120 (1992), which added a new heading, subsection a. and a new b.
NOTE: Sections 185 b. and 189 a.(1)(b) of the Atomic Energy Act of 1954, as added by P.L. 102–486, § 2801 and 2802, 106 Stat. 3120 (1992), shall apply to all proceedings involving a combined license for which an application was filed after May 8, 1991, under such sections.

procedures provided under the Administrative Procedures Act. Just compensation shall be paid for the use of the facility.

Sec. 187. Modification of License

The terms and conditions of all licenses shall be subject to amendment, revision, or modification, by reason of amendments of this Act, or by reason of rules and regulations issued in accordance with the terms of this Act.

Sec. 188. Continued Operation of Facilities

Whenever the Commission finds that the public convenience and necessity or the production program of the Commission requires continued operation of a production facility or utilization facility the license for which has been revoked pursuant to section 186, the Commission may, after consultation with the appropriate regulatory agency, State or Federal, having jurisdiction, order that possession be taken of and such facility be operated for such period of time as the public convenience and necessity or the production program of the Commission may, in the judgment of the Commission, require, or until a license for the operation of the facility shall become effective. Just compensation shall be paid for the use of the facility.

Sec. 189. Hearings and Judicial Review

a. (1)(A) In any proceeding under this Act, for the granting, suspending, revoking, or amending of any license or construction permit, or application to transfer control, and in any proceeding for the issuance or modification of rules and regulations dealing with the activities of licensees, and in any proceeding for the payment of compensation, an award, or royalties under section 153, 157, 186c., or 188, the Commission shall grant a hearing upon the request of any person whose interest may be affected by the proceeding, and shall admit any such person as a party to such proceeding. The Commission shall hold a

hearing after thirty days' notice and publication once in the Federal Register, on each application under section 103 or 104b. for a construction permit for a facility, and on any application under section 104c. for a construction permit for a testing facility. In cases where such a construction permit has been issued following the holding of such a hearing, the Commission may, in the absence of a request therefor by any person whose interest may be affected, issue an operating license or an amendment to a construction permit or an amendment to an operating license without a hearing, but upon thirty days' notice and publication once in the Federal Register of its intent to do so. The Commission may dispense with such thirty days' notice and publication with respect to any application for an amendment to a construction permit or an amendment to an operating license upon a determination by the Commission that the amendment involves no significant hazards consideration.[316]

---

[316]   Amended by P.L. 87–615 § 2, 76 Stat. 409 (1962). Before amendment, it read:
SEC. 189. HEARINGS AND JUDICIAL REVIEW.–
    a. In any proceeding under this Act, for the granting, suspending, revoking, or amending of any license or construction permit, or application to transfer control, and in any proceeding for the issuance or modification of rules and regulations dealing with the activities of licensees, and in any proceeding for the payment of compensation, an award or royalties under section 153, 157, 186c., or 188, the Commission shall grant a hearing upon the request of any person whose interest may be affected by the proceeding, and shall admit any such person as a party to such proceeding. The Commission shall hold a hearing after thirty days notice and publication once in the Federal Register on each application under section 103 or 104b. for a license for a facility, and on any application under section 104c. for a license for a testing facility.

(B)(i) Not less than 180 days before the date scheduled for initial loading of fuel into a plant by a licensee that has been issued a combined construction permit and operating license under section 185b., the Commission shall publish in the Federal Register notice of intended operation. That notice shall provide that any person whose interest may be affected by operation of the plant, may within 60 days request the Commission to hold a hearing on whether the facility as constructed complies, or on completion will comply, with the acceptance criteria of the license.

(ii) A request for hearing under clause (i) shall show, prima facie, that one or more of the acceptance criteria in the combined license have not been, or will not be met, and the specific operational consequences of nonconformance that would be contrary to providing reasonable assurance of adequate protection of the public health and safety.

(iii) After receiving a request for a hearing under clause (i), the Commission expeditiously shall either deny or grant the request. If the request is granted, the Commission shall determine, after considering petitioners' prima facie showing and any answers thereto, whether during a period of interim operation, there will be reasonable assurance of adequate protection of the public health and safety. If the Commission determines that there is such reasonable assurance, it shall allow operation during an interim period under the combined license.

(iv) The Commission, in its discretion, shall determine appropriate hearing procedures, whether informal or formal adjudicatory, for any hearing under clause (i), and shall state its reasons therefor.

(v) The Commission shall, to the maximum possible extent, render a decision on issues raised by the hearing request within 180 days of the publication of the notice provided by clause (i) or the anticipated date for initial loading of fuel into the reactor, whichever is later. Commencement of operation under a combined license is not subject to subparagraph (A).[317]

(2)(A) The Commission may issue and make immediately effective any amendment to an operating license or any amendment to a combined construction and operating license, upon a determination by the Commission that such amendment involves no significant hazards consideration, notwithstanding the pendency before the Commission of a request for a hearing from any person. Such amendment may be issued and made immediately effective in advance of the holding and completion of any required hearing. In determining under this section whether such amendment involves no significant hazards consideration, the Commission shall consult with

P.L. 85–256, § 7, 71 Stat. 576 (1957), had previously amended section 189 a. by adding the last sentence. P.L. 102–486, 106 Stat. 3120 (1992), added a subparagraph designator (A), to section 189 a.(1) and added a new subsection (B)(i).
[317]   Amended by P.L. 102–486, 106 Stat. 3121 (1992).
NOTE: Sections 185 b. and 189 a.(1)(b) of the Atomic Energy Act of 1954, as added by P.L. 102–486, § 2801 and 2802, 106 Stat. 3120 (1992), shall apply to all proceedings involving a combined license for which an application was filed after May 8, 1991, under such sections.

the State in which the facility involved is located. In all other respects such amendment shall meet the requirements of this Act.

Notice
publication.

(B) The Commission shall periodically (but not less frequently than once every thirty days) publish notice of any amendments issued, or proposed to be issued, as provided in subparagraph (A). Each such notice shall include all amendments issued, or proposed to be issued, since the date of publication of the last such periodic notice. Such notice shall, with respect to each amendment or proposed amendment (i) identify the facility involved; and (ii) provide a brief description of such amendment. Nothing in this subsection shall be construed to delay the effective date of any amendment.

Regulations
establishing
standards,
criteria,
and procedures.

(C) The Commission shall, during the ninety-day period following the effective date of this paragraph, promulgate regulations establishing (i) standards for determining whether any amendment to an operating license involves no significant hazards consideration; (ii) criteria for providing or, in emergency situations, dispensing with prior notice and reasonable opportunity for public comment on any such determination, which criteria shall take into account the exigency of the need for the amendment involved; and (iii) procedures for consultation on any such determination with the State in which the facility involved is located.[318]

42 USC
2239(b).

b. The following Commission actions shall be subject to judicial review in the manner prescribed in chapter 158 of title 28, United States Code, and chapter 7 of title 5, United States Code:

(1) Any final order entered in any proceeding of the   kind specified in subsection (a).

(2) Any final order allowing or prohibiting a facility to begin operating under a combined construction and operating license.

(3) Any final order establishing by regulation standards to govern the Department of Energy's gaseous diffusion uranium enrichment plants, including any such facilities leased to a corporation established under the USEC Privatization Act.

(4) Any final determination under section 1701(c) relating to whether the gaseous diffusion plants, including any such facilities leased to a corporation established under the USEC Privatization Act, are in compliance with the Commission's standards governing the gaseous diffusion plants and all applicable laws.[319]

42 USC 2240.

Sec. 190. Licensee Incident Reports[320]

No report by any licensee of any incident arising out of or in connection with a licensed activity made pursuant to any requirement of the Commission shall be admitted as evidence in any suit or action for damages growing out of any matter mentioned in such report.

---

[318]   Amended by P.L. 97–415, § 12, 96 Stat. 2067 (1983), by inserting (1) after subsection a. and by adding at end thereof new paragraphs (2)(A), (B), and (C).
[319] Amended by P.L. 104–134, Title III, Ch. 1, Subch. A, § 3116(c), 110 Stat. 1321–349 (1996), which substituted subsection b. for one which read:
   b. Any final order entered in any proceeding of the kind specified in
   subsection (a) above or any final order allowing or prohibiting a facility to
   begin operating under a combined construction and operating license shall be
   subject to judicial review in the manner prescribed in the Act of
   December 29, 1950, as amended (Ch. 1189, 64 Stat. 1129), and to the
   provisions of section 10 of the Administrative Procedure Act, as amended.
[320]   Added by P.L. 87–206, § 16, 75 Stat. 475 (1961).

Sec. 191. Atomic Safety and Licensing Board[321]

a. Notwithstanding the provisions of sections 7(a) and 8(a) of the Administrative Procedure Act, the Commission is authorized to establish one or more atomic safety and licensing boards, each comprised of three members, one of whom shall be qualified in the conduct of administrative proceedings and two of whom shall have such technical or other qualifications as the Commission deems appropriate to the issues to be decided, to conduct such hearings as the Commission may direct and make such intermediate or final decisions as the Commission may authorize with respect to the granting, suspending, revoking or amending of any license or authorization under the provisions of this Act, any other provision of law, or any regulation of the Commission issued thereunder.[322]

The Commission may delegate to a board such other regulatory functions as the Commission deems appropriate. The Commission may appoint a panel of qualified persons from which board members may be selected.

b. Board members may be appointed by the Commission from private life, or designated from the staff of the Commission or other Federal agency. Board members appointed from private life shall receive a per diem compensation for each day spent in meetings or conferences, and all members shall receive their necessary traveling or other expenses while engaged in the work of a board. The provisions of section 163 shall be applicable to board members appointed from private life.

Sec. 192. Temporary Operating License[323]

a. In any proceeding upon an application for an operating license for a utilization facility required to be licensed under section 103 or 104b. of this Act, in which a hearing is otherwise required pursuant to section 189a., the applicant may petition the Commission for a temporary operating license for such facility authorizing fuel loading, testing, and operation at a specific power level to be determined by the Commission, pending final action by the Commission on the application. The initial petition for a temporary operating license for each such facility, and any temporary operating license issued for such facility based upon the initial petition, shall be limited to power levels not to exceed 5 percent of rated full thermal power. Following issuance by the Commission of the temporary operating license for each such facility, the licensee may file petitions with the Commission to amend the license to allow facility operation in staged increases at specific power levels, to be determined by the Commission, exceeding 5 percent of rated full thermal power. The initial petition for a temporary operating license for each such facility may be filed at any time after the filing of: (1) the report of the Advisory

---

[321]   Added by P.L. 87–615, § 1, 76 Stat. 409 (1962).
[322]   Amended by P.L. 91–560, § 10, 84 Stat. 1472 (1970). Before amendment, it read as follows:

> Notwithstanding the provisions of sections 7(a) and 8(a) of the Administrative Procedure Act, the Commission is authorized to establish one or more atomic safety and licensing boards, each composed of three members, two of whom shall be technically qualified and one of whom shall be qualified in the conduct of administrative proceedings, to conduct such hearings as the Commission may direct and make such intermediate or final decisions as the Commission may authorize with respect to the granting, suspending, revoking or amending of any license or authorization under the provisions of this Act, any other provision of law, or any regulation of the Commission issued hereunder.

[323]   Added by P.L. 92–307, 86 Stat. 191 (1972).

Committee on Reactor Safeguards required by section 182b.; (2) the filing of the initial Safety Evaluation Report by the Nuclear Regulatory Commission staff and the Nuclear Regulatory Commission staff's first supplement to the report prepared in response to the report of the Advisory Committee on Reactor Safeguards for the facility; (3) the Nuclear Regulatory Commission staff's final detailed statement on the environmental impact of the facility prepared pursuant to section 102(2)(C) of the National Environmental Policy Act of 1969 (42 USC 4332(2)(C)); and (4) a State, local, or utility emergency preparedness plan for the facility. Petitions for the issuance of a temporary operating license, or for an amendment to such a license allowing operation at a specific power level greater than that authorized in the initial temporary operating license, shall be accompanied by an affidavit or affidavits setting forth the specific fact upon which the petitioner relies to justify issuance of the temporary operating license or the amendment thereto. The Commission shall publish notice of each such petition in the Federal Register and in such trade or news publications as the Commission deems appropriate to give reasonable notice to persons who might have a potential interest in the grant of such temporary operating license or amendment thereto. Any person may file affidavits or statements in support of, or in opposition to, the petition within thirty days after the publication of such notice in the Federal Register.

Affidavits.

Publication in Federal Register*.*

b. With respect to any petition filed pursuant to subsection a. of this section, the Commission may issue a temporary operating license, or amend the license to authorize temporary operation at each specific power level greater than that authorized in the initial temporary operating license, as determined by the Commission, upon finding that–

(1) in all respects other than the conduct or completion of any required hearing, the requirements of law are met;

(2) in accordance with such requirements, there is reasonable assurance that operation of the facility during the period of the temporary operating license in accordance with its terms and conditions will provide adequate protection to the public health and safety and the environment during the period of temporary operation; and

(3) denial of such temporary operating license will result in delay between the date on which construction of the facility is sufficiently completed, in the judgment of the Commission, to permit issuance of the temporary operating license, and the date when such facility would otherwise receive a final operating license pursuant to this Act.

Final order, *transmittal to* *congressional* *committees.*

The temporary operating license shall become effective upon issuance and shall contain such terms and conditions as the Commission may deem necessary, including the duration of the license and any provision for the extension thereof. Any final order authorizing the issuance or amendment of any temporary operating license pursuant to this section shall recite with specificity the facts and reasons justifying the findings under this subsection, and shall be transmitted upon such issuance to the Committee on Natural Resources and on Energy and Commerce of the House of Representatives and the Committee on Environment and Public Works of the Senate. The final order of the Commission with respect to the issuance or amendment of a temporary operating license shall be subject to judicial review pursuant to chapter 158 of title 28, United States Code. The requirements of section 189a. of this Act with respect to the issuance or amendment of facility licenses shall not apply to the issuance or amendment of a temporary operating license under this section.

Judicial review. 28 USC 2341 et seq*.*

Hearing.

c. Any hearing on the application for the final operating license for a facility required pursuant to section 189a. shall be concluded as promptly as practicable. The Commission shall suspend the temporary operating license if it finds that the applicant is not prosecuting the application for the final operating license with due diligence. Issuance of a temporary operating license under subsection b. of this section shall be without prejudice to the right of any party to raise any issue in a hearing required pursuant to section 189a; and failure to assert any ground for denial or limitation of a temporary operating license shall not bar the assertion of such ground in connection with the issuance of a subsequent final operating license. Any party to a hearing required pursuant to section 189a. on the final operating license for a facility for which a temporary operating license has been issued under subsection b., and any member of the Atomic Safety and Licensing Board conducting such hearing, shall promptly notify the Commission of any information indicating that the terms and conditions of the temporary operating license are not being met, or that such terms and conditions are not sufficient to comply with the provisions of paragraph (2) of subsection b.

d. The Commission is authorized and directed to adopt such administrative remedies as the Commission deems appropriate to minimize the need for issuance of temporary operating licenses pursuant to this section.

Expiration date.

42 USC 2243.

e. The authority to issue new temporary operating licenses under this section shall expire on December 31, 1983.[324]

Sec. 193. Licensing of Uranium Enrichment Facilities[325]

(a) Environmental Impact Statement.–

(1) Major Federal Action.–The issuance of a license under sections 53 and 63 for the construction and operation of any uranium enrichment facility shall be considered a major Federal action significantly affecting the quality of the human environment for purposes of the National Environmental Policy Act of 1969 (42 USC 4321 *et seq.*).

(2) Timing.–An environmental impact statement prepared under paragraph (1) shall be prepared before the hearing on the issuance of a license for the construction and operation of a uranium enrichment facility is completed.

(b) Adjudicatory Hearing.–

(1) In General.–The Commission shall conduct a single adjudicatory hearing on the record with regard to the licensing of the construction and operation of a uranium enrichment facility under sections 53 and 63.

(2) Timing.–Such hearing shall be completed and a decision issued before the issuance of a license for such construction and operation.

(3) Single Proceeding.–No further Commission licensing action shall be required to authorize operation.

---

[324]   Amended by P.L. 103–437, § 15(f)(5), 108 Stat. 4581 (1994); P.L. 97–415, 96 Stat. 2067 (1983), before which section 192 read as follows:

Sec. 192. Temporary Operating Licenses.–

a. In an proceeding upon an application for an operating license for a nuclear power reactor, in which a hearing is otherwise required pursuant to section 189a , the applicant may petition the Commission for a temporary operating license authorizing operation of the facility pending final action by the Commission on the application. Such petition may be filed at any time after filing of: (1) the report of the Advisory Committee on Reactor Safeguards required by subsection 182b.; (2) the safety evaluation of the application by the Commission's regulatory staff; and (3) the regulatory staff's final detailed statement on the environmental impact of the facility prepared pursuant to section 102(2)(C) of the National Environmental Policy Act of 1969 (83 Stat. 853) or, in the case of an application for operating license filed on or before September 9, 1971, if the regulatory staff's final detailed statement required under section 102(2)(C) is not completed, the Commission must satisfy the applicable requirements of the National Environmental Policy Act prior to issuing any temporary operating license under this section 192. The petition shall be accompanied by an affidavit or affidavits setting forth the facts upon which the petitioner relies to justify issuance of the temporary operating license. Any party to the proceeding may file affidavits in support of, or opposition to, the petition within fourteen days subject to judicial review pursuant to the Act of December 29, 1950, as amended (Ch. 1189, 64 Stat. 1129).

c. The hearing on the application for the final operating license otherwise required pursuant to section 189a. shall be concluded as promptly as practicable. The Commission shall vacate the temporary operating license if it finds that the applicant is not prosecuting the application for the final operating license with due diligence. Issuance of a temporary operating license pursuant to subsection b. of this section shall be without prejudice to the position of any party to the proceeding in which a hearing is otherwise required pursuant to section 189a.; and failure to assert any ground for denial or limitation of a temporary operating license shall not bar the assertion of such ground in connection with the issuance of a subsequent final operating license.

d. The authority under this section shall expire on October 30, 1973.

[325]   Added by P.L. 101–575, 104 Stat. 2835 (1990).

Federal Register publication.
Claims.
Nuclear materials.

(c) Inspection and Operation.–Prior to commencement of operation of a uranium enrichment facility licensed hereunder, the Commission shall verify through inspection that the facility has been constructed in accordance with the requirements of the license for construction and operation. The Commission shall publish notice of the inspection results in the Federal Register.

(d) Insurance and Decommissioning.–"(1) The Commission shall require, as a condition of the issuance of a license under sections 53 and 63 for a uranium enrichment facility, that the licensee have and maintain liability insurance of such type and in such amounts as the Commission judges appropriate to cover liability claims arising out of any occurrence within the United States, causing, within or outside the United States, bodily injury, sickness, disease, or death, or loss of or damage to property, or loss of use of property, arising out of or resulting from the radioactive, toxic, explosive, or other hazardous properties of chemical compounds containing source or special nuclear material.

(2) The Commission shall require, as a condition for the issuance of a license under sections 53 and 63 for a uranium enrichment facility, that the licensee provide adequate assurance of the availability of funds for the decommissioning (including decontamination) of such facility using funding mechanisms that may include, but are not necessarily limited to, the following:

(A) Prepayment (in the form of a trust, escrow account, government fund, certificate of deposit, or deposit of government securities).

(B) Surety (in the form of a surety or performance bond, letter of credit, or line of credit), insurance, or other guarantee (including parent company guarantee) method.

(C) External sinking fund in which deposits are made at least annually.

(e) No Price-Anderson Coverage.–Section 170 of this Act shall not apply to any license under section 53 or 63 for a uranium enrichment facility constructed after the date of enactment of this section.

(f)[326] LIMITATION.–No license or certificate of compliance may be issued to the United States Enrichment Corporation or its successor under this section or section 53, 63, or 1701, if the Commission determines that–

(1) the Corporation is owned, controlled, or dominated by an alien, a foreign corporation, or a foreign government; or

(2) the issuance of such a license or certificate of compliance would be inimical to–

(A) the common defense and security of the United States; or

(B) the maintenance of a reliable and economical domestic source of enrichment services.

Chapter 17–Joint Committee on Atomic Energy

(Repealed[327])

---

[326]   Added by P.L. 104–134, Title III, Ch. 1, Subch. A, § 3116(b)(2), 110 Stat. 1321–349 (1996).
[327]   Chapter 17 repealed by P.L. 95–110, 91 Stat. 884 (1977). Chapter 17 read as follows:
JOINT COMMITTEE ON ATOMIC ENERGY
Sec. 201. Membership.–There is hereby established a Joint Committee on Atomic Energy to be composed of nine Members of the Senate to be appointed by the President of the Senate, and nine Members of the House of

Representatives to be appointed by the Speaker of the House of Representatives. In each instance not more than five Members shall be members of the same political party.

Sec. 202. Authority and Duty.–

a. The Joint Committee shall make continuing studies of the activities of the Atomic Energy Commission and of problems relating to the development, use, and control of atomic energy. During the first ninety[a] days of each session of the Congress, the Joint Committee may conduct hearings in either open or executive session for the purpose of receiving information concerning the development, growth, and state of the atomic energy industry[b] The Commission shall keep the Joint Committee fully and currently informed with respect to all of the Commission's activities. The Department of Defense shall keep the Joint Committee fully and currently informed with respect to all matters within the Department of Defense relating to the development, utilization, or application of atomic energy. Any Government agency shall furnish any information requested by the Joint Committee with respect to the activities or responsibilities of that agency in the field of atomic energy. All bills, resolutions, and other matters in the Senate or the House of Representatives relating primarily to the Commission or to the development, use, or control of atomic energy shall be referred to the Joint Committee. The members of the Joint Committee who are Members of the Senate shall from time to time report to the Senate, and the members of the Joint Committee who are Members of the House of Representatives shall from time to time report to the House, by bill or otherwise, their recommendations with respect to matters within the jurisdiction of their respective Houses which are referred to the Joint Committee or otherwise within the jurisdiction of the Joint Committee.

b. The members of the Joint Committee who are Members of the Senate and the Members of the Joint Committee who are Members of the House of Representatives shall, on or before June 30 of each year, report to their respective Houses on the development, use, and control of nuclear energy for the common defense and security and for peaceful purposes. Each report shall provide facts and information available to the Joint Committee concerning nuclear energy which will assist the appropriate committees of the Congress and individual members in the exercise of informed judgment on matters of weaponry; foreign policy; defense; international trade; and in respect to the expenditure and appropriation of Government revenues. Each report shall be presented formally under circumstances which provide for clarification and discussion by the Senate and the House of Representatives. In recognition of the need for public understanding, presentations of the reports shall be made to the maximum extent possible in open session and by means of unclassified written materials.

Sec. 203. Chairman.–Vacancies in the membership of the Joint Committee shall not affect the power of the remaining members to execute the functions of the Joint Committee, and shall be filled in the same manner as in the case of the original selection. The Joint Committee shall select a Chairman and a Vice Chairman from among its members at the beginning of each Congress. The Vice Chairman shall act in the place and stead of the Chairman in the absence of the Chairman. The Chairmanship shall alternate between the Senate and the House of Representatives with each Congress, and the Chairman shall be selected by the Members from that House entitled to the Chairmanship. The Vice Chairman shall be chosen from the House other than that of the Chairman by the Members from that House.

Sec. 204 Powers.–In carrying out its duties under this Act, the Joint Committee, or any duly authorized subcommittee thereof, is authorized to hold such hearings or investigations, to sit and act at such places and times to require, by subpoena or otherwise, the attendance of such witnesses and the production of such books, papers, and documents, to administer such oaths, to take such testimony, to procure such printing and binding, and to make such expenditures as it deems advisable. The Joint Committee may make such rules respecting its organization and procedures as it deems necessary: *Provided, however,* That no measure or recommendation shall be reported from the Joint Committee unless a majority of the committee assent. Subpoenas may be issued over the signature of the Chairman of the Joint Committee or by any member designated by him or by the Joint Committee, and may be served by such person or persons as may be designated by such Chairman or member. The Chairman of the Joint Committee or any member thereof may administer

Chapter 18–Enforcement

42 USC 2271.
General
provisions.

Sec. 221. General Provisions

a. To protect against the unlawful dissemination of Restricted Data and to safeguard facilities, equipment, materials, and other property of the Commission, the President shall have authority to utilize the services of any Government agency to the extent he may deem necessary or desirable.

b. The Federal Bureau of Investigation of the Department of Justice shall investigate all alleged or suspected criminal violations of this Act.

c. No action shall be brought against any individual or person for any violation under this Act unless and until the Attorney General of the United States has advised the Commission with respect to such action and no such action shall be commenced except by the Attorney General of the United States. And provided further, that nothing in this subsection shall be construed as applying to administrative action taken by the Commission.[328]

---

oaths to witnesses. The Joint Committee may use a committee seal. The provisions of sections 102 to 104, inclusive, of the Revised Statutes, as amended, shall apply in case of any failure of any witness to comply with a subpoena or to testify when summoned under authority of this section. The expenses of the Joint Committee shall be paid from the contingent fund of the Senate from funds appropriated for the Joint Committee upon vouchers approved by the Chairman. The cost of stenographic service to report public hearings shall not be in excess of the amounts prescribed by law for reporting the hearings of standing committees of the Senate. The cost of stenographic service to report executive hearings shall be fixed at an equitable rate by the Joint Committee. Members of the Joint Committee, and its employees and consultants, while traveling on official business for the Joint Committee, may receive either the per diem allowance authorized to be paid to Members of Congress or its employees, or their actual and necessary expenses provided an itemized statement of such expenses is attached to the voucher.

Sec. 205. Staff and Assistance.– The Joint Committee is empowered to appoint and fix the compensation of such experts, consultants, technicians, and staff employees as it deems necessary and advisable. The Joint Committee is authorized to utilize the services, information, facilities, and personnel of the departments and establishments of the Government. The Joint Committee is authorized to permit such of its members, employees, and consultants as it deems necessary in the interest of common defense and security to carry firearms while in the discharge of their official duties for the committee.

Sec. 206. Classification of Information.– The Joint Committee may classify information originating within the committee in accordance with standards used generally by the executive branch for classifying Restricted Data or defense information.

Sec. 207. Records.– The Joint Committee shall keep a complete record of all committee actions, including a record of the votes on any question on which a record vote is demanded. All committee records, data, charts, and files shall be the property of the Joint Committee and shall be kept in the offices of the Joint Committee or other places as the Joint Committee may direct under such security safeguards as the Joint Committee shall determine in the interest of the common defense and security.

Chapter previously amended by P.L. 87–206, § 17, 75 Stat. 475 (1961); P.L. 88–294, 78 Stat. 172 (1964), amended the second sentence of section 202. Before amendment, this sentence read:

During the first ninety days of each session of the Congress, the Joint Committee shall conduct hearings in either open or executive session for the purpose of receiving information concerning the development, growth, and state of the atomic energy industry.

Subsection 202b. was added by P.L. 93–514, 88 Stat. 1611 (1974).
[328]   Amended by P.L. 101–647, § 1211, 104 Stat. 4789.

42 USC 2272.
Violation of
specific
sections.

Sec. 222. Violations of Specific Sections

a. Whoever willfully violates, attempts to violate, or conspires to violate, any provisions of section 57 or 101, or whoever unlawfully interferes, attempts to interfere, or conspires to interfere with any recapture or entry under section 108, shall, upon conviction thereof, be punished by a fine of not more than $10,000 or by imprisonment for not more than ten years, or both, except that whoever commits such an offense with intent to injure the United States or with intent to secure an advantage to any foreign nation shall, upon conviction thereof, be punished by imprisonment for life, or by imprisonment for any term of years or a fine of not more than $20,000 or both.[329]

b.[330] Any person who violates, or attempts or conspires to violate, section 92 shall be fined not more than $2,000,000 and sentenced to a term of imprisonment not less than 25 years or to imprisonment for life. Any person who, in the course of a violation of section 92, uses, attempts or conspires to use, or possesses and threatens to use, any atomic weapon shall be fined not more than $2,000,000 and imprisoned for not less than 30 years or imprisoned for life. If the death of another results from a person's violation of section 92, the person shall be fined not more than $2,000,000 and punished by imprisonment for life.

42 USC 2273.
Violation of
sections
generally.

Sec. 223. Violation of Sections Generally

a.[331] Whoever willfully violates, attempts to violate, or conspires to violate, any provision of this Act for which no criminal[332] penalty is specifically provided or of any regulation or order prescribed or issued under section 65 or subsection 161b., i., or o.,[333] shall, upon conviction thereof, be punished by a fine of not more than $5,000 or by imprisonment for not more than two years, or both, except that whoever commits such as offense with intent to injure the United States or with intent to secure an advantage to any foreign nation, shall, upon conviction thereof, be punished by a fine of not more than $20,000 or by imprisonment for not more than twenty years, or both.

b.[334] Any individual director, officer, or employee of a firm constructing, or supplying the components of any utilization facility required to be licensed under section 103 or 104b. of this Act who by act or omission, in connection with such construction or supply, knowingly and willfully violates or causes to be violated, any section of this Act, any rule, regulation, or order issued thereunder, or any license condition, which violation results, or if undetected could have resulted, in a significant impairment of a basic component of such a facility shall, upon conviction, be subject to a fine of not more than $25,000 for each day of violation, or to imprisonment not to exceed two years, or both. If the conviction is for a violation committed after a first conviction under this subsection, punishment shall be a fine of not more than $50,000 per day of violation, or imprisonment for not more than two years, or both. For
Basic
component.
the purposes of this subsection, the term "basic component" means a facility structure, system, component or part thereof necessary to assure–

(1) the integrity of the reactor coolant pressure boundary,

---

[329]   Amended by P.L. 91–161, § 2, 3(a), 83 Stat. 444 (1969); P.L. 108–458, 118 Stat 3771 (2004).
[330]   Added by P.L. 108–458, 118 Stat. 3771 (2004).
[331]   Designated as subsection a. by P.L. 96–295, § 203, 94 Stat. 786 (1980).
[332]   Amended by P.L. 91–161, § 6, 83 Stat. 444 (1969).
[333]   Amended by P.L. 90–190, § 12, 81 Stat. 575 (1967).
[334]   Added by P.L. 96–295, § 203, 94 Stat. 786 (1980).

(2) the capability to shut-down the facility and maintain it in a safe shut-down condition, or

(3) the capability to prevent or mitigate the consequences of accidents which could result in an unplanned offsite release of quantities of fission products in excess of the limits established by the Commission.

42 USC 2133.
42 USC 2134.

The provisions of this subsection shall be prominently posted at each site where a utilization facility required to be licensed under section 103 or 104b. of the Act is under construction and on the premises of each plant where components for such a facility are fabricated.

Contracts.

c.[335] Any individual director, officer or employee of a person indemnified under an agreement of indemnification under section 170d. (or of a subcontractor or supplier thereto) who, by act or omission, knowingly and willfully violates or causes to be violated any section of this Act or any applicable nuclear safety-related rule, regulation or order issued thereunder by the Secretary of Energy (or expressly incorporated by reference by the Secretary for purposes of nuclear safety, except any rule, regulation, or order issued by the Secretary of Transportation), which violation results in or, if undetected, would have resulted in a nuclear incident as defined in subsection 11q. shall, upon conviction, notwithstanding section 3571 of title 18, United States Code, be subject to a fine of not more than $25,000, or to imprisonment not to exceed two years, or both. If the conviction is for a violation committed after the first conviction under this subsection, notwithstanding section 3571 of title 18, United States Code, punishment shall be a fine of not more than $50,000, or imprisonment for not more than five years, or both.

42 USC 2274.
Communication of restricted data.

Sec. 224. Communication of Restricted Data

Whoever, lawfully or unlawfully, having possession of, access to, control over, or being entrusted with any document, writing, sketch, photograph, plan, model, instrument, appliance, note, or information involving or incorporating Restricted Data–

a. communicates, transmits, or discloses the same to any individual or person, or attempts or conspires to do any of the foregoing, with intent to injure the United States or with intent to secure an advantage to any foreign nation, upon conviction thereof, shall be punished by imprisonment for life, or by imprisonment for any term of years or a fine of not more than $100,000 or both;

b. communicates, transmits, or discloses the same to any individual or person, or attempts or conspires to do any of the foregoing, with reason to believe such data will be utilized to injure the United States or to secure an advantage to any foreign nation, shall, upon conviction, be punished by a fine of not more than $50,000 or imprisonment for not more than ten years, or both.[336]

42 USC 2275.
Receipt of restricted data.

Sec. 225. Receipt of Restricted Data

Whoever, with intent to injure the United States or with intent to secure an advantage to any foreign nation, acquires, or attempts or conspires to acquire any document, writing, sketch, photograph, plan, model, instrument, appliance, note, or information involving or

---

[335]   Added by P.L. 100–408, § 18, 102 Stat. 1066 (1988).
[336]   Amended by P.L. 79–585, § 10, 60 Stat. 766, (1946); P.L. 83–703, § 1, 68 Stat. 958 (1954); P.L. 91–161, § 3(b), 83 Stat. 444 (1969); P.L. 102–486, Title IX, § 902(a)(8), 106 Stat. 2944 (1992) (renumbered Title I); P.L. 106–65, Div. C, Title XXXI, § 3148(a), 113 Stat. 938, October 5, 1991; P.L. 106–398, § 1, 114 Stat. 1654, October 30, 2000 (Div. A, Title X, § 1087(g)(9)).

incorporating Restricted Data shall, upon conviction thereof, be punished by imprisonment for life, or by imprisonment for any term of years or a fine of not more than $100,000 or both.[337]

42 USC 2276. Tampering with restricted data.

Sec. 226. Tampering with Restricted Data

Whoever, with intent to injure the United States or with intent to secure an advantage to any foreign nation, removes, conceals, tampers with, alters, mutilates, or destroys any document, writing, sketch, photograph, plan, model, instrument, appliance, or note involving or incorporating Restricted Data and used by any individual or person in connection with the production of special nuclear material, or research or development relating to atomic energy, conducted by the United States, or financed in whole or in part by Federal funds, or conducted with the aid of special nuclear material, shall be punished by imprisonment for life, or by imprisonment for any term of years or a fine of not more than $20,000 or both.

42 USC 2277. Disclosure of restricted data.

Sec. 227. Disclosure of Restricted Data

Whoever, being or having been an employee or member of the Commission, a member of the Armed Forces, an employee of any agency of the United States, or being or having been a contractor of the Commission or of an agency of the United States, or being or having been an employee of a contractor of the Commission or of an agency of the United States, or being or having been a licensee of the Commission, or being or having been an employee of a licensee of the Commission, knowingly communicates, or whoever conspires to communicate or to receive, any Restricted Data, knowing or having reason to believe that such data is Restricted Data, to any person not authorized to receive Restricted Data pursuant to the provisions of this Act or under rule or regulation of the Commission issued pursuant thereto, knowing or having reason to believe such person is not so authorized to receive Restricted Data shall, upon conviction thereof, be punishable by a fine of not more than $12,500.[338]

42 USC 2278. Statute of limitations.

Sec. 228. Statute of Limitations

Except for a capital offense, no individual or person shall be prosecuted, tried, or punished for any offense prescribed or defined in sections 224 to 226, inclusive, of this Act, unless the indictment is found or the information is instituted within ten years next after such offense shall have been committed.

42 USC 2278a. Trespass on Commission installations.

Sec. 229. Trespass on Commission Installations[339]

a.(1) The Commission is authorized to issue regulations relating to the entry upon or carrying, transporting, or otherwise introducing or causing to be introduced any dangerous weapon, explosive, or other dangerous instrument or material likely to produce substantial injury or damage to persons or property, into or upon any facility, installation, or real property subject to the jurisdiction, administration, in the custody of the Commission, or subject to the licensing authority of the Commission or certification by the Commission under this chapter or any other Act,

(2) Every such regulation of the Commission shall be posted conspicuously at the location involved.[340]

---

[337] Amended by P.L. 106–65, Div. C, Title XXXI, Subtitle G, § 3148(b), 113 Stat. 938, (2000).

[338] Amended by P.L. 106–65, Div. C., Title XXXI, Subtitle G, § 3148(a), 113 Stat. 938 (1999), by substituting $12,500 for $2,500.

[339] Amended by P.L. 84–1006, § 6, 70 Stat. 1069 (1956), added a new section 229.

[340] Amended by P.L. 109–58, § 654(3)(B), 119 Stat. 812 (2005), redesignating the second sentence of subsection a. as paragraph (2).

b. Whoever shall willfully violate any regulation of the Commission issued pursuant to subsection a. shall, upon conviction thereof, be punishable by a fine of not more than $1,000.

c. Whoever shall willfully violate any regulation of the Commission issued pursuant to subsection a. with respect to any installation or other property which is enclosed by a fence, wall, floor, roof, or other structural barrier shall be guilty of a misdemeanor and upon conviction thereof shall be punished by a fine of not to exceed $5,000 or to imprisonment for not more than one year, or both.

42 USC 2278b.

Photographing
of Commission
installations.

Sec. 230. Photographing, etc., of Commission Installations[341]

It shall be an offense, punishable by a fine of not more than $1,000 or imprisonment for not more than one year, or both–

(1) to make any photograph, sketch, picture, drawing, map or graphical representation, while present on property subject to the jurisdiction, administration or in the custody of the Commission, of any installations or equipment designated by the President as requiring protection against the general dissemination of information relative thereto, in the interest of the common defense and security, without first obtaining the permission of the Commission, and promptly submitting the product obtained to the Commission for inspection or such other action as may be deemed necessary; or

(2) to use or permit the use of an aircraft or any contrivance used, or designed for navigation or flight in air, for the purpose of making a photograph, sketch, picture, drawing, map or graphical representation of any installation or equipment designated by the President as provided in the preceding paragraph, unless authorized by the Commission.

42 USC 2279.
Other laws.

Sec. 231. Other Laws

Sections 224 to 230 shall not exclude the applicable provisions of any other laws.[342]

42 USC 2280.
Injunction
proceedings.

Sec. 232. Injunction Proceedings

Whenever in the judgment of the Commission any person has engaged or is about to engage in any acts or practices which constitute or will constitute a violation of any provision of this Act, or any regulation or order issued thereunder, the Attorney General on behalf of the United States may make application to the appropriate court for an order enjoining such acts or practices, or for an order enforcing compliance with such provision, and upon a showing by the Commission that such person has engaged or is about to engage in any such acts or practices, a permanent or temporary injunction, restraining order, or other order may be granted.[343]

42 USC 2281.
Contempt
proceedings.

Sec. 233. Contempt Proceedings

In case of failure or refusal to obey a subpoena served upon any person pursuant to subsection 161c., the district court for any district in which such person is found or resides or transacts business, upon application by the Attorney General on behalf of the United States, shall have jurisdiction to issue an order requiring such person to appear and give testimony or to appear and produce documents, or both, in

---

[341]   Added by P.L. 84–1006, § 6, 70 Stat. 1069 (1956).
[342]   Amended by P.L. 84–1006, § 7, 70 Stat 1069 (1956), redesignating former section 229 as section 231. Before redesignation, section 229 read: "Sec. 229. Other Laws.– Sections 224 to 228 shall not exclude the applicable provisions of any other laws."
[343]   Amended by P.L. 84–1006, § 6, 70 Stat. 1069 (1956), which renumbered former sections 230 and 231 to sections 232 and 233, respectively.

accordance with the subpoena; and any failure to obey such order of the court may be punished by such court as a contempt thereof.[344]

Sec. 234. Civil Monetary Penalties for Violations of Licensing Requirements[345]

a.[346] Any person who (1) violates any licensing or certification[347] provision of section 53, 57, 62, 63, 81, 82, 101, 103, 104, 107, 109, or 1701 or any rule, regulation, or order issued thereunder, or any term, condition, or limitation of any license or certification issued thereunder, or (2) commits any violation for which a license may be revoked under section 186, shall be subject to a civil penalty, to be imposed by the Commission, of not to exceed $100,000[348] for each such violation.[349] If any violation is a continuing one, each day of such violation shall constitute a separate violation for the purpose of computing the applicable civil penalty. The Commission shall have the power to compromise, mitigate, or remit such penalties.

b. Whenever the Commission has reason to believe that a person has become subject to the imposition of a civil penalty under the provisions of this section, it shall notify such person in writing (1) setting forth the date, facts, and nature of each act or omission with which the person is charged, (2) specifically identifying the particular provision or provisions of the section, rule, regulation, order, or license involved in the violation, and (3) advising of each penalty which the Commission proposes to impose and its amount. Such written notice shall be sent by registered or certified mail by the Commission to the last known address of such person. The person so notified shall be granted an opportunity to show in writing, within such reasonable period as the Commission shall by regulation prescribe, why such penalty should not be imposed. The notice shall also advise such person that upon failure to pay the civil penalty subsequently determined by the Commission, if any, the penalty may be collected by civil action.

c. On the request of the Commission, the Attorney General is authorized to institute a civil action to collect a penalty imposed pursuant to this section. The Attorney General shall have the exclusive power to compromise, mitigate, or remit such civil penalties as are referred to him for collection.

Sec. 234A. Civil Monetary Penalties for Violations of Department of Energy Safety Regulations[350]

a. Any person who has entered into an agreement of indemnification under subsection 170d. (or any subcontractor or supplier thereto) who violates (or whose employee violates) any applicable rule, regulation or order related to nuclear safety prescribed or issued by the Secretary of Energy pursuant to this Act (or expressly incorporated by reference by the Secretary for purposes of nuclear safety, except any rule, regulation, or order issued by the Secretary of Transportation) shall be subject to a civil penalty of not to exceed $100,000 for each such violation. If any

Margin notes: 42 USC 2282. / 42 USC 2073. / 42 USC 2077. / 42 USC 2092. / 42 USC 2093. / 42 USC 2111. / 42 USC 2112. / 42 USC 2131. / 42 USC 2133. / 42 USC 2134. / 42 USC 2137. / 42 USC 2139. / 42 USC 2236. / Civil penalties. / Written notification. / 42 USC 2282a. / Contracts.

---

[344] Amended by P.L. 84–1006, § 6, 70 Stat. 1069 (1956), which renumbered former sections 230 and 231 to sections 232 and 233, respectively.
[345] Added by P.L. 91–161, § 4, 83 Stat. 444 (1969).
[346] Added by P.L. 100–408, § 17, 102 Stat. 1066 (1988).
[347] Amended by P.L. 104–134, § 416, 110 Stat. 1321 (1996).
[348] The NRC by rulemaking periodically enlarges this number to account for inflation pursuant to the Federal Civil Penalties Act of 1990, as amended. The current number is $140,000 (73 Fed. Reg. 54671; September 23, 2008).
[349] Amended by P.L. 96–295, § 206, 94 Stat. 787 (1980).
[350] Amended by P.L. 106–65, § 3147(c), 113 Stat. 938 (1999).

violation under this subsection is a continuing one, each day of such violation shall constitute a separate violation for the purpose of computing the applicable civil penalty.

b. (1) The Secretary shall have the power to compromise, modify or remit, with or without conditions, such civil penalties and to prescribe regulations as he may deem necessary to implement this section.

(2) In determining the amount of any civil penalty under this subsection, the Secretary shall take into account the nature, circumstances, extent, and gravity of the violation or violations and, with respect to the violator, ability to pay, effect on ability to continue to do business, any history of prior such violations, the degree of culpability, and such other matters as justice may require.[351]

c. (1) Before issuing an order assessing a civil penalty against any person under this section, the Secretary shall provide to such person notice of the proposed penalty. Such notice shall inform such person of his opportunity to elect in writing within thirty days after the date of receipt of such notice to have the procedures of paragraph (3) (in lieu of those paragraph (2)) apply with respect to such assessment.

(2) (A) Unless an election is made within thirty calendar days after receipt of notice under paragraph (1) to have paragraph (3) apply with respect to such penalty, the Secretary shall assess the penalty, by order, after a determination of violation has been made on the record after an opportunity for an agency hearing pursuant to section 554 of title 5, United States Code, before an administrative law judge appointed under section 3105 of such title 5. Such assessment order shall include the administrative law judge's findings and the basis for such assessment.

Courts, U.S.

(B) Any person against whom a penalty is assessed under this paragraph may, within sixty calendar days after the date of the order of the Secretary assessing such penalty, institute an action in the United States court of appeals for the appropriate judicial circuit for judicial review of such order in accordance with chapter 7 of title 5, United States Code. The court shall have jurisdiction to enter a final judgment affirming, modifying or setting aside in whole or in part, the order of the Secretary, or the court may remand the proceeding to the Secretary for such further action as the court may direct.

(3) (A) In the case of any civil penalty with respect to which the procedures of this paragraph have been elected, the Secretary shall promptly assess such penalty, by order, after the date of the election under paragraph (1).

Courts, U.S.

(B) If the civil penalty has not been paid within sixty calendar days after the assessment order has been made under subparagraph (A), the Secretary shall institute an action in the appropriate district court of the United States for an order affirming the assessment of the civil penalty. The court shall have authority to review de novo the law and facts involved, and shall have jurisdiction to enter a judgment enforcing, modifying, and enforcing as so modified, or setting aside in whole or in part, such assessment.

(C) Any election to have this paragraph apply may not be revoked except with consent of the Secretary.

---

[351]   Amended by P.L. 109–58, § 610(a), 119 Stat. 781 (2005).

Courts, U.S.

(4) If any person fails to pay an assessment of a civil penalty after it has become a final and unappealable order under paragraph (2), or after the appropriate district court has entered final judgment in favor of the Secretary under paragraph (3), the Secretary shall institute an action to recover the amount of such penalty in any appropriate district court of the United States. In such action, the validity and appropriateness of such final assessment order or judgment shall not be subject to review.

Schools and
colleges.
Corporations.

d. (1) Notwithstanding subsection a., in the case of any not-for-profit contractor, subcontractor, or supplier, the total amount of civil penalties paid under subsection a. may not exceed the total amount of fees paid within any 1-year period (as determined by the Secretary) under the contract under which the violation occurs.

(2) For purposes of this section, the term 'not-for-profit' means that no part of the net earnings of the contractor, subcontractor, or supplier inures to the benefit of any natural person or for-profit artificial person.[352]

42 USC 2282b.

Sec. 234B. Civil Monetary Penalties for Violations of Department of Energy Regulations Regarding Security of Classified or Sensitive Information or Data[353]

a. Any person who has entered into a contract or agreement with the Department of Energy, or a subcontract or subagreement thereto, and who violates (or whose employee violates) any applicable rule, regulation, or order prescribed or otherwise issued by the Secretary pursuant to this chapter relating to the safeguarding or security of Restricted Data or other classified or sensitive information shall be subject to a civil penalty of not to exceed $100,000 for each such violation.

b. The Secretary shall include in each contract with a contractor of the Department provisions which provide an appropriate reduction in the fees or amounts paid to the contractor under the contract in the event of a violation by the contractor or contractor employee of any rule, regulation, or order relating to the safeguarding or security of Restricted Data or other classified or sensitive information. The provisions shall specify

---

[352]   Amended by P.L. 109–58, § 610(a), (b), 119 Stat. 781 (2005), which struck out the last sentence: "In implementing this section, the Secretary shall determine by rule whether nonprofit educational institutions should receive aut matic remission of any penalty under this section"; and section 610(b), rewrote subsection (d), which formerly read:
    (d) The provisions of this section shall not apply to:
    (1) The University of Chicago (and any subcontractors or suppliers thereto) for activities associated with Argonne National Laboratory;
    (2) The University of California (and any subcontractors or suppliers thereto) for activities associated with Los Alamos National Laboratory, Lawrence Livermore National Laboratory, and Lawrence Berkeley National Laboratory;
    (3) American Telephone an [sic.; probably should read 'and'] Telegraph Company and its subsidiaries (and any subcontractors or suppliers thereto) for activities associated with Sandia National Laboratories;
    (4) Universities Research Association, Inc. (and any subcontractors or suppliers thereto) for activities associated with FERMI National Laboratory;
    (5) Princeton University (and any subcontractor or suppliers thereto) for activities associated with Princeton Plasma Physics Laboratory;
    (6) The Associated Universities, Inc. (and any subcontractors or suppliers thereto) for activities associated with Pacific Northwest Laboratory.
NOTE: Effective and Applicability Provisions: P.L. 109–58, 119 Stat. 782 (2005), Title VI, § 610(c), provided that: "The amendments made by this section shall not apply to any violation of the Atomic Energy Act of 1954 (42 U.S.C. 2011 et seq.) occurring under a contract entered into before the date of enactment of this section."
[353]   Added by P.L. 106–65, Div. C, Title XXXI, § 3147(a), 113 Stat. 937 (1999).

various degrees of violations and the amount of the reduction attributable to each degree of violation.

c. The powers and limitations applicable to the assessment of civil penalties under section 2282a of this title, except for subsection (d) of that section, shall apply to the assessment of civil penalties under this section.

d. In the case of an entity specified in subsection (d) of section 2282a of this title--

(1) the assessment of any civil penalty under subsection (a) of this section against that entity may not be made until the entity enters into a new contract with the Department of Energy or an extension of a current contract with the Department; and

(2) the total amount of civil penalties under subsection (a) of this section in a fiscal year may not exceed the total amount of fees paid by the Department of Energy to that entity in that fiscal year.

42 USC 2283.

Sec. 235. Protection of Nuclear Inspectors[354]

a. Whoever kills any person who performs any inspections which–

(1) are related to any activity or facility licensed by the Commission, and

42 USC 2133.
42 USC 2134.

(2) are carried out to satisfy requirements under this Act or under any other Federal law governing the safety of utilization facilities required to be licensed under section 103 or 104b, or the safety of radioactive materials, shall be punished as provided under sections 1111 and 1112 of title 18, United States Code. The preceding sentence shall be applicable only if such person is killed while engaged in the performance of such inspection duties or on account of the performance of such duties.

b. Whoever forcibly assaults, resists, opposes, impedes, intimidates, or interferes with any person who performs inspections as described under subsection a. of this section, while such person is engaged in such inspection duties or on account of the performance of such duties, shall be punished as provided under section 111 of title 18, United States Code.

42 USC 2284.

Sec. 236. Sabotage of Nuclear Facilities or Fuel

a. Any person who knowingly[355] destroys or causes physical damage to–

(1) any production facility or utilization facility licensed under this Act;

(2) any nuclear waste treatment, storage, or disposal facility licensed under this Act;

(3) any nuclear fuel for a utilization facility licensed under this Act,[356] or any spent nuclear fuel from such a facility;

(4) any uranium enrichment, uranium conversion, or nuclear fuel fabrication facility licensed or certified[357] by the Nuclear Regulatory Commission;

---

[354]   Added by P.L. 96–295, § 202(a), 94 Stat. 786 (1980).

[355]   Amended by P.L. 109–58, Title VI, § 655, 119 Stat. 813 (2005).

[356]   Amended by P.L. 109–58, Title VI, § 655, 119 Stat. 813 (2005), amended paragraph 3 by deleting "such a utilization facility" and inserting "a utilization facility licensed under this act."

[357]   Amended by P.L. 109–58, Title VI, section 655, 119 Stat. 813 (2005); P.L. 96–295, 94 Stat. 787 (1980); P.L. 97–415, Title II, § 204(a), 96 Stat. 2076 (1983); P.L. 101–575, § 16, 104 Stat. 2835 (1990); P.L. 102–486, § 5(d), 106 Stat. 2944 (1992); P.L. 107–56, Title IX, § 902(a)(8), 115 Stat. 380, 381 (2001).

(5) any production, utilization, waste storage, waste treatment, waste disposal, uranium enrichment, uranium conversion, or nuclear fuel fabrication facility subject to licensing or certification under this Act during construction of the facility, if the destruction or damage caused or attempted to be caused could adversely affect public health and safety during the operation of the facility;

(6) any primary facility or backup facility from which a radiological emergency preparedness alert and warning system is activated; or

(7) any radioactive material or other property subject to regulation by the Commission that, before the date of the offense, the Commission determines, by order or regulation published in the Federal Register, is of significance to the public health and safety or to common defense and security;[358]

Penalties.

or attempts or conspires to do such an act, shall be fined not more than $10,000 or imprisoned for not more than 20 years, or both, and, if death results to any person, shall be imprisoned for any term of years or for life.

b. Any person who knowingly[359] causes an interruption of normal operation of any such facility through the unauthorized use of or tampering with the machinery, components, or controls of any such facility, or attempts or conspires to do such an act, shall be fined not more than $10,000 or imprisoned for not more than 20 years, or both, and, if death results to any person, shall be imprisoned for any term of years or for life.[360]

Chapter 19–Miscellaneous

42 USC 2015.
Transfer of property.

Sec. 241. Transfer of Property

Nothing in this Act shall be deemed to repeal, modify, amend, or alter the provisions of section 9(a) of the Atomic Energy Act of 1946, as heretofore amended.[361]

42 USC 2015a.

Sec. 242. Cold Standby

The Secretary is authorized to expend such funds as may be necessary for the purposes of maintaining enrichment capability at the Portsmouth, Ohio, facility.[362]

42 USC 2015b.

Sec. 243. Scholarship and Fellowship Program

a. SCHOLARSHIP PROGRAM.–To enable students to study, for at least 1 academic semester or equivalent term, science, engineering, or another field of study that the Commission determines is in a critical skill area related to the regulatory mission of the Commission, the Commission may carry out a program to–

(1) award scholarships to undergraduate students who–

(A) are United States citizens; and

(B) enter into an agreement under subsection c. to be employed by the Commission in the area of study for which the scholarship is awarded.

b. FELLOWSHIP PROGRAM.–To enable students to pursue education in science, engineering, or another field of study that the

---

[358]   Paragraphs (5), (6), and (7) added by P.L. 109–58, Title VI, § 655, 119 Stat. 813 (2005).

[359]   Amended by P.L. 109–58, Title VI, § 655, 119 Stat. 813 (2005).

[360]   Amended by P.L. 107–56, Title VIII, §§ 810(f), 811(h), 115 Stat. 380, 381 (2001).

[361]   Added by P.L. 83–703, 68 Stat. 960 (1954); Renumbered by P.L. 102–486, 106 Stat. 2944 (1992).

[362]   Amended by P.L. 107–222, 116 Stat. 1336, (2002).

Commission determines is in a critical skill area related to its regulatory mission, in a graduate or professional degree program offered by an institution of higher education in the United States, the Commission may carry out a program to–

(1) award fellowships to graduate students who–

(A) are United States citizens; and

(B) enter into an agreement under subsection c. to be employed by the Commission in the area of study for which the fellowship is awarded.

c. REQUIREMENTS.–

(1) IN GENERAL.–As a condition of receiving a scholarship or fellowship under subsection a. or b., a recipient of the scholarship or fellowship shall enter into an agreement with the Commission under which, in return for the assistance, the recipient shall–

(A) maintain satisfactory academic progress in the studies of the recipient, as determined by criteria established by the Commission;

(B) agree that failure to maintain satisfactory academic progress shall constitute grounds on which the Commission may terminate the assistance;

(C) on completion of the academic course of study in connection with which the assistance was provided, and in accordance with criteria established by the Commission, engage in employment by the Commission for a period specified by the Commission, that shall be not less than 1 time and not more than 3 times the period for which the assistance was provided; and

(D) if the recipient fails to meet the requirements of subparagraph (A), (B), or (C), reimburse the United States Government for–

(i) the entire amount of the assistance provided the recipient under the scholarship or fellowship; and

(ii) interest at a rate determined by the Commission.

(2) WAIVER OR SUSPENSION.–The Commission may establish criteria for the partial or total waiver or suspension of any obligation of service or payment incurred by a recipient of a scholarship or fellowship under this section.

d. COMPETITIVE PROCESS.–Recipients of scholarships or fellowships under this section shall be selected through a competitive process primarily on the basis of academic merit and such other criteria as the Commission may establish, with consideration given to financial need and the goal of promoting the participation of individuals identified in section 33 or 34 of the Science and Engineering Equal Opportunities Act (42 U.S.C. 1885a, 1885b.).

e. DIRECT APPOINTMENT.–The Commission may appoint directly, with no further competition, public notice, or consideration of any other potential candidate, an individual who has–

(1) received a scholarship or fellowship awarded by the Commission under this section; and

(2) completed the academic program for which the scholarship or fellowship was awarded.[363]

---

[363]   Amended by P.L. 109–58, § 622, 119 Stat. 182 (2005).

42 USC 2015c.

Sec. 244. Partnership Program with Institutions of Higher Education
a. DEFINITIONS.–In this section:

(1) HISPANIC-SERVING INSTITUTION.–The term 'Hispanic-serving institution' has the meaning given the term in section 502(a) of the Higher Education Act of 1965 (20 U.S.C. 1101a(a)).

(2) HISTORICALLY BLACK COLLEGE AND UNIVERSITY.– The term 'historically Black college or university' has the meaning given the term 'part B institution' in section 322 of the Higher Education Act of 1965 (20 U.S.C. 1061).

(3) TRIBAL COLLEGE.–The term 'Tribal college' has the meaning given the term 'tribally controlled college or university' in section 2(a) of the Tribally Controlled College or University Assistance Act of 1978 (25 U.S.C. 1801(a)).

b. PARTNERSHIP PROGRAM.–The Commission may establish and participate in activities relating to research, mentoring, instruction, and training with institutions of higher education, including Hispanic-serving institutions, historically Black colleges or universities, and Tribal colleges, to strengthen the capacity of the institutions–

(1) to educate and train students (including present or potential employees of the Commission); and

(2) to conduct research in the field of science, engineering, or law, or any other field that the Commission determines is important to the work of the Commission.[364]

42 USC 2016.
Report to
Congress.

Sec. 251. Report to Congress [Repealed][365]
The Commission shall submit to the Congress, in January of each year, a report concerning the activities of the Commission. The Commission shall include in such report, and shall at such other times as it deems desirable submit to the Congress, such recommendations *for* additional legislation as the Commission deems necessary or desirable.

42 USC 2017.

Sec. 261. Appropriations
a.[366] No appropriation shall be made to the Commission, nor shall the Commission waive charges for the use of materials under the Cooperative

---

[364]   Amended by P.L. 109–58, Title VI, § 651(c)(4), 119 Stat. 802 (2005); P.L. 110–315, § 941(k)(2)(L), made technical corrections resulting in no change in text.

[365]   Repealed by P.L. 105–85, Div. C, Title XXI, Subtitle D, § 3152(a), 111 Stat. 2042 (1997). It provided for a report to Congress on the activities of the Atomic Energy Commission.

[366]   Amended by P.L. 88–72, section 107, 77 Stat. 84 (1963). Before amendment, this section read as follows:
    Sec. 261. APPROPRIATIONS—
    a. There are hereby authorized to be appropriated such sums as may be necessary and appropriate to carry out the provisions and purposes of this Act, except—
    (1) Such as may be necessary for acquisition or condemnation of any real property or any facility or for plant or facility acquisition, construction or expansion: *Provided,* That for the purposes of this subsection a., any nonmilitary experimental reactor which is designed to produce more than 10,000 thermal kilowatts of heat (except for intermittent excursions) or which is designed to be used in the production of electric power shall be deemed to be a facility.
    (2) Such as may be necessary to carry out cooperative programs with persons for the development and construction of reactors for the demonstration of their use, in whole or in part, in the production of electric power or process heat, or for propulsion, or solely or principally for the commercial provision of byproduct material, irradiation, or other special services, for civilian use, by arrangements (including contracts, agreements, and loans) or amendments thereto, providing for the payment of funds, the rendering of services, and the undertaking of research and development without full reimbursement, the waiver of charges accompanying such arrangement, or the provision by the

Power Reactor Demonstration Program, unless previously authorized by legislation enacted by the Congress.[367]

b. Any Act appropriating funds to the Commission may appropriate specified portions thereof to be accounted for upon the certification of the Commission only.

c. Notwithstanding the provisions of subsection a., funds are hereby authorized to be appropriated for the restoration or replacement of any plant or facility destroyed or otherwise seriously damaged, and the Commission is authorized to use available funds for such purposes.

d. Funds authorized to be appropriated for any construction project to be used in connection with the development or production of special nuclear material or atomic weapons may be used to start another construction project not otherwise authorized if the substituted construction project is within the limit of cost of the construction project for which substitution is to be made, and the Commission certifies that–

(1) the substituted project is essential to the common defense and security;

(2) the substituted project is required by changes in weapon characteristics or weapon logistics operations; and

(3) the Commission is unable to enter into a contract with any person on terms satisfactory to it to furnish from a privately owned plant or facility the product or services to be provided by the new project.

Sec. 271. Agency Jurisdiction

42 USC 2018.
Agency
jurisdiction.

Nothing in this Act shall be construed to affect the authority or regulations of any Federal, State, or Local agency with respect to the generation, sale, or transmission of electric power produced through the use of nuclear facilities licensed by the Commission: *Provided,* That this section shall not be deemed to confer upon any Federal, State, or local agency any authority to regulate, control, or restrict any activities of the Commission.[368]

Commission of any other financial assistance pursuant to such arrangement, or which involves the acquisition or condemnation of any real property or any facility or for plant or facility acquisition, construction or expansion undertaken by the Commission as a part of such arrangements.

b. The acts appropriating such sums may appropriate specified portions thereof to be accounted for upon the certification of the Commission only.

c. Funds are hereby authorized to be appropriated for advance planning, construction design, and architectural services in connection with any plant or facility not otherwise authorized, and for the restoration or replacement of any plant or facility destroyed or otherwise seriously damaged, and the Commission is authorized to use available funds for such purposes.

d. Funds hereafter authorized to be appropriated for any project to be used in connection with the development or production of special nuclear material or atomic weapons may be used to start another project not otherwise authorized. If the substituted project is within the limit of cost of the project for which substitution is to be made, and the Commission certifies that–

(1) the substituted project is essential to the common defense and security;

(2) the substituted project is required by changes in weapon characteristics or weapon logistic operations; and

(3) the Commission is unable to enter into a contract with any person on terms satisfactory to it to furnish from a privately owned plant or facility the product or services to be provided by the new project.

[367] Excerpts from legislation appropriating funds to the Atomic Energy Commission and the U.S. Nuclear Regulatory Commission are set forth in NUREG 0980, Volume 2, Section 4.

[368] Amended by P.L. 89–135, 79 Stat. 551 (1965). Before amendment, this section read as follows: "Sec. 271. AGENCY JURISDICTION–Nothing in this Act shall be construed to

42 USC 2019.
Applicability of
Federal Power
Act.

Sec. 272. Applicability of Federal Power Act

Every licensee under this Act who holds a license from the Commission for a utilization of production facility for the generation of commercial electric energy under section 103 and who transmits such electric energy in interstate commerce or sells it as wholesale in interstate commerce shall be subject to the regulatory provisions of the Federal Power Act.

42 USC 2020.
Licensing of
Government
agencies.

Sec. 273. Licensing of Government Agencies

Nothing in this Act shall preclude any Government agency now or hereafter authorized by law to engage in the production, marketing, or distribution of electric energy from obtaining a license under section 103, if qualified under the provisions of section 103, for the construction and operation of production of utilization facilities for the primary purpose of producing electric energy for disposition for ultimate public consumption.

42 USC 2021.
Cooperation
with States.

Sec. 274. Cooperation with States[369]

a. It is the purpose of this section–

(1) to recognize the interests of the States in the peaceful uses of atomic energy, and to clarify the respective responsibilities under this Act of the States and the Commission with respect to the regulation of byproduct, source, and special nuclear materials;

(2) to recognize the need, and establish programs for cooperation between the States and the Commission with respect to control of radiation hazards associated with use of such materials;

(3) to promote an orderly regulatory pattern between the Commission and State governments with respect to nuclear development and use and regulation of byproduct, source, and special nuclear materials;

(4) to establish procedures and criteria for discontinuance of certain of the Commission's regulatory responsibilities with respect to byproduct, source, and special nuclear materials, and the assumption thereof by the States;

(5) to provide for coordination of the development of radiation standards for the guidance of Federal agencies and cooperation with the States; and

(6) to recognize that, as the States improve their capabilities to regulate effectively such materials, additional legislation may be desirable.

Agreements
with States.

b. Except as provided in subsection c., the Commission is authorized to enter into agreements with the Governor of any State providing for discontinuance of the regulatory authority of the Commission under Chapters 6, 7, and 8, and section 161 of this Act, with respect to any one or more of the following materials within the State:

(1) Byproduct materials (as defined in section 11e.).

(2) Source materials;

(3) Special nuclear materials in quantities not sufficient to form a critical mass.[370]

---

affect the authority or regulations of any Federal, State, or local agency with respect to the generation, sale, or transmission of electric power."

[369] Added by P.L. 86–373, § 1, 73 Stat. 688 (1959).

[370] Amended by P.L. 109–58, Title VI, § 651(e)(2), 119 Stat. 807 (2005), which revised paragraph (1), and renumbered paragraphs (3) and (4) of section 274 b. as paragraphs (2) and (3).

During the duration of such an agreement it is recognized that the State shall have authority to regulate the materials covered by the agreement for the protection of the public health and safety from radiation hazards.

c. No agreement entered into pursuant to subsection b. shall provide for discontinuance of any authority and the Commission shall retain authority and responsibility with respect to regulation of–

    (1) the construction and operation of any production or utilization facility or any uranium enrichment facility;[371]

    (2) the export from or import into the United States of byproduct, source, or special nuclear material, or of any production or utilization facility;

    (3) the disposal into the ocean or sea of byproduct, source, or special nuclear waste materials as defined in regulations or orders of the Commission;

    (4) the disposal of such other byproduct, source, or special nuclear material as the Commission determines by regulation or order should, because of the hazards or potential hazards thereof, not be so disposed of without a license from the Commission.

The Commission shall also retain authority under any such agreement to make a determination that all applicable standards and requirements have been met prior to termination of a license for byproduct material, as defined in section 11e.(2).[372] Notwithstanding any agreement between the Commission and any State pursuant to subsection b., the Commission is authorized by rule, regulation, or order to require that the manufacturer, processor, or producer of any equipment, device, commodity, or other product containing source, byproduct, or special nuclear material shall not transfer possession or control of such product except pursuant to a license issued by the Commission.

d. The Commission shall enter into an agreement under subsection b. of this section with any State if–

    (1) The Governor of that State certifies that the State has a program for the control of radiation hazards adequate to protect the public health and safety with respect to the materials within the State covered by the proposed agreement, and that the State desires to assume regulatory responsibility for such materials; and

    (2) the Commission finds that the State program is in accordance with the requirements of subsection o. and in all other respects[373] compatible with the Commission's program for regulation of such materials, and that the State program is adequate to protect the public health and safety with respect to the materials covered by the proposed agreement.

e. (1) Before any agreement under subsection b. is signed by the Commission, the terms of the proposed agreement and of proposed exemptions pursuant to subsection f. shall be published once each week for four consecutive weeks in the Federal Register; and such opportunity for comment by interested persons on the proposed agreement and exemptions shall be allowed as the Commission determines by regulation or order to be appropriate.

    (2) Each proposed agreement shall include the proposed effective date of such proposed agreement or exemptions. The agreement and

Margin notes:
42 USC 2014.
Conditions.
Publication in *Federal Register.*

---

[371]   Amended by P.L. 102–486, 106 Stat. 2944 (1992).
[372]   Amended by P.L. 95–604, section 204(f), 92 Stat. 3038 (1978).
[373]   Amended by P.L. 95–604, section 904(b), 92 Stat. 3037 (1978).

Exemptions.
Licensing
requirements.

exemptions shall be published in the Federal Register within thirty days after signature by the Commission and the Governor.

f. The Commission is authorized and directed, by regulation or order, to grant such exemptions from the licensing requirements contained in Chapters 6, 7, and 8, and from its regulations applicable to licensees as the Commission finds necessary or appropriate to carry out any agreement entered into pursuant to subsection b. of this section.

g. The Commission is authorized and directed to cooperate with the States in the formulation of standards for protection against hazards of radiation to assure that State and Commission programs for protection against hazards of radiation will be coordinated and compatible.

h. The Administrator of the Environmental Protection Agency shall consult qualified scientists and experts in radiation matters, including the President of the National Academy of Sciences, the Chairman of the National Committee on Radiation Protection and Measurement, and qualified experts in the field of biology and medicine and in the field of health physics. The Special Assistant to the President for Science and Technology, or his designee, is authorized to attend meetings with, participate in the deliberations of, and to advise the Administrator. The Administrator shall advise the President with respect to radiation matters, directly or indirectly affecting health, including guidance for all Federal agencies in the formulation of radiation standards and in the establishment and execution of programs of cooperation with States. The Administrator shall also perform such other functions as the President may assign to him by Executive Order.[374]

Inspections.

i. The Commission in carrying out its licensing and regulatory responsibilities under this Act is authorized to enter into agreements with any State, or group of States, to perform inspections or other functions on a cooperative basis as the Commission deems appropriate. The Commission is also authorized to provide training, with or without charge, to employees of, and such other assistance to, any such State or political subdivision thereof or group of States as the Commission deems appropriate. Any such provision or assistance by the Commission shall take into account the additional expenses that may be incurred by a State as a consequence of the State's entering into an agreement with the Commission pursuant to subsection b.

Termination of
agreement.

j. (1) The Commission, upon its own initiative after reasonable notice and opportunity for hearing to the State with which an agreement under subsection b. has become effective, or upon request of the Governor of such State, may terminate or suspend all or part of its agreement with the State and reassert the licensing and regulatory authority vested in it under this Act, if the Commission finds that (1) such termination or suspension is required to protect the public health and safety, or (2) the State has not complied with one or more of the requirements of this section. The Commission shall periodically review such agreements and actions taken by the States under the agreements to insure compliance with the provisions of this section.[375]

(2)[376] The Commission, upon its own motion or upon request of the Governor of any State, may, after notifying the Governor,

---

[374] Reorganization Plan No. 3 of 1970 abolished the Federal Radiation Council and transferred its functions to the Administrator of the Environmental Protection Agency.
[375] Amended by P.L. 96–295, § 205, 94 Stat. 787 (1980); P.L. 95–604, § 204(d)(1), (2), and (3), 92 Stat. 3037 (1978), amended section 274j.
[376] Added by P.L. 96–295, § 205, 94 Stat. 787 (1980).

temporarily suspend all or part of its agreement with the State without notice or hearing if, in the judgment of the Commission:

(A) an emergency situation exists with respect to any material covered by such an agreement creating danger which requires immediate action to protect the health or safety of persons either within or outside of the State, and

(B) the State has failed to take steps necessary to contain or eliminate the cause of the danger within a reasonable time after the situation arose.

A temporary suspension under this paragraph shall remain in effect only for such time as the emergency situation exists and shall authorize the Commission to exercise its authority only to the extent necessary to contain or eliminate the danger.

k. Nothing in this section shall be construed to affect the authority of any State or local agency to regulate activities for purposes other than protection against radiation hazards.

Notice of filing.

l. With respect to each application for Commission license authorizing an activity as to which the Commission's authority is continued pursuant to subsection c., the Commission shall give prompt notice to the State or States in which the activity will be conducted of the filing of the license application; and shall afford reasonable opportunity for State representatives to offer evidence, interrogate witnesses, and advise the Commission as to the application without requiring such representatives to take a position for or against the granting of the application.

m. No agreement entered into under subsection b., and no exemption granted pursuant to subsection f., shall affect the authority of the Commission under subsection 161b. or i.; to issue rules, regulations, or orders to protect the common defense and security, to protect restricted data or to guard against the loss or diversion of special nuclear material. For purposes of subsection 161i., activities covered by exemptions granted pursuant to subsection f. shall be deemed to constitute activities authorized pursuant to this Act; and special nuclear material acquired by any person pursuant to such an exemption shall be deemed to have been acquired pursuant to section 53.

Definition.

n. As used in this section, the term "State" means any State, Territory, or possession of the United States, the Canal Zone, Puerto Rico, and the District of Columbia. As used in this section, the term "agreement" includes any amendment to any agreement.[377]

Agreement.

o.[378] In the licensing and regulation of byproduct material, as defined in section 11e. (2) of this Act, or of any activity which results in the production of byproduct material as so defined under an agreement entered into pursuant to subsection b., a State shall require–

(1) compliance with the requirements of subsection b. of section 83 (respecting ownership of byproduct material and land), and

(2) compliance with standards which shall be adopted by the State for the protection of the public health, safety, and the environment from hazards associated with such material which are equivalent, to the extent practicable, or more stringent than, standards adopted and enforced by the Commission for the same purpose, including requirements and standards promulgated by the Commission and the

---

[377]   Amended by P.L. 95–604, § 204(c), 92 Stat. 3037 (1978).
[378]   Added by P.L. 95–604, § 204(e), 92 Stat. 3037 (1978).

Administrator of the Environmental Protection Agency pursuant to sections 83, 84, and 275, and

(3) procedures which–

(A) in the case of licenses, provide procedures under State law which include–

(i) an opportunity, after public notice, for written comments and a public hearing, with a transcript,

(ii) an opportunity for cross examination, and

(iii) a written determination which is based upon findings included in such determination and upon the evidence presented during the public comment period and which is subject to judicial review;

(B) in the case of rulemaking, provide an opportunity for public participation through written comments or a public hearing and provide for judicial review of the rule;

(C) require for each license which has a significant impact on the human environment a written analysis (which shall be available to the public before the commencement of any such proceedings) of the impact of such license, including any activities conducted pursuant thereto, on the environment, which analysis shall include–

(i) an assessment of the radiological and nonradiological impacts to the public health of the activities to be conducted pursuant to such license;

(ii) an assessment of any impact on any waterway and groundwater resulting from such activities;

(iii) consideration of alternatives, including alternative sites and engineering methods, to the activities to be conducted pursuant to such license; and

(iv) consideration of the long-term impacts, including decommissioning, decontamination, and reclamation impacts, associated with activities to be conducted pursuant to such license, including the management of any byproduct material, as defined by section 11e.(2); and

(D) prohibit any major construction activity with respect to such material prior to complying with the provisions of subparagraph (C).

If any State under such agreement imposes upon any licensee any requirement for the payment of funds to such State for the reclamation or long-term maintenance and monitoring of such material, and if transfer to the United States of such material is required in accordance with section 83b. of this Act, such agreement shall be amended by the Commission to provide that such State shall transfer to the United States upon termination of the license issued to such licensee the total amount collected by such State from such licensee for such purpose. If such payments are required, they must be sufficient to ensure compliance with the standards established by the Commission pursuant to section 161x. of this Act. No State shall be required under paragraph (3) to conduct proceedings concerning any license or regulation which would duplicate proceedings conducted by the Commission.

42 USC 2201.

In adopting requirements pursuant to paragraph (2) of this subsection with respect to sites at which ores are processed primarily for their source material content or which are used for the disposal of byproduct material as defined in section 11e.(2), the State may adopt alternatives (including, where appropriate, site-specific alternatives) to the requirements adopted and enforced by the Commission for the same purpose if, after notice and

42 USC 2014.

opportunity for public hearing, the Commission determines that such alternatives will achieve a level of stabilization and containment of the sites concerned, and a level of protection for public health, safety, and the environment from radiological and nonradiological hazards associated with such sites, which is equivalent to, to the extent practicable, or more stringent than the level which would be achieved by standards and requirements adopted and enforced by the Commission for the same purpose and any final standards promulgated by the Administrator of the Environmental Protection Agency in accordance with section 275. Such alternative State requirements may take into account local or regional conditions, including geology, topography, hydrology and meteorology.[379]

42 USC 2022.

42 USC 2022.

Rule.

42 USC 7911.

42 USC 2014.
42 USC 6901
note.

Sec. 275. Health and Environmental Standards for Uranium Mill Tailings[380]

a. As soon as practicable, but not later than October 1, 1982,[381] the Administrator of the Environmental Protection Agency (hereinafter referred to in this section as the "Administrator") shall, by rule, promulgate standards of general application (including standards applicable to licenses under section 104(h) of the Uranium Mill Tailings Radiation Control Act of 1978) for the protection of the public health, safety, and the environment from radiological and nonradiological hazards associated with residual radioactive materials (as defined in section 101 of the Uranium Mill Tailings Radiation Control Act of 1978) located at inactive uranium mill tailings sites and depository sites for such materials selected by the Secretary of Energy, pursuant to title I of the Uranium Mill Tailings Radiation Control Act of 1978. Standards promulgated pursuant to this subsection shall, to the maximum extent practicable, be consistent with the requirements of the Solid Waste Disposal Act, as amended. In establishing such standards, the Administrator shall consider the risk to the public health, safety, and the environment, the environmental and economic costs of applying such standards, and such other factors as the Administrator determines to be appropriate.[382] The Administrator may periodically revise any standard promulgated pursuant to this subsection. After October 1, 1982, if the Administrator has not promulgated standards in final form under this subsection, any action of the Secretary of Energy under title I of the Uranium Mill Tailings Radiation Control Act of 1978 which is required to comply with, or be taken in accordance with, standards of the Administrator shall comply with, or be taken in accordance with, the standards proposed by the Administrator under this subsection until such time as the Administrator promulgates such standards in final form.[383]

b.(1) As soon as practicable, but not later than October 31, 1982, the Administrator shall, by rule, propose and within 11 months thereafter promulgate in final form, standards, general application for the protection of the public health, safety, and the environment from radiological and non-radiological hazards associated with processing and with the possession, transfer, and disposal of byproduct material, as defined in section 11e.(2) of this Act, at sites at which ores are processed primarily for their source material content or which are used for the disposal of

---

[379]   This paragraphs was added by P.L. 97–415, § 19, 96 Stat. 2067 (1983).
[380]   Added by P.L. 95–604, § 206(a), 92 Stat. 3039 (1978), added section 275.
[381]   Amended by P.L. 97–415, § 18, 96 Stat. 2067 (1983).
[382]   Amended by P.L. 97–415, § 22, 96 Stat. 2067 (1983).
[383]   Amended by P.L. 97–415, § 18, 96 Stat. 2067 (1983).

Promulgation
authority.

such byproduct material. If the Administrator fails to promulgate standards in final form under this subsection by October 1, 1983, the authority of the Administrator to promulgate such standards shall terminate, and the Commission may take actions under this Act without regard to any provision of this Act requiring such actions to comply with, or be taken in accordance with, standards promulgated by the

42 USC 2014.

Administrator. In any such case, the Commission shall promulgate, and from time to time revise, any such standards of general application which the Commission deems necessary to carry out its responsibilities in the conduct of its licensing activities under this Act. Requirements established by the Commission under this Act with respect to byproduct material as defined in section 11e.(2) shall confirm to such standards. Any requirements adopted by the Commission respecting such byproduct material before promulgation by the Commission of such standards shall be amended as the Commission deems necessary to conform to such standards in the same manner as provided in subsection f.(3). Nothing in this subsection shall be construed to prohibit or suspend the implementation or enforcement by the Commission of any requirement of the Commission respecting byproduct material as defined in section 11e.(2) pending promulgation by the Commission of any such standard of general application. In establishing such standards, the Administrator shall consider the risk to the public health, safety, and the environment, the environmental and economic costs of applying such standards, and such other factors as the Administrator determines to be appropriate.[384]

(2) Such generally applicable standards promulgated pursuant to this subsection for nonradiological hazards shall provide for the protection of human health and the environment consistent with the standards required under subtitle C of the Solid Waste Disposal Act, as amended, which are applicable to such hazards: *Provided, however,* That no permit issued by the Administrator is required under this Act or the Solid Waste Disposal Act, as amended, for the processing, possession, transfer, or disposal of byproduct material, as defined in section 11e.(2) of this Act. The

42 USC 2021.

Administration may periodically revise any standard promulgated pursuant to this subsection. Within three years after such revision of any such standard, the Commission and any State permitted to exercise authority under section 274b.(2) shall apply such revised standard in the case of any license for byproduct material as defined in section 11e.(2) or any revision thereof.

Consultation.
Notice, hearing
opportunity.
Publication in
*Federal
Register.*

c. (1) Before the promulgation of any rule pursuant to this section, the Administrator shall publish the proposed rule in the Federal Register, together with a statement of the research, analysis, and other available information in support of such proposed rule, and provide a period of public comment of at least thirty days for written comments thereon and an opportunity, after such comment period and after public notice, for any interested person to present oral data, views, and arguments at a public hearing. There shall be a transcript of any such hearing. The Administrator shall consult with the Commission and the Secretary of Energy before promulgation of any such rule.

Judicial review.

(2) Judicial review of any rule promulgated under this section may be obtained by any interested person only upon such person filing a petition for review within sixty days after such promulgation in the United States

---

[384]   Amended by P.L. 97–415, §§ 18, 22, 96 Stat. 2067 (1983); added this language at the end of subsection b.

court of appeals for the Federal judicial circuit in which such person resides or has his principal place of business. A copy of the petition shall be forthwith transmitted by the clerk of the court to the Administrator. The Administrator thereupon shall file in the court the written submission to, and transcript of, the written or oral proceedings on which such rule was based as provided in section 2112 of title 28, United States Code. The court shall have jurisdiction to review the rule in accordance with chapter 7 of title 5, United States Code, and to grant appropriate relief as provided in such chapter. The judgment of the court affirming, modifying, or setting aside, in whole or in part, any such rule shall be final, subject to judicial review by the Supreme Court of the United States upon certiorari or certification as provided in section 1254 of title 28, United States Code.

(3) Any rule promulgated under this section shall not take effect earlier than sixty calendar days after such promulgation.

d. Implementation and enforcement of the standards promulgated pursuant to subsection b. of this section shall be the responsibility of the Commission in the conduct of its licensing activities under this Act. States exercising authority pursuant to section 274b.(2) of this Act shall implement and enforce such standards in accordance with subsection o. of such section.

e. Nothing in this Act applicable to byproduct material, as defined in section 11e.(2) of this Act, shall affect the authority of the Administrator under the Clean Air Act of 1970, as amended, or the Federal Water Pollution Control Act, as amended.

f.[385](1) Prior to January 1, 1983, the Commission shall not implement or enforce the provisions of the Uranium Mill Licensing Requirements published as final rules at 45 Federal Register 65521 to 65538 on October 3, 1980 (hereinafter in this subsection referred to as the "October 3 regulations") After December 31, 1982, the Commission is authorized to implement and enforce the provisions of such October 3 regulations (and any subsequent modifications or additions to such regulations which may be adopted by the Commission), except as otherwise provided in paragraphs (2) and (3) of this subsection.

(2) Following the proposal by the Administrator of standards under subsection b., the Commission shall review the October 3 regulations, and, not later than 90 days after the date of such proposal, suspend implementation and enforcement of any provision of such regulations which the Commission determines after notice and opportunity for public comment to require a major action or major commitment by licensees which would be unnecessary if–

(A) the standards proposed by the Administrator are promulgated in final form without modification, and

(B) the Commission's requirements are modified to conform to such standards.

Such suspension shall terminate on the earlier of April 1, 1984 or the date on which the Commission amends the October 3 regulations to conform to final standards promulgated by the Administrator under subsection b. During the period of such suspension, the Commission shall continue to regulate byproduct material (as defined in section 11e.(2)) under this Act on a licensee-by-licensee basis as the Commission deems necessary to protect public health, safety, and the environment.

5 USC *et seq.*

42 USC 2021.

42 USC 2014.
42 USC 7401
note.

Uranium mill licensing requirement regulations. Implementation and enforcement.

Review, public comment, and suspension.

---

[385]   Added by P.L. 97–415, § 18, 96 Stat. 2067 (1983).

(3) Not later than 6 months after the date on which the Administrator promulgates final standards pursuant to subsection b. of this section, the Commission shall, after notice and opportunity for public comment, amend the October 3 regulations, and adopt such modifications, as the Commission deems necessary to conform to such final standards of the Administrator.

42 USC 2114.

(4) Nothing in this subsection may be construed as affecting the authority or responsibility of the Commission under section 84 to promulgate regulations to protect the public health and safety and the environment.

42 USC 2023.

Sec. 276. State Authority to Regulate Radiation Below Level of Regulatory Concern of Nuclear Regulatory Commission[386]

(a) IN GENERAL.–No provision of this Act, or of the Low-Level Radioactive Waste Policy Act, may be construed to prohibit or otherwise restrict the authority of any State to regulate, on the basis of radiological hazard, the disposal or off-site incineration of low-level radioactive waste, if the Nuclear Regulatory Commission, after the date of the enactment of the Energy Policy Act of 1992 exempts such waste from regulation.

(b) RELATION TO OTHER STATE AUTHORITY.–This section may not be construed to imply preemption of existing State authority. Except as expressly provided in subsection (a), this section may not be construed to confer on any State any additional authority to regulate activities licensed by the Nuclear Regulatory Commission.

(c) DEFINITIONS.–For purposes of this section:

(1) The term "low-level radioactive waste" means radioactive material classified by the Nuclear Regulatory Commission as low-level radioactive waste on the date of the enactment of the Energy Policy Act of 1992.

(2) The term "off-site incineration" means any incineration of radioactive materials at a facility that is located off the site where such materials were generated.

(3) The term "State" means each of the several States, the District of Columbia, and any commonwealth, territory, or possession of the United States.

Sec. 281. Separability

Separability.

If any provision of this Act or the application of such provision to any person or circumstances, is held invalid, the remainder of this Act or the application of such provision to persons or circumstances other than those as to which it is held invalid, shall not be affected thereby.

Sec. 291. Short Title

Short title.

This Act may be cited as the "Atomic Energy Act of 1954 ."

---

[386]   Added by P.L. 102–486, 106 Stat. 3122 (1992). P.L. 109–58 states:
(b) REVOCATION OF RELATED NRC POLICY STATEMENTS.–The policy statements of the Nuclear Regulatory Commission published in the Federal Register on July 3, 1990 (55 Fed. Reg. 27522) and August 29, 1986 (51 Fed. Reg. 30839), relating to radioactive waste below regulatory concern, shall have no effect after the date of the enactment of this Act.

Chapter 20–Joint Committee on Atomic Energy Abolished; Functions and Responsibilities Reassigned[387]

42 USC 2258.

Sec. 301. Joint Committee on Atomic Energy Abolished

a. The Joint Committee on Atomic Energy is abolished.

b. Any reference in any rule, resolution, or order of the Senate or the House of Representatives or in any law, regulation, or Executive order to the Joint Committee on Atomic Energy shall, on and after the date of enactment of this section, be considered as referring to the committees of the Senate and the House of Representatives which, under the rules of the Senate and the House, have jurisdiction over the subject matter of such reference.

Records, transfer.

c. All records, data, charts, and files of the Joint Committee on Atomic Energy are transferred to the committees of the Senate and House of Representatives which, under the rules of the Senate and the House, have jurisdiction over the subject matters to which such records, data, charts, and files relate. In the event that any record, data, chart, or file shall be within the jurisdiction of more than one committee, duplicate copies shall be provided upon request.

Sec. 302. Transfer of Certain Functions of the Joint Committee on Atomic Energy and Conforming Amendments to Certain Other Laws

Repeal.
42 USC 2251 *et seq.*
Repealed.
42 USC 2315.

a. Effective on the date of enactment of this section, chapter 17 of this Act is repealed.

b. Section 103 of the Atomic Energy Community Act of 1955, as amended, is repealed.

c. Section 3 of the Congressional Budget and Impoundment Control Act of 1974 is amended by–

(1) striking the subsection designation "(a)"; and

Repealed.
2 USC 190j.
Repeal.

(2) repealing subsection (b).

d. Section 252(a)(3) of the Legislative Reorganization Act of 1970 is repealed.

42 USC 2259.

Sec. 303. Information and Assistance to Congressional Committees

a. The Secretary of Energy and the Nuclear Regulatory Commission shall keep the committees of the Senate and the House of Representatives which, under the rules of the Senate and the House, have jurisdiction over the functions of the Secretary or the Commission, fully and currently informed with respect to the activities of the Secretary and the Commission.

b. The Department of Defense and Department of State shall keep the committees of the Senate and the House of Representatives which, under the rules of the Senate and the House, have jurisdiction over national security considerations of nuclear energy, fully and currently informed with respect to such matters within the Department of Defense and Department of State relating to national security considerations of nuclear technology which are within the jurisdiction of such committees.

c. Any Government agency shall furnish any information requested by the committees of the Senate and the House of Representatives which, under the rules of the Senate and the House, have jurisdiction over the development, utilization, or application of nuclear energy, with respect to the activities or responsibilities of such agency in the field of nuclear energy which are within the jurisdiction of such committees.

---

[387]   Chapter 20 added by P.L. 95–110, 91 Stat. 884 (1977).

d. The committees of the Senate and the House of Representatives which, under the rules of the Senate and the House, have jurisdiction over the development, utilization or application of nuclear energy, are authorized to utilize the services, information, facilities, and personnel of any Government agency which has activities or responsibilities in the field of nuclear energy which are within the jurisdiction of such committees: *Provided, however,* That any utilization of personnel by such committees shall be on a reimbursable basis and shall require, with respect to committees of the Senate, the prior written consent of the Committee on Rules and Administration, and with respect to committees of the House of Representatives, the prior written consent of the Committee on House Oversight.[388]

Chapter 21–Defense Nuclear Facilities Safety Board[389]

42 USC 2286.

Sec. 311. Establishment

(a) Establishment.–There is hereby established an independent establishment in the executive branch, to be known as the "Defense Nuclear Facilities Safety Board" (hereafter in this chapter referred to as the "Board").

President of U.S.

(b) Membership.–(1) The Board shall be composed of five members appointed from civilian life by the President, by and with the advice and consent of the Senate, from among United States citizens who are respected experts in the field of nuclear safety with a demonstrated competence and knowledge relevant to the independent investigative and oversight functions of the Board. Not more than three members of the Board shall be of the same political party.

(2) Any vacancy in the membership of the Board shall be filled in the same manner in which the original appointment was made.

(3) No member of the Board may be an employee of, or have any significant financial relationship with, the Department of Energy or any contractor of the Department of Energy.

42 USC 2286.
President of U.S.
Reports.

(4) Not later than 180 days after the date of the enactment of this chapter, the President shall submit to the Senate nominations for appointment to the Board. In the event that the President is unable to submit the nominations within such 180-day period, the President shall submit to the Committees on Armed Services and on Appropriations of the Senate and to the Speaker of the House of Representatives a report describing the reasons for such inability and a plan for submitting the nominations within the next 90 days. If the President is unable to submit the nominations within that 90-day period, the President shall again submit to such committees and the Speaker such a report and plan. The President shall continue to submit to such committees and the Speaker such a report and plan every 90 days until the nominations are submitted.

President of U.S.

(c) Chairman and Vice Chairman.–(1) The President shall designate a Chairman and Vice Chairman of the board from among members of the Board.

(2) The Chairman shall be the chief executive officer of the Board and, subject to such policies as the Board may establish, shall exercise the functions of the Board with respect to–

---

[388]   Amended by P.L. 104–186, § 222(1), 110 Stat. 1749 (1996).
[389]   Chapter 21 added by P.L. 100–456, 102 Stat. 2084 (1988).

(A) the appointment and supervision of employees of the Board;

(B) the organization of any administrative units established by the Board; and

(C) the use and expenditure of funds.

(3) The Chairman may delegate any of the functions under this paragraph to any other member or to any appropriate officer of the Board.

(4) The Vice Chairman shall act as Chairman in the event of the absence or incapacity of the Chairman or in case of a vacancy in the office of Chairman.

(d) Terms.–(1) Except as provided under paragraph (2), the members of the Board shall serve for terms of five years. Members of the Board may be reappointed.

(2) of the members first appointed–

(A) one shall be appointed for a term of one year;

(B) one shall be appointed for a term of two years;

(C) one shall be appointed for a term of three years;

(D) one shall be appointed for a term of four years; and

(E) one shall be appointed for a term of five years, as designated by the President at the time of appointment.

(3) Any member appointed to fill a vacancy occurring before the expiration of the term of office for which such member's predecessor was appointed shall be appointed only for the remainder of such term. A member may serve after the expiration of that member's term until a successor has taken office.

(e) Quorum.–Three members of the Board shall constitute a quorum, but a lesser number may hold hearings.

42 USC 2286a.   **Sec. 312. Functions of the Board**

The Board shall perform the following functions:

(1) Review and Evaluation of Standards.–The Board shall review and evaluate the content and implementation of the standards relating to the design, construction, operation, and decommissioning of defense nuclear facilities of the Department of Energy (including all applicable Department of Energy orders, regulations, and requirements) at each Department of Energy defense nuclear facility. The Board shall recommend to the Secretary of Energy those specific measures that should be adopted to ensure that public health and safety are adequately protected. The Board shall include in its recommendations necessary changes in the content and implementation of such standards, as well as matters on which additional data or additional research is needed.

(2) Investigations.–(A) The Board shall investigate any event or practice at a Department of Energy defense nuclear facility which the Board determines has adversely affected, or may adversely affect, public health and safety.

(B) The purpose of any Board investigation under subparagraph (A) shall be–

(i) to determine whether the Secretary of Energy is adequately implementing the standards described in paragraph (1) of the Department of Energy (including all applicable Department of Energy orders, regulations, and requirements) at the facility;

(ii) to ascertain information concerning the circumstances of such event or practice and its implications for such standards;

(iii) to determine whether such event or practice is related to other events or practices at other Department of Energy defense nuclear facilities; and

(iv) to provide to the Secretary of Energy such recommendations for changes in such standards or the implementation of such standards (including Department of Energy orders, regulations, and requirements) and such recommendations relating to data or research needs as may be prudent or necessary.

(3) Analysis of Design and Operational Data.–The Board shall have access to and may systematically analyze design and operational data, including safety analysis reports, from any Department of Energy defense nuclear facility.

(4) Review of Facility Design and Construction.–The Board shall review the design of a new Department of Energy defense nuclear facility before construction of such facility begins and shall recommend to the Secretary, within a reasonable time, such modifications of the design as the Board considers necessary to ensure adequate protection of public health and safety. During the construction of any such facility, the Board shall periodically review and monitor the construction and shall submit to the Secretary, within a reasonable time, such recommendations relating to the construction of that facility as the Board considers necessary to ensure adequate protection of public health and safety. An action of the Board, or a failure to act, under this paragraph may not delay or prevent the Secretary of Energy from carrying out the construction of such a facility.

(5) Recommendations.–The Board shall make such recommendations to the Secretary of Energy with respect to Department of Energy defense nuclear facilities, including operations of such facilities, standards, and research needs, as the Board determines are necessary to ensure adequate protection of public health and safety. In making its recommendations the Board shall consider the technical and economic feasibility of implementing the recommended measures.

42 USC 2286b.   **Sec. 313. Powers of Board**

(a) Hearings.–(1) The Board or a member authorized by the Board may, for the purpose of carrying out this chapter, hold such hearings and sit and act at such times and places, and require, by subpoena or otherwise, the attendance and testimony of such witnesses and the production of such evidence as the Board or an authorized member may find advisable.

(2) (A) Subpoenas may be issued only under the signature of the Chairman or any member of the Board designated by him and shall be served by any person designated by the Chairman, any member, or any person as otherwise provided by law. The attendance of witnesses and the production of evidence may be required from any place in the United States at any designated place of hearing in the United States.

(B) Any member of the Board may administer oaths or affirmations to witnesses appearing before the Board.

(C) If a person issued a subpoena under paragraph (1) refuses to obey such subpoena or is guilty of contumacy, any court of the United States within the judicial district within which the hearing is conducted or within the judicial district within which such person is found or resides or transacts business may (upon application by the Board) order such person to appear before the

Board to produce evidence or to give testimony relating to the matter under investigation. Any failure to obey such order of the court may be punished by such court as a contempt of the court.

(D) The subpoenas of the Board shall be served in the manner provided for subpoenas issued by a United States district court under the Federal Rules of Civil Procedure for the United States district courts.

(E) All process of any court to which application may be made under this section may be served in the judicial district in which the person required to be served resides or may be found.

(b) Staff.–The Board may, for the purpose of performing its responsibilities under this Chapter–

(1) hire such staff as it considers necessary to perform the functions of the Board, but not more than the equivalent of 100 full-time employees; and

(2) procure the temporary and intermittent services of experts and consultants to the extent authorized by section 3109(b) of title 5, United States Code, at rates the Board determines to be reasonable.

(c) Regulations.–The Board may prescribe regulations to carry out the

(d) Reporting Requirements.–The Board may establish reporting requirements for the Secretary of Energy which shall be binding upon the Secretary. The information which the Board may require the Secretary of Energy to report under this subsection may include any information designated as classified information, or any information designated as safeguards information and protected from disclosure under section 147 or 148 of this Act.

Classified information.

42 USC 2286b.

(e) Use of Government Facilities, Etc.–The Board may, for the purpose of carrying out its responsibilities under this chapter, use any facility, contractor, or employee of any other department or agency of the Federal Government with the consent of and under appropriate support arrangements with the head of such department or agency and, in the case of a contractor, with the consent of the contractor.

(f) Assistance From Certain Agencies of The Federal Government.–With the consent of and under appropriate support arrangements with the Nuclear Regulatory Commission, the Board may obtain the advice and recommendations of the staff of the Commission on matters relating to the Board's responsibilities and may obtain the advice and recommendations of the Advisory Committee on Reactor Safeguards on such matters.

(g) Assistance From Organizations Outside The Federal Government–The Board may enter into an agreement with the National Research Council of the National Academy of Sciences or any other appropriate group or organization of experts outside the Federal Government chosen by the Board to assist the Board in carrying out its responsibilities under this chapter.

(h) Resident Inspectors.–The Board may assign staff to be stationed at any Department of Energy defense nuclear facility to carry out the functions of the Board.

(i) Special Studies.–The Board may conduct special studies pertaining to adequate protection of public health and safety at any Department of Energy defense nuclear facility.

(j) Evaluation of Information.–The Board may evaluate information received from the scientific and industrial communities, and from the interested public, with respect to–

(1) events or practices at any Department of Energy defense nuclear facility; or

(2) suggestions for specific measures to improve the content of standards described in section 312(1), the implementation of such standards, or research relating to such standards at Department of Energy defense nuclear facilities.

Sec. 314. Responsibilities of the Secretary of Energy

(a) Cooperation.–The Secretary of Energy shall fully cooperate with the Board and provide the Board with ready access to such facilities, personnel, and information as the Board considers necessary to carry out its responsibilities under this chapter. Each contractor operating a Department of Energy defense nuclear facility under a contract awarded by the Secretary shall, to the extent provided in such contract or otherwise with the contractor's consent, fully cooperate with the Board and provide the Board with ready access to such facilities, personnel, and information of the contractor as the Board considers necessary to carry out its responsibilities under this chapter.

(b) Access To Information.–The Secretary of Energy may deny access to information provided to the Board to any person who–

(1) has not been granted an appropriate security clearance or access authorization by the Secretary of Energy; or

(2) does not need such access in connection with the duties of such person.

Sec. 315. Board Recommendations

(a) Public Availability and Comment.–Subject to subsections (g) and (h) and after receipt by the Secretary of Energy of any recommendations from the Board under section 312, the Board promptly shall make such recommendations available to the public in the Department of Energy's regional public reading rooms and shall publish in the Federal Register such recommendations and a request for the submission to the Board of public comments on such recommendations. Interested persons shall have 30 days after the date of the publication of such notice in which to submit comments, data, views, or arguments to the Board concerning the recommendations.

(b) Response By Secretary.–

(1) The Secretary of Energy shall transmit to the Board, in writing, a statement on whether the Secretary accepts or rejects, in whole or in part, the recommendations submitted to him by the Board under section 312, a description of the actions to be taken in response to the recommendations, and his views on such recommendations. The Secretary of Energy shall transmit his response to the Board within 45 days after the date of the publication, under subsection (a), of the notice with respect to such recommendations or within such additional period, not to exceed 45 days, as the Board may grant.

(2) At the same time as the Secretary of Energy transmits his response to the Board under paragraph (1), the Secretary, subject to subsection (h), shall publish such response, together with a request for public comment on his response, in the Federal Register.

(3) Interested persons shall have 30 days after the date of the publication of the Secretary of Energy's response in which to submit comments, data, views, or arguments to the Board concerning the Secretary's response.

(4) The Board may hold hearings for the purpose of obtaining public comments on its recommendations and the Secretary of Energy's response.

(c) Provision of Information To Secretary.–The Board shall furnish the Secretary of Energy with copies of all comments, data, views, and arguments submitted to it under subsection (a) or (b).

Federal Register, publication. Reports.

(d) Final Decision.–If the Secretary of Energy, in a response under subsection (b)(1), rejects (in whole or part) any recommendation made by the Board under section 312, the Board shall either reaffirm its original recommendation or make a revised recommendation and shall notify the Secretary of its action. Within 30 days after receiving the notice of the Board's action under this subsection, the Secretary shall consider the Board's action and make a final decision on whether to implement all or part of the Board's recommendations. Subject to subsection (h), the Secretary shall publish the final decision and the reasoning for such decision in the Federal Register and shall transmit to the Committees on Armed Services and on Appropriations of the Senate, and to the Speaker of the House of Representatives a written report containing that decision and reasoning.

(e) Implementation Plan.–The Secretary of Energy shall prepare a plan for the implementation of each Board recommendation, or part of a recommendation, that is accepted by the Secretary in his final decision. The Secretary shall transmit the implementation plan to the Board within 90 days after the date of the publication of the Secretary's final decision on such recommendation in the Federal Register. The Secretary may have an additional 45 days to transmit the plan if the Secretary submits to the Board and to the Committees on Armed Services and on Appropriations of the Senate and to the Speaker of the House of Representatives a notification setting forth the reasons for the delay and describing the actions the Secretary is taking to prepare an implementation plan under this subsection. The Secretary may implement any such recommendation (or part of any such recommendation) before, on, or after the date on which the Secretary transmits the implementation plan to the Board under this subsection.

Reports.

(f) Implementation.–

(1) Subject to paragraph (2), not later than one year after the date on which the Secretary of Energy transmits an implementation plan with respect to a recommendation (or part thereof) under subsection (e), the Secretary shall carry out and complete the implementation plan. If complete implementation of the plan takes more than 1 year, the Secretary of Energy shall submit a report to the Committees on Armed Services and on Appropriations of the Senate and to the Speaker of the House of Representatives setting forth the reasons for the delay and when implementation will be completed.

Reports.

(2) If the Secretary of Energy determines that the implementation of a Board recommendation (or part thereof) is impracticable because of budgetary considerations, or that the implementation would affect the Secretary's ability to meet the annual nuclear weapons stockpile requirements established pursuant to section 91 of this Act, the Secretary shall submit to the President, to the Committees on Armed Services and on Appropriations of the Senate, and to the Speaker of the House of Representatives a report containing the recommendation and the Secretary's determination.

Public health and safety.

(g) Imminent Or Severe Threat.–

(1) In any case in which the Board determines that a recommendation submitted to the Secretary of Energy under section 312 relates to an imminent or severe threat to public health and safety, the Board and the Secretary of Energy shall proceed under this subsection in lieu of subsections (a) through (d).

President of
U.S.

(2) At the same time that the Board transmits a recommendation relating to an imminent or severe threat to the Secretary of Energy, the Board shall also transmit the recommendation to the President and for information purposes to the Secretary of Defense. The Secretary of Energy shall submit his recommendation to the President. The President shall review the Secretary of Energy's recommendation and shall make the decision concerning acceptance or rejection of the Board's recommendation.

Public
information.

(3) After receipt by the President of the recommendation from the Board under this subsection, the Board promptly shall make such recommendation available to the public and shall transmit such recommendation to the Committees on Armed Services and on Appropriations of the Senate and to the Speaker of the House of Representatives. The President shall promptly notify such committees and the Speaker of his decision and the reasons for that decision.

(h) Limitation.–Notwithstanding any other provision of this section, the requirements to make information available to the public under this section–

(1) shall not apply in the case of information that is classified; and

(2) shall be subject to the orders and regulations issued by the Secretary of Energy under section 147 and 148 of this Act to prohibit dissemination of certain information.

42 USC 2286e.

Sec. 316. Reports

(a) Board Report.–

(1) The Board shall submit to the Committees on Armed Services and on Appropriations of the Senate and to the Speaker of the House of Representatives each year, at the same time that the President submits the budget to Congress pursuant to section 1105(a) of title 31, United States Code, a written report concerning its activities under this chapter, including all recommendations made by the Board, during the year preceding the year in which the report is submitted. The Board may also issue periodic unclassified reports on matters within the Board's responsibilities.

(2) The annual report under paragraph (1) shall include an assessment of–

(A) The improvements in the safety of Department of Energy defense nuclear facilities during the period covered by the report;

(B) the improvements in the safety of Department of Energy defense nuclear facilities resulting from actions taken by the Board or taken on the basis of the activities of the Board; and

(C) the outstanding safety problems, if any, of Department of Energy defense nuclear facilities.

(b) DOE Report.–The Secretary of Energy shall submit to the Committees on Armed Services and on Appropriations of the Senate and to the Speaker of the House of Representatives each year, at the same time that the President submits the budget to Congress pursuant to section 1105(a) of title 31, United State Code, a written report concerning the activities of the Department of Energy under this chapter during the year preceding the year in which the report is submitted.

42 USC 2286f.

Sec. 317. Judicial Review

Chapter 7 of title 5, United States Code, shall apply to the activities of the Board under this chapter.

42 USC 2286g.

Sec. 318. Definition

As used in this chapter, the term "Department of Energy defense nuclear facility" means any of the following:

(1) A production facility or utilization facility (as defined in section 11 of this Act) that is under the control or jurisdiction of the Secretary of Energy and that is operated for national security purposes, but the term does not include–

(A) any facility or activity covered by Executive Order No. 12344, dated February 1, 1982, pertaining to the Naval nuclear propulsion program;

(B) any facility or activity involved with the assembly or testing of nuclear explosives or with the transportation of nuclear explosives or nuclear material;

(C) any facility that does not conduct atomic energy defense activities; or

(D) any facility owned by the United States Enrichment Corporation.

(2) A nuclear waste storage facility under the control or jurisdiction of the Secretary of Energy, but the term does not include a facility developed pursuant to the Nuclear Waste Policy Act of 1982 (42 USC 10101 *et seq.*) and licensed by the Nuclear Regulatory Commission.

42 USC 2286h.

Sec. 319. Contract Authority Subject to Appropriations

The authority of the Board to enter into contracts under this chapter is effective only to the extent that appropriations (including transfers of appropriations) are provided in advance for such purpose.

42 USC 2286h-1.

Sec. 320. Transmittal of Certain Information to Congress[390]

Whenever the Board submits or transmits to the President or the Director of the Office of Management and Budget any legislative recommendation, or any statement or information in preparation of a report to be submitted to the Congress pursuant to section 316(a), the Board shall submit at the same time a copy thereof to the Congress.

42 USC 2286i.

Sec. 321. Annual Authorization of Appropriations[391]

Authorizations of appropriations for the Board for fiscal years beginning after fiscal year 1989 shall be provided annually in authorizations Acts.

Negotiated Rulemaking on Financial Protection for Radiopharmaceutical Licensees

42 USC 2210
note.
Contracts.

(A) Rulemaking Proceeding.–

(1) Purpose.–The Nuclear Regulatory Commission (hereafter in this section referred to as the "Commission") shall initiate a proceeding, in accordance with the requirements of this section, to determine whether to enter into indemnity agreements under section 170 of the Atomic Energy Act of 1954 (42 USC 2210) with persons licensed by the Commission under section 81, 104(a), or 104(c) of the Atomic Energy Act of 1954 (42 USC 2111, 2134(a), and 2134(c)) or by a State under section 274(b) of the Atomic Energy Act of 1954 (42 USC 2021(b)) for the manufacture, production, possession, or use of radioisotopes or radiopharmaceutical for medical purposes (hereafter in this section referred to as "radiopharmaceutical licensees")

(2) Final Determination.–A final determination with respect to whether radiopharmaceutical licensees, or any class of such licensees,

---

[390]   Added by P.L. 103–160, Div C., Title XXXII, § 3202(a)(2), 107 Stat 1959 (1993).

[391]   Amended by P.L. 103–160, Div C., Title XXXII, § 3202(a)(1), 107 Stat 1959 (1993), which renumbered this section [formerly section 320] as section 321.

shall be indemnified pursuant to section 170 of the Atomic Energy Act of 1954 (42 USC2210) and if so, the terms and conditions of such indemnification, shall be rendered by the Commission within 18 months of the date of the enactment of this Act.

(b) Negotiated Rulemaking.–

(1) Administrative Conference Guidelines.–For the purpose of making the determination required under subsection (a), the Commission shall, to the extent consistent with the provisions of this Act, conduct a negotiated rulemaking in accordance with the guidance provided by the Administrative Conference of the United States in Recommendation 82-4, "Procedures for Negotiating Proposed Regulations" (42 Fed. Reg. 30708, July 15, 1982).

(2) Designation of Convener.–Within 30 days of the date of the enactment of this Act, the Commission shall designate an individual or individuals recommended by the Administrative Conference of the United States to serve as a convener for such negotiations.

Contracts.

(3) Submission Recommendations of The Convener.–The convener shall, not later than seven months after the date of the enactment of this Act, submit to the Commission recommendations for a proposed rule regarding whether the Commission should enter into indemnity agreements under section 170 of the Atomic Energy Act of 1954 (42 USC 2210) with radiopharmaceutical licensees and, if so, the terms and conditions of such indemnification. If the convener recommends that such indemnity be provided for radiopharmaceutical licensees, the proposed rule submitted by the convener shall set forth the procedures for the execution if indemnification agreements with radiopharmaceutical licensees.

(4) Publication of Recommendations and Proposed Rule.–If the convener recommends that such indemnity be provided for radiopharmaceutical licensees, the Commission shall publish the recommendations of the convener submitted under paragraph (3) as a notice of proposed rulemaking within 30 days of the submission of such recommendations under such paragraph.

(5) Administrative Procedures.–To the extent consistent with the provisions of this Act, the Commission shall conduct the proceeding required under subsection (a) in accordance with section 553 of title 5, United States Code.[392]

Title II–United States Enrichment Corporation

Chapter 22–General Provisions[393]

Sec. 1201. Definitions

For purposes of this title:

---

[392]   Amended by P.L. 100–408, § 19, 102 Stat. 1066 (1988), to provide for this rulemaking, which is not part of the Atomic Energy Act.
[393]   Chapter 22 added by P.L. 102–486, § 901, 106 Stat. 2924 (1992). Section will be repealed on privatization date. P.L. 104–134, Title III, Chapter 1, Subchapter A, § 3116(a), 110 Stat. 1321–349 (1996), provides: "Chapters 22 through 26 of the Atomic Energy Act of 1954 (42 U.S.C. 2297–2297e–7) are repealed as of the privatization date." [The "privatization date" is defined at 42 U.S.C. section 2297h(9) as the date on which 100 percent of the ownership of the United States Enrichment Corporation has been transferred to private investors.]

42 USC 2297.

(1) The term "alternative technologies for uranium enrichment" means technologies to enrich uranium by methods other than the gaseous diffusion process.

(2) The term "AVLIS" means atomic vapor laser isotope separation technology.

(3) The term "Board" means the Board of Directors of the Corporation established under section 1304.

(4) The term "Corporation" means the United States Enrichment Corporation.

(5) The term "corrective actions" has the meaning given such term by the Administrator of the Environmental Protection Agency under section 3004(u) of the Solid Waste Disposal Act (42 USC 6924(u)).

(6) The term "decontamination and decommissioning" means those activities, other than response actions or corrective actions, undertaken to decontaminate and decommission inactive uranium enrichment facilities that have residual radioactive or mixed radioactive and hazardous chemical contamination, including depleted tailings.

(7) The term "Department" means the Department of Energy.

(8) The term "highly enriched uranium" means uranium enriched to 20 percent or more of the uranium-235 isotope.

(9) The term "low-enriched uranium" means uranium enriched to less than 20 percent of the uranium-235 isotope.

(10) The term "releases" has the meaning given the term "release" in section 101(22) of the Comprehensive Environmental response, Compensation, and Liability Act of 1980 (42 USC 9601(22)).

(11) The term "remedial action" has the meaning given such term in section 101(24) of the Comprehensive Environmental Response, Compensation, and Liability Act of 1980 (42 USC 9601(24)).

(12) the term "response actions" has the meaning given the term "response" in section 101(25) of the Comprehensive Environmental Response, Compensation, and Liability Act of 1980 (42 USC 9601(25)).

(13) The term "Secretary" means the Secretary of Energy.

(14) The term "uranium enrichment" means the separation of uranium of a given isotopic content into 2 components, 1 having a higher percentage of a fissile isotope and 1 having a lower percentage.[394]

42 USC 2297a.

Sec. 1202. Purposes

The Corporation is created for the following purposes:

---

[394]   Severability provisions for Title IX of P.L. 102–486, § 904, 106 Stat. 2946 (1992), provide the following:

If any provision of this title [42 U.S.C. sections 2297 et seq., generally, for full classification, consult U.S.C. Tables volumes], or the amendments made by this title [42 U.S.C. sections 2297 et seq., generally, for full classification, consult U.S.C. Tables volumes], or the application of any provision to any entity, person, or circumstance, is for any reason adjudged by a court of competent jurisdiction to be invalid, the remainder of this title, and the amendments made by this title [adding 42 U.S.C. sections 2297 et seq., generally, for full classification, consult U.S.C. Tables volumes], or its application shall not be affected.

P.L. 104–134, Title III, Chapter 1, subchapter A, § 3116(e), 110 Stat. 1321–350, (1996), provides: "Following the privatization date, all references in the Atomic Energy Act of 1954 [42 U.S.C. sections 2011 et seq.] to the United States Enrichment Corporation shall be deemed to be references to the private corporation."

(1) To operate as a business enterprise on a profitable and efficient basis.

(2) To maximize the long-term value of the Corporation to the Treasury of the United States.

(3) To lease Department uranium enrichment facilities, as needed.

(4) To acquire uranium for uranium enrichment, low-enriched uranium for resale, and highly enriched uranium for conversion into low-enriched uranium, as needed.

(5) To market and sell its enriched uranium and uranium enrichment and related services to–

(A) the Department for governmental purposes; and

(B) domestic and foreign persons, as provided in section 1303(6).

(6) To conduct research and development as required to meet business objectives for the purposes of identifying, evaluating, improving, and testing alternative technologies for uranium enrichment.

(7) To conduct the business as a self-financing corporation and eliminate the need for Federal Government appropriations or sources of Federal financing other than those provided in this title.

(8) To help maintain a reliable and economical domestic source of uranium enrichment services.

(9) To comply with laws, and regulations promulgated thereunder, to protect the public health, safety, and the environment.

(10) To continue at all times to meet the objectives of ensuring the Nation's common defense and security, including abiding by United States laws and policies concerning special nuclear materials and nonproliferation of atomic weapons and other nonpeaceful uses of atomic energy.

(11) To take all other lawful actions in furtherance of these purposes.

Chapter 23–Establishment, Powers, and Organization of Corporation[395]

Sec. 1301. Establishment of the Corporation

(a)[396] IN GENERAL.–There is established a body corporate to be known as the United States Enrichment Corporation.

42 USC 2297b.

(b) GOVERNMENT CORPORATION.–The Corporation shall be established as a wholly owned Government corporation subject to chapter 91 of title 31, United States Code (commonly referred to as the Government Corporation Control Act), except as otherwise provided in this title.

(c) FEDERAL AGENCY.–The Corporation shall be an agency and instrumentality of the United States.

---

[395]   Section 1301 to 1316 repealed by P.L. 104–134, Title III, § 3116(a)(1), 110 Stat. 1321–349 (1996). (USEC's privatization was completed July 28, 1998). It provides: "Chapters 22 through 26 of the Atomic Energy Act of 1954 (42 U.S.C. 2297–2297e–7) are repealed as of the privatization date." [The "privatization date" is defined at 42 U.S.C. section 2297h(9) as the date on which 100 percent of the ownership of the United States Enrichment Corporation has been transferred to private investors.]

[396]   Added by P.L. 102–486, 106 Stat. 2925.

42 USC 2297b-1.

Sec. 1302. Corporate Offices

The Corporation shall maintain an office for the service of process and papers in the District of Columbia, and shall be deemed, for purposes of venue in civil actions, to be a resident thereof. The Corporation may establish offices in such other place or places as it may deem necessary or appropriate in the conduct of its business.

42 USC 2297b-2.

Sec. 1303. Powers of the Corporation

In order to accomplish its purposes, the Corporation–

(1) shall, except as provided in this title or applicable Federal law, have all the powers of a private corporation incorporated under the District of Columbia Business Corporation Act;

(2) shall have the priority of the United States with respect to the payment of debts out of bankrupt, insolvent, and decedents' estates;

(3) may obtain from the Administrator of General Services the services the Administrator is authorized to provide agencies of the United States, on the same basis as those services are provided to other agencies of the United States;

(4) shall enrich uranium, provide for uranium to be enriched by others, or acquire enriched uranium (including low-enriched uranium derived from highly enriched uranium provided under section 1408);

(5) may conduct, or provide for conducting, those research and development activities related to uranium enrichment and related processes and activities the Corporation considers necessary or advisable to maintain the Corporation as a commercial enterprise operating on a profitable and efficient basis;

(6) may enter into transactions regarding uranium, enriched uranium, or depleted uranium with–

(A) persons licensed under section 53, 63, 103, or 104 in accordance with the licenses held by those persons;

(B) persons in accordance with, and within the period of, an agreement for cooperation arranged under section 123; or

(C) persons otherwise authorized by law to enter into such transactions;

(7) may enter into contracts with persons licensed under section 53, 63, 103, or 104, for as long as the Corporation considers necessary or desirable, to provide uranium or uranium enrichment and related services;

(8) may enter into contracts to provide uranium or uranium enrichment and related services in accordance with, and within the period of, an agreement for cooperation arranged under section 123 or as otherwise authorized by law; and

(9) shall sell to the Department as provided in this title, without regard to section 57e., the amounts of uranium enrichment and related services that the Department determines from time to time are required for it to–

(A) carry out Presidential directions and authorizations under section 91; and

(B) conduct other Department programs.

42 USC 2297b-3.

Sec. 1304. Board of Directors

(a) IN GENERAL.–The powers of the Corporation are vested in the Board of Directors.

(b) APPOINTMENT.–The Board of Directors shall consist of five individuals, to be appointed by the President by and with the advice and consent of the Senate. The President shall designate a Chairman of the Board from among members of the Board.

(c) QUALIFICATIONS.–Members of the Board shall be citizens of the United States. No member of the Board shall be an employee of the Corporation or have any direct financial relationship with the Corporation other than that of being a member of the Board.

(d) TERMS.–

(1) IN GENERAL.–Except as provided in paragraph (2), members of the Board shall serve five-year terms or until the election of a new Board of Directors under section 1704, whichever comes first.

(2) INITIAL MEMBERS.–of the members first appointed to the Board–

(A) 1 shall be appointed for a 1-year term;

(B) 1 shall be appointed for a 2-year term;

(C) 1 shall be appointed for a 3-year term; and

(D) 1 shall be appointed for a 4-year term.

(3) REAPPOINTMENT.–Members of the Board may be reappointed by the President, by and with the advice and consent of the Senate.

(e) VACANCIES.–Upon the occurrence of a vacancy on the Board, the President by and with the advice and consent of the Senate shall appoint an individual to fill such vacancy for the remainder of the applicable term.

(f) MEETINGS AND QUORUM.–The Board shall meet at any time pursuant to the call of the Chairman and as provided by the bylaws of the Corporation, but not less than quarterly. Three voting members of the Board shall constitute a quorum. A majority of the Board shall adopt and from time to time may amend bylaws for the operation of the Board.

(g) POWERS.–The Board shall be responsible for general management of the Corporation and shall have the same authority, privileges, and responsibilities as the board of directors of a private corporation incorporated under the District of Columbia Business Corporation Act.

(h) COMPENSATION.–Members of the Board shall serve on a part-time basis and shall receive per diem, when engaged in the actual performance of Corporation duties, plus reimbursement for travel, subsistence, and other necessary expenses incurred in the performance of their duties.

(i) MEMBERSHIP OF SECRETARY OF TREASURY.–The President may appoint the Secretary of the Treasury or his designee to serve as a member of the Board or as a nonvoting, ex officio member of the Board.

(j) CONFLICT OF INTEREST REQUIREMENTS.–No director, officer, or other management level employee of the Corporation may have a financial interest in any customer, contractor, or competitor of the Corporation or in any business that may be adversely affected by the success of the Corporation.

42 USC 2297b-4. **Sec. 1305. Employees of the Corporation**

(a) APPOINTMENT.–The Board shall appoint such officers and employees as are necessary for the transaction of its business.

(b) COMPENSATION, DUTIES, AND REMOVAL.–The Board shall, without regard to section 5301 of title 6, United States Code, fix the compensation of all officers and employees of the Corporation, define their duties, and provide a system of organization to fix responsibility and promote efficiency. Any officer or employee of the Corporation may be removed in the discretion of the Board.

(c) APPLICABLE CRITERIA.–The Board shall ensure that the personnel function and organization is consistent with the principles of section 2301(b) of title 5, United States Code, relating to merit system principles. Officers and employees shall be appointed, promoted, and assigned on the basis of merit and fitness, and other personnel actions shall be consistent with the principles of fairness and due process but without regard to those provisions of title 5 of the United States Code governing appointments and other personnel actions in the competitive service.

(d) TREATMENT OF PERSONS EMPLOYED PRIOR TO TRANSITION DATE.–Compensation, benefits, and other terms and conditions of employment in effect immediately prior to the transition date, whether provided by statute or by rules of the Department or the executive branch, shall continue to apply to officers and employees who transfer to the Corporation from other Federal employment until changed by the Board.

(e) PROTECTION OF EXISTING EMPLOYEES.–

(1) IN GENERAL.–It is the purpose of this subsection to ensure that the establishment of the Corporation pursuant to this chapter shall not result in any adverse effects on the employment rights, wages, or benefits of employees at facilities that are operated, directly or under contract, in the performance of the functions vested in the Corporation.

(2) APPLICABILITY OF EXISTING COLLECTIVE BARGAINING AGREEMENT.–Any employer (including the Corporation) at a facility described in paragraph (1) shall abide by the terms of a collective bargaining agreement in effect on April 30, 1991, at each individual facility until–

(A) the earlier of the date on which a new bargaining agreement is signed; or

(B) the end of the 2-year period beginning on the date of the enactment of this title.

(3) APPLICABILITY OF NLRA.–Except as specifically provided in this subsection, the Corporation is subject to the provisions of the National Labor Relations Act (29 USC 151 *et seq.*).

(4) BENEFITS OF TRANSFEREES AND DETAILEES.–At the request of the Board and subject to the approval of the Secretary, an employee of the Department may be transferred or detailed as provided for in section 1315, to the Corporation without any loss in accrued benefits or standing within the Civil Service System. For those employees who accept transfer to the Corporation, it shall be their option as to whether to have any accrued retirement benefits transferred to a retirement system established by the Corporation or to retain their coverage under either the Civil Service Retirement System or the Federal Employees' Retirement System, as applicable, in lieu of coverage by the Corporation's retirement system. For those employees electing to remain with one of the Federal retirement systems, the Corporation shall withhold pay and make such payments as are required under the Federal retirement system. For those Department employees detailed, the Department shall offer those employees a position of like grade, compensation, and proximity to their official duty station after their services are no longer required by the Corporation.

42 USC 2297b-5.   **Sec. 1306. Audits**

(a) INDEPENDENT AUDITS.–

(1) IN GENERAL.–The financial statements of the Corporation shall be prepared in accordance with generally accepted accounting principles and shall be audited annually by an independent certified

public accountant in accordance with auditing standards issued by the Comptroller General. Such auditing standards shall be consistent with the private sector's generally accepted auditing standards.

(2) REVIEW BY GAO.–The Comptroller General may review any audit of the Corporation's financial statements conducted under paragraph (1). The Comptroller General shall report to the Congress and the Corporation the results of any such review and shall include in such report appropriate recommendations.

(b) GAO AUDITS.–

(1) IN GENERAL.–The Comptroller General may audit the financial statements of the Corporation for any year in the manner provided in subsection (a)(1).

(2) REIMBURSEMENT BY CORPORATION.–The Corporation shall reimburse the Comptroller General for the full cost of any audit conducted under this subsection, as determined by the Comptroller General.

(c) AVAILABILITY OF BOOKS AND RECORDS.–All books, accounts, financial records, reports, files, papers, and other property belonging to or in use by the Corporation and its auditor that the Comptroller General considers necessary to the performance of any audit or review under this section shall be made available to the Comptroller General, subject to section 1314.

(d) TREATMENT OF GAO AUDITS.–Activities the Comptroller General conducts under this section shall be in lieu of any other audit of the financial transactions of the Corporation the Comptroller General is required to make under chapter 91 of title 31, United States Code, or other law.

42 USC 2297b-6.   **Sec. 1307. Annual Reports**

(a) IN GENERAL.–The Corporation shall prepare and submit an annual report of its activities to the President and the Congress. This report shall contain–

(1) a general description of the Corporation's operations;

(2) a summary of the Corporation's operating and financial performance, including an explanation of the decision to pay or not pay dividends;

(3) copies of audit reports prepared under section 1305;

(4) the information required under regulations issued under section 13 of the Securities Exchange Act of 1934 (15 USC 78m); and

(5) an identification and assessment of any impairment of capital or ability of the Corporation to comply with this title.

(b) DEADLINE.–The report shall be completed not later than 150 days following the close of each of the Corporation's fiscal years and shall accurately reflect the financial position of the Corporation at fiscal year end.

42 USC 2297b-7.   **Sec. 1308. Accounts**

(a) ESTABLISHMENT OF UNITED STATES ENRICHMENT CORPORATION FUND.–There is established in the Treasury of the United States a revolving fund, to be known as the "United States Enrichment Corporation Fund", which shall be available to the Corporation, without need for further appropriation and without fiscal year limitation, for carrying out its purposes, functions, and powers, and which shall not be subject to apportionment under subchapter II of chapter 15 of title 31, United States Code.

(b) TRANSFER OF UNEXPENDED BALANCES.–On the transfer date, the Secretary shall, without need of further appropriation, transfer to the Corporation the unexpended balance of appropriations and other monies available to the Department (inclusive of funds set aside for accounts payable), and accounts receivable which are related to functions and activities acquired by the Corporation from the Department pursuant to this title, including all advance payments.

42 USC 2297b-8.   **Sec. 1309.- Obligations**

(a) ISSUANCE.–

(1) IN GENERAL.–The Corporation may issue and sell bonds, notes, and other evidences of indebtedness (collectively referred to in this title as "bonds"), except that the Corporation may not issue or sell bonds for the purpose of constructing new uranium enrichment facilities or conducting directly related preconstruction activities. Borrowing under this paragraph during any fiscal year ending before October 1, 1996, shall be subject to approval in appropriation Acts.

(2) USE OF REVENUES.–The Corporation may pledge and use its revenues for payment of the principal of and interest on its bonds, for their purchase or redemption, and for other purposes incidental to these functions, including creation of reserve funds and other funds that may be similarly pledged and used.

(3) AGREEMENTS WITH HOLDERS AND TRUSTEES.– The Corporation may enter into binding covenants with the holders and trustees of its bonds with respect to

(A) the establishment of reserve and other funds;

(B) stipulations concerning the subsequent issuance of bonds; and

(C) other matters not inconsistent with this title; that the Corporation determines necessary or desirable to enhance the marketability of the bonds.

(b) NOT OBLIGATIONS OF UNITED STATES.–Bonds issued by the Corporation under this section shall not be obligations of, or guaranteed as to principal or interest by, the United States, and the bonds shall so plainly state.

(c) TERMS AND CONDITIONS.–

(1) NEGOTIABLE; MATURITY.–Bonds issued by the Corporation under this section shall be negotiable instruments unless otherwise specified in the bond and shall mature not more than 50 years after their date of issuance.

(2) ROLE OF SECRETARY OF THE TREASURY.–

(A) RIGHT OF DISAPPROVAL.–The Corporation may set the terms and conditions of bonds issued under this section, subject to disapproval of such terms and conditions by the Secretary of the Treasury within 5 days after the Secretary of the Treasury is notified of the following terms and conditions of the bonds:

(i) Their forms and denominations.

(ii) The times, amounts, and prices at which they are sold.

(iii) Their rates of interest.

(iv) The terms at which they may be redeemed by the Corporation before maturity.

(v) The priority of their claims on the Corporation's net revenues with respect to principal and interest payments.

(vi) Any other terms and conditions.

(B) INAPPLICABILITY OF RIGHT TO PRESCRIBE TERMS.–Section 9108(a) of title 31, United States Code, shall not apply to the Corporation.

(d) INAPPLICABILITY OF SECURITIES REQUIREMENTS.– The Corporation shall be considered an executive department of the United

States for purposes of section 3(c) of the Securities Exchange Act of 1934 (15 USC 78c(c)).

(e) INAPPLICABILITY OF FEE.–The Corporation shall not issue or sell any bonds to the Federal Financing Bank.

42 USC 2297b-9.

Sec. 1310. Exemption from Taxation and Payments in Lieu of Taxes

(a) EXEMPTION FROM TAXATION.–In order to render financial assistance to those States and localities in which the facilities of the Corporation are located, the Corporation shall, beginning in fiscal year 1998, make payments to State and local governments as provided in this section. These payments shall be in lieu of any and all State and local taxes on the real and personal property of the Corporation. All property of the Corporation is expressly exempted from taxation in any manner or form by any State, county, or other local government entity including State, county, or other local government sales tax.

(b) PAYMENTS IN LIEU OF TAXES.–Beginning in fiscal year 1998, the Corporation shall make annual payments, in amounts determined by the Corporation to be fair and reasonable, to the State and local governmental agencies having tax jurisdiction in any area where facilities of the Corporation are located. In making these determinations, the Corporation shall be guided by the following criteria:

(1) The Corporation shall take into account the customs and practices prevailing in the area with respect to appraisal, assessment, and classification of industrial property and any special considerations extended to large-scale industrial operations.

(2) The payment made to any taxing authority for any period shall not be less than the payments that would have been made to the taxing authority for the same period by the Department and its cost-type contractors on behalf of the Department with respect to property that has been transferred to the Corporation under section 1404 and that would have been attributable to the ownership, management, operation, and maintenance of the Department's uranium enrichment facilities, applying the laws and policies prevailing immediately prior to the transition date.

(c) TIME OF PAYMENTS.–Payments shall be made by the Corporation at the time when payments of taxes by taxpayers to each taxing authority are due and payable.

(d) DETERMINATION OF AMOUNT DUE.–The determination by the Corporation of the amounts due under this section shall be final and conclusive.

42 USC 2297b-10.

Sec. 1311. Cooperation with Other Agencies

The Corporation may request to use on a reimbursable basis the available services, equipment, personnel, and facilities of agencies of the United States, and on a similar basis may cooperate with such agencies in the establishment and use of services, equipment, and facilities of the Corporation. Further, the Corporation may confer with and avail itself of the cooperation, services, records, and facilities of State, territorial, municipal, or other local agencies.

Sec. 1312. Applicability of Certain Federal Laws

42 USC 2297b-11.

(a) ANTITRUST LAWS.–The Corporation shall conduct its activities in a manner consistent with the policies expressed in the following antitrust laws:

(1) The Sherman Act (15 USC 1-7).

(2) The Clayton Act (15 USC 12-27).

(3) Sections 73 and 74 of the Wilson Tariff Act (15 USC 8 and 9).

(b) ENVIRONMENTAL LAWS.–The Corporation shall be subject to, and comply with, all Federal and State, interstate, and local environmental laws and requirements, both substantive and procedural, in the same manner, and to the same extent, as any person who is subject to such laws and requirements. For purposes of enforcing any such law or substantive or procedural requirements (including any injunctive relief, administrative order, or civil or administrative penalty or fine) against the Corporation, the United States expressly waives any immunity otherwise applicable to the Corporation. For the purposes of this subsection, the term "person" means an individual, trust, firm, joint stock company, corporation, partnership, association, State, municipality, or political subdivision of a State.

(c) OSHA REQUIREMENTS.–Notwithstanding sections 3(5), 4(b)(1), and 19 of the Occupational Safety and Health Act of 1970 (29 USC 652(5), 653(b) (1), and 668)), the Corporation shall be subject to, and comply with, such Act and all regulations and standards promulgated thereunder in the same manner, and to the same extent, as an employer is subject to such Act. For the purposes of enforcing such Act (including any injunctive relief, administrative order, or civil, administrative, or criminal penalty or fine) against the Corporation, the United States expressly waives any immunity otherwise applicable to the Corporation.

(d) LABOR STANDARDS.–The Act of March 3, 1931 (known as the Davis-Bacon Act) (40 USC 276a *et seq*.) and the Service Contract Act of 1965 (41 USC 351 *et seq*.) shall apply to the Corporation. All laborers and mechanics employed on the construction, alteration, or repair of projects funded, in whole or in part, by the Corporation shall be paid wages at rates not less than those prevailing on projects of a similar character in the locality as determined by the Secretary of Labor in accordance with such Act of March 3, 1931. The Secretary of Labor shall have, with respect to the labor standards specified in this subsection, the authority and functions set forth in Reorganization Plan Numbered 14 of 1950 (15 F.R. 3176, 64 Stat. 1267) and the Act of June 13, 1934 (40 USC 276c).

(e) ENERGY REORGANIZATION ACT REQUIREMENTS.– The Corporation is subject to the provisions of section 210 of the Energy Reorganization Act of 1974 (42 USC 5850) to the same extent as an employer subject to such section, and, with respect to the operation of the facilities leased by the Corporation, section 206 of the Energy Reorganization Act of 1974 (42 USC 5846) shall apply to the directors and officers of the Corporation.

(f) EXEMPTION FROM FEDERAL PROPERTY REQUIREMENTS.–The Corporation shall not be subject to the Federal Property and Administrative Services Act of 1949 (41 USC 471 *et seq*.).

42 USC 2297b-12. **Sec. 1313. Security**

Any references to the term "Commission" or to the Department in sections 161k., 221a., and 230 shall be considered to include the Corporation.

42 USC 2297b-13. **Sec. 1314. Control of Information**

(a) IN GENERAL.–Except as provided in subsection (b), the Corporation may protect trade secrets and commercial or financial information to the same extent as a privately owned corporation.

(b) OTHER APPLICABLE LAWS.–Section 552(d) of title 5, United States Code, shall apply to the Corporation, and such information shall be subject to the applicable provisions of law protecting the confidentiality of trade secrets and business and financial information, including section 1905 of title 18, United States Code.

42 USC 2297b-14.

President.

Sec. 1315. Transition

(a) TRANSITION MANAGER.–Within 30 days after the date of the enactment of this title, the President shall appoint a Transition Manager, who shall serve at the pleasure of the President until a quorum of the Board has been appointed and confirmed in accordance with section 1304.

(b) POWERS.–

(1) IN GENERAL.–Until a quorum of the Board has qualified, the Transition Manager shall exercise the powers and duties of the Board and shall be responsible for taking all actions needed to effect the transfer of the uranium enrichment enterprise from the Secretary to the Corporation on the transition date.

(2) CONTINUATION UNTIL BOARD HAS QUORUM.–In the event that a quorum of the Board has not qualified by the transition date, the Transition Manager shall continue to exercise the powers and duties of the Board until a quorum has qualified.

(c) RATIFICATION OF TRANSITION MANAGER'S ACTIONS.– All actions taken by the Transition Manager before the qualification of a quorum of the Board shall be subject to ratification by the Board.

(d) RESPONSIBILITIES OF SECRETARY.–Before the transition date, the Secretary shall–

(1) continue to be responsible for the management and operation of the uranium enrichment plants;

(2) provide funds, to the extent provided in appropriations Acts, to the Transition Manager to pay salaries and expenses;

(3) delegate Department employees to assist the Transition Manager in meeting his responsibilities under this section; and

(4) assist and cooperate with the Transition Manager in preparing for the transfer of the uranium enrichment enterprise to the Corporation on the transition date.

(e) TRANSITION DATE.–The transition date shall be July 1, 1993.

(f) DETAIL OF PERSONNEL.–For the purpose of continuity of operations, maintenance, and authority, the Department shall detail, for up to 18 months after the date of the enactment of this title, appropriate Department personnel as may be required in an acting capacity, until such time as a Board is confirmed and top officers of the Corporation are hired. The Corporation shall reimburse the Department and its contractors for the detail of such personnel.

42 USC 2297b-15.

Sec. 1316. Working Capital Account

There shall be established within the Corporation a Working Capital Account in which the Corporation may retain all revenue necessary for legitimate business expenses, or investments, related to carrying out its purposes.

Chapter 24–Rights, Privileges, and Assets of the Corporation[397]

42 USC 2297c.

Sec. 1401. Marketing and Contracting Authority

(a)[398] EXCLUSIVE MARKETING AGENT.–The Corporation shall act as the exclusive marketing agent on behalf of the United States Government for entering into contracts for providing enriched uranium (including low-enriched uranium derived from highly enriched uranium) and uranium enrichment and related services. The Department may not market enriched uranium (including low-enriched uranium derived from highly enriched uranium), or uranium enrichment and related services, after the transition date.

(b) TRANSFER OF CONTRACTS.

(1) IN GENERAL.–Except as provided in paragraph (2), all contracts, agreements, and leases with the Department, including all uranium enrichment contracts and power purchase contracts, that have been executed by the Department before the transition date and that relate to uranium enrichment and related services shall transfer to the Corporation.

(2) EXCEPTIONS.

(A) TVA SETTLEMENT.–The rights and responsibilities of the Department under the settlement agreement with the Tennessee Valley Authority, filed on December 18, 1987, with the United States Court of Federal Claims,[399] shall not transfer to the Corporation.

(B) NONTRANSFERABLE POWER CONTRACTS.–If the Secretary determines that a power purchase contract executed by the Department prior to the transition date cannot be transferred under its terms, the Secretary may continue to receive power under the contract and resell such power to the Corporation at cost.

(C) NONPOWER APPLICATIONS.–Contracts for enriched uranium and uranium services in existence as of the date of the enactment of this title for research and development or other nonpower applications shall remain with the Department. At the request of the Department, the Corporation, in consultation with the Department, may enter into such contracts it determines to be appropriate.

42 USC 2297c-1.

Sec. 1402. Pricing

(a) SERVICES PROVIDED TO COMMERCIAL CUSTOMERS. – The Corporation shall establish prices for its products, materials, and services provided to customers other than the Department on a basis that will allow it to attain the normal business objectives of a profitmaking corporation.

(b) SERVICES PROVIDED TO DOE.–The Corporation shall charge prices to the Department for uranium enrichment services provided under section 1303(9) on a basis that will allow it to recover its costs, on a

---

[397]   Section 1401 to 1408 repealed by P.L. 104–134, Title III, Chapter 1, subchapter A, section 3116(a), 110 Stat. 1321–349 (1996), provides the following: "Chapters 22 through 26 of the Atomic Energy Act of 1954 (42 U.S.C. 2297–2297e–7) are repealed as of the privatization date." [The "privatization date" is defined at 42 U.S.C. section 2297h(9) as the date on which 100 percent of the ownership of the United States Enrichment Corporation has been transferred to private investors.]

[398]   Added by P.L. 102–486, 106 Stat. 2934.

[399]   Amended by P.L. 102–572, Title IX, section 902(b)(1), 106 Stat. 4516 (1992).

yearly basis, for providing products, materials, and services, and provide for a reasonable profit.

42 USC 2297c-2. **Sec. 1403. Leasing of Gaseous Diffusion Facilities of Department**

(a) IN GENERAL.–The Corporation shall lease the Paducah Gaseous Diffusion Plant in Paducah, Kentucky, the Portsmouth Gaseous Diffusion Plant in Piketon, Ohio, and related property of the Department, for a period of 6 years from the transition date. Thereafter, the Corporation shall have the exclusive option to lease such facilities and related property for additional periods.

(b) TERMS OF LEASE.–The Corporation and the Department shall set mutually agreeable terms for a lease under subsection (a), including specifying annual payments to the Department by the Corporation to be made. The amount of annual payments shall be equal to the cost incurred by the Department in administering the lease and providing services related to the lease to the Corporation (excluding depreciation and imputed interest on original plant investments in the Department's gaseous diffusion plants and costs under subsection (d)).

(c) EXCLUSION OF FACILITIES FOR PRODUCTION OF HIGHLY ENRICHED URANIUM.–Subsection (a) shall not apply to Department facilities necessary for the production of highly enriched uranium. The Secretary may grant to the Corporation access to such facilities for purposes other than the production of highly enriched uranium.

(d) DOE RESPONSIBILITY FOR PREEXISTING CONDITIONS.–The payment of any costs of decontamination and decommissioning, response actions, or corrective actions with respect to conditions existing before the transition date, in connection with property of the Department leased under subsection (a), shall remain the sole responsibility of the Department.

(e) ENVIRONMENTAL AUDIT.–The Secretary, in consultation with the Administrator of the Environmental Protection Agency, shall conduct a comprehensive environmental audit identifying environmental conditions that will remain the responsibility of the Department pursuant to subsection (d) after the transition date. Such audit shall be completed no later than the transition date.

(f) TREATMENT UNDER PRICE-ANDERSON PROVISIONS. – Any lease executed between the Secretary and the Corporation under this section shall be deemed to be a contract for purposes of section 170d.

(g) WAIVER OF EIS REQUIREMENT.–The execution of the lease by the Corporation and the Department shall not be considered a major Federal action significantly affecting the quality of the human environment for purposes of section 102 of the National Environmental Policy Act of 1969 (42 USC 4332).

42 USC 2297c-3. **Sec. 1404. Capital Structure of Corporation**

(a) CAPITAL STOCK.–

(1) ISSUANCE TO SECRETARY OF THE TREASURY.– The Corporation shall issue capital stock representing an equity investment equal to the greater of–

(A) $3,000,000,000; or

(B) the book value of assets transferred to the Corporation, as reported in the Uranium Enrichment Annual Report for fiscal year 1991, modified to reflect continued depreciation and other usual changes that occur up to the transfer date.

The Secretary of the Treasury shall hold such stock for the United States, except that all rights and duties pertaining to management of the Corporation shall remain vested in the Board.

(2) RESTRICTION ON TRANSFERS OF STOCK BY UNITED STATES.–The capital stock of the Corporation shall not be sold, transferred, or conveyed by the United States, except to carry out the privatization of the Corporation under section 1502.

(3) ANNUAL ASSESSMENT.–The Secretary of the Treasury shall annually assess the value of the stock held by the Secretary under paragraph (1) and submit to the Congress a report setting forth such value. The annual assessment of the Secretary shall be subject to review by an independent auditor.

(b) PAYMENT OF DIVIDENDS.–The Corporation shall pay into miscellaneous receipts of the Treasury of the United States or such other fund as is provided by law, dividends on the capital stock, out of earnings of the Corporation, as a return on the investment represented by such stock. Until privatization occurs under section 1502, the Corporation shall pay as dividends to the Treasury of the United States all net revenues remaining at the end of each fiscal year not required for operating expenses or for deposit into the Working Capital Account established in section 1316.

(c) PROHIBITION ON ADDITIONAL FEDERAL ASSISTANCE. – Except as otherwise specifically provided in this title, the Corporation shall receive no appropriations, loans, or other financial assistance from the Federal Government.

(d) SOLE RECOVERY OF UNRECOVERED COSTS.–Receipt by the United States of the proceeds from the sale of stock issued by the Corporation under subsection (a)(1), and the dividends paid under subsection (b), shall constitute the sole recovery by the United States of previously unrecovered costs (including depreciation and imputed interest on original plant investments in the Department's gaseous diffusion plants) that have been incurred by the United States for uranium enrichment activities prior to the transition date.

42 USC 2297c-4.   **Sec. 1405. Patents and Inventions**

The Corporation may at any time apply to the Department for a patent license for the use of an invention or discovery useful in the production or utilization of special nuclear material or atomic energy covered by a patent when the patent has not been declared to be affected with the public interest under section 153a. and when use of the patent is within the Corporation's authority. An application shall constitute an application under section 153c. subject to section 153c., d., e., f., g., and h.

42 USC 2297c-5.   **Sec. 1406. Liabilities**

(a) LIABILITIES BASED ON OPERATIONS BEFORE TRANSITION.–Except as otherwise provided in this title, all liabilities attributable to operation of the uranium enrichment enterprise before the transition date shall remain direct liabilities of the Department.

(b) JUDGMENTS BASED ON OPERATIONS BEFORE TRANSITION.– Any judgment entered against the Corporation imposing liability arising out of the operation of the uranium enrichment enterprise before the transition date shall be considered a judgment against and shall be payable solely by the Department.

(c) REPRESENTATION.–With regard to any claim seeking to impose liability under subsection (a) or (b), the United States shall be represented by the Department of Justice.

(d) JUDGMENTS BASED ON OPERATIONS AFTER TRANSITION.–Any judgment entered against the Corporation arising from operations of the Corporation on or after the transition date shall be payable solely by the Corporation from its own funds. The Corporation shall not be considered a Federal agency for purposes of chapter 171 of title 28, United States Code.

42 USC 2297c-6.   **Sec. 1407. Transfer of Uranium Inventories**

The Secretary shall transfer to the Corporation without charge all raw and low-enriched uranium inventories of the Department necessary for the fulfillment of contracts transferred under section 1401(b).

42 USC 2297c-7.   **Sec. 1408. Purchase of Highly Enriched Uranium from Former Soviet Union**

(a) IN GENERAL.–The Corporation is authorized to negotiate the purchase of all highly enriched uranium made available by any State of the former Soviet Union under a government-to-government agreement or shall assume the obligations of the Department under any contractual agreement that has been reached with any such State or any private entity before the transition date. The Corporation may only purchase this material so long as the quality of the material can be made suitable for use in commercial reactors.

(b) ASSESSMENT OF POTENTIAL USE.–The Corporation shall prepare an assessment of the potential use of highly enriched uranium in the business operations of the Corporation.

(c) PLAN FOR BLENDING AND CONVERSION.–In the event that the agreement under subsection (a) provides for the Corporation to provide for the blending and conversion the assessment shall include a plan for such blending and conversion. The plan shall determine the least-cost approach to providing blending and conversion services, compatible with environmental, safety, security, and nonproliferation requirements. The plan shall include a competitive process that the Corporation shall use for selecting a provider of such services, including the public solicitation of proposals from the private sector to allow a determination of the least-cost approach.

(d) MINIMIZATION OF IMPACT ON DOMESTIC INDUSTRIES.– The Corporation shall seek to minimize the impact on domestic industries (including uranium mining) of the sale of low-enriched uranium derived from highly enriched uranium.

Chapter 25–Privatization of the Corporation[400]

42 USC 2297d.   **Sec. 1501. Strategic Plan for Privatization**

(a) IN GENERAL.–Within 2 years after the transition date, the Corporation shall prepare a strategic plan for transferring ownership of the Corporation to private investors. The Corporation shall revise the plan as needed.

(b) CONSIDERATION OF ALTERNATIVE MEANS OF TRANSFERRING OWNERSHIP.–The plan shall include consideration of alternative means for transferring ownership of the Corporation to private investors, including public stock offering, private placement, or merger or acquisition. The plan may call for the phased transfer of ownership or for complete transfer at a single point of time. If the plan calls for phased transfer of ownership, then–

---

[400]   Added by P.L. 102–486, 106 Stat. 2937 (1992). Sections 1501 and 1502 are repealed by P.L. 104–134, Title III, § 3116(a)(1), 110 Stat. 1321–349 (1996). It provides the following: "Chapters 22 through 26 of the Atomic Energy Act of 1954 (42 U.S.C. 2297-2297e–7) are repealed as of the privatization date." [The "privatization date" is defined at 42 U.S.C. 2297h(9) as the date on which 100 percent of the ownership of the United States Enrichment Corporation has been transferred to private investors.]

(1) privatization shall be deemed to occur when 100 percent of ownership has been transferred to private investors;

(2) prior to privatization, such stock shall be nonvoting stock; and

(3) at the time of privatization, such stock shall convert to voting stock.

(c) EVALUATION AND RECOMMENDATION.–The plan shall evaluate the relative merits of the alternatives considered and the estimated return on the Government's investment in the Corporation achievable through each alternative. The plan shall include the Corporation's recommendation on its preferred means of privatization.

(d) TRANSMITTAL.–The Corporation shall transmit copies of the strategic plan for privatization to the President and Congress upon completion.

42 USC 2297d-1.   **Sec. 1502. Privatization**

(a) IMPLEMENTATION.–Subsequent to transmitting a plan for privatization pursuant to section 1501, and subject to subsections (b) and (c), the Corporation may implement the privatization plan if the Corporation determines, in consultation with appropriate agencies of the United States, that privatization will–

(1) result in a return to the United States at least equal to the net present value of the Corporation;

(2) not result in the Corporation being owned, controlled, or dominated by an alien, a foreign corporation, or a foreign government;

(3) not be inimical to the health and safety of the public or the common defense and security; and

(4) provide reasonable assurance that adequate enrichment capacity will remain available to meet the domestic electric utility industry.

(b) REQUIREMENT OF PRESIDENTIAL APPROVAL.–The Corporation may not implement the privatization plan without the approval of the President.

(c) NOTIFICATION OF CONGRESS AND GAO EVALUATION.– The Corporation shall notify the Congress of its intent to implement the privatization plan. Within 30 days of notification, the Comptroller General shall submit a report to Congress evaluating the extent to which–

(1) the privatization plan would result in any ongoing obligation or undue cost to the Federal Government; and

(2) the revenues gained by the Federal Government under the privatization plan would represent at least the net present value of the Corporation.

(d) PERIOD FOR CONGRESSIONAL REVIEW.–The Corporation may not implement the privatization plan less than 60 days after notification of the Congress.

(e) DEPOSIT OF PROCEEDS.–Proceeds from the sale of capital stock of the Corporation under this section shall be deposited in the general fund of the Treasury.

Chapter 26–AVLIS and Alternative Technologies for Uranium Enrichment

42 USC 2297e.

Sec. 1601. Assessment by United States Enrichment Corporation[401]

(a) IN GENERAL.–The Corporation shall prepare an assessment of the economic viability of proceeding with the commercialization of AVLIS and alternative technologies for uranium enrichment in accordance with this chapter. The assessment shall include–

(1) an evaluation of market conditions together with a marketing strategy;

(2) an analysis of the economic viability of competing enrichment technologies;

(3) an identification of predeployment and capital requirements for the commercialization of AVLIS and alternative technologies for uranium enrichment;

(4) an estimate of potential earnings from the licensing of AVLIS and alternative technologies for uranium enrichment to a private government sponsored corporation;

(5) an analysis of outstanding and potential patent and related claims with respect to AVLIS and alternative technologies for uranium enrichment, and a plan for resolving such claims; and

(6) a contingency plan for providing enriched uranium and related services in the event that deployment of AVLIS and alternative technologies for uranium enrichment is determined not to be economically viable.

(b) DETERMINATION BY CORPORATION TO PROCEED WITH COMMERCIALIZATION OF AVLIS OR ALTERNATIVE TECHNOLOGIES FOR URANIUM ENRICHMENT.–The succeeding sections of this chapter shall apply only to the extent the Corporation determines in its business judgment, on the basis of the assessment prepared under subsection (a), to proceed with the commercialization of AVLIS or alternative technologies for uranium enrichment.

42 USC 2297e-1.

Sec. 1602. Transfer of Rights and Property to United States Enrichment Corporation

(a) EXCLUSIVE RIGHT TO COMMERCIALIZE.–The Corporation shall have the exclusive commercial right to deploy and use any AVLIS patents, processes, and technical information owned or controlled by the Government, upon completion of a royalty agreement with the Department.

(b) TRANSFER OF RELATED PROPERTY TO CORPORATION.–

(1) IN GENERAL.–TO the extent requested by the Corporation, the President shall transfer without charge to the Corporation all of the Department's right, title, or interest in and to property owned by the Department, or by the United States but under control or custody of the Department, that is directly related to and materially useful in the performance of the Corporation's purposes regarding AVLIS and alternative technologies for uranium enrichment, including–

---

[401]   Added by P.L. 102–486, 106 Stat. 2939 (1992). Sections 1601 and 1608 are repealed by P.L. 104–134, Title III, Chapter 1, Subchapter A, § 3116(a), 110 Stat. 1321–349 (1996), provides: "Chapters 22 through 26 of the Atomic Energy Act of 1954 (42 U.S.C. 2297–2297e–7) are repealed as of the privatization date." [The "privatization date" is defined at 42 U.S.C. 2297h(9) as the date on which 100 percent of the ownership of the United States Enrichment Corporation has been transferred to private investors.]

(A) facilities, equipment, and materials for research, development, and demonstration activities; and

(B) all other facilities, equipment, materials, processes, patents, technical information of any kind, contracts, agreements, and leases.

(2) EXCEPTION.–Facilities, real estate, improvements, and equipment related to the gaseous diffusion, and gas centrifuge, uranium enrichment programs of the Department shall not transfer under paragraph (1)(B).

(3) EXPIRATION OF TRANSFER AUTHORITY.–The President's authority to transfer property under this subsection shall expire upon privatization under section 1502.

(c) LIABILITY FOR PATENT AND RELATED CLAIMS.–With respect to any right, title, or interest provided to the Corporation under subsection (a) or (b), the Corporation shall have sole liability for any payments made or awards under section 157b.(3), or any settlements or judgments involving claims for alleged patent infringement. Any royalty agreement under subsection (a) shall provide for a reduction of royalty payments to the Department to offset any payments, awards, settlements, or judgments under this subsection.

42 USC 2297e-2.

Sec. 1603. Predeployment Activities by United States Enrichment Corporation

The Corporation may begin activities necessary to prepare AVLIS or alternative technologies for uranium enrichment for commercialization including–

(1) completion of preapplication activities with the Nuclear Regulatory Commission;

(2) preparation of a transition plan to move AVLIS or alternative technologies for uranium enrichment from the laboratory to the marketplace;

(3) confirmation of technical performance;

(4) validation of economic projections;

(5) completion of feasibility and risk studies;

(6) initiation of preliminary plant design and engineering; and

(7) site selection, site characterization, and environmental documentation activities on the basis of site evaluations and recommendations prepared for the Department by the Argonne National Laboratory.

42 USC 2297e-3.

Sec. 1604. United States Enrichment Corporation Sponsorship of Private For-Profit Corporation to Construct AVLIS and Alternative Technologies for Uranium Enrichment

(a) ESTABLISHMENT.–

(1) IN GENERAL.–If the Corporation determines to proceed with the commercialization of AVLIS or alternative technologies for uranium enrichment under this chapter, the Corporation may provide for the establishment of a private for-profit corporation, which shall have as its initial purpose the construction of a uranium enrichment facility using AVLIS technology or alternative technologies for uranium enrichment.

(2) PROCESS OF ORGANIZATION.–For purposes of the establishment of the private corporation under paragraph (1), the Corporation shall appoint not less than 3 persons to be incorporators. The incorporators so appointed shall each sign the articles of incorporation and shall serve as the initial board of directors until the members of the 1st regular board of directors shall have been appointed and elected. Such incorporators shall take whatever actions

are necessary or appropriate to establish the private corporation, including the filing of articles of incorporation in such jurisdiction as the incorporators determine to be appropriate. The incorporators shall also develop a plan for the issuance by the private corporation of voting common stock to the public, which plan shall be subject to the approval of the Secretary of the Treasury.

(b) LEGAL STATUS OF PRIVATE CORPORATION.–

(1) NOT FEDERAL AGENCY.–The private corporation established under subsection (a) shall not be an agency, instrumentality, or establishment of the United States Government and shall not be a Government corporation or Government controlled corporation.

(2) NO RECOURSE AGAINST UNITED STATES.–Obligations of the private corporation established under subsection (a) shall not be obligations of, or guaranteed as to principal or interest by, the Corporation or the United States, and the obligations shall so plainly state.

(3) NO CLAIMS COURT JURISDICTION.–NO action under section 1491 of title 28, United States Code, shall be allowable against the United States based on the actions of the private corporation established under subsection (a).

(c) TRANSACTIONS BETWEEN UNITED STATES ENRICHMENT CORPORATION AND PRIVATE CORPORATION;–

(1) GRANTS FROM USEC.–The Corporation may make grants to the private corporation established under subsection (a) from amounts available in the AVLIS Commercialization Fund. Such grants shall be used by the private corporation to carry out any remaining predeployment activity assigned to the private corporation by the Corporation. Such grants may not be used for the costs of constructing an AVLIS, or alternative technologies for uranium enrichment, production facility or engaging in directly related preconstruction activities (other than such assigned predeployment activities). The aggregate amount of such grants shall not exceed $364,000,000.

(2) LICENSING AGREEMENT.–The Corporation shall license to the private corporation established under subsection (a) the rights, titles, and interests provided to the Corporation under section 1602. The licensing agreement shall require the private corporation to make periodic payments to the Corporation in an amount that is not less than the aggregate amounts paid by the Corporation during the period involved under subsections (a) and (c) of section 1602.

(3) PURCHASE AGREEMENT.–The Corporation may enter into a commitment to purchase all enriched uranium produced at an AVLIS, or alternative technologies for uranium enrichment, facility of the private corporation established under subsection (a) at a price negotiated by the 2 corporations that–

(A) provides the private corporation with a reasonable return on its investment; and

(B) is less costly than enriched uranium available from other sources.

(4) ADDITIONAL ASSISTANCE.–The Corporation may provide to the private corporation established under subsection (a), on a reimbursable basis, such additional personnel, services, and equipment as the 2 corporations may determine to be appropriate.

42 USC 2297e-4. **Sec. 1605. AVLIS Commercialization Fund Within United States Enrichment Corporation**

(a) ESTABLISHMENT.–The Corporation may establish within the Corporation an AVLIS Commercialization Fund, which shall consist of not more than $364,000,000 paid into the Fund by the Corporation from amounts provided in appropriation Acts for such purposes and from the retained earnings of the Corporation.

(b) EXPENDITURES FROM FUND.–Amounts in the AVLIS Commercialization Fund shall be available for–

(1) expenses of the Corporation in preparing the assessment under section 1601;

(2) expenses of predeployment activities under section 1603; and

(3) grants to the private corporation under section 1604.

(c) LIMITATIONS.–

(1) EXCLUSIVE SOURCE OF FUNDS.–The Corporation may not incur any obligation, or expend any amount, with respect to AVLIS or alternative technologies for uranium enrichment, except from amounts available in the AVLIS Commercialization Fund.

(2) UNAVAILABLE FOR CONSTRUCTION COSTS.–No amount may be used from the AVLIS Commercialization Fund for the costs of constructing an AVLIS, or alternative technologies for uranium enrichment, production facility or engaging in directly related preconstruction activities (other than activities specified in subsection (b)).

(d) AUTHORIZATION OF APPROPRIATIONS.–There is authorized to be appropriated $364,000,000 from the Uranium Enrichment Special Fund for purposes of this section.

(e) COST REPORT.–On the basis of the assessment under section 1601(a)(3), the Corporation shall submit to the Congress a report on the capital requirements for commercialization of AVLIS.

42 USC 2297e-5. **Sec. 1606. Department Research and Development Assistance**

If requested by the Corporation, the Secretary shall provide, on a reimbursable basis, research and development of AVLIS and alternative technologies for uranium enrichment.

42 USC 2297e-6. **Sec. 1607. Site Selection**

This chapter shall not prejudice consideration of the site of an existing uranium enrichment facility as a candidate site for future expansion or replacement of uranium enrichment capacity through AVLIS or alternative technologies for uranium enrichment. Selection of a site for the AVLIS, or alternative technologies for uranium enrichment, facility shall be made on a competitive basis, taking into consideration economic performance, environmental compatibility, and use of any existing uranium enrichment facilities.

42 USC 2297e-7. **Sec. 1608. Exclusion from Price-Anderson Coverage**

Section 170 shall not apply to any license under section 53, 63, or 103 for a uranium enrichment facility constructed after the date of the enactment of this title.

Chapter 27–Licensing and Regulation of Uranium Enrichment Facilities

42 USC 2297f.

Sec. 1701. Gaseous Diffusion Facilities

(a)[402] ISSUANCE OF STANDARDS.–Within 2 years after the date of the enactment of this title [enacted October 24, 1992], the Nuclear Regulatory Commission shall establish by regulation such standards as are necessary to govern the gaseous diffusion uranium enrichment facilities of the Department in order to protect the public health and safety from radiological hazard and provide for the common defense and security. Regulations promulgated pursuant to this subsection shall, among other things, require that adequate safeguards (within the meaning of section 147) are in place.

(b) ANNUAL REPORT.–

(1) IN GENERAL.–Not later than the date on which a certificate of compliance is issued under subsection (c), the Nuclear Regulatory Commission in consultation with the Department and the Environmental Protection Agency, shall report to the Congress on the status of health, safety, and environmental conditions at the gaseous diffusion uranium enrichment facilities of the Department.

(2) REQUIRED DETERMINATION.–Such report shall include a determination regarding whether the gaseous diffusion uranium enrichment facilities of the Department are in compliance with the standards established under subsection (a) and all applicable laws.

(c) CERTIFICATION PROCESS.–

(1) ESTABLISHMENT.–The Nuclear Regulatory Commission shall establish a certification process to ensure that the Corporation complies with standards established under subsection (a).

(2) PERIODIC APPLICATION FOR CERTIFICATE OF COMPLIANCE.–The Corporation shall apply to the Nuclear Regulatory Commission for a certificate of compliance under paragraph (1) periodically, as determined by the Commission, but not less than every 5 years. The Commission shall review any such application and any determination made under subsection (b)(2) shall be based on the results of any such review.[403]

(3) TREATMENT OF CERTIFICATE OF COMPLIANCE.–The requirement for a certificate of compliance under paragraph (1) shall be in lieu of any requirement for a license for any gaseous diffusion facility of the Department leased by the Corporation.

(4) NRC REVIEW.–

(A) IN GENERAL.–The Nuclear Regulatory Commission, in consultation with the Environmental Protection Agency, shall review the operations of the Corporation with respect to any gaseous diffusion uranium enrichment facilities of the Department leased by the Corporation to ensure that public health and safety are adequately protected.

(B) ACCESS TO FACILITIES AND INFORMATION.– The Corporation and the Department shall cooperate fully with the Nuclear Regulatory Commission and the Environmental

---

[402]   Added by P.L. 102–486, 106 Stat. 2951 (1992), as amended by P.L. 104–134, Title III, Chapter 1, Subchapter A, § 3116(b)(3), 110 Stat. 1321–349 (1996); P.L. 105–362, Title II, § 1202, 112 Stat. 3292 (1998).

[403]   Amended by P.L. 104–134, Title III, Chapter 1, Subchapter A, § 3116(b)(3), 110 Stat. 1321–349 (1996); P.L. 105–362, Title XII, § 1202, 112 Stat. 3292 (1998).

Protection Agency and shall provide the Nuclear Regulatory Commission and the Environmental Protection Agency with the ready access to the facilities, personnel, and information the Nuclear Regulatory Commission and the Environmental Protection Agency consider necessary to carry out their responsibilities under this subsection. A contractor operating a Corporation facility for the Corporation shall provide the Nuclear Regulatory Commission and the Environmental Protection Agency with ready access to the facilities, personnel, and information of the contractor as the Nuclear Regulatory Commission and the Environmental Protection Agency consider necessary to carry out their responsibilities under this subsection.

(C) LIMITATION.–The Nuclear Regulatory Commission shall limit its finding under subsection (b)(2) to a determination of whether the facilities are in compliance with the standards established under subsection (a).

(d) REQUIREMENT FOR OPERATION.–The gaseous diffusion uranium enrichment facilities of the Department may not be operated by the Corporation unless the Nuclear Regulatory Commission, in consultation with the Environmental Protection Agency, makes a determination of compliance under subsection (b) or approves a plan prepared by the Department for achieving compliance required under subsection (b).

42 USC 2297f-1.

Sec. 1702. Licensing of Other Technologies

(a) IN GENERAL.–Corporation facilities using alternative technologies for uranium enrichment, including than AVLIS, shall be licensed under sections 53, 63, and 193.[404]

(b) COSTS FOR DECONTAMINATION AND DECOMMISSIONING.–The Corporation shall provide for the costs of decontamination and decommissioning of any Corporation facilities described in subsection (a) in accordance with the requirements of the amendments made by section 5 of the Solar, Wind, Waste, and Geothermal Power Production Act of 1990.

42 USC 2297f-2.

Sec. 1703. Regulation of Restricted Data

The Corporation shall be subject to this Act with respect to the use of, or access to, Restricted Data to the same extent as any private corporation.

Chapter 28–Decontamination and Decommissioning

42 USC 2297g.

Sec. 1801. Uranium Enrichment Decontamination and Decommissioning Fund

(a)[405] ESTABLISHMENT.–There is established in the Treasury of the United States an account to be known as the Uranium Enrichment Decontamination and Decommissioning Fund (referred to in this chapter as the "Fund"). The Fund, and any amounts deposited in it, including any interest earned thereon, shall be available to the Secretary subject to appropriations for the exclusive purpose of carrying out this chapter.

(b) ADMINISTRATION.–

(1) IN GENERAL.–The Secretary of the Treasury shall hold the Fund and, after consultation with the Secretary, annually report to the

---

[404]   Amended by P.L. 104–134, Title III, Chapter 1, Subchapter A, § 3116(b)(4), 110 Stat. 1321–349 (1996).

[405]   Added by P.L. 102–486, 106 Stat. 2953 (1992).

Congress on the financial condition and operations of the Fund during the preceding fiscal year.

(2) INVESTMENTS.–The Secretary of the Treasury shall invest amounts contained within the Fund in obligations of the United State–

(A) having maturities determined by the Secretary of the Treasury to be appropriate for what the Department determines to be the needs of the Fund; and

(B) bearing interest at rates determined to be appropriate by the Secretary of the Treasury, taking into consideration the current average market yield on outstanding marketable obligations of the United States with remaining periods to maturity comparable to these obligations.

42 USC 2297g-1.   **Sec. 1802. Deposits**

(a) AMOUNT.–The Fund shall consist of deposits in the amount of $518,233,333 per fiscal year (to be annually adjusted for inflation beginning on the date of the enactment of the Energy Policy Act of 1992 [October 24, 1992][406] using the Consumer Price Index for all-urban consumers published by the Department of Labor) as provided in this section.

(b) SOURCE.–Deposits described in subsection (a) shall be from the following sources:

(1) Sums collected pursuant to subsection (c).

(2) Appropriations made pursuant to subsection (d).

(c) SPECIAL ASSESSMENT.–The Secretary shall collect a special assessment from domestic utilities. The total amount collected for a fiscal year shall not exceed $160,000,000 (to be annually adjusted for inflation using the Consumer Price Index for all-urban consumers published by the Department of Labor). The amount collected from each utility pursuant to this subsection on for a fiscal year shall be in the same ratio to the amount required under subsection (a) to be deposited for such fiscal year as the total amount of separative work units such utility has purchased from the Department of Energy for the purpose of commercial electricity generation, before the date of the enactment of this title, bears to the total amount of separative work units purchased from the Department of Energy for all purposes (including units purchased or produced for defense purposes) before the date of the enactment of this title. For purposes of this subsection–

(1) a utility shall be considered to have purchased a separative work unit from the Department if such separative work unit was produced by the Department, but purchased by the utility from another source; and

(2) a utility shall not be considered to have purchased a separative work unit from the Department if such separative work unit was purchased by the utility, but sold to another source.

(d) AUTHORIZATION OF APPROPRIATIONS.–There are authorized to be appropriated to the Fund, for the period encompassing 15 years after the date of the enactment of this title, such sums as are necessary to ensure that the amount required under subsection (a) is deposited for each fiscal year.

(e) TERMINATION OF ASSESSMENTS.–The collection of amounts under subsection (c) shall cease after the earlier of–

(1) 16 years after the date of the enactment of this title; or

---

[406]   Amended by P.L. 105–388, 112 Stat. 3485 (1998).

(2) the collection of $2,260,000,000 (to be annually adjusted for inflation using the Consumer Price Index for all-urban consumers published by the Department of Labor) under such subsection.

(f) CONTINUATION OF DEPOSITS.–Except as provided in subsection (e), deposits shall continue to be made into the Fund under subsection (d) for the period specified in such subsection.

(g) TREATMENT OF ASSESSMENT.–Any special assessment levied under this section on domestic utilities for the decontamination and decommissioning of the Department's gaseous diffusion enrichment facilities shall be deemed a necessary and reasonable current cost of fuel and shall be fully recoverable in rates in all jurisdictions in the same manner as the utility's other fuel cost.

42 USC 2297g-2.

Sec. 1803. Department Facilities

(a) STUDY BY NATIONAL ACADEMY OF SCIENCES.–The National Academy of Sciences shall conduct a study and provide recommendations for reducing costs associated with decontamination and decommissioning, and shall report its findings to the Congress within 3 years after the date of the enactment of this title. Such report shall include a determination of the decontamination and decommissioning required for each facility shall identify alternative methods, using different technologies, shall include sit-specific surveys of the actual contamination, and shall provide estimated costs of those activities.

(b) PAYMENT OF DECONTAMINATION AND DECOMMISSIONING COSTS.–The costs of all decontamination and decommissioning activities of the Department shall be paid from the Fund until such time as the Secretary certifies and the Congress concurs, by law, that such activities are complete.

(c) PAYMENT OF REMEDIAL ACTION COSTS.–The annual cost of remedial action at the Department's gaseous diffusion facilities shall be paid from the Fund to the extent the amount available in the Fund is sufficient. To the extent the amount in the Fund is insufficient, the Department shall be responsible for the cost of remedial action. No provision of this title may be construed to relieve in any way the responsibility or liability of the Department for remedial action under applicable Federal and State laws and regulations.

42 USC 2297g-3.

Sec. 1804. Employee Provisions

All laborers and mechanics employed by contractors or subcontractors in the performance of decontamination or decommissioning of uranium enrichment facilities of the Department shall be paid wages at rates not less than those prevailing on projects of a similar character in the locality as determined by the Secretary of Labor in accordance with the Act of March 3, 1931 (known as the Davis-Bacon Act) (40 USC 276a et seq.). The Secretary of Labor shall have, with respect to the labor standards specified in this section, the authority and functions set forth in Reorganization Plan Numbered 14 of 1950 (15 FR 3176, 64 Stat. 1267) and the Act of June 13, 1934 (40 USC 276c). This section may not be construed to require the contracting out of activities associated with the decontamination or decommissioning of uranium enrichment facilities.

42 USC 2297g-4.

Sec. 1805. Reports to Congress

Within 3 years after the date of the enactment of this title, and at least once every 3 years thereafter, the Secretary shall report to the Congress on progress under this chapter. The 5th report submitted under this section shall contain recommendations of the Secretary for the reauthorization of the program and Fund under this title.

Title III–Rescissions and Offsets

Chapter 1–Energy and Water Development Uranium Enrichment Capacity

Subchapter A–United States Enrichment Corporation Privatization

42 USC 2011 note.

42 USC 2297h.

Sec. 3101. Short Title

This subchapter may be cited as the "USEC Privatization Act."[407]

Sec. 3102. Definitions

Except as provided in section 3112A. [42 USC 2297h-10a], for purposes of this subchapter:[408]

(1) The term "AVLIS" means atomic vapor laser isotope separation technology.

(2) The term "Corporation" means the United States Enrichment Corporation and, unless the context otherwise requires, includes the private corporation and any successor thereto following privatization.

(3) The term "gaseous diffusion plants" means the Paducah Gaseous Diffusion Plant at Paducah, Kentucky and the Portsmouth Gaseous Diffusion Plant at Piketon, Ohio.

(4) The term "highly enriched uranium" means uranium enriched to 20 percent or more of the uranium-235 isotope.

(5) The term "low-enriched uranium" means uranium enriched to less than 20 percent of the uranium-235 isotope, including that which is derived from highly enriched uranium.

(6) The term "low-level radioactive waste" has the meaning given such term in section 2(9) of the Low-Level Radioactive Waste Policy Act (42 USC 2021b(9)).

(7) The term "private corporation" means the corporation established under section 3105.

(8) The term "privatization" means the transfer of ownership of the Corporation to private investors.

(9) The term "privatization date" means the date on which 100 percent of the ownership of the Corporation has been transferred to private investors.

(10) The term "public offering" means an underwritten offering to the public of the common stock of the private corporation pursuant to section 3104.

(11) The "Russian HEU Agreement" means the Agreement Between the Government of the United States of America and the Government of the Russian Federation Concerning the Disposition of Highly Enriched Uranium Extracted from Nuclear Weapons, dated February 18, 1993.

(12) The term "Secretary" means the Secretary of Energy.

(13) The "Suspension Agreement" means the Agreement to Suspend the Antidumping Investigation on Uranium from the Russian Federation, as amended.

(14) The term "uranium enrichment" means the separation of uranium of a given isotopic content into two components, one having

[407] Sections 3102 through 3115 were enacted as part of the USEC Privatization Act, P.L. 104–134, Title III, Chapter 1, Subchapter A, 110 Stat. 1321–335 (1996) and not as part of the Atomic Energy Act of 1954, which generally comprises this Chapter.
[408] Amended by P.L. 110–329, 122 Stat. 3647 (2008).

a higher percentage of a fissile isotope and one having a lower percentage.

42 USC 2297h-1.

Sec. 3103. Sale of the Corporation

(a) Authorization.–The Board of Directors of the Corporation, with the approval of the Secretary of the Treasury, shall transfer the interest of the United States in the United States Enrichment Corporation to the private sector in a manner that provides for the long-term viability of the Corporation, provides for the continuation by the Corporation of the operation of the Department of Energy's gaseous diffusion plants, provides for the protection of the public interest in maintaining a reliable and economical domestic source of uranium mining, enrichment and conversion services, and, to the extent not inconsistent with such purposes, secures the maximum proceeds to the United States.

(b) Proceeds.–Proceeds from the sale of the United States' interest in the Corporation shall be deposited in the general fund of the Treasury.

42 USC 2297h-2.

Sec. 3104. Method of Sale

(a) Authorization.–The Board of Directors of the Corporation, with the approval of the Secretary of the Treasury, shall transfer ownership of the assets and obligations of the Corporation to the private corporation established under section 3105 (which may be consummated through a merger or consolidation effected in accordance with, and having the effects provided under, the law of the State of incorporation of the private corporation, as if the Corporation were incorporated thereunder).

(b) Board Determination.–The Board, with the approval of the Secretary of the Treasury, shall select the method of transfer and establish terms and conditions for the transfer that will provide the maximum proceeds to the Treasury of the United States and will provide for the long-term viability of the private corporation, the continued operation of the gaseous diffusion plants, and the public interest in maintaining reliable and economical domestic uranium mining and enrichment industries.

(c) Adequate Proceeds.–The Secretary of the Treasury shall not allow the privatization of the Corporation unless before the sale date the Secretary of the Treasury determines that the method of transfer will provide the maximum proceeds to the Treasury consistent with the principles set forth in section 3103(a).

(d) Application of Securities Laws.–Any offering or sale of securities by the private corporation shall be subject to the Securities Act of 1933 (15 USC 77a *et seq.*), the Securities Exchange Act of 1934 (15 USC 78a *et seq.*), and the provisions of the Constitution and laws of any State, territory, or possession of the United States relating to transactions in securities.

(e) Expenses.–Expenses of privatization shall be paid from Corporation revenue accounts in the United States Treasury.

42 USC 2297h-3.

Sec. 3105. Establishment of Private Corporation

(a) Incorporation.–

(1) The directors of the Corporation shall establish a private for-profit corporation under the laws of a State for the purpose of receiving the assets and obligations of the Corporation at privatization and continuing the business operations of the Corporation following privatization.

(2) The directors of the Corporation may serve as incorporators of the private corporation and shall take all steps necessary to establish the private corporation, including the filing of articles of incorporation consistent with the provisions of this subchapter.

(3) Employees and officers of the Corporation (including members of the Board of Directors) acting in accordance with this section on behalf of the private corporation shall be deemed to be acting in their official capacities as employees or officers of the Corporation for purposes of section 205 of title 18, United States Code.

(b) Status of the Private Corporation.–

(1) The private corporation shall not be an agency, instrumentality, or establishment of the United States, a Government corporation, or a Government-controlled corporation.

(2) Except as otherwise provided by this subchapter, financial obligations of the private corporation shall not be obligations of, or guaranteed as to principal or interest by, the Corporation or the United States, and the obligations shall so plainly state.

(3) No action under section 1491 of title 28, United States Code, shall be allowable against the United States based on actions of the private corporation.

(c) Application of Post-Government Employment Restrictions.–

Beginning on the privatization date, the restrictions stated in section 207(a), (b), (c), and (d) of title 18, United States Code, shall not apply to the acts of an individual done in carrying out official duties as a director, officer, or employee of the private corporation, if the individual was an officer or employee of the Corporation (including a director) continuously during the 45 days prior to the privatization date.

(d) Dissolution.–In the event that the privatization does not occur, the Corporation will provide for the dissolution of the private corporation within one year of the private corporation's incorporation unless the Secretary of the Treasury or his delegate, upon the Corporation's request, agrees to delay any such dissolution for an additional year.

42 USC 2297h-4.

Sec. 3106. Transfers to the Private Corporation

Concurrent with privatization, the Corporation shall transfer to the private corporation–

(1) the lease of the gaseous diffusion plants in accordance with section 3107,

(2) all personal property and inventories of the Corporation,

(3) all contracts, agreements, and leases under section 3108(a),

(4) the Corporation's right to purchase power from the Secretary under section 3108(b),

(5) such funds in accounts of the Corporation held by the Treasury or on deposit with any bank or other financial institution as approved by the Secretary of the Treasury, and

Records.

(6) all of the Corporation's records, including all of the papers and other documentary materials, regardless of physical form or characteristics, made or received by the Corporation.

42 USC 2297h-5.

Sec. 3107. Leasing of Gaseous Diffusion Facilities

(a) Transfer of Lease.–Concurrent with privatization, the Corporation shall transfer to the private corporation the lease of the gaseous diffusion plants and related property for the remainder of the term of such lease in accordance with the terms of such lease.

(b) Renewal.–The private corporation shall have the exclusive option to lease the gaseous diffusion plants and related property for additional periods following the expiration of the initial term of the lease.

(c) Exclusion of Facilities for Production of Highly Enriched Uranium.–The Secretary shall not lease to the private corporation any facilities necessary for the production of highly enriched uranium but may, subject to the requirements of the Atomic Energy Act of 1954 (42

USC 2011 *et seq.*), grant the Corporation access to such facilities for purposes other than the production of highly enriched uranium.

(d) DOE Responsibility for Preexisting Conditions.–The payment of any costs of decontamination and decommissioning, response actions, or corrective actions with respect to conditions existing before July 1, 1993, at the gaseous diffusion plants shall remain the sole responsibility of the Secretary.

(e) Environmental Audit.–For purposes of subsection (d), the conditions existing before July 1, 1993, at the gaseous diffusion plants shall be determined from the environmental audit conducted pursuant to section 1403(e) of the Atomic Energy Act of 1954 (42 USC 2297c-2(e)).

(f) Treatment Under Price-Anderson Provisions.–Any lease executed between the Secretary and the Corporation or the private corporation, and any extension or renewal thereof, under this section shall be deemed to be a contract for purposes of section 170d. of the Atomic Energy Act of 1954 (42 USC 2210(d)).

(g) Waiver of EIS Requirement.–The execution or transfer of the lease between the Secretary and the Corporation or the private corporation, and any extension or renewal thereof, shall not be considered to be a major Federal action significantly affecting the quality of the human environment for purposes of section 102 of the National Environmental Policy Act of 1969 (42 USC 4332).

(h) Maintenance of Security

(1) In General–With respect to the Paducah Gaseous Diffusion Plant, Kentucky, and the Portsmouth Gaseous Diffusion Plant, Ohio, the guidelines relating to the authority of the Department of Energy's contractors (including any Federal agency, or private entity operating a gaseous diffusion plant under a contract or lease with the Department of Energy) and any subcontractor (at any tier) to carry firearms and make arrests in providing security at Federal installations, issued under section 161k. of the Atomic Energy Act of 1954 (42 USC 2201k.) shall require, at a minimum, the presence of all security police officers[409] carrying sidearms at all times to ensure maintenance of security at the gaseous diffusion plants (whether a gaseous diffusion plant is operated directly by a Federal agency or by a private entity under a contract or lease with a Federal agency).

(2)[410] Funding

(A) The costs of arming and providing arrest authority to the security policy officers required under paragraph (1) shall be paid as follows:

(i) the Department of Energy (the "Department") shall pay the percentage of the costs equal to the percentage of the total number of employees at the gaseous diffusion plant who are: (I) employees of the Department or the contractor or subcontractors of the Department; or (II) employees of the private entity leasing the gaseous diffusion plant who perform work on behalf of the Department (including employees of a contractor or subcontractor of the private entity); and

(ii) the private entity leasing the gaseous diffusion plant shall pay the percentage of the costs equal to the percentage of the total number of employees at the gaseous diffusion plant

---

[409]   Amended by P.L. 105–245, Title III, § 310, 112 Stat. 1853 (1998).
[410]   Added by P.L. 105–245, Title III, § 310, 112 Stat. 1853 (1998).

who are employees of the private entity (including employees of a contractor or subcontractor) other than those employees who perform work for the Department.

(B) Neither the private entity leasing the gaseous diffusion plant nor the Department shall reduce its payments under any contract or lease or take other action to offset its share of the costs referred to in subparagraph (A), and the Department shall not reimburse the private entity for the entity's share of these costs.

(C) Nothing in this subsection shall alter the Department's responsibilities to pay the safety, safeguards and security costs associated with the Department's highly enriched uranium activities.[411]

42 USC 2297h-6.

Sec. 3108. Transfer of Contracts

(a) Transfer of Contracts.–Concurrent with privatization, the Corporation shall transfer to the private corporation all contracts, agreements, and leases, including all uranium enrichment contracts, that were–

(1) transferred by the Secretary to the Corporation pursuant to section 1401(b) of the Atomic Energy Act of 1954 (42 USC 2297c(b)), or

(2) entered into by the Corporation before the privatization date.

(b) Nontransferable Power Contracts.–The Corporation shall transfer to the private corporation the right to purchase power from the Secretary under the power purchase contracts for the gaseous diffusion plants executed by the Secretary before July 1, 1993. The Secretary shall continue to receive power for the gaseous diffusion plants under such contracts and shall continue to resell such power to the private corporation at cost during the term of such contracts.

(c) Effect of Transfer.–(1) Notwithstanding subsection (a), the United States shall remain obligated to the parties to the contracts, agreements, and leases transferred under subsection (a) for the performance of its obligations under such contracts, agreements, or leases during their terms. Performance of such obligations by the private corporation shall be considered performance by the United States.

(2) If a contract, agreement, or lease transferred under subsection (a) is terminated, extended, or materially amended after the privatization date–

(A) the private corporation shall be responsible for any obligation arising under such contract, agreement, or lease after any extension or material amendment, and

(B) the United States shall be responsible for any obligation arising under the contract, agreement, or lease before the termination, extension, or material amendment.

(3) The private corporation shall reimburse the United States for any amount paid by the United States under a settlement agreement entered into with the consent of the private corporation or under a judgment, if the settlement or judgment–

(A) arises out of an obligation under a contract, agreement, or lease transferred under subsection (a), and

---

[411]   Amended by P.L. 104–134, Title III, Chapter 1, Subchapter A, § 3107, 110 Stat. 1321–338 (1996); P.L. 105–62, Title V, § 511, 111 Stat. 1341 (1997); P.L. 105–245, Title III, § 310, 112 Stat. 1853 (1998).

(B) arises out of actions of the private corporation between the privatization date and the date of a termination, extension, or material amendment of such contract, agreement, or lease.

(d) Pricing.–The Corporation may establish prices for its products, materials, and services provided to customers on a basis that will allow it to attain the normal business objectives of a profit making corporation.

42 USC 2297h-7. **Sec. 3109. Liabilities**

(a) Liability of the United States.–

(1) Except as otherwise provided in this subchapter, all liabilities arising out of the operation of the uranium enrichment enterprise before July 1, 1993, shall remain the direct liabilities of the Secretary.

(2) Except as provided in subsection (a)(3) or otherwise provided in a memorandum of agreement entered into by the Corporation and the Office of Management and Budget prior to the privatization date, all liabilities arising out of the operation of the Corporation between July 1, 1993, and the privatization date shall remain the direct liabilities of the United States.

(3) All liabilities arising out of the disposal of depleted uranium generated by the Corporation between July 1, 1993, and the privatization date shall become the direct liabilities of the Secretary.

(4) Any stated or implied consent for the United States, or any agent or officer of the United States, to be sued by any person for any legal, equitable, or other relief with respect to any claim arising from any action taken by any agent or officer of the United States in connection with the privatization of the Corporation is hereby withdrawn.

(5) To the extent that any claim against the United States under this section is of the type otherwise required by Federal statute or regulation to be presented to a Federal agency or official for adjudication or review, such claim shall be presented to the Department of Energy in accordance with procedures to be established by the Secretary. Nothing in this paragraph shall be construed to impose on the Department of Energy liability to pay any claim presented pursuant to this paragraph.

(6) The Attorney General shall represent the United States in any action seeking to impose liability under this subsection.

(b) Liability of the Corporation.–Notwithstanding any provision of any agreement to which the Corporation is a party, the Corporation shall not be considered in breach, default, or violation of any agreement because of the transfer of such agreement to the private corporation under section 3108 or any other action the Corporation is required to take under this subchapter.

(c) Liability of the Private Corporation.–Except as provided in this subchapter, the private corporation shall be liable for any liabilities arising out of its operations after the privatization date.

(d) Liability of Officers and Directors.–

(1) No officer, director, employee, or agent of the Corporation shall be liable in any civil proceeding to any party in connection with any action taken in connection with the privatization if, with respect to the subject matter of the action, suit, or proceeding, such person was acting within the scope of his employment.

(2) This subsection shall not apply to claims arising under the Securities Act of 1933 (15 USC 77a. *et seq*.), the Securities Exchange Act of 1934 (15 USC 78a. *et seq*.), or under the Constitution or laws of any State, territory, or possession of the United States relating to transactions in securities.

42 USC 2297h-8. **Sec. 3110. Employee Protections**

(a) Contractor Employees.–

(1) Privatization shall not diminish the accrued, vested pension benefits of employees of the Corporation's operating contractor at the two gaseous diffusion plants.

(2) In the event that the private corporation terminates or changes the contractor at either or both of the gaseous diffusion plants, the plan sponsor or other appropriate fiduciary of the pension plan covering employees of the prior operating contractor shall arrange for the transfer of all plan assets and liabilities relating to accrued pension benefits of such plan's participants and beneficiaries from such plant to a pension plan sponsored by the new contractor or the private corporation or a joint labor-management plan, as the case may be.

(3) In addition to any obligations arising under the National Labor Relations Act (29 USC 151 *et seq.*), any employer (including the private corporation if it operates a gaseous diffusion plant without a contractor or any contractor of the private corporation) at a gaseous diffusion plant shall–

(A) abide by the terms of any unexpired collective bargaining agreement covering employees in bargaining units at the plant and in effect on the privatization date until the stated expiration or termination date of the agreement; or

(B) in the event a collective bargaining agreement is not in effect upon the privatization date, have the same bargaining obligations under section 8(d) of the National Labor Relations Act (29 USC 158(d)) as it had immediately before the privatization date.

(4) If the private corporation replaces its operating contractor at a gaseous diffusion plant, the new employer (including the new contractor or the private corporation if it operates a gaseous diffusion plant without a contractor) shall–

(A) offer employment to non-management employees of the predecessor contractor to the extent that their jobs still exist or they are qualified for new jobs, and

(B) abide by the terms of the predecessor contractor's collective bargaining agreement until the agreement expires or a new agreement is signed.

(5) In the event of a plant closing or mass layoff (as such terms are defined in section 2101(a) (2) and (3) of title 29, United States Code) at either of the gaseous diffusion plants, the Secretary of Energy shall treat any adversely affected employee of an operating contractor at either plant who was an employee at such plant on July 1, 1993, as a Department of Energy employee for purposes of sections 3161 and 3162 of the National Defense Authorization Act for Fiscal Year 1993 (42 USC 7274h-7274i).

(6)(A) The Secretary and the private corporation shall cause the post-retirement health benefits plan provider (or its successor) to continue to provide benefits for eligible persons, as described under subparagraph (B), employed by an operating contractor at either of the gaseous diffusion plants in an economically efficient manner and at substantially the same level of coverage as eligible retirees are entitled to receive on the privatization date.

(B) Persons eligible for coverage under subparagraph (A) shall be limited to:

(i) persons who retired from active employment at one of the gaseous diffusion plants on or before the privatization date as vested participants in a pension plan maintained either by the Corporation's operating contractor or by a contractor employed prior to July 1, 1993, by the Department of Energy to operate a gaseous diffusion plant; and

(ii) persons who are employed by the Corporation's operating contractor on or before the privatization date and are vested participants in a pension plan maintained either by the Corporation's operating contractor or by a contractor employed prior to July 1, 1993, by the Department of Energy to operate a gaseous diffusion plant.

(C) The Secretary shall fund the entire cost of post-retirement health benefits for persons who retired from employment with an operating contractor prior to July 1, 1993.

(D) The Secretary and the Corporation shall fund the cost of post-retirement health benefits for persons who retire from employment with an operating contractor on or after July 1, 1993, in proportion to the retired person's years and months of service at a gaseous diffusion plant under their respective management.

(7)(A) Any suit under this subsection alleging a violation of an agreement between an employer and a labor organization shall be brought in accordance with section 301 of the Labor Management Relations Act (29 USC 185).

(B) Any charge under this subsection alleging an unfair labor practice violative of section 8 of the National Labor Relations Act (29 USC 158) shall be pursued in accordance with section 10 of the National Labor Relations Act (29 USC 160).

(C) Any suit alleging a violation of any provision of this subsection, to the extent it does not allege a violation of the National Labor Relations Act, may be brought in any district court of the United States having jurisdiction over the parties, without regard to the amount in controversy or the citizenship of the parties.

Deadline.
Ohio.
Kentucky.

(8)[412] CONTINUITY OF BENEFITS–To the extend appropriations are provided in advance for this purpose or are otherwise available, not later than 30 days after the date of enactment of this paragraph, the Secretary shall implement such actions as are necessary to ensure that any employee who–

(A) is involved in providing infrastructure or environmental remediation services at the Portsmouth, Ohio, or the Paducah, Kentucky, Gaseous Diffusion Plant;

(B) has been an employee of the Department of Energy's predecessor management and integrating contractor (or its first or second tier subcontractors), or of the Corporation, at the Portsmouth, Ohio, or the Paducah, Kentucky, facility; and

(C) was eligible as of April 1, 2005, to participate in or transfer into the Multiple Employer Pension Plan or the associated multiple employer retiree health care benefit plans, as defined in those plans,

shall continue to be eligible to participate in or transfer into such pension or health care benefit plans.

(b) Former Federal Employees.–

---

[412]   Added by P.L. 109–58, § 633, 119 Stat. 790 (2005).

(1)(A) An employee of the Corporation that was subject to either the Civil Service Retirement System (referred to in this section as "CSRS") or the Federal Employees' Retirement System (referred to in this section as "FERS") on the day immediately preceding the privatization date shall elect–

(i) to retain the employee's coverage under either CSRS or FERS, as applicable, in lieu of coverage by the Corporation's retirement system, or

(ii) to receive a deferred annuity or lump-sum benefit payable to a terminated employee under CSRS or FERS, as applicable.

(B) An employee that makes the election under subparagraph (A)(ii) shall have the option to transfer the balance in the employee's Thrift Savings Plan account to a defined contribution plan under the Corporation's retirement system, consistent with applicable law and the terms of the Corporation's defined contribution plan.

(2) The Corporation shall pay to the Civil Service Retirement and Disability Fund–

(A) such employee deductions and agency contributions as are required by sections 8334, 8422, and 8423 of title 5, United States Code, for those employees who elect to retain their coverage under either CSRS or FERS pursuant to paragraph (1);

(B) such additional agency contributions as are determined necessary by the Office of Personnel Management to pay, in combination with the sums under subparagraph (A), the "normal cost" (determined using dynamic assumptions) of retirement benefits for those employees who elect to retain their coverage under CSRS pursuant to paragraph (1), with the concept of "normal cost" being used consistent with generally accepted actuarial standards and principles; and

(C) such additional amounts, not to exceed two percent of the amounts under subparagraphs (A) and (B), as are determined necessary by the Office of Personnel Management to pay the cost of administering retirement benefits for employees who retire from the Corporation after the privatization date under either CSRS or FERS, for their survivors, and for survivors of employees of the Corporation who die after the privatization date (which amounts shall be available to the Office of Personnel Management as provided in section 8348(a)(1)(B) of title 5, United States Code).

(3) The Corporation shall pay to the Thrift Savings Fund such employee and agency contributions as are required or authorized by sections 8432 and 8351 of Title 5, for employees who elect to retain their coverage under CSRS or FERS pursuant to paragraph (1).[413]

(4) Any employee of the Corporation who was subject to the Federal Employee Health Benefits Program (referred to in this section as "FEHBP") on the day immediately preceding the privatization date and who elects to retain coverage under either CSRS or FERS pursuant to paragraph (1) shall have the option to receive health benefits from a health benefit plan established by the Corporation or

---

[413]   Amended by P.L. 104–206, Title III, 110 Stat. 2995 (1996).

to continue without interruption coverage under the FEHBP, in lieu of coverage by the Corporation's health benefit system.

(5) The Corporation shall pay to the Employees Health Benefits Fund–

(A) such employee deductions and agency contributions as are required by section 8906(a)-(f) of title 5, United States Code, for those employees who elect to retain their coverage under FEHBP pursuant to paragraph (4); and

(B) such amounts as are determined necessary by the Office of Personnel Management under paragraph (6) to reimburse the Office of Personnel Management for contributions under section 8906(g)(1) of title 5, United States Code, for those employees who elect to retain their coverage under FEHBP pursuant to paragraph (4).

(6) The amounts required under paragraph (5)(B) shall pay the Government contributions for retired employees who retire from the Corporation after the privatization date under either CSRS or FERS, for survivors of such retired employees, and for survivors of employees of the Corporation who die after the privatization date, with said amounts prorated to reflect only that portion of the total service of such employees and retired persons that was performed for the Corporation after the privatization date.

42 USC 2297h-9.   ### Sec. 3111. Ownership Limitations

(a) Securities Limitations.–No director, officer, or employee of the Corporation may acquire any securities, or any rights to acquire any securities of the private corporation on terms more favorable than those offered to the general public–

(1) in a public offering designed to transfer ownership of the Corporation to private investors,

(2) pursuant to any agreement, arrangement, or understanding entered into before the privatization date, or

(3) before the election of the directors of the private corporation.

(b) Ownership Limitation.–Immediately following the consummation of the transaction or series of transactions pursuant to which 100 percent of the ownership of the Corporation is transferred to private investors, and for a period of three years thereafter, no person may acquire, directly or indirectly, beneficial ownership of securities representing more than 10 percent of the total votes of all outstanding voting securities of the Corporation. The foregoing limitation shall not apply to–

(1) any employee stock ownership plan of the Corporation,

(2) members of the underwriting syndicate purchasing shares in stabilization transactions in connection with the privatization, or

(3) in the case of shares beneficially held in the ordinary course of business for others, any commercial bank, broker-dealer, or clearing agency.

42 USC 2297h-10.   ### Sec. 3112. Uranium Transfers and Sales

(a) Transfers and Sales by the Secretary.–The Secretary shall not provide enrichment services or transfer or sell any uranium (including natural uranium concentrates, natural uranium hexafluoride, or enriched uranium in any form) to any person except as consistent with this section.

(b) Russian HEU.

(1) On or before December 31, 1996, the United States Executive Agent under the Russian HEU Agreement shall transfer to the Secretary without charge title to an amount of uranium hexafluoride equivalent to the natural uranium component of low-enriched uranium derived from at least 18 metric tons of highly enriched uranium

purchased from the Russian Executive Agent under the Russian HEU Agreement. The quantity of such uranium hexafluoride delivered to the Secretary shall be based on a tails assay of 0.30 U^{235}. Uranium hexafluoride transferred to the Secretary pursuant to this paragraph shall be deemed under United States law for all purposes to be of Russian origin.

(2) Within seven years of the date of enactment of this Act, the Secretary shall sell, and receive payment for, the uranium hexafluoride transferred to the Secretary pursuant to paragraph (1). Such uranium hexafluoride shall be sold–

(A) at any time for use in the United States for the purpose of overfeeding;

(B) at any time for end use outside the United States;

(C) in 1995 and 1996 to the Russian Executive Agent at the purchase price for use in matched sales pursuant to the Suspension Agreement; or,

(D) in calendar year 2001 for consumption by end users in the United States not prior to January 1, 2002, in volumes not to exceed 3,000,000 pounds U_3O_8 equivalent per year.

(3) With respect to all enriched uranium delivered to the United States Executive Agent under the Russian HEU Agreement on or after January 1, 1997, the United States Executive Agent shall, upon request of the Russian Executive Agent, enter into an agreement to deliver concurrently to the Russian Executive Agent an amount of uranium hexafluoride equivalent to the natural uranium component of such uranium. An agreement executed pursuant to a request of the Russian Executive Agent, as contemplated in this paragraph, may pertain to any deliveries due during any period remaining under the Russian HEU Agreement. The quantity of such uranium hexafluoride delivered to the Russian Executive Agent shall be based on a tails assay of 0.30 U^{235}. Title to uranium hexafluoride delivered to the Russian Executive Agent pursuant to this paragraph shall transfer to the Russian Executive Agent upon delivery of such material to the Russian Executive Agent, with such delivery to take place at a North American facility designated by the Russian Executive Agent. Uranium hexafluoride delivered to the Russian Executive Agent pursuant to this paragraph shall be deemed under U.S. law for all purposes to be of Russian origin. Such uranium hexafluoride may be sold to any person or entity for delivery and use in the United States only as permitted in subsections (b)(5), (b)(6) and (b)(7) of this section.

(4) In the event that the Russian Executive Agent does not exercise its right to enter into an agreement to take delivery of the natural uranium component of any low-enriched uranium, as contemplated in paragraph (3), within 90 days of the date such low-enriched uranium is delivered to the United States Executive Agent, or upon request of the Russian Executive Agent, then the United States Executive Agent shall engage an independent entity through a competitive selection process to auction an amount of uranium hexafluoride or U_3O_8 (in the event that the conversion component of such hexafluoride has previously been sold) equivalent to the natural uranium component of such low-enriched uranium. An agreement executed pursuant to a request of the Russian Executive Agent, as contemplated in this paragraph, may pertain to any deliveries due during any period remaining under the Russian HEU Agreement. Such independent entity shall sell such uranium hexafluoride in one

or more lots to any person or entity to maximize the proceeds from such sales, for disposition consistent with the limitations set forth in this subsection. The independent entity shall pay to the Russian Executive Agent the proceeds of any such auction less all reasonable transaction and other administrative costs. The quantity of such uranium hexafluoride auctioned shall be based on a tails assay of 0.30 U^{235}. Title to uranium hexafluoride auctioned pursuant to this paragraph shall transfer to the buyer of such material upon delivery of such material to the buyer. Uranium hexafluoride auctioned pursuant to this paragraph shall be deemed under United States law for all purposes to be of Russian origin.

(5) Except as provided in paragraphs (6) and (7), uranium hexafluoride delivered to the Russian Executive Agent under paragraph (3) or auctioned pursuant to paragraph (4), may not be delivered for consumption by end users in the United States either directly or indirectly prior to January 1, 1998, and thereafter only in accordance with the following schedule:

Annual Maximum Deliveries to End Users

Year:	(millions lbs. U_3O_8 equivalent)
1998	2
1999	4
2000	6
2001	8
2002	10
2003	12
2004	14
2005	16
2006	17
2007	18
2008	19
2009 and each year thereafter	20.

(6) Uranium hexafluoride delivered to the Russian Executive Agent under paragraph (3) or auctioned pursuant to paragraph (4) may be sold at any time as Russian-origin natural uranium in a matched sale pursuant to the Suspension Agreement, and in such case shall not be counted against the annual maximum deliveries set forth in paragraph (5).

(7) Uranium hexafluoride delivered to the Russian Executive Agent under paragraph (3) or auctioned pursuant to paragraph (4) may be sold at any time for use in the United States for the purpose of overfeeding in the operations of enrichment facilities.

(8) Nothing in this subsection (b) shall restrict the sale of the conversion component of such uranium hexafluoride.

(9) The Secretary of Commerce shall have responsibility for the administration and enforcement of the limitations set forth in this subsection. The Secretary of Commerce may require any person to provide any certifications, information, or take any action that may be necessary to enforce these limitations. The United States Customs Service shall maintain and provide any information required by the Secretary of Commerce and shall take any action requested by the Secretary of Commerce which is necessary for the administration and

enforcement of the uranium delivery limitations set forth in this section.

(10) The President shall monitor the actions of the United States Executive Agent under the Russian HEU Agreement and shall report to the Congress not later than December 31 of each year on the effect the low-enriched uranium delivered under the Russian HEU Agreement is having on the domestic uranium mining, conversion, and enrichment industries, and the operation of the gaseous diffusion plants. Such report shall include a description of actions taken or proposed to be taken by the President to prevent or mitigate any material adverse impact on such industries or any loss of employment at the gaseous diffusion plants as a result of the Russian HEU Agreement.

(c) Transfers to the Corporation.–(1) The Secretary shall transfer to the Corporation without charge up to 50 metric tons of enriched uranium and up to 7,000 metric tons of natural uranium from the Department of Energy's stockpile, subject to the restrictions in subsection (c)(2).

(2) The Corporation shall not deliver for commercial end use in the United States–

(A) any of the uranium transferred under this subsection before January 1, 1998;

(B) more than 10 percent of the uranium (by uranium hexafluoride equivalent content) transferred under this subsection or more than 4,000,000 pounds, whichever is less, in any calendar year after 1997; or

(C) more than 800,000 separative work units contained in low-enriched uranium transferred under this subsection in any calendar year.

(d) Inventory Sales.–(1) In addition to the transfers authorized under subsections (c) and (e), the Secretary may, from time to time, sell natural and low-enriched uranium (including low-enriched uranium derived from highly enriched uranium) from the Department of Energy's stockpile.

(2) Except as provided in subsections (b), (c), and (e), no sale or transfer of natural or low-enriched uranium shall be made unless–

(A) the President determines that the material is not necessary for national security needs,

(B) the Secretary determines that the sale of the material will not have an adverse material impact on the domestic uranium mining, conversion, or enrichment industry, taking into account the sales of uranium under the Russian HEU Agreement and the Suspension Agreement, and

(C) the price paid to the Secretary will not be less than the fair market value of the material.

(e) Government Transfers.–Notwithstanding subsection (d)(2), the Secretary may transfer or sell enriched uranium–

(1) to a Federal agency if the material is transferred for the use of the receiving agency without any resale or transfer to another entity and the material does not meet commercial specifications;

(2) to any person for national security purposes, as determined by the Secretary; or

(3) to any State or local agency or nonprofit, charitable, or educational institution for use other than the generation of electricity for commercial use.

(f) Savings Provision.–Nothing in this subchapter shall be read to modify the terms of the Russian HEU Agreement.

42 USC 2297h-10a.

Sec. 3112A. Incentives for Additional Downblending of Highly Enriched Uranium by the Russian Federation[414]

(a) DEFINITIONS.–In this section:

(1) COMPLETION OF THE RUSSIAN HEU AGREEMENT.– The term 'completion of the Russian HEU Agreement' means the importation into the United States from the Russian Federation pursuant to the Russian HEU Agreement of uranium derived from the downblending of not less than 500 metric tons of highly enriched uranium of weapons origin.

(2) DOWNBLENDING.–The term 'downblending' means processing highly enriched uranium into a uranium product in any form in which the uranium contains less than 20 percent uranium-235.

(3) HIGHLY ENRICHED URANIUM.–The term 'highly enriched uranium' has the meaning given that term in section 3102(4).

(4) HIGHLY ENRICHED URANIUM OF WEAPONS ORIGIN.– The term 'highly enriched uranium of weapons origin' means highly enriched uranium that–

(A) contains 90 percent or more uranium-235; and

(B) is verified by the Secretary of Energy to be of weapons origin.

(5) LOW-ENRICHED URANIUM.–The term 'low-enriched uranium' means a uranium product in any form, including uranium hexafluoride (UF6) and uranium oxide (UO2), in which the uranium contains less than 20 percent uranium-235, including natural uranium, without regard to whether the uranium is incorporated into fuel rods or complete fuel assemblies.

(6) RUSSIAN HEU AGREEMENT.–The term 'Russian HEU Agreement' has the meaning given that term in section 3102(11).

(7) URANIUM-235.–The term 'uranium-235' means the isotope 235U.

(b) STATEMENT OF POLICY.–It is the policy of the United States to support the continued downblending of highly enriched uranium of weapons origin in the Russian Federation in order to protect the essential security interests of the United States with respect to the nonproliferation of nuclear weapons.

(c) PROMOTION OF DOWNBLENDING OF RUSSIAN HIGHLY ENRICHED URANIUM.–

(1) COMPLETION OF THE RUSSIAN HEU AGREEMENT.– Prior to the completion of the Russian HEU Agreement, the importation into the United States of low-enriched uranium, including low-enriched uranium obtained under contracts for separative work units, that is produced in the Russian Federation and is not imported pursuant to the Russian HEU Agreement, may not exceed the following amounts:

(A) In the 4-year period beginning with calendar year 2008, 16,559 kilograms.

(B) In calendar year 2012, 24,839 kilograms.

(C) In calendar year 2013 and each calendar year thereafter through the calendar year of the completion of the Russian HEU Agreement, 41,398 kilograms.

---

[414]   Added by P.L. 110–329, § 8118, 122 Stat. 3647 (2008).

(2) INCENTIVES TO CONTINUE DOWNBLENDING RUSSIAN HIGHLY ENRICHED URANIUM AFTER THE COMPLETION OF THE RUSSIAN HEU AGREEMENT.–

(A) IN GENERAL.–After the completion of the Russian HEU Agreement, the importation into the United States of low-enriched uranium, including low-enriched uranium obtained under contracts for separative work units, that is produced in the Russian Federation, whether or not such low-enriched uranium is derived from highly enriched uranium of weapons origin, may not exceed–

(i) in calendar year 2014, 485,279 kilograms;

(ii) in calendar year 2015, 455,142 kilograms;

(iii) in calendar year 2016, 480,146 kilograms;

(iv) in calendar year 2017, 490,710 kilograms;

(v) in calendar year 2018, 492,731 kilograms;

(vi) in calendar year 2019, 509,058 kilograms; and

(vii) in calendar year 2020, 514,754 kilograms.

(B) ADDITIONAL IMPORTS IN EXCHANGE FOR A COMMITMENT TO DOWNBLEND AN ADDITIONAL 300 METRIC TONS OF HIGHLY ENRICHED URANIUM.–

(i) IN GENERAL.–In addition to the amount authorized to be imported under subparagraph (A) and except as provided in clause (ii), if the Russian Federation enters into a bilateral agreement with the United States under which the Russian Federation agrees to downblend an additional 300 metric tons of highly enriched uranium after the completion of the Russian HEU Agreement, 4 kilograms of low-enriched uranium, whether or not such low-enriched uranium is derived from highly enriched uranium of weapons origin and including low-enriched uranium obtained under contracts for separative work units, may be imported in a calendar year for every 1 kilogram of Russian highly enriched uranium of weapons origin that was downblended in the preceding calendar year, subject to the verification of the Secretary of Energy under paragraph (10).

(ii) MAXIMUM ANNUAL IMPORTS.–Not more than 120,000 kilograms of low-enriched uranium may be imported in a calendar year under clause (i).

(3) EXCEPTIONS.–The import limitations described in paragraphs (1) and (2) shall not apply to low-enriched uranium produced in the Russian Federation that is imported into the United States–

(A) for use in the initial core of a new nuclear reactor;

(B) for processing and to be certified for reexportation and not for consumption in the United States; or

(C) to be added to the inventory of the Department of Energy.

(4) LIMITED WAIVER AUTHORITY.–

Deadline.

(A) IN GENERAL.–Notwithstanding paragraph (1)(C), if the completion of the Russian HEU Agreement does not occur before December 31, 2013, the import limitations under paragraph (1)(C) shall be waived, and low-enriched uranium may be imported into the United States in the quantities specified in paragraph (2) in a calendar year after 2013, if–

(i) the Secretary of Energy and the Secretary of State jointly determine that–

(I) the failure of the completion of the Russian HEU Agreement arises from causes beyond the control and without the fault or negligence of the Government of the Russian Federation; and

(II) the Government of the Russian Federation has made reasonable efforts to avoid and mitigate the effects of the failure of the completion of the Russian HEU Agreement; and

(ii) the Secretary of Energy and the Secretary of State jointly notify Congress of, and publish in the Federal Register, the determination under clause (i) and the reasons for the determination.

(B) NOTICE AND WAIT.–A waiver under subparagraph (A) may not take effect until the date that is 180 days after the date on which Secretary of Energy and the Secretary of State notify Congress under subparagraph (A)(ii).

(C) TERMINATION.–A waiver under subparagraph (A) shall terminate on December 31 of the calendar year with respect to which the Secretary makes the determination under subparagraph (A)(i).

(5) ADJUSTMENTS TO IMPORT LIMITATIONS.–

(A) IN GENERAL.–The import limitations described in paragraph (2)(A) are based on the reference data in the 2005 Market Report on the Global Nuclear Fuel Market Supply and Demand 2005–2030 of the World Nuclear Association. In each of calendar years 2016 and 2019, the Secretary of Commerce shall review the projected demand for uranium for nuclear reactors in the United States and adjust the import limitations described in paragraph (2)(A) to account for changes in such demand in years after the year in which that report or a subsequent report is published.

(B) INCENTIVE ADJUSTMENT.–Beginning in the second calendar year after the calendar year of the completion of the Russian HEU Agreement, the Secretary of Energy shall increase or decrease the amount of low-enriched uranium that may be imported in a calendar year under paragraph (2)(B) (including the amount of low-enriched uranium that may be imported for each kilogram of highly enriched uranium downblended under paragraph (2)(B)(i)) by a percentage equal to the percentage increase or decrease, as the case may be, in the average amount of uranium loaded into nuclear power reactors in the United States in the most recent 3-calendar-year period for which data are available, as reported by the Energy Information Administration of the Department of Energy, compared to the average amount of uranium loaded into such reactors during the 3-calendar-year period beginning on January 1, 2011, as reported by the Energy Information Administration.

(C) PUBLICATION OF ADJUSTMENTS.–As soon as practicable, but not later than July 31 of each calendar year, the Secretary of Energy shall publish in the Federal Register the amount of low-enriched uranium that may be imported in the current calendar year after the adjustments under subparagraph (B).

(6) AUTHORITY FOR ADDITIONAL ADJUSTMENT.–In addition to the adjustment under paragraph (5)(A), the Secretary of Commerce may adjust the import limitations under paragraph (2)(A) for a calendar year if the Secretary–

(A) in consultation with the Secretary of Energy, determines that the available supply of low-enriched uranium and the available stockpiles of uranium of the Department of Energy are insufficient to meet demand in the United States in the following calendar year; and

(B) notifies Congress of the adjustment not less than 45 days before making the adjustment.

(7) EQUIVALENT QUANTITIES OF LOW-ENRICHED URANIUM IMPORTS.–

(A) IN GENERAL.–The import limitations described in paragraphs (1) and (2) are expressed in terms of uranium containing 4.4 percent uranium-235 and a tails assay of 0.3 percent.

(B) ADJUSTMENT FOR OTHER URANIUM.–Imports of low-enriched uranium under paragraphs (1) and (2), including low-enriched uranium obtained under contracts for separative work units, shall count against the import limitations described in such paragraphs in amounts calculated as the quantity of low-enriched uranium containing 4.4 percent uranium-235 necessary to equal the total amount of uranium-235 contained in such imports.

(8) DOWNBLENDING OF OTHER HIGHLY ENRICHED URANIUM.–

(A) IN GENERAL.–The downblending of highly enriched uranium not of weapons origin may be counted for purposes of paragraph (2)(B), subject to verification under paragraph (10), if the Secretary of Energy determines that the highly enriched uranium to be downblended poses a risk to the national security of the United States.

(B) EQUIVALENT QUANTITIES OF HIGHLY ENRICHED URANIUM.– For purposes of determining the additional lowenriched uranium imports allowed under paragraph (2)(B), highly enriched uranium not of weapons origin downblended pursuant to subparagraph (A) shall count as downblended highly enriched uranium of weapons origin in amounts calculated as the quantity of highly enriched uranium containing 90 percent uranium-235 necessary to equal the total amount of uranium-235 contained in the highly enriched uranium not of weapons origin downblended pursuant to subparagraph (A).

(9) TERMINATION OF IMPORT RESTRICTIONS.–The provisions of this subsection shall terminate on December 31, 2020.

(10) TECHNICAL VERIFICATIONS BY SECRETARY OF ENERGY.–

(A) IN GENERAL.–The Secretary of Energy shall verify the origin, quantity, and uranium-235 content of the highly enriched uranium downblended for purposes of paragraphs (2)(B) and (8).

(B) METHODS OF VERIFICATION.–In conducting the verification required under subparagraph (A), the Secretary of Energy shall employ the transparency measures and access provisions agreed to under the Russian HEU Agreement for monitoring the downblending of Russian highly enriched uranium of weapons origin and such other methods as the Secretary determines appropriate.

(11) ENFORCEMENT OF IMPORT LIMITATIONS.–The Secretary of Commerce shall be responsible for enforcing the import limitations imposed under this subsection and shall enforce such

import limitations in a manner that imposes a minimal burden on the commercial nuclear industry.

(12) EFFECT ON OTHER AGREEMENTS.–

(A) RUSSIAN HEU AGREEMENT.–Nothing in this section shall be construed to modify the terms of the Russian HEU Agreement, including the provisions of the Agreement relating to the amount of low-enriched uranium that may be imported into the United States.

(B) OTHER AGREEMENTS.–If a provision of any agreement between the United States and the Russian Federation, other than the Russian HEU Agreement, relating to the importation of low-enriched uranium, including lowenriched uranium obtained under contracts for separative work units, into the United States conflicts with a provision of this section, the provision of this section shall supersede the provision of the agreement to the extent of the conflict.

42 USC 2297h-11. **Sec. 3113. Low-Level Waste**

(a) Responsibility of DOE.–

(1) The Secretary, at the request of the generator, shall accept for disposal low-level radioactive waste, including depleted uranium if it were ultimately determined to be low-level radioactive waste, generated by–

(A) the Corporation as a result of the operations of the gaseous diffusion plants or as a result of the treatment of such wastes at a location other than the gaseous diffusion plants, or

(B) any person licensed by the Nuclear Regulatory Commission to operate a uranium enrichment facility under section 53, 63, and 193 of the Atomic Energy Act of 1954 (42 USC 2073, 2093, and 2243).

(2) Except as provided in paragraph (3), the generator shall reimburse the Secretary for the disposal of low-level radioactive waste pursuant to paragraph (1) in an amount equal to the Secretary's costs, including a pro rata share of any capital costs, but in no event more than an amount equal to that which would be charged by commercial, State, regional, or interstate compact entities for disposal of such waste.

(3) In the event depleted uranium were ultimately determined to be low-level radioactive waste, the generator shall reimburse the Secretary for the disposal of depleted uranium pursuant to paragraph (1) in an amount equal to the Secretary's costs, including a pro rata share of any capital costs.

(4)[415] In the event that a licensee requests the Secretary to accept for disposal depleted uranium pursuant to this subsection, the Secretary shall be required to take title to and possession of such depleted uranium at an existing DUF6 storage facility.

(b) Agreements With Other Persons.–The generator may also enter into agreements for the disposal of low-level radioactive waste subject to subsection (a) with any person other than the Secretary that is authorized by applicable laws and regulations to dispose of such wastes.

(c) State or Interstate Compacts.–Notwithstanding any other provision of law, no State or interstate compact shall be liable for the treatment, storage, or disposal of any low-level radioactive waste (including mixed

---

[415]   Added by P.L. 108–447, Div. C, § 311, 118 Stat. 2959 (2004).

waste) attributable to the operation, decontamination, and decommissioning of any uranium enrichment facility.

42 USC 2297h-12.

Sec. 3114. AVLIS

(a) Exclusive Right to Commercialize.–The Corporation shall have the exclusive commercial right to deploy and use any AVLIS patents, processes, and technical information owned or controlled by the Government, upon completion of a royalty agreement with the Secretary.

(b) Transfer of Related Property to Corporation.–

President.

(1) In general.–To the extent requested by the Corporation and subject to the requirements of the Atomic Energy Act of 1954 (42 USC 2011, *et seq.*), the President shall transfer without charge to the Corporation all of the right, title, or interest in and to property owned by the United States under control or custody of the Secretary that is directly related to and materially useful in the performance of the Corporation's purposes regarding AVLIS and alternative technologies for uranium enrichment, including–

(A) facilities, equipment, and materials for research, development, and demonstration activities; and

(B) all other facilities, equipment, materials, processes, patents, technical information of any kind, contracts, agreements, and leases.

(2) Exception.–Facilities, real estate, improvements, and equipment related to the gaseous diffusion, and gas centrifuge, uranium enrichment programs of the Secretary shall not transfer under paragraph (1)(B).

(3) Expiration of transfer authority.–The President's authority to transfer property under this subsection shall expire upon the privatization date.

(c) Liability for Patent and Related Claims.–With respect to any right, title, or interest provided to the Corporation under subsection (a) or (b), the Corporation shall have sole liability for any payments made or awards under section 157b.(3) of the Atomic Energy Act of 1954 (42 USC 2187(b)(3)), or any settlements or judgments involving claims for alleged patent infringement. Any royalty agreement under subsection (a) of this section shall provide for a reduction of royalty payments to the Secretary to offset any payments, awards, settlements, or judgments under this subsection.

42 USC 2297h-13.

Sec. 3115. Application of Certain Laws

(a) OSHA.–

(1) As of the privatization date, the private corporation shall be subject to and comply with the Occupational Safety and Health Act of 1970 (29 USC 651 *et seq.*).

Contracts.

(2) The Nuclear Regulatory Commission and the Occupational Safety and Health Administration shall, within 90 days after the date of enactment of this Act, enter into a memorandum of agreement to govern the exercise of their authority over occupational safety and health hazards at the gaseous diffusion plants, including inspection, investigation, enforcement, and rulemaking relating to such hazards.

(b) Antitrust Laws.–For purposes of the antitrust laws, the performance by the private corporation of a "matched import" contract under the Suspension Agreement shall be considered to have occurred prior to the privatization date, if at the time of privatization, such contract had been agreed to by the parties in all material terms and confirmed by the Secretary of Commerce under the Suspension Agreement.

(c) Energy Reorganization Act Requirements.–

(1) The private corporation and its contractors and subcontractors shall be subject to the provisions of section 211 of the Energy Reorganization Act of 1974 (42 USC 5851) to the same extent as an employer subject to such section.

(2) With respect to the operation of the facilities leased by the private corporation, section 206 of the Energy Reorganization Act of 1974 (42 USC 5846) shall apply to the directors and officers of the private corporation.

3. Energy Reorganization Act of 1974, as Amended

3. Energy Reorganization Act of 1974, as Amended
Contents

A. ENERGY REORGANIZATION ACT OF 1974, AS AMENDED

Public Law 93–438   **88 STAT. 1233**

October 11, 1974

An Act

Energy
Reorganization
Act
of 1974.

To reorganize and consolidate certain functions of the Federal Government in a new Energy Research and Development Administration and in a new Nuclear Regulatory Commission in order to promote more efficient management of such functions.

Be it enacted by the Senate and House of Representatives of the United States of America in Congress assembled,

That the Energy Reorganization Act of 1974, as amended, is amended to read as follows:

42 USC 5801
note.

Sec. 1. Short Title

The Act may be cited as the "Energy Reorganization Act of 1974".

42 USC 5801.

Sec. 2. Declaration of Purpose

(a) The Congress hereby declares that the general welfare and the common defense and security require effective action to develop, and increase the efficiency and reliability of use of, all energy sources to meet the needs of present and future generations, to increase the productivity of the national economy and strengthen its position in regard to international trade, to make the Nation self-sufficient in energy, to advance the goals of restoring, protecting, and enhancing environmental quality, and to assure public health and safety.

Energy
Research and
Development
Administration,
establishment.

(b) The Congress finds that, to best achieve these objectives, improve Government operations, and assure the coordinated and effective development of all energy sources, it is necessary to establish an Energy Research and Development Administration to bring together and direct Federal activities relating to research and development on the various sources of energy, to increase the efficiency and reliability in the use of energy, and to carry out the performance of other functions, including but not limited to the Atomic Energy Commission's military and production activities and its general basic research activities. In establishing an Energy Research and Development Administration to achieve these objectives, the Congress intends that all possible sources of energy be developed consistent with warranted priorities.

Separation of
AEC licensing
and regulatory
functions.

(c) The Congress finds that it is in the public interest that the licensing and related regulatory functions of the Atomic Energy Commission be separated from the performance of the other functions of the Commission, and that this separation be effected in an orderly manner, pursuant to this Act, assuring adequacy of technical and other resources necessary for the performance of each.

Small business
participation.

(d) The Congress declares that it is in the public interest and the policy of Congress that small business concerns be given a reasonable opportunity to participate, insofar as is possible, fairly and equitably in grants, contracts, purchases, and other Federal activities relating to research, development, and demonstration of sources of energy efficiency, and utilization and conservation of energy. In carrying out this policy, to the extent practicable, the Administrator shall consult with the Administrator of the Small Business Administration.

Priorities.

(e) Determination of priorities which are warranted should be based on such considerations as power-related values of an energy source, preservation of material resources, reduction of pollutants, export market potential (including reduction of imports), among others. On such a basis, energy sources warranting priority might include, but not be limited to, the various methods of utilizing solar energy.

Title I–Energy Research and Development Administration

42 USC 5811.

Sec. 101. Establishment

There is hereby established an independent executive agency to be known as the Energy Research and Development Administration (hereinafter in this Act referred to as the "Administration").[1]

42 USC 5812.
Administrator.

Sec. 102. Officers

(a) There shall be at the head of the Administration an Administrator of Energy Research and Development (hereinafter in this Act referred to as the "Administrator"), who shall be appointed from civilian life by the President by and with the advice and consent of the Senate. A person may not be appointed as Administrator within two years after release from active duty as a commissioned officer of a regular component of an Armed Force. The Administration shall be administered under the supervision and direction of the Administrator, who shall be responsible for the efficient and coordinated management of the Administration.

Deputy
Administrator.

(b) There shall be in the Administration a Deputy Administrator, who shall be appointed by the President, by and with the advice and consent of the Senate.

(c) The President shall appoint the Administrator and Deputy Administrator from among individuals who, by reason of their general background and experience are specially qualified to manage a full range of energy research and development programs.

Assistant
Administrators.

(d) There shall be in the Administration six Assistant Administrators, one of whom shall be responsible for fossil energy, another for nuclear energy, another for environment and safety, another for conservation, another for solar, geothermal, and advanced energy systems, and another for national security. The Assistant Administrators shall be appointed by the President, by and with the advice and consent of the Senate. The President shall appoint each Assistant Administrator from among individuals who, by reason of general background and experience, are specially qualified to manage the energy technology area assigned to such Assistant Administrator.

General
Counsel.

(e) There shall be in the Administration a General Counsel who shall be appointed by the Administrator and who shall serve at the pleasure of and be removable by the Administrator.

Additional
officers.

(f) There shall be in the Administration not more than eight additional officers appointed by the Administrator. The positions of such officers shall be considered career positions and be subject to subsection 161d. of the Atomic Energy Act.

Director of
Military
Application.

(g) The Division of Military Application transferred to and established in the Administration by section 104(d) of this Act shall be under the direction of a Director of Military Application, who shall be appointed by the Administrator and who shall serve at the pleasure of and

---

[1]   This title established the Energy Research and Development Administration. The Administration was terminated, and its functions were transferred to the U.S. Department of Energy, by the Department of Energy Organization Act, P.L. 95–91, 91 Stat. 565 (1977).

42 USC 2011
note.

International
cooperation.

Order of
succession.

42 USC 5813.

be removable by the Administrator and shall be an active commissioned officer of the Armed Forces serving in general or flag officer rank or grade The functions, qualifications, and compensation of the Director of Military Application shall be the same as those provided under the Atomic Energy Act of 1954, as amended, for the Assistant General Manager for Military Application.

(h) Officers appointed pursuant to this section shall perform such functions as the Administrator shall delegate to one such officer the special responsibility for international cooperation in all energy and related environmental research and development.

(i) The Deputy Administrator (or in the absence or disability of the Deputy Administrator, or in the event of a vacancy in the office of the Deputy Administrator, an Assistant Administrator, the General Counsel or such other official, determined according to such order as the Administrator shall prescribe) shall act for and perform the functions of the Administrator during any absence or disability of the Administrator or in the event of a vacancy in the office of the Administrator.

Sec. 103. Responsibilities of the Administrator

The responsibilities of the Administrator shall include, but not be limited to–

(1) exercising central responsibility for policy planning, coordination, support, and management of research and development programs respecting all energy sources, including assessing the requirements for research and development in regard to various energy sources in relation to near-term and long-range needs, policy planning in regard to meeting those requirements, undertaking programs for the optimal development of the various forms of energy sources, managing such programs, and disseminating information resulting therefrom;

(2) encouraging and conducting research and development, including demonstration of commercial feasibility and practical applications of the extraction, conversion , storage, transmission, and utilization phases related to the development and use of energy from fossil, nuclear, solar, geothermal, and other energy sources;

(3) engaging in and supporting environmental, biomedical, physical, and safety research related to the development of energy sources and utilization technologies;

(4) taking into account the existence, progress, and results of other public and private research and development activities, including those activities of the Federal Energy Administration relating to the development of energy resources using currently available technology in promoting increased utilization of energy resources, relevant to the Administration's mission in formulating its own research and development programs;

(5) participating in and supporting cooperative research and development projects which may involve contributions by public or private persons or agencies, of financial or other resources to the performance of the work;

(6) developing, collecting, distributing, and making available for distribution, scientific and technical information concerning the manufacture or development of energy and its efficient extraction, conversion, transmission, and utilization;

(7) creating and encouraging the development of general information to the public on all energy conservation technologies and energy sources as they become available for general use, and the Administrator, in conjunction with the Administrator of the Federal Energy Administration

shall, to the extent practicable, disseminate such information through the use of mass communications;

(8) encouraging and conducting research and development in energy conservation, which shall be directed toward the goals of reducing total energy consumption to the maximum extent practicable, and toward maximum possible improvement in the efficiency of energy use. Development of new and improved conservation measures shall be conducted with the goal of the most expeditious possible application of these measures;

(9) encouraging and participating in international cooperation in energy and related environmental research and development;

(10) helping to assure an adequate supply of manpower for the accomplishment of energy research and development programs, by sponsoring and assisting in education and training activities in institutions of higher education, vocational schools, and other institutions, and by assuring the collection, analysis, and dissemination of necessary manpower supply and demand data;

(11) encouraging and conducting research and development in clean and renewable energy sources.[2]

42 USC 5814.
Atomic Energy
Commission.

Sec. 104. Abolition and Transfers

(a) The Atomic Energy Commission is hereby abolished. Sections 21 and 22 of the Atomic Energy Act of 1954, as amended (42 USC 2031 and 2032) are repealed.

(b) All other functions of the Commission, the Chairman and members of the Commission, and the officers and components of the Commission are hereby transferred or allowed to lapse pursuant to the provisions of this Act.

(c) There are hereby transferred to and vested in the Administrator all functions of the Atomic Energy Commission, the Chairman and members of the Commission, and the officers and components of the Commission, except as otherwise provided in this Act.

(d) The General Advisory Committee established pursuant to section 26 of the Atomic Energy Act of 1954, as amended (42 USC 2036), the Patent Compensation Board established pursuant to section 157 of the Atomic Energy Act of 1954, as amended (42 USC 2187) and the Divisions of Military Application and Naval Research established pursuant to section 25 of the Atomic Energy Act of 1954, as amended (42 USC 2035), are transferred to the Energy Research and Development Administration and the functions of the Commission with respect thereto, and with respect to relations with the Military Liaison Committee established by section 27 of the Atomic Energy Act of 1954, as amended (42 USC 2037), are transferred to the Administrator.

---

[2]   Amended by P.L. 102–486, § 143(b), 1406 Stat. 2843 (1992), which redesignated former paragraphs (8) through (12) as paragraphs (7) through (11), respectively, and struck out former paragraph (7), which included establishment of an Energy Extension Service as the responsibility of the Administrator. P.L. 95–39, § 510(a), 91 Stat. 200 (1977), amended section 103 by redesignating paragraphs (7) through (11) as paragraphs (8) through (12), respectively, and inserted a new paragraph (7), which read as follows:
"(7) establishing, in accordance with the National Energy Extension Service Act, an Energy Extension Service to provide technical assistance, instruction, and practical demonstration on energy conservation measures and alternative energy systems to individuals, businesses, and State and local government officials."

Interior
Department
functions.

(e) There are hereby transferred to and vested in the Administrator such functions of the Secretary of the Interior, the Department of the Interior, and officers and components of such department–

(1) as relate to or are utilized by the Office of Coal Research established pursuant to the Act of July 1, 1960[3];

(2) as relate to or are utilized in connection with fossil fuel energy research and development programs and related activities conducted by the Bureau of Mines "energy centers" and synthane plant to provide greater efficiency in the extraction, processing, and utilization of energy resources for the purpose of conserving those resources, developing alternative energy resources such as oil and gas secondary and tertiary recovery, oil shale and synthetic fuels, improving methods of managing energy-related wastes and pollutants, and providing technical guidance needed to establish and administer national energy policies; and

(3) as relate to or are utilized for underground electric power transmission research.

Helium
applications study.
Report to
President and
Congress.

The Administrator shall conduct a study of the potential energy applications of helium and, within six months from the date of the enactment of this Act, report to the President and Congress his recommendations concerning the management of the Federal helium programs, as they relate to energy.

National
Science
Foundation
functions.

(f) There are hereby transferred to and vested in the Administrator such functions of the National Science Foundation as relate to or are utilized in connection with–

(1) solar heating and cooling development; and

(2) geothermal power development.

Environmental
Protection
Agency
functions.

(g) There are hereby transferred to and vested in the Administrator such functions of the Environmental Protection Agency and the officers and components thereof as relate to or are utilized in connection with research, development, and demonstration, but not assessment or monitoring for regulatory purposes, of alternative automotive power systems.

(h) To the extent necessary or appropriate to perform functions and carry out programs transferred by this Act, the Administrator and Commissions may exercise, in relation to the functions so transferred, any authority or part thereof available by law, including appropriation Acts, to the official or agency from which such functions were transferred.

Use of other
agencies'
capabilities.

(i) In the exercise of his responsibilities under section 103, the Administrator shall utilize, with their consent, to the fullest extent he determines advisable the technical and management capabilities of other executive agencies having facilities, personnel, or other resources which can assist or advantageously be expanded to assist in carrying out such responsibilities. The Administrator shall consult with the head of each agency with respect to such facilities, personnel, or other resources, and may assign, with their consent, specific programs or projects in energy research and development as appropriate. In making such assignments under this subsection, the head of each such agency shall insure that–

(1) such assignments shall be in addition to and not detract from the basic mission responsibilities of the agency, and

---

[3]   The "Act" is probably referencing P.L. 86–599, 74 Stat. 336 (1960), which is classified, to Chapter 18 of Title 30 U.S.C. 661–668, "Mineral Lands and Mining."

(2) such assignments shall be carried out under such guidance as the Administrator deems appropriate.

Sec. 105. Administrative Provisions

42 USC 5815.
Regulations.

(a) The Administrator is authorized to prescribe such policies, standards, criteria, procedures, rules, and regulations as he may deem to be necessary or appropriate to perform functions now or hereafter vested in him.

Policy planning and evaluation.

(b) The Administrator shall engage in such policy planning, and perform, such program evaluation analyses and other studies, as may be necessary to promote the efficient and coordinated administration of the Administration and properly assess progress toward the achievement of its missions.

Delegation of functions.

(c) Except as otherwise expressly provided by law, the Administrator may delegate any of his functions to such officers and employees of the Administration as he may designate, and may authorize such successive re-delegations of such functions as he may deem to be necessary or appropriate.

Organization.

(d) Except as provided in section 102 and in section 104(d), the Administrator may organize the Administration as he may deem to be necessary or appropriate.

Field offices.

(e) The Administrator is authorized to establish, maintain, alter, or discontinue such State, regional, district, local, or other field offices as he may deem to be necessary or appropriate to perform functions now or hereafter vested in him.

Seal.

(f) The Administrator shall cause a seal of office to be made for the Administration of such device as he shall approve, and judicial notice shall be taken of such seal.

Working capital fund.

(g) The Administrator is authorized to establish a working capital fund, to be available without fiscal year limitation, for expenses necessary for the maintenance and operation of such common administrative services as he shall find to be desirable in the interests of economy and efficiency. There shall be transferred to the fund the stocks of supplies, equipment, assets other than real property, liabilities, and unpaid obligations relating to the services which he determines will be performed through the fund. Appropriations to the fund, in such amounts as may be necessary to provide additional working capital, are authorized. The working capital fund shall recover from the appropriations and funds for which services are performed, either in advance or by way of reimbursement, amounts which will approximate the costs incurred, including the accrual of annual leave and the depreciation of equipment. The fund shall also be credited with receipts from the sale or exchange of its property, and receipts in payment for loss or damage to property owned by the fund.

Information from other agencies.

(h) Each department, agency, and instrumentality of the executive branch of the Government is authorized to furnish to the Administrator, upon his request, any information or other data which the Administrator deems necessary to carry out his duties under this title.

Sec. 106. Personnel and Services

42 USC 5816.
Appointment and pay.

(a) The Administrator is authorized to select, appoint, employ, and fix the compensation of such officers and employees, including attorneys, pursuant to section 161d. of the Atomic Energy Act of 1954, as amended

(42 USC 2201(d)) as are necessary to perform the functions now or hereafter vested in him and to prescribe their functions.[4]

Experts and consultants. Military personnel.

(b) The Administrator is authorized to obtain services as provided by section 3109 of title 5 of the United States Code.

(c) The Administrator is authorized to provide for participation of military personnel in the performance of his functions. Members of the Army, the Navy, the Air Force, or the Marine Corps may be detailed for service in the Administration by the appropriate military Secretary, pursuant to cooperative agreements with the Secretary, for service in the Administration in positions other than a position the occupant of which must be approved by and with the advice and consent of the Senate.

(d) Appointment, detail, or assignment to, acceptance of, and service in, any appointive or other position in the Administration under this section shall in no way affect the status, office, rank, or grade which such officers or enlisted men may occupy or hold, or any emolument, prerequisite, right, privilege, or benefit incident to or arising out of any such status, office, rank, or grade. A member so appointed, detailed, or assigned shall not be subject to direction or control by his Armed Force, or any officer thereof, directly or indirectly, with respect to the responsibilities exercised in the position to which appointed, detailed, or assigned.

Transportation and per diem.

(e) The Administrator is authorized to pay transportation expenses, and per diem in lieu of subsistence expenses, in accordance with chapter 57 of title 5 of the United States Code for travel between places of recruitment and duty, and while at places of duty, of persons appointed for emergency, temporary, or seasonal services in the field service of the Administration.

Personnel of other agencies.

(f) The Administrator is authorized to utilize, on a reimbursable basis, the services of any personnel made available by any department, agency, or instrumentality, including any independent agency of the Government.

5 USC App. I. Advisory boards.

(g) The Administrator is authorized to establish advisory boards, in accordance with the provisions of the Federal Advisory Committee Act (Public Law 92-463), to advise with and make recommendations to the Administrator on legislation, policies, administration, research, and other matters.

Noncitizens.

(h) The Administrator is authorized to employ persons who are not citizens of the United States in expert, scientific, technical, or professional capacities whenever he deems it in the public interest.

Sec. 107. Powers

42 USC 5817. Research and development.

(a) The Administrator is authorized to exercise his powers in such manner as to insure the continued conduct of research and development and related activities in areas or fields deemed by the Administrator to be pertinent to the acquisition of an expanded fund of scientific, technical, and practical knowledge in energy matters. To this end, the Administrator is authorized to make arrangements (including contracts, agreements, and loans) for the conduct of research and development activities with private or public institutions or persons, including participation in joint or cooperative projects of a research, development, or experimental nature; to make payments (in lump sum or installments,

Contracts, etc.

---

[4]   42 U.S.C. 5816a has been repealed. P.L. 95–39, Title III, § 308, 91 Stat. 189, (1977); P.L. 96–470, Title II, § 203(d), 94 Stat. 2243) (1980), was repealed by P.L. 104–106, Div. D, Title XLIII, Subtitle A, § 4304(b)(7), 110 Stat. 664, (1996), which became effective February 10, 1996. This section provided for financial statements of Department of Energy officers and employees.

42 USC 2011
note.
and in advance or by way of reimbursement, with necessary adjustments on account of overpayments or underpayments); and generally to take such steps as he may deem necessary or appropriate to perform functions now or hereafter vested in him such functions of the Administrator under this Act as are applicable to the nuclear activities transferred pursuant to this title shall be subject to the provisions of the Atomic Energy Act of 1954, as amended, and to other authority applicable to such nuclear activities. The non-nuclear responsibilities and functions of the Administrator referred to in sections 103 and 104 of this Act shall be carried out pursuant to the provisions of this Act, applicable authority existing immediately before the effective date of this Act, or in accordance with the provisions of chapter 4 of the Atomic Energy Act of 1954, as amended (42 USC 2051-2053).

5 USC App. II.
40 USC 601
note.
Facilities and
real property.
(b) Except for public buildings as defined in the Public Buildings Act of 1959, as amended, and with respect to leased space subject to the provisions of Reorganization Plan Numbered 18 of 1950, the Administrator is authorized to acquire (by purchase, lease, condemnation, or otherwise), construct, improve, repair, operate, and maintain facilities and real property as the Administrator deems to be necessary in and outside of the District of Columbia. Such authority shall apply only to facilities required for the maintenance and operation of laboratories, research and testing sites and facilities, quarters, and related accommodations for employees and dependents of employees of the Administration, and such other special-purpose real property as the Administrator deems to be necessary in and outside the District of Columbia. Title to any property or interest therein, real, personal, or mixed, acquired pursuant to this section, shall be in the United States.

Services for
employees at
remote
locations.
(c)(1) The Administrator is authorized to provide, construct, or maintain, as necessary and when not otherwise available, the following for employees and their dependents stationed at remote locations:

(A) Emergency medical services and supplies.

(B) Food and other subsistence supplies.

(C) Messing facilities.

(D) Audiovisual equipment, accessories, and supplies for recreation and training.

(E) Reimbursement for food, clothing, medicine, and other supplies furnished by such employees in emergencies for the temporary relief of distressed persons.

(F) Living and working quarters and facilities.

(G) Transportation for school-age dependents of employees to the nearest appropriate education facilities.

(2) The furnishing of medical treatment under sub-paragraph (A) of paragraph (1) and the furnishing of services and supplies under paragraphs (B) and (C) of paragraph (1) shall be at prices reflecting reasonable value as determined by the Administrator.

(3) Proceeds from reimbursements under this section shall be deposited in the Treasury and may be withdrawn by the Administrator to pay directly the cost of such work or services, to repay or make advances to appropriations or funds which do or will bear all or a part of such cost, or to refund excess sums when necessary; except that such payments may be credited to a service or working capital fund otherwise established by law, and used under the law governing such funds, if the fund is available for use by the Administrator for performing the work or services for which payment is received.

Acquisition of copyrights, patents, etc.

(d) The Administrator is authorized to acquire any of the following described rights if the property acquired thereby is for use in, or is useful to, the performance of functions vested in him:

(1) Copyrights, patents, and applications for patents, designs, processes, specifications, and data.

(2) Licenses under copyrights, patents, and applicants for patents.

(3) Releases, before suit is brought, for past infringement of patents or copyrights.

Dissemination of information.

(e) Subject to the provisions of chapter 12 of the Atomic Energy Act of 1954, as amended (42 USC 2161-2166), and other applicable law, the Administrator shall disseminate scientific, technical, and practical information acquired pursuant to this title through information programs and other appropriate means, and shall encourage the dissemination of scientific, technical, and practical information relating to energy so as to enlarge the fund of such information and to provide that free interchange of ideas and criticism which is essential to scientific and industrial progress and public understanding.

Gifts and bequests.

(f) The Administrator is authorized to accept, hold, administer, and utilize gifts, and bequests of property, both real and personal, for the purpose of aiding or facilitating the work of the Administration. Gifts and bequests of money and proceeds from sales of other property received as gifts or bequests shall be deposited in the Treasury and shall be disbursed upon the order of the Administrator. For the purposes of Federal income, estate, and gift taxes, property accepted under this section shall be considered as a gift or bequest to the United States.

42 USC 5818.

Sec. 108. (Repealed)
(Repealed[5])

---

[5]   Repealed by P.L. 95–91, 91 Stat. 608 (1977). Previously, section read as follows:
(a) There is established in the Executive Office of the President an Energy Resources Council. The Council shall be composed of the Secretary of the Interior, the Administrator of the Federal Energy Administration, the Administrator of the Energy Research and Development Administration, the Secretary of State, the Director, Office of Management and Budget, and such other officials of the Federal Government as the President may designate. The President shall designate one of the members of the Council to serve as Chairman.
(b) It shall be the duty and function of the Council to—
(1) insure communication and coordination among the agencies of the Federal Government which have responsibilities for the development and implementation of energy policy or for the management of energy resources;
(2) make recommendations to the President and to the Congress for measures to improve the implementation of Federal energy policies or the management of energy resources with particular emphasis upon policies and activities involving two or more Departments or independent agencies; (See I)
(3) advise the President in the preparation of the reorganization recommendations required by section 110 of this Act; and (See II)
(4) insure that Federal agencies fully discharge their responsibilities under section 507 and 508 of the National Energy Extension Service Act for coordinating and planning of their related activities under such Act and any other law, including but not limited to the Energy Policy and Conservation Act. (See III)
(5) prepare a report on national energy conservation activities which shall be submitted to the President and the Congress annually, beginning on July 1, 1977, and which shall include—
(A) a review of all Federal energy conservation expenditures and activities, the purpose of each such activity, the relation of the activity to national conservation targets and plans, and the success of the activity and the plans for the activity in future years;
(B) an analysis of all conservation targets established for industry, residential, transportation, and public sectors of the economy, whether the targets can be achieved or whether they can be further improved, and the progress toward their achievement in the past year;
(C) a review of the progress made pursuant to the State energy conservation plans under section 361 through 366 of the Energy Policy and Conservation Act and other similar efforts at the State and local level, and whether further conservation can be carried on by the States or by local governments, and whether further Federal assistance is required;
(D) a review of the principal conservation efforts in the private sector, the potential for more widespread implementation of such efforts and the Federal Government's efforts to promote more widespread use of private energy conservation initiatives; and
(E) an assessment of whether existing conservation targets and goals are sufficient to bridge the gap between domestic energy production capacity and domestic energy needs, whether additional incentives or programs are necessary or useful to close that gap further, and a discussion of what mandatory measures might be useful to further bring domestic demand into harmony with domestic supply.
The Chairman of the Energy Resources Council shall coordinate the preparation of the report required under paragraph (5). (See IV)
(c) The President through the Energy Resources Council shall—
(1) prepare a plan for the reorganization of the Federal Government's activities in energy and natural resources, including, but not limited to, a study of—
(A) the principal laws and directives that constitute the energy and natural resource policy of the United States;
(B) prospects of developing a consolidated national energy policy;
(C) the major problems and issues of existing energy and natural resource organizations;
(D) the options for Federal energy and natural resource organizations;
(E) an overview of available resources pertinent to energy and natural resource organization;

42 USC 5819.
Report to
Congress.

Sec. 109. Future Reorganization

(a) The President shall transmit to the Congress as promptly as possible, but not later than June 30, 1975, such additional recommendations as he deems advisable for organization of energy and related functions in the Federal Government, including, but not limited to, whether or not there shall be established (1) a Department of Energy and Natural Resources, (2) an Energy Policy Council, and (3) a consolidation in whole or in part of regulatory functions concerning energy.

(b) This report shall replace and serve the purposes of the report required by section 15(a)(4) of the Federal Energy Administration Act.

42 USC 5820.

Sec. 110. Coordination with Environmental Efforts

The Administrator is authorized to establish programs to utilize research and development performed by other Federal agencies to minimize the adverse environmental effects of energy projects. The Administrator of the Environmental Protection Agency, as well as other affected agencies and departments, shall cooperate fully with the Administrator in establishing and maintaining such programs, and in establishing appropriate interagency agreements to develop cooperative programs and to avoid unnecessary duplication.

42 USC 5821.

Sec. 111. Provisions Applicable to Annual Authorization Acts[6]

42 USC 2017.

(a) All appropriations made to the Energy Research and Development Administration or the Administrator shall, except as otherwise provided by law, be subject to annual authorization in accordance with section 261 of the Atomic Energy Act of 1954, section 16 of the Federal Nonnuclear Energy Research and Development Act of 1974, and section 305 of this Act. The provisions of this section shall apply with respect to

---

(F) recent proposals for a national energy and natural resource policy for the United States; and

(G) the relationship between energy policy goals and other national objectives;

(2) submit to Congress —

(A) no later than December 31, 1976, the plan prepared pursuant to subsection (c)(1) and a report containing his recommendations for the reorganization of the Federal Government's responsibility for energy and natural resource matters together with such proposed legislation as he deems necessary or appropriate for the implementation of such plans or recommendations; and

(B) not later than April 15, 1977, such revisions to the plan and report described in subparagraph (A) of this paragraph as he may consider appropriate; and

(3) provide interim and transitional policy planning for energy and natural resource matters in the Federal Government. (See V)

(d) The Chairman of the Council may not refuse to testify before the Congress or any duly authorized committee thereof regarding the duties of the Council or other matters concerning interagency coordination of energy policy and activities.

(e) There is hereby established an Energy Conservation Subcommittee within the Council which shall be chaired by the Administrator of the Energy Research and Development Administration to discharge the responsibilities specified in subsection (b)(4) of this section and other related functions associated with the coordination and management of Federal efforts in the areas of energy conservation and energy conservation research, development and demonstration.(See VI)

(f) This section shall be effective no later than sixty days after the enactment of this Act or such earlier date as the President shall prescribe and publish in the Federal Register, and shall terminate upon enactment of a permanent department responsible for energy and natural resources or not later than September 30, 1977, whichever shall occur first. (See VII)

[6]   Added by P.L. 95–238, § 201, 92 Stat. 56 (1978).

appropriations made pursuant to the Act providing such authorization (hereinafter in this section referred to as "annual authorization Acts").

(b)(1) Funds appropriated pursuant to an annual authorization Act for "Operating expenses" may be used for–

> (A) the construction or acquisition of any facilities, or major items of equipment, which may be required at locations other than installations of the Administration, for the performance of research, development, and demonstration activities, and

> (B) grants to any organization for purchase or construction of research facilities.

No such funds shall be used under this subsection for the acquisition of land. Fee title to all such facilities and items of equipment shall be vested in the United States, unless the Administrator or his designee determines in writing that the research, development, and demonstration authorized by such Act would best be implemented by permitting fee title or any other property interest to be vested in an entity other than the United States; but before approving the vesting of such title or interest in such entity, the Administrator shall (i) transmit such determination, together with all pertinent data, to the Committee on Science and Technology of the House of Representatives and the Committee on Energy and Natural Resources of the Senate and (ii) wait a period of thirty calendar days (not including any day in which either House of Congress is not in session because of adjournment of more than three calendar days to a day certain), unless prior to the expiration of such period each such committee has transmitted to the Administrator written notice to the effect that such committee has no objection to the proposed action.

(2) No funds shall be used under paragraph (1) for any facility or major item of equipment, including collateral equipment, if the estimated cost to the Federal Government exceeds $5,000,000 in the case of such a facility or $2,000,000 in the case of such an item of equipment, unless such facility or item has been previously authorized by the appropriate committees of the House of representatives and the Senate, or the Administrator–

> (A) transmit to the appropriate committees of the House of Representatives and the Senate a report on such facility or item showing its nature, purpose, and estimated cost, and

> (B) waits a period of thirty calendar days (not including any day in which either House of Congress is not in session because of adjournment of more than three calendar days to a day certain), unless prior to the expiration of such period each such committee has transmitted to the Administrator written notice to the effect that such committee has no objection to the proposed action.

(c)(1) Not to exceed 1 per centum of all funds appropriated pursuant to any annual authorization Act for "Operating expenses" may be used by the Administrator to construct, expand, or modify laboratories and other facilities, including the acquisition of land, at any location under the control of the Administrator, if the Administrator determines that (A) such action would be necessary because of changes in the national programs authorized to be funded by such Act or because the new scientific or engineering developments, and (B) deferral of such action until the enactment of the next authorization Act would be inconsistent with the policies established by Congress for the Administration.

Notice.

(2) No funds may be obligated for expenditure or expended under paragraph (1) for activities described in such paragraph unless–

(A) a period of thirty calendar days (not including any day in which either House of Congress is not in session because of adjournment of more than three calendar days to a day certain) has passed after the Administrator has transmitted to the appropriate committees of the House of Representatives and the Senate a written report containing a full and complete statement concerning (i) the nature of the construction, expansion, or modification involved, (ii) the cost thereof, including the cost of any real estate action pertaining thereto, and (iii) the reason why such construction, expansion, or modification is necessary and in the national interest, or

(B) each such committee before the expiration of such period has transmitted to the Administrator a written notice to the effect that such committee has no objection to the proposed action;

except that this paragraph shall not apply to any project the estimated total cost of which does not exceed $50,000.

Report, transmittal to congressional committees. Notice.

(d)(1) Except as otherwise provided in the authorization Act involved–

(A) no amount appropriated pursuant to any annual authorization Act may be used for any program in excess of the amount actually authorized for that particular program by such Act, and

(B) no amount appropriated pursuant to any annual authorization Act may be used for any program which has not been presented to, or requested of the Congress, unless (i) a period of thirty calendar days (not including any day in which either House of Congress is not in session because of adjournment of more than three calendar days to a day certain) has passed after the receipt by the appropriate committees of the House of Representatives and the Senate of notice given by the Administrator containing a full and complete statement of the action proposed to be taken and the facts and circumstances relied upon in support of such proposed action, or (ii) each such committee before the expiration of such period has transmitted to the Administrator written notice to the effect that such committee has no objection to the proposed action.

(2) Notwithstanding any other provision of this section or the authorization Act involved, the aggregate amount available for use within the categories of coal, petroleum and natural gas, oil shale, solar, geothermal nuclear energy (non-weapons), environment and safety, and conservation from sums appropriated pursuant to an annual authorization Act may not, as a result of reprogramming, be decreased by more than 10 per centum of the total of the sums appropriated pursuant to such Act for those categories.

Funds merger, limitations.

(e) Subject to the applicable requirements and limitations of this section and the authorization Act involved, when so specified in an appropriation Act, amounts appropriated pursuant to any annual authorization Act for "Operating expenses" or for "Plant and capital equipment" may be merged with any other amounts appropriated for like purposes pursuant to any other Act authorizing appropriations for the Administration: *Provided,* That no such amounts appropriated for "Plant and capital equipment" may be merged with amounts appropriated for "Operating expenses."

(f) When so specified in an appropriation Act, amounts appropriated pursuant to any annual authorization Act for "Operating expenses" or for "Plant and capital equipment" may remain available until expended.

Construction design services.

(g) The Administrator is authorized to perform construction design services for any administration construction project whenever (1) such construction project has been included in a proposed authorization bill transmitted to the Congress by the Administration, and (2) the Administration determines that the project is of such urgency in order to meet the needs of national defense or protection of life and property or health and safety that construction of the project should be initiated promptly upon enactment of legislation appropriating funds for its construction.

(h) When so specified in appropriation Acts, any moneys received by the Administration may be retained and used for operating expenses, and may remain available until expended, notwithstanding the provisions of section 3617 of the Revised Statutes (31 USC 484); except that–

    (1) this subsection shall not apply with respect to sums received from disposal of property under the Atomic Energy Community Act of 1955 or the Strategic and Critical Materials Stockpiling Act, as amended, or with respect to fees received for tests or investigations under the Act of May 16, 1910, as amended (42 USC 2301; 50 USC 98h; 30 USC 7); and

    (2) revenues received by the Administration from the enrichment of uranium shall (when so specified) be retained and used for the specific purpose of offsetting costs incurred by the Administration in providing uranium enrichment service activities.

Funds transfer.

(i) When so specified in an appropriation Act, transfers of sums from the "Operating expenses" appropriation made pursuant to an annual authorization Act may be made to other agencies of the Government for the performance of the work for which the appropriation is made, and in such cases the sums so transferred may be merged with the appropriations to which they are transferred.[7]

Title II–Nuclear Regulatory Commission; Nuclear Whistleblower Protection[8]

42 USC 5841.
Members and Chairman.

Sec. 201. Establishment and Transfers

(a)(1) There is established an independent regulatory commission to be known as the Nuclear Regulatory Commission which shall be composed of five members, each of whom shall be a citizen of the United States. The President shall designate one member of the Commission as Chairman thereof to serve as such during the pleasure of the President. The Chairman may from time to time designate any other member of the Commission as Acting Chairman to act in the place and stead of the Chairman during his absence. The Chairman (or the Acting Chairman in the absence of the Chairman) shall preside at all meetings of the Commission and a quorum for the transaction of business shall consist of at least three members present. Each member of the Commission, including the Chairman, shall have equal responsibility and authority in all decisions and actions of the Commission, shall have full access to all information relating to the

---

[7]   Amended amended, P.L. 103–437, § 15(c)(7), 108 Stat. 4592 (1994).
[8]   Amended by P.L. 102–486, 106 Stat. 3124 (1992).

performance of his duties or responsibilities, and shall have one vote. Action of the Commission shall be determined by a majority vote of the members present. The Chairman (or Acting Chairman in the absence of the Chairman) shall be the official spokesman of the Commission in its relations with the Congress, Government agencies, persons, or the public, and on behalf of the Commission, shall see to the faithful execution of the policies and decisions of the Commission, and shall report thereon to the Commission from time to time or as the Commission may direct. The Commission shall have an official seal which shall be judicially noticed.[9]

(2) The Chairman of the Commission shall be the principal executive officer of the Commission, and he shall exercise all of the executive and administrative functions of the Commission, including functions of the commission with respect to (a) the appointment and supervision of personnel employed under the Commission (other than personnel employed regularly and full time in the immediate offices of Commissioners other than the Chairman, and except as otherwise provided in the Energy Reorganization Act of 1974), (b) the distribution of business among such personnel and among administrative units of the Commission, and (c) the use and expenditure of funds.

(3) In carrying out any of his functions under the provisions of this section the Chairman shall be governed by general policies of the Commission and by such regulatory decisions, findings, and determinations as the Commission may by law be authorized to make.

(4) The appointment by the Chairman of the heads of major administrative units under the Commission shall be subject to the approval of the Commission.

(5) There are hereby reserved to the Commission its functions with respect to revising budget estimates and with respect to determining upon the distribution of appropriated funds according to major programs and purposes.[10]

(b) (1) Members of the Commission shall be appointed by the President, by and with the advice and consent of the Senate.

(2) Appointments of members pursuant to this subsection shall be made in such a manner that not more than three members of the Commission shall be members of the same political party.

(c) Each member shall serve for a term of five years, each such term to commence on July 1, except that of the five members first appointed to the Commission, one shall serve for one year, one for two years, one for three years, one for four years, and one for five years, to be designated by the President at the time of appointment; and except that any member appointed to fill a vacancy occurring prior to the expiration of the term for which his predecessor was appointed, shall be appointed for the remainder of such term. For the purpose of determining the expiration date of the terms of office of the five members first appointed to the Nuclear Regulatory Commission, each such term "shall" be deemed to have begun July 1, 1975.[11]

Seal. Commission Chairman. Functions.

42 USC 5801 note.

42 USC 5841. Term of Office.

---

[9]   Amended by P.L. 94–79, § 201, 89 Stat. 413 (1975).
[10]   Amended by P.L. 94–79, § 201, 89 Stat. 413 (1975), which added new subparagraphs (2) through (5).
[11]   Amended by P.L. 94–79, §§ 202, 203, 89 Stat. 413 (1975). Before the amendment, this subsection read as follows:
    (c) Each member shall serve for a term of five years, each such term to commence on July 1, except that of the five members first appointed to the

(d) Such initial appointments shall be submitted to the Senate within sixty days of the signing of this Act. Any individual who is serving as a member of the Atomic Energy commission at the time of the enactment of this Act, and who may be appointed by the President to the Commission, shall be appointed for a term designated by the President, but which term shall terminate not later than the end of his present term as a member of the Atomic Energy Commission, without regard to the requirements of subsection (b)(2) of this section. Any subsequent appointment of such individuals shall be subject to the provisions of this section.

(e) Any member of the Commission may be removed by the President for inefficiency, neglect of duty, or malfeasance in office. No member of the Commission shall engage in any business, vocation, or employment other than that of serving as member of the Commission.

(f) There are hereby transferred to the Commission all the licensing and related regulatory functions of the Atomic Energy Commission, the Chairman and member of the Commission, the General Counsel, and other officers and components of the Commission—which functions, officers, components, and personnel are excepted from the transfer to the Administrator by section 104(c) of this Act.

(g) In addition to other functions and personnel transferred to the Commission, there are also transferred to the Commission—

(1) the functions of the Atomic Safety and Licensing Board Panel and the Atomic Safety and Licensing Appeal Board;

(2) such personnel as the Director of the Office of Management and Budget determines are necessary for exercising responsibilities under section 205, relating to, research, for the purpose of confirmatory assessment relating to licensing and other regulation under the provisions of the Atomic Energy Act of 1954, as amended, and of this Act.[12]

Sec. 202. Licensing and Related Regulatory Functions Respecting Selected Administration Facilities

Notwithstanding the exclusions provided for in section 110 a. or any other provisions of the Atomic Energy Act of 1954, as amended (42 USC 2140(a)), the Nuclear Regulatory Commission shall, except as otherwise specifically provided by section 110 b. of the Atomic Energy Act of 1954, as amended (42 USC 2140(b)), or other law, have licensing and related regulatory authority pursuant to chapters 6, 7, 8, and 10 of the Atomic Energy Act of 1954, as amended, as to the following facilities of the Administration:

(1) Demonstration Liquid Metal Fast Breeder reactors when operated as part of the power generation facilities of an electric utility system, or when operated in any other manner for the purpose of demonstrating the suitability for commercial application of such a reactor.

(2) Other demonstration nuclear reactors—except those in existence on the effective date of this Act—when operated as part of the power generation facilities of an electric utility system, or when

---

Commission, one shall serve for one year, one for two years, one for three years, one for four years, and one for five years, to be designated by the President at the time of appointment.

[12]   P.L. 95–209, § 2, 91 Stat. 1482 (1977), added a new subsection h, which was subsequently deleted by P.L. 99–386, 100 Stat. 822 (1986).

operated in any other manner for the purpose of demonstrating the suitability for commercial application of such a reactor.

(3) Facilities used primarily for the receipt and storage of high-level radioactive wastes resulting from activities licensed under such Act.

(4) Retrievable Surface Storage Facilities and other facilities authorized for the express purpose of subsequent long-term storage of high-level radioactive waste generated by the Administration, which are not used for, or are part of, research and development activities.

(5) Any facility under a contract with and for the account of the Department of Energy that is utilized for the express purpose of fabricating mixed plutonium-uranium oxide nuclear reactor fuel for use in a commercial nuclear reactor licensed under such Act other than any such facility that is utilized for research, development, demonstration, testing, or analysis purposes.[13]

42 USC 5843.
Director.
Establishment.

Sec. 203. Office of Nuclear Reactor Regulation

(a) There is hereby established in the Commission an Office of Nuclear Reactor Regulation under the direction of a Director of Nuclear Reactor Regulation, who shall be appointed by the Commission, who may report directly to the Commission, as provided in section 209, and who shall serve at the pleasure of and be removable by the Commission.

Functions.

(b) Subject to the provisions of this Act, the Director of Nuclear Reactor Regulation shall perform such functions as the Commission shall delegate including:

42 USC 2011
note.

(1) Principal licensing and regulation involving all facilities, and materials licensed under the Atomic Energy Act of 1954, as amended, associated with the construction and operation of nuclear reactors licensed under the Atomic Energy Act of 1954, as amended;

(2) Review the safety and safeguards of all such facilities, materials, and activities, and such review functions shall include, but not be limited to–

(A) monitoring, testing and recommending upgrading of systems designed to prevent substantial health or safety hazards; and

(B) evaluating methods of transporting special nuclear and other nuclear materials and of transporting and storing high-level radioactive wastes to prevent radiation hazards to employees and the general public.

(3) Recommend research necessary for the discharge of the functions of the Commission.

(c) Nothing in this section shall be construed to limit in any way the functions of the Administration relating to the safe operation of all facilities resulting from all activities within the jurisdiction of the Administration pursuant to this Act.

42 USC 5844.
Director.
Establishment.

Sec. 204. Office of Nuclear Material Safety and Safeguards

(a) There is hereby established in the Commission an Office of Nuclear Material Safety and Safeguards under the direction of a Director of Nuclear Material Safety and Safeguards, who shall be appointed by the Commission, who may report directly to the Commission as provided in section 209, and who shall serve at the pleasure of and be removable by the Commission.

---

[13]   Amended by P.L. 105–261, Div. C, Title XXXI, Subtitle C, § 3134(a), 112 Stat. 2247 (1998).

Functions.

42 USC 2011
note.

Report to
Congress.

42 USC 5845.
Director.
Establishment.

Functions.

Cooperation of
Federal agencies.

(b) Subject to the provisions of this Act, the Director of Nuclear Material Safety and Safeguards shall perform such functions as the Commission shall delegate including:

(1) Principal licensing and regulation involving all facilities and materials, licensed under the Atomic Energy Act of 1954, as amended, associated with the processing, transport, and handling of nuclear materials, including the provision and maintenance of safeguards against threats, thefts, and sabotage of such licensed facilities, and materials.

(2) Review safety and safeguards of all such facilities and materials licensed under the Atomic Energy Act of 1954, as amended, and such review shall include, but not be limited to–

(A) monitoring, testing, and recommending upgrading of internal accounting systems for special nuclear and other nuclear materials licensed under the Atomic Energy Act of 1954, as amended:

(B) developing, in consultation and coordination with the Administration, contingency plans for dealing with threats, thefts, and sabotage relating to special nuclear materials, high-level radioactive wastes and nuclear facilities resulting from all activities licensed under the Atomic Energy Act of 1954, as amended;

(C) assessing the need for, and the feasibility of, establishing a security agency within the office for the performance of the safeguards functions, and a report with recommendations on this matter shall be prepared within one year of the effective date of this Act and promptly transmitted to the Congress by the Commission.

(3) Recommending research to enable the Commission to more effectively perform its functions.

(c) Nothing in this section shall be construed to limit in any way the functions of the Administration relating to the safeguarding of special nuclear materials, high-level radioactive wastes and nuclear facilities resulting from all activities within the jurisdiction of the Administration pursuant to this Act.

Sec. 205. Office of Nuclear Regulatory Research

(a) There is hereby established in the Commission an Office of Nuclear Regulatory Research under the direction of a Director of Nuclear Regulatory research, who shall be appointed by the Commission, who may report directly to the Commission as provided in section 209, and who shall serve at the pleasure of and be removable by the Commission.

(b) Subject to the provisions of this Act, the Director of Nuclear Regulatory Research shall perform such functions as the Commission shall delegate including:

(1) Developing recommendations for research deemed necessary for performance by the Commission of its licensing and related regulatory functions.

(2) Engaging in or contracting for research which the Commission deems necessary for the performance of its licensing and related regulatory functions.

(c) The Administrator of the Administration and the head of every other Federal agency shall–

(1) cooperate with respect to the establishment of priorities for the furnishing of such research services as requested by the Commission for the conduct of its functions;

(2) furnish to the Commission, on a reimbursable basis, through their own facilities or by contract or other arrangement, such research services as the Commission deems necessary and requests for the performance of its functions; and

(3) consult and cooperate with the Commission on research and development matters of mutual interest and provide such information and physical access to its facilities as will assist the Commission in acquiring the expertise necessary to perform its licensing and related regulatory functions.

(d) Nothing in subsections (a) and (b) of this section or section 201 of this Act shall be construed to limit in any way the functions of the Administration relating to the safety of activities within the jurisdiction of the Administration.

Information and research services. 42 USC 5845. Improved safety systems research. Long-term plan development. **42 USC 5846.**

42 USC 2011 note.

(e) Each Federal agency, subject to the provisions of existing law, shall cooperate with the Commission and provide such information and research services, on a reimbursable basis, as it may have or be reasonably able to acquire.

(f)[14] The Commission shall develop a long-term plan for projects for the development of new or improved safety systems for nuclear power plants.

Sec. 206. Noncompliance

(a) Any individual director, or responsible officer of a firm constructing, owning, operation, or supplying the components of any facility or activity which is licensed or otherwise regulated pursuant to the Atomic Energy Act of 1954 as amended, or pursuant to this Act, who obtains information reasonably indicating that such facility or activity or basic components supplied to such facility or activity–

(1) fails to comply with the Atomic Energy Act of 1954, as amended, or any applicable rule, regulation, order, or license of the Commission relating to substantial safety hazards, or

(2) contains a defect which could create a substantial safety hazard, as defined by regulations which the Commission shall promulgate, shall immediately notify the Commission of such failure to comply, or of such defect, unless such person has actual knowledge that the Commission has been adequately informed of such defect or failure to comply.

42 USC 2282. Penalty.

(b) Any person who knowingly and consciously fails to provide the notice required by subsection (a) of this section shall be subject to a civil penalty in an amount equal to the amount provided by section 234 of the Atomic energy Act of 1954, as amended.

42 USC 2011 note. Posting of enforcement.

(c) The requirements of this section shall be prominently posted on the premises of any facility licensed or otherwise regulated pursuant to the Atomic Energy Act of 1954, as amended.

(d) The Commission is authorized to conduct such reasonable inspections and other enforcement activities as needed to insure compliance with the provisions of this section.

42 USC 5847. Federal-State-local cooperation.

Solicitation of views.

Sec. 207. Nuclear Energy Center Site Survey

(a)(1) The Commission is authorized and directed to make or cause to be made under its direction, a national survey, which shall include consideration of each of the existing or future electric reliability regions, or other appropriate regional areas, to locate and identify possible nuclear energy center sites. This survey shall be conducted in cooperation with

---

[14]   Added by P.L. 95–209, § 4, 91 Stat. 1482 (1977).

other interested Federal, State, and local agencies, and the views of interested persons, including electric utilities, citizens' groups, and others, shall be solicited and considered.

Definition.

(2) For purposes of this section, the term "nuclear energy center site" means any site, including a site not restricted to land, large enough to support utility operations or other elements of the total nuclear fuel cycle, or both including, if appropriate, nuclear fuel reprocessing facilities, nuclear fuel fabrication plants, retrievable nuclear waste storage facilities, and uranium enrichment facilities.

(3) The survey shall include–

(a) a regional evaluation of natural resources, including land, air, and water resources, available for use in connection with nuclear energy center sites; estimates of future electric power requirements that can be served by each nuclear energy center site; an assessment of the economic impact of each nuclear energy site; and consideration of any other relevant factors, including but not limited to population distribution, proximity to electric load centers and to other elements of the fuel cycle, transmission line rights-of-way, and the availability of other fuel resources;

(b) an evaluation of the environmental impact likely to result from construction and operation of such nuclear energy centers, including an evaluation whether such nuclear energy centers will result in greater or lesser environmental impact than separate siting of the reactors and/or fuel cycle facilities; and

(c) consideration of the use of federally owned property and other property designated for public use, but excluding national parks, national forests, national wilderness areas, and national historic monuments.

Report to
Congress and
Council on
Environmental
Quality; public
availability.

(4) A report of the results of the survey shall be published and transmitted to the Congress and the Council on Environmental Quality not later than one year from the date of the enactment of this Act and shall be made available to the public, and shall be updated from time to time thereafter as the Commission, in its discretion, deems advisable. The report shall include the Commission's evaluation of the results of the survey and any conclusions and recommendations, including recommendations for legislation, which the Commission may have concerning the feasibility and practicality of locating nuclear power reactors and/or other elements of the nuclear fuel cycle or nuclear energy center sites. The Commission is authorized to adopt policies which will encourage the location of nuclear power reactors and related fuel cycle facilities on nuclear energy center sites insofar as practicable.

42 USC 5848.
Reports to
Congress.
42 USC 2011
note.

Sec. 208. Abnormal Occurrence Reports

The Commission shall submit to the Congress an annual report listing for the previous fiscal year any abnormal occurrences at or associated with any facility which is licensed or otherwise regulated pursuant to the Atomic Energy Act of 1954 as amended, or pursuant to this Act. For the purposes of this section an abnormal occurrence is an unscheduled incident or event which the Commission determines is significant from the standpoint of public health or safety. Nothing in the preceding sentence shall limit the authority of a court to review the determination of the Commission. Each such report shall contain–

(1) the date and place of each occurrence;

(2) the nature and probable consequence of each occurrence;

(3) the cause or causes of each; and

(4) any action taken to prevent reoccurrence;

Public
dissemination
of information.

the Commission shall also provide as wide dissemination to the public of the information specified in clauses (1) and (2) of this section as reasonably possible within fifteen days of its receiving information of each abnormal occurrence and shall provide as wide dissemination to the public as reasonably possible of the information specified in clauses (3) and (4) as soon as such information becomes available to it.[15]

42 USC 5849.
Executive
Director.

Sec. 209. Other Officers

(a) The Commission shall appoint an Executive Director for Operations, who shall serve at the pleasure of and be removable by the Commission.

Functions.

(b) The Executive Director shall perform such functions as the Commission may direct, except that the Executive Director shall not limit the authority of the director of any component organization provided in this Act to communicate with or report directly to the Commission when such director of a component organization deems it necessary to carry out his responsibilities. Notwithstanding the preceding sentence, each such director shall keep the Executive Director fully and currently informed concerning the content of all such direct communications with the Commission.[16]

Equal
employment
opportunity.
Annual status
report.

(c) The Executive Director shall report to the Commission at semiannual public meetings on the problems, progress, and status of the Commission's equal employment opportunity efforts.[17]

(d)[18] The Executive Director shall prepare and forward to the Commission an annual report (for the fiscal year 1978 and each succeeding fiscal year) on the status of the Commission's programs concerning domestic safeguards matters including an assessment of the effectiveness and adequacy of safeguards at facilities and activities licensed by the Commission The Commission shall forward to the

Report to
Congress.

Congress a report under this section prior to February 1, 1979, as a separate document, and prior to February 1 of each succeeding year as a separate chapter of the Commission's annual report (required under section 307(c) of the Energy Reorganization Act of 1974) following the fiscal year to which such report applies.

42 USC 5877.
Other officers.

(e)[19] There shall be in the Commission not more than five additional officers appointed by the Commission. The positions of such officers shall be considered career positions and be subject to subsection 161 d. of the Atomic Energy Act.

42 USC 5850.
Progress
reports.
Submittal to
Congress.

Sec. 210. Unresolved Safety Issues Plan[20]

The Commission shall develop a plan providing for the specification and analysis of unresolved safety issues relating to nuclear reactors and shall take such action as may be necessary to implement corrective measures with respect to such issues. Such plans shall be submitted to the Congress on or before January 1, 1978, and progress reports shall be included in the annual report of the Commission thereafter.

---

[15]   Amended by P.L. 104–66, Title II, Subtitle Q, § 2171, 109 Stat. 731 (1995).
[16]   Amended by P.L. 95–601, § 4(a), 92 Stat. 2949 (1978).
[17]   Amended by P.L. 95–601, § 4(b), 92 Stat. 2949 (1978), by adding a new subsection (c) and redesignating it as subsection (e).
[18]   Added by P.L. 95–601, § 6, 92 Stat. 2949 (1978). As a result of P.L. 104–66, § 3003, 109 Stat. 734 (1995), reporting requirements ceased to be effective after December 21, 1999.
[19]   Amended by P.L. 95–601, § 4(b), 92 Stat. 2949 (1978). Formerly, subsection (c). See footnote 17.
[20]   Added by P.L. 95–209, § 3, 91 Stat. 1482 (1977).

42 USC 5851.

Sec. 211. Employee Protection[21]

(a)(1) No employer may discharge any employee or otherwise discriminate against any employee with respect to his compensation, terms, conditions, or privileges of employment because the employee (or any person acting pursuant to a request of the employee)–

(A) notified his employer of an alleged violation of this Act or the Atomic Energy Act of 1954 (42 USC 2011 *et seq.*);

(B) refused to engage in any practice made unlawful by this act or the Atomic Energy Act of 1954, if the employee has identified the alleged illegality to the employer;

(C) testified before Congress or at any Federal or State proceeding regarding any provision (or proposed provision) of this Act or the Atomic Energy Act of 1954;

42 USC 2011 note.

(D) commenced, caused to be commenced, or is about to commence or cause to be commenced a proceeding under this Act or the Atomic Energy Act of 1954, as amended, or a proceeding for the administration or enforcement of any requirement imposed under this Act or the Atomic Energy Act of 1954, as amended;

(E) testified or is about to testify in any such proceeding or;

(F) assisted or participated or is about to assist or participate in any manner in such a proceeding or in any other manner in such a proceeding or in any other action to carry out the purposes of this Act or the Atomic Energy Act of 1954, as amended.

(2) For purposes of this section, the term "employer" includes–

(A) a licensee of the Commission or of an Agreement State under section 274 of the Atomic Energy Act of 1954 (42 USC 2021);

(B) an applicant for a license from the Commission or such an Agreement State;

(C) a contractor or subcontractor of such a licensee or applicant;

(D) a contractor or subcontractor of the Department of Energy that is indemnified by the Department under section 170d. of the Atomic Energy Act of 1954 (42 USC 2210(d)), but such term shall not include any contractor or subcontractor covered by Executive Order No. 12344;

(E) a contractor or subcontractor of the Commission;

(F) the Commission; and

(G) the Department of Energy.[22]

Complaint, filing and notification.

(b)(1) Any employee who believes that he has been discharged or otherwise discriminated against by any person in violation of subsection (a) may, within 180 days after such violation occurs, file (or have any person file on his behalf) a complaint with the Secretary of Labor (in this section referred to as the "Secretary") alleging such discharge or discrimination. Upon receipt of such a complaint, the Secretary shall notify the person named in the complaint of the filing of the complaint, the Commission and the Department of Energy.

---

[21]   Sec. 211, formerly Sec. 210, was added by P.L. 95–601, 92 Stat. 2951 (1978). It was renumbered as Sec. 211 by P.L. 102–486, 106 Stat 3123 (1992).
[22]   Sections (E), (F), and (G) added by P.L. 109–58, § 629, 119 Stat. 785 (2005).

Investigation
and notification.

Order.

Notice and
hearing.
Settlement.

Relief.

(2)(A) Upon receipt of a complaint filed under paragraph (1), the Secretary shall conduct an investigation of the violation alleged in the complaint. Within thirty days of the receipt of such complaint, the Secretary shall complete such investigation and shall notify in writing the complainant (and any person acting in his behalf) and the person alleged to have committed such violation of the results of the investigation conducted pursuant to this subparagraph. Within ninety days of the receipt of such complaint the Secretary shall, unless the proceeding on the complaint is terminated by the Secretary on the basis of a settlement entered into by the Secretary and the person alleged to have committed such violation, issue an order either providing the relief prescribed by subparagraph (B) or denying the complaint. An order of the Secretary shall be made on the record after notice and notice and opportunity for public hearing. Upon the conclusion of such hearing and the issuance of a recommended decision that the complaint has merit, the Secretary shall issue a preliminary order providing the relief prescribed in subparagraph (B), but may not order compensatory damages pending a final order. The Secretary may not enter into a settlement terminating a proceeding on a complaint without the participation and consent of the complainant.

(B) If, in response to a complaint filed under paragraph (1), the Secretary determines that a violation of subsection (a) has occurred, the Secretary shall order the person who committed such violation to (i) take affirmative action to abate the violation, and (ii) reinstate the complainant to his former position together with the compensation (including back pay), terms, conditions, and privileges of his employment, and the Secretary may order such person to provide compensatory damages to the complainant. If an order is issued under this paragraph, the Secretary, at the request of the complainant shall assess against the person against whom the order is issued a sum equal to the aggregate amount of all costs and expenses (including attorneys' and expert witness fees) reasonably incurred, as determined by the Secretary, by the complainant for, or in connection with, the bringing of the complaint upon which the order was issued.

(3) (A) The Secretary shall dismiss a complaint filed under paragraph (1), and shall not conduct the investigation required under paragraph (2), unless the complainant has made a prima facie showing that any behavior described in subparagraphs (A) through (F) of subsection (a)(1) was a contributing factor in the unfavorable personnel action alleged in the complaint.

(B) Notwithstanding a finding by the Secretary that the complainant has made the showing required by subparagraph (A), no investigation required under paragraph (2) shall be conducted if the employer demonstrates, by clear and convincing evidence, that it would have taken the same unfavorable personnel action in the absence of such behavior.

(C) The Secretary may determine that a violation of subsection (a) has occurred only if the complainant has demonstrated that any behavior described in subparagraphs (A) through (F) of subsection (a)(1) was a contributing factor in the unfavorable personnel action alleged in the complaint.

(D) Relief may not be ordered under paragraph (2) if the employer demonstrates by clear and convincing evidence that it would have taken the same unfavorable personnel action in the absence of such behavior.

Review.
5 USC 701 *et seq.*

(c)(1) Any person adversely affected or aggrieved by an order issued under subsection (b) may obtain review of the order in the United States court of appeals for the circuit in which the violation, with respect to which the order was issued, allegedly occurred. The petition for review must be filed within sixty days from the issuance of the Secretary's order. Review shall conform to chapter 7 of title 5 of the United States Code. The commencement of proceedings under this subparagraph shall not, unless order by the court, operate as a stay of the Secretary's order.

(2) An order of the Secretary with respect to which review could have been obtained under paragraph (1) shall not be subject to judicial review in any criminal or other civil proceeding.

Jurisdiction.

(d) Whenever a person has failed to comply with an order issued under subsection (b) (2), the Secretary may file a civil action in the United States district court for the district in which the violation was found to occur to enforce such order. In actions brought under this subsection, the district courts shall have jurisdiction to grant all appropriate relief including, but not limited to, injunctive relief, compensatory, and exemplary damages.

(e)(1) Any person on whose behalf an order was issued under paragraph (2) of subsection (b) may commence a civil action against the person to whom such order was issued to require compliance with such order. The appropriate United States district court shall have jurisdiction, without regard to the amount in controversy or the citizenship of the parties, to enforce such order.

Litigative costs.

(2) The court, in issuing any final order under this subsection, may award costs of litigation (including reasonable attorney and expert witness fees) to any party whenever the court determines such award is appropriate.

(f) Any nondiscretionary duty imposed by this section shall be enforceable in a mandamus proceeding brought under section 1361 of title 28 of the United States Code.

42 USC 2011.

(g) Subsection (a) shall not apply with respect to any employee who, acting without direction from his or her employer (or the employer's agent), deliberately causes a violation of any requirement of this Act or of the Atomic Energy Act of 1954, as amended.[23]

(h) This section may not be construed to expand, diminish, or otherwise affect any right otherwise available to an employee under Federal or State law to redress the employee's discharge or other discriminatory action taken by the employer against the employee.

(i) The provisions of this section shall be prominently posted in any place of employment to which this section applies.

(j)(1) The Commission or the Department of Energy shall not delay taking appropriate action with respect to an allegation of a substantial safety hazard on the basis of–

(A) the filing of a complaint under subsection (b)(1) arising from such allegation; or

(B) any investigation by the Secretary, or other action, under this section in response to such complaint.

---

[23]   Formerly Section 201, as added by P.L. 95–601, § 10, 92 Stat. 2951 (1978). Renumbered and amended by Pub.L. 102–486, Title XXIX, § 2902(a) to (g), (h)(2), (3), 106 Stat. 3123, 3124 (1992); Pub.L. 109–58, Title VI, § 629, 119 Stat. 785 (2005).

(2) A determination by the Secretary under this section that a violation of subsection (a) has not occurred shall not be considered by the Commission or the Department of Energy in its determination of whether a substantial safety hazard exists.

Deadline.

(4)[24] If the Secretary has not issued a final decision within 1 year after the filing of a complaint under paragraph (1), and there is no showing that such delay is due to the bad faith of the person seeking relief under this paragraph, such person may bring an action at law or equity for de novo review in the appropriate district court of the United States, which shall have jurisdiction over such an action without regard to the amount of controversy.

42 USC 5853.

Sec. 212. Limitation on Legal Fee Reimbursement[25]

The Department of Energy shall not, except as required under a contract entered into before the date of enactment of this section, reimburse any contractor or subcontractor of the Department for any legal fees or expenses incurred with respect to a complaint subsequent to–

(1) an adverse determination on the merits with respect to such complaint against the contractor or subcontractor by the Director of the Department of Energy's Office of Hearings and Appeals pursuant to part 708 of title 10, Code of Federal Regulations, or by a Department of Labor Administrative Law Judge pursuant to section 211 of this Act; or

(2) an adverse final judgment by any State or Federal court with respect to such complaint against the contractor or subcontractor for wrongful termination or retaliation due to the making of disclosures protected under chapter 12 of title 5, United States Code, section 211 of this Act, or any comparable State law, unless the adverse determination or final judgment is reversed upon further administrative or judicial review.

Title III–Miscellaneous and Transitional Provisions

42 USC 5871.

Sec. 301. Transitional Provisions

(a) Except as otherwise provided in this Act, whenever all of the functions or programs of an agency, or other body, or any component thereof, affected by this Act, have been transferred from that agency, or other body, or any component thereof by this Act, the agency, or other body, or component thereof shall lapse. If an agency, or other body, or any component thereof, lapses pursuant to the preceding sentence, each position and office therein which was expressly authorized by law, or the incumbent of which was authorized to receive compensation at the rate prescribed for an officer or position at level II, III, IV, or V of the Executive Schedule (5 USC 5313–5316), shall lapse.

Lapses of agencies and positions.

Savings clauses.

(b) All orders, determinations, rules, regulations, permits, contracts, certificates, licenses, and privileges–

(1) which have been issued, made, granted, or allowed to become effective by the President, any Federal department or agency or official thereof, or by a court of competent jurisdiction, in the performance of functions which are transferred under this Act, and

(2) which are in effect at the time this Act takes effect, shall continue in effect according to their terms until modified, terminated,

---

24   Added by P.L. 109–58, § 629(b), 119 Stat. 785 (2005).
25   Added byP.L. 109–58, § 627, 119 Stat. 784 (2005).

superseded, set aside, or revoked by the President, the Administrator, the Commission, or other authorized officials, a court of competent jurisdiction, or by operation of law.

(c) The provisions of this Act shall not affect any proceeding pending, at the time this section takes effect, before the Atomic Energy Commission or any department or agency (or component thereof) functions of which are transferred by this Act; but such proceedings, to the extent that they relate to functions so transferred, shall be continued. Orders shall be issued in such proceedings, appeals shall be taken therefrom, and payments shall be made pursuant to such orders, as if this Act had not been enacted; and orders issued in any such proceedings shall continue in effect until ;modified, terminated, superseded, or revoked by a duly authorized official, by a court of competent jurisdiction, or by operation of law. Nothing in this subsection shall be deemed to prohibit the discontinuance or modification of any such proceeding under the same terms and conditions and to the same extent that such proceeding could have been disconnected if this Act had not been enacted.

(d) Except as provided in subsection (f)–

(1) the provisions of this Act shall not affect suits commenced prior to the date this Act takes effect, and

(2) in all such suits proceedings shall be had, appeals taken, and judgments rendered, in the same manner and effect as if this Act had not been enacted.

(e) No suit, action, or other proceeding commenced by or against any officer in his official capacity as an officer of any department or agency, functions of which are transferred by this Act, shall abate by reason of the enactment of this Act. No cause of action by or against any department or agency, functions of which are transferred by this Act, or by or against any officer thereof in his official capacity shall abate by reason of the enactment of this Act. Causes of actions, suits, actions, or other proceedings may be asserted by or against the United States or such official as may be appropriate and, in any litigation pending when this section takes effect, the court may at any time, on its own motion or that of any party, enter any order which will give effect to the provisions of this section.

(f) If, before the date on which this Act takes effect, any department or agency, or officer thereof in his official capacity, is a party to a suit, and under this Act any function of such department, agency, or officer is transferred to the Administrator or Commission, or any other official, then such suit shall be continued as if this Act had not been enacted, with the Administrator of Commission, or other official, as the case may be, substituted.

(g) Final orders and actions of any official or component in the performance of functions transferred by this Act shall be subject to judicial review to the same extent and in the same manner as if such orders or actions had been make or taken by the officer, department, agency, or instrumentality in the performance of such functions immediately preceding the effective date of the Act. Any statutory requirements relating to notices, hearings, action upon the record, or administrative review that apply to any function transferred by this Act shall apply to the performance of those functions by the Administrator or Commission, or any officer or component.

(h) With respect to any function transferred by this Act and performed after the effective date of this Act, reference in any other law to any department or agency, or any officer or office, the functions of which are so transferred, shall be deemed to refer to the Administration, the Administrator or Commission, or other office or official in which this Act vests such functions.

(i) Nothing contained in this Act shall be construed to limit, curtail, abolish, or terminate any function of the President which he had immediately before the effective date of this Act; or to limit, curtail, abolish, or terminate his authority to perform such function; or to limit, curtail, abolish, or terminate his authority to delegate, redelegate, or terminate any delegation of functions.

(j) Any reference in this Act to any provision if law shall be deemed to include, as appropriate, references thereto as now or hereafter amended or supplemented.

(k) Except as may be otherwise expressly provided in this Act, all functions expressly conferred by this Act shall be in addition to and not in substitution for functions existing immediately before the effective date of this Act and transferred by this Act.

42 USC 5872.

Sec. 302. Transfer of Personnel and Other Matters

(a) Except as provided in the next sentence, the personnel employed in connection with, and the personnel positions, assets, liabilities, contracts, property, records, and unexpended balances of appropriations, authorizations, allocations, and other funds employed, held, used, arising from, available to or to be made available in connection with the functions and programs transferred by this Act, are, subject to section 202 of the Budget and Accounting Procedures Act of 1950 (31 USC 581c), correspondingly transferred for appropriate allocation. Personnel positions expressly created by law, personnel occupying those positions on the effective date of this Act, and personnel authorized to receive compensation at the rate prescribed for offices and positions at levels II, III, IV, or V of the Executive Schedule (5 USC 5313-5316) on the effective date of this Act shall be subject to the provisions of subsection (c) of this section and section 301 of this Act.

(b) Except as provided in subsection (c), transfer of nontemporary personnel pursuant to this Act shall not cause any such employee to be separated or reduced in grade or compensation for one year after such transfer.

(c) Any person who, on the effective date of this Act, held a position compensated in accordance with the Executive Schedule prescribed in chapter 52 of title 5 of the United States Code, and who, without a break in service, is appointed in the Administration to a position having duties comparable to those performed immediately preceding his appointment shall continue to be compensated in his new position at not less than the rate provided for his previous position.

42 USC 5873.

Sec. 303. Incidental Dispositions

The Director of the Office of Management and Budget is authorized to make such additional incidental dispositions of personnel, personnel positions, assets, liabilities, contracts, property, records, and unexpended balances of appropriations, authorizations, allocations, and other funds held, used, arising from, available to or to be made available in connection with functions transferred by this Act, as he may deem necessary or appropriate to accomplish the intent and purpose of this Act.

42 USC 5874.

Sec. 304. Definitions

As used in this Act–

(1) any reference to "function" or "functions" shall be deemed to include references to duty, obligation, power, authority, responsibility, right, privilege, and activity, or the plural thereof, as the case may be; and

(2) any reference to "perform" or "performance", when used in relation to functions, shall be deemed to include the exercise of power, authority, rights, and privileges.

42 USC 5875.

Sec. 305. Authorizations of Appropriations

(a) Except as otherwise provided by law, appropriations made under this Act shall be subject to an annual authorization.

(b) Authorization of appropriations to the Commission shall reflect the need for effective licensing and other regulation of the nuclear power industry in relation to the growth of such industry.

42 USC 5876.

Sec. 306. Comptroller General Audit

42 USC 2206.

(a) Section 166. "Comptroller General Audit" of the Atomic Energy Act of 1954, as amended, shall be deemed to be applicable, respectively, to the nuclear and nonnuclear activities under title I and to the activities under title II.

Report to
Congress.

(b) The Comptroller General of the United States shall audit, review, and evaluate the implementation of the provisions of title II of this Act by the Nuclear Safety and Licensing Commission not later than sixty months after the effective date of this Act, the Comptroller General shall prepare and submit to the Congress a report on his audit, which shall contain, but not be limited to–

(1) an evaluation of the effectiveness of the licensing and related regulatory activities of the Commission and the operations of the Office of Nuclear Safety Research and the Bureau of Nuclear Materials Security;

(2) an evaluation of the effect of such Commission activities on the efficiency, effectiveness, and safety with which the activities licensed under the Atomic Energy Act of 1954, as amended, are carried out;

(3) recommendations concerning any legislation he deems necessary, and the reasons therefor, for improving the implementation of title II.

42 USC 5877.
Administration
activities and
progress.
Reports to the
President and
Congress.

Sec. 307. Reports[26]

(a) The Administrator shall, as soon as practicable after the end of each fiscal year, make a report to the President for submission to the Congress on the activities of the Administration during the preceding fiscal year. Such report shall include a statement of the short-range and long-range goals, priorities, and plans of the Administration together with an assessment of the progress made toward the attainment of objectives and toward the more effective and efficient management of the Administration and the coordination of its functions.

Feasibility of
transferring
military
application
functions.

(b) During the first year of operation of the Administration, the Administrator, in collaboration with the Secretary of Defense, shall conduct a thorough review of the desirability and feasibility of transferring to the Department of Defense or other Federal agencies the functions of the Administrator respecting military application and restricted data, and within one year after the Administrator first takes office, the Administrator shall make a report to the President, for submission to the Congress, setting forth his comprehensive analysis, the principal alternatives, and the specific recommendations of the Administrator and the Secretary of Defense.

---

[26]   The requirements of this section are included in the reporting provisions under section 657 of the Department of Energy Reorganization Act, P.L. 95–91, 42 U.S.C. 7267.

Commission
activities and
findings.

(c) The Commission shall, as soon as practicable after the end of each fiscal year, make a report to the President for submission to the Congress on the activities of the Commission during the preceding fiscal year. Such report shall include a clear statement of the short-range and long-range goals, priorities, and plans of the Commission as they relate to the benefits, costs, and risks of commercial nuclear power. Such report shall also include a clear description of the Commission's activities and findings in the following areas–

(1) insuring the safe design of nuclear power plants and other licensed facilities;

(2) investigating abnormal occurrences and defects in nuclear power plants and other licensed facilities;

(3) safeguarding special nuclear materials at all stages of the nuclear fuel cycle;

(4) investigating suspected, attempted, or actual thefts of special nuclear materials in the licensed sector and developing contingency plans for dealing with such incidents;

(5) insuring the safe, permanent disposal of high-level radioactive wastes through the licensing of nuclear activities and facilities;

(6) protecting the public against the hazards of low-level radioactive emissions from licensed nuclear activities and facilities.

42 USC 5878.

Sec. 308. Information to Committees

The Administrator shall keep the appropriate congressional committees fully and currently informed with respect to all of the Administration's activities.

42 USC 5879.

Sec. 309. Transfer of Funds

The Administrator, when authorized in an appropriation Act, may, in any fiscal year, transfer funds from one appropriation to another within the Administration; except, that no appropriation shall be either increased or decreased pursuant to this section by more than 5 per centum of the appropriation for such fiscal year.

Sec. 310. Conforming Amendments to Certain Other Laws

Subchapter II (relating to Executive Schedule pay rates) of chapter 53 of title 5, United States Code, is amended as follows:

5 USC 5313.

(1) Section 5313 is amended by striking out "(8) Chairman, Atomic Energy Commission," and inserting in lieu thereof "(8) Chairman, Nuclear Regulatory Commission," and by adding at the end thereof the following:

(22) Administrator of Energy Research and Development Administration.

5 USC 5314.

(2) Section 5314 is amended by striking out "(42) Members, Atomic Energy Commission." and inserting in lieu thereof "(42) Members, Nuclear Regulatory Commission.", and by adding at the end thereof the following:

(60) Deputy Administrator, Energy Research and Development Administration.

5 USC 5315.

(3) Section 5315 is amended by striking out paragraph (50), and by adding at the end thereof the following:

(100) Assistant Administrator, Energy Research and Development Administration (6).

(101) Director of Nuclear Reactor Regulation, Nuclear Regulatory Commission.

(102) Director of Nuclear Material Safety and Safeguards, Nuclear Regulatory Commission.

(103) Director of Nuclear Regulatory Research, Nuclear Regulatory Commission.

(104) Executive Director for Operations, Nuclear Regulatory Commission.

5 USC 5316.

(4) Section 5316 is amended by striking out paragraphs (29), (62), (69), and (102), by striking out "(81), General Counsel of the Atomic Energy Commission," and inserting in lieu thereof "(81) General Counsel of the Nuclear Regulatory Commission.", and by adding at the end thereof the following:

(134) General Counsel, Energy Research and Development Administration.

(135) Additional officers, Energy Research and Development Administration (8).

(136) Additional officers, Nuclear Regulatory Commission (5).

42 USC 5801 note.

Sec. 311. Separability

If any provision of this Act, or the application thereof to any person or circumstance, is held invalid, the remainder of this Act, and the application of such provision to other persons or circumstances, shall not be affected thereby.

42 USC 5801 note.

Publication in Federal Register.

Sec. 312. Effective Date and Interim Appointments

(a) This Act shall take effect one hundred and twenty days after the date of its enactment, or on such earlier date the President may prescribe and publish in the Federal Register; except that any of the officers provided for in title I of this Act may be nominated and appointed, as provided by this Act, at any time after the date of enactment of this Act. Funds available to any department or agency (or any official or component thereof), any functions of which are transferred to the Administrator and the Commission by this Act, may, with the approval of the President, be used to pay the compensation and expenses of any officer appointed pursuant to this subsection until such time as funds for that purpose are otherwise available.

(b) In the event that any officer required by this Act to be appointed by and with the advice and consent of the Senate shall not have entered upon office on the effective date of this Act, the President may designate any officer, whose appointment was required to be made by and with the advice and consent of the Senate and who was such an officer immediately prior to the effective date of this Act, to act in such office until the office is filled as provided in this Act. While so acting, such persons shall receive compensation at the rates provided by this Act for the respective offices in which they act.

Title IV–Sex Discrimination

42 USC 5891.

Sec. 401. Sex Discrimination Prohibited

No person shall on the ground of sex be excluded from participation in, be denied a license under, be denied the benefits of, or be subjected to discrimination under any program or activity carried on or receiving Federal assistance under any title of this Act. This provision will be enforced through agency provisions and rules similar to those already established, with respect to racial and other discrimination, under title VI of the Civil Rights Act of 1964. However, this remedy is not exclusive and will not prejudice or cut off any other legal remedies available to a discriminate.

42 USC 2000d.

4. Reorganization Plan of 1980 and Other Documents Pertaining to NRC Jurisdiction

4. Reorganization Plan of 1980 and Other Documents Pertaining to NRC Jurisdiction
Contents

A. REORGANIZATION PLAN NO. 1 OF 1980

5 USC App. 1.   Prepared by the President and submitted to the Senate and the House of Representatives in Congress assembled March 27, 1980,[1] pursuant to the provisions of chapter 9 of title 5 of the United States Code.[2]

Nuclear Regulatory Commission

Sec. 1. (a) Those functions of the Nuclear Regulatory Commission, hereinafter referred to as the "Commission," concerned with:

(1) policy formulation;

(2) rulemaking, as defined in section 553 of title 5 of the United States Code, except that those matters set forth in 553(a)(2) and (b) which do not pertain to policy formulation orders or adjudications shall be reserved to the Chairman of the Commission;

(3) orders and adjudications, as defined in section 551 (6) and (7) of title 5 of the United States Code;

shall remain vested in the Commission. The Commission may determine by majority vote, in an area of doubt, whether any matter, action, question or area of inquiry pertains to one of these functions. The performance of any portion of these functions may be delegated by the Commission to a member of the Commission, including the Chairman of the Nuclear Regulatory Commission, hereinafter referred to as the "Chairman", and to the staff through the Chairman.

(b)(1) With respect to the following officers or successor officers duly established by statute or by the Commission, the Chairman shall initiate the appointment, subject to the approval of the Commission; and the Chairman or a member of the Commission may initiate an action for removal, subject to the approval of the Commission:

(i) Executive Director for Operations,

(ii) General Counsel,

(iii) Secretary of the Commission,

(iv) Director of the Office of Policy Evaluation,

(v) Director of the Office of Inspector and Auditor,

(vi) Chairman, Vice Chairman, Executive Secretary, and Members of the Atomic Safety and Licensing Board Panel,

(vii) Chairman, Vice Chairman and Members of the Atomic Safety and Licensing Appeal Panel.

(2) With respect to the following officers or successor officers duly established by statute or by the Commission, the Chairman, after consultation with the Executive Director for Operations, shall initiate the appointment, subject to the approval of the Commission, and the Chairman, or a member of the Commission may initiate an action for removal, subject to the approval of the Commission:

(i) Director of Nuclear Reactor Regulation,

(ii) Director of Nuclear Material Safety and Safeguards,

(iii) Director of Nuclear Regulatory Research,

(iv) Director of Inspection and Enforcement.

---

[1]   As amended May 5, 1980. See House Document No. 96–307, Amendments to Reorganization Plan No. 1 of 1980, Message from the President of the United States.
[2]   This Reorganization Plan was originally approved under special Congressional procedures; the Supreme Court decision in *Immigration & Naturalization Service vs. Chadha* (462 US 919 (1983)) called into question the legality of this plan. Congress responded by enacting this Reorganization Plan in P.L. 98–614.

(v) Director of Standards Development.

(3) The Chairman or a member of the Commission shall initiate the appointment of the Members of the Advisory Committee on Reactor Safeguards, subject to the approval of the Commission. The provisions for appointment of the Chairman of the Advisory Committee on Reactor Safeguards and the term of the members shall not be affected by the provisions of this Reorganization Plan.

(4) The Commission shall delegate the function of appointing, removing and supervising the staff of the following offices or successor offices to the respective heads of such offices: General Counsel, Secretary of the Commission, Office of Policy Evaluation, Office of Inspector and Auditor. The Commission shall delegate the functions of appointing, removing and supervising the staff of the following panels and committee to the respective Chairman thereof: Atomic Safety and Licensing Board Panel, Atomic Safety and Licensing Appeal Panel and Advisory Committee on Reactor Safeguards.

(c) Each member of the Commission shall continue to appoint, remove and supervise the personnel employed in his or her immediate office.

(d) The Commission shall act as provided by subsection 201(a)(1) of the Energy Reorganization Act of 1974, as amended (42 USC 5841 (a)(1), as amended) in the performance of its functions as described in subsections (a) and (b) of this section.

Sec. 2. (a) All other functions of the Commission, not specified by section 1 of this Reorganization Plan, are hereby transferred to the Chairman. The Chairman shall be the official spokesman for the Commission, and shall appoint, supervise, and remove, without further action by the Commission, the Directors and staff of the Office of Public Affairs and the Office of Congressional Relations. The Chairman may consult with the Commission as he deems appropriate in exercising this appointment function.

(b) The Chairman shall also be the principal executive officer of the Commission, and shall be responsible to the Commission for developing policy planning and guidance for consideration by the Commission; shall be responsible for the Commission for assuring that the Executive Director for Operations and the staff of the Commission (other than the officers and staff referred to in sections (1)(b)(4), (1)(c) and (2)(a) of this Reorganization Plan) are responsive to the requirements of the Commission in the performance of its functions; shall determine the use and expenditure of funds of the Commission, in accordance with the distribution of appropriated funds according to major programs and purposes approved by the Commission; shall present to the Commission for its consideration the proposals and estimates set forth in subsection (3) of this paragraph; and shall be responsible for the following functions, which he shall delegate, subject to his direction and supervision, to the Executive Director for Operations unless otherwise provided by this Reorganization Plan:

(1) administrative functions of the Commission;

(2) distribution of business among such personnel and among administrative units and offices of the Commission;

(3) preparation of

(i) proposals for the reorganization of the major offices within the Commission;

(ii) the budget estimate for the Commission; and

(iii) the proposed distribution of appropriated funds according to major programs and purposes.

(4) appointing and removing without any further action by the Commission, all officers and employees under the Commission other than those whose appointment and removal are specifically provided for by subsections 1 (b), (c) and 2(a) of this Reorganization Plan.

(c) The Chairman as principal executive officer and the Executive Director for Operations shall be governed by the general policies of the Commission and by such regulatory decisions, findings, and determinations, including those for reorganization proposals, budget revisions and distribution of appropriated funds, as the Commission may by law, including this Plan, be authorized to make. The Chairman and the Executive Director for Operations, through the Chairman, shall be responsible for insuring that the Commission is fully and currently informed about matters within its functions.

Sec. 3. (a) Notwithstanding sections 1 and 2 of this Reorganization Plan, there are hereby transferred to the Chairman all the functions vested in the Commission pertaining to an emergency concerning a particular facility or materials licensed or regulated by the Commission, including the functions of declaring, responding, issuing orders, determining specific policies, advising the civil authorities, and the public, directing, and coordinating actions relative to such emergency incident.

(b) The Chairman may delegate the authority to perform such emergency functions, in whole or in part, to any of the other members of the Commission. Such authority may also be delegated or re-delegated, in whole or in part to the staff of the Commission.

(c) In acting under this section, the Chairman, or other member of the Commission delegated authority under subsection (b), shall conform to the policy guidelines of the Commission. To the maximum extent possible under the emergency conditions, the Chairman or other member of the Commission delegated authority under subsection (b), shall inform the Commission of actions taken relative to the emergency.

(d) Following the conclusion of the emergency, the Chairman, or the member of the Commission delegated the emergency functions under subsection (b), shall render a complete and timely report to the Commission on the actions taken during the emergency.

Sec. 4. (a) The Chairman may make such delegations and provide for such reporting as the Chairman deems necessary, subject to provisions of law and this Reorganization Plan. Any officer or employee under the Commission may communicate directly to the Commission, or to any member of the Commission, whenever in the view of such officer or employee a critical problem or public health and safety or common defense and security is not being properly addressed.

(b) The Executive Director for Operations shall report for all matters to the Chairman.

(c) The function of the Director of Nuclear Reactor Regulation, Nuclear Material Safety and Safeguards, and Nuclear Regulatory Research of reporting directly to the Commission is hereby transferred so that such officers report to the Executive Director for Operations. The function of receiving such reports is hereby transferred from the Commission to the Executive Director for Operations.

(d) The heads of the Commission level offices or successor offices, of General Counsel, Secretary to the Commission, Office of Policy Evaluation, Office of Inspector and Auditor, the Atomic Safety and Licensing Board Panel and Appeal Panel, and Advisory Committee on

Reactor Safeguards shall continue to report directly to the Commission and the Commission shall continue to receive such reports.

Sec. 5. The provisions of this Reorganization Plan shall take effect October 1, 1980, or at such earlier time or times as the President shall specify, but no sooner than the earliest time allowable under section 906 of title 5 of the United States Code.[3]

---

[3]   See 45 FR 40561.

B. REORGANIZATION PLAN NO. 3 OF 1970

Title III–The President

5 USC App. 1. Prepared by the President and transmitted to the Senate and the House of Representatives in Congress assembled July 9, 1970, pursuant to the provisions of chapter 9 of title 5 of the United States Code.[1] [2]

Environmental Protection Agency

Sec. 1. Establishment of Agency

(a) There is hereby established the Environmental Protection Agency, hereinafter referred to as the "Agency."

(b) There shall be at the head of the Agency the Administrator of the Environmental Protection Agency, hereinafter referred to as the "Administrator." The Administrator shall be appointed by the President, by and with the advice and consent of the Senate, and shall be compensated at the rate now or hereafter provided for Level II of the Executive Schedule Pay rates (5 USC 5313).

(c) There shall be in the Agency a Deputy Administrator of the Environmental Protection Agency who shall be appointed by the President, by and with the advice and consent of the Senate, and shall be compensated at the rate now or hereafter provided for Level III of the Executive Schedule Pay Rates (5 USC 5314). The Deputy Administrator shall perform such functions as the Administrator shall from time to time assign or delegate, and shall act as Administrator during the absence or disability of the Administrator or in the event of a vacancy in the office of Administrator.

(d) There shall be in the Agency not to exceed five Assistant Administrators of the Environmental Protection Agency who shall be appointed by the President, by and with the advice and consent of the Senate, and shall be compensated at the rate now or hereafter provided for Level IV of the Executive Schedule Pay Rates (5 USC 5315). Each Assistant Administrator shall perform such functions as the Administrator shall from time to time assign or delegate.

Sec. 2. Transfers to Environmental Protection Agency

(a) There are hereby transferred to the Administrator:

(1) All functions vested by law in the Secretary of the Interior and the Department of the Interior which are administered through the Federal Water Quality Administration, all functions which were transferred to the Secretary of the Interior by Reorganization Plan No. 2 of 1966 (80 Stat. 1608), and all functions vested in the Secretary of the Interior or the Department of the Interior by the Federal Water Pollution Control Act or by provisions of law amendatory or supplementary thereof.

(2)(i) The functions vested in the Secretary of the Interior by the Act of August 1, 1958, 72 Stat. 479, 16 USC 742d-1 (being an Act relating to studies on the effects of insecticides, herbicides, fungicides, and pesticides upon the fish and wildlife resources of the

---

[1]   Effective December 2, 1970, under the provisions of section 7 of the plan.

[2]   This Reorganization Plan was originally approved under special Congressional procedures; the Supreme Court decision in *Immigration & Naturalization Service vs. Chadha* (462 US 919 (1983)) called into question the legality of this plan. Congress responded by enacting this Reorganization Plan in P.L. 98–614.

United States), and (ii) the functions vested by law in the Secretary of the Interior and the Department of the Interior which are administered by the Gulf Breeze Biological Laboratory of the Bureau of Commercial Fisheries at Gulf Breeze, Florida.

(3) The functions vested by law to the Secretary of Health, Education, and Welfare or in the Department of Health, Education, and Welfare which are administered through the Environmental Health Service, including the functions exercised by the following components thereof:

      (i) The National Air Pollution Control Administration,
      (ii) The Environmental Control Administration
    (A) Bureau of Solid Waste Management,
    (B) Bureau of Water Hygiene,
    (C) Bureau of Radiological Health,

except that functions carried out by the following components of the Environmental Control Administration of the Environmental Health Service are not transferred: (i) Bureau of Community Environmental Management, (ii) Bureau of Occupational Safety and Health, and (iii) Bureau of Radiological Health, insofar as the functions carried out by the latter Bureau pertain to (A) regulation of radiation from consumer products, including electronic product radiation, (B) radiation as used in the healing arts, (C) occupational exposures to radiation, and (D) research, technical assistance, and training related to clauses (A), (B), and (C).

(4) The functions vested in the Secretary of Health, Education, and Welfare of establishing tolerances for pesticide chemicals under the Federal Food, Drug, and Cosmetic Act as amended, 21 USC 346, 346a, and 348, together with authority, in connection with the functions transferred, (i) to monitor compliance with the tolerances and the effectiveness of surveillance and enforcement, and (ii) to provide technical assistance to the States and conduct research under the Federal Food, Drug, and Cosmetic Act, as amended, and the Public Health Service Act, as amended.

(5) So much of the functions of the Council on Environmental Quality under section 204(5) of the National Environmental Policy Act of 1969 (Public Law 91-190, approved January 1, 1970, 83 Stat. 855), as pertains to ecological systems.

(6) The functions of the Atomic Energy Commission under the Atomic Energy Act of 1954, as amended, administered through its Division of Radiation Protection Standards, to the extent that such functions of the Commission consist of establishing generally applicable environmental standards for the protection of the general environment from radioactive material. As used herein, standards mean limits on radiation exposures or levels, or concentrations or quantities of radioactive material, in the general environment outside the boundaries of locations under the control of persons possessing or using radioactive material.

(7) All functions of the Federal Radiation Council (42 USC 2021(h)).

(8)(i) The functions of the Secretary of Agriculture and the Department of Agriculture under the Federal Insecticide, Fungicide, and the Rodenticide Act, as amended (7 USC 135-135k), (ii) the functions of the Secretary of Agriculture and the Department of Agriculture under section 408 (1) of the Federal Food, Drug, and Cosmetic Act, as amended (21 USC 346a (1)), and (iii) the functions vested by law in the Secretary of Agriculture and the Department of

Agriculture which are administered through the Environmental Quality Branch of the Plant Protection Division of the Agricultural Research Service.

(9) So much of the functions of the transferor officers and agencies referred to in or affected by the foregoing provisions of this section as is incidental to or necessary for the performance by or under the Administrator of the functions transferred by those provisions or relates primarily to those functions. The transfers to the Administrator made by this section shall be deemed to include the transfer of (1) authority, provided by law, to prescribe regulations relating primarily to the transferred functions, and (2) the functions vested in the Secretary of the Interior and the Secretary of Health, Education, and Welfare by section 169(d)(1)(b) and (3) of the Internal Revenue Code of 1954 (as enacted by section 704 of the Tax Reform Act of 1969, 83 Stat. 668); but shall be deemed to exclude the transfer of the functions of the Bureau of Reclamation under section 3(b)(1) of the Water Pollution Control Act (33 USC 466a(b)(1)).

(b) There are hereby transferred to the Agency:

(1) From the Department of the Interior, (i) the Water Pollution Control Advisory Board (33 USC 466f), together with its functions, and (ii) the hearing boards provided for in section 10(c)(4) and 10(f) of the Federal Water Pollution Control Act, as amended (33 USC 466g(c)(4): 466g(f)). The functions of the Secretary of the Interior with respect to being or designating the Chairman of the Water Pollution Control Advisory Board are hereby transferred to the Administrator.

(2) From the Department of Health, Education, and Welfare, the Air Quality Advisory Board (42 USC 1857e), together with its functions. The functions of the Secretary of Health, Education, and Welfare with respect to being a member and the Chairman of that Board are hereby transferred to the Administrator.

Sec. 3. Performance of Transferred Functions

The Administrator may from time to time make such provisions as he shall deem appropriate authorizing the performance of any of the functions transferred to him by the provisions of this reorganization plan by any other officer or by any organizational entity or employee, of the Agency.

Sec. 4. Incidental Transfers

(a) So much of the personnel, property, records, and unexpended balances of appropriations, allocations, and other funds employed, used, held, available, or to be made available in connection with the functions transferred to the Administrator or the Agency by this reorganization plan as the Director of the Office of Management and Budget shall determine shall be transferred to the Agency at such time or times as the Director shall direct.

(b) Such further measures and dispositions as the Director of Office of Management and Budget shall deem to be necessary in order to effectuate the transfers referred to in subsection (a) of this section shall be carried out in such manner as he shall direct and by such agencies as he shall designate.

Sec. 5. Interim Officers

(a) The President may authorize any person who immediately prior to the effective date of this reorganization plan held a position in the executive branch of the Government to act as Administrator until the office of Administrator is for the first time filled pursuant to the provisions of this reorganization plan or by recess appointment, as the case may be.

(b) The President may similarly authorize any such person to act as Deputy Administrator, authorize any such person to act as Assistant Administrator, and authorize any such person to act as the head of any principal constituent organizational entity of the Administration.

(c) The President may authorize any person who serves in an acting capacity under the foregoing provisions of this section to receive the compensation attached to the office in respect of which he so serves. Such compensation, if authorized, shall be in lieu of, but not in addition to, other compensation from the United States to which such person may be entitled.

Sec. 6. Abolitions

(a) Subject to the provisions of this reorganization plan, the following, exclusive of any functions, are hereby abolished:

(1) The Federal Water Quality Administration in the Department of the Interior (33 USC 466-1).

(2) The Federal Radiation Council (73 Stat. 690; 42 USC 2021(h)).

(b) Such provisions as may be necessary with respect to terminating any outstanding affairs shall be made by the Secretary of the Interior in the case of the Federal Water Quality Administration and by the Administrator of General Services in the case of the Federal Radiation Council.

Sec. 7. Effective Date

The provisions of this reorganization plan shall take effect sixty days after the date they would take effect under 5 USC 906(a) in the absence of this section.

C. EXECUTIVE ORDER 11834—ACTIVATION OF THE ENERGY RESEARCH AND DEVELOPMENT ADMINISTRATION AND THE NUCLEAR REGULATORY COMMISSION

THE WHITE HOUSE

Activation of the Energy Research and Development Administration and the Nuclear Regulatory Commission

By virtue of the authority vested in me by the Energy Reorganization Act of 1974 (Public Law 93-438; 88 Stat. 1233), section 301 of title 3 of the United States Code, and as President of the United States of America, it is hereby ordered:

Sec. 1. Pursuant to section 312(a) of the Energy Reorganization Act of 1974, I hereby prescribe January 19, 1975, as the effective date of that Act. This action shall not impair in any way the activation of the Energy Resources Council by Executive Order No. 11814 of October 11, 1974.

Sec. 2. The Director of the Office of Management and Budget shall take all steps necessary or appropriate to ensure or effectuate the transfers provided for in the Energy Reorganization Act of 1974, the Solar Heating and Cooling Demonstration Act of 1974 (Public Law 93-409; 88 Stat. 1069), the Geothermal Energy Research, Development, and Demonstration Act of 1974 (Public Law 93-410; 88 Stat. 1079), the Solar Energy Research, Development, and Demonstration Act of 1974 (Public Law 93-473; 88 Stat. 1431), to the extent required or permitted by law, including transfers of funds, personnel and positions, assets liabilities, contracts, property, records, and other items related to the transfer of functions, programs, or authorities.

Sec. 3. As required by the Energy Reorganization Act of 1974, this Order shall be published in the Federal Register.

GERALD R. FORD

THE WHITE HOUSE,

January 15, 1975.

D. OMB MEMORANDUM REGARDING RESPONSIBILITY FOR SETTING RADIATION PROTECTION STANDARDS

Washington, D.C. 20503

December 7, 1973

MEMORANDUM FOR: ADMINISTRATOR TRAIN
                  CHAIRMAN RAY

SUBJECT: Responsibility for Setting Radiation Protection Standards

FROM: Roy L. Ash

Thank you for providing position papers which outline the background and the current difference of views between your two agencies as to which should have the responsibility for issuing standards to define permissible limits on radioactivity that may be emitted from facilities in the nuclear power industry.

It is clear, as your paper indicated, that a decision is needed on this matter so that the nuclear power industry and the general public will know where the responsibility lies for developing (including public participation in development), promulgating and enforcing radiation protection standards for various types of facilities in the nuclear power industry. We must, in the national interest, avoid confusion in this area, particularly since nuclear power is expected to supply a growing share of the Nation's energy requirements; and it must be clear that we are assuring continued full protection of the public health and the environment from radiation hazards.

It is also clear from the information which you provided that:

> the area of responsibility now in controversy is intimately related to the direct regulatory responsibilities and capabilities of the Atomic Energy Commission, responsibilities about which there is no dispute.

> EPA has construed too broadly its responsibilities, as set forth in Reorganization Plan No. 3 of 1970, to set "generally applicable environmental standards for the protection of the general environment from radioactive material."

On behalf of the President, this memorandum is to advise you that the decision is that AEC should proceed with its plans for issuing uranium fuel cycle standards, taking into account the comments received from all sources, including EPA; that EPA should discontinue its preparations for issuing, now or in the future, any standards for types of facilities; and that EPA should continue, under its current authority, to have responsibility for setting standards for the total amount of radiation in the general environment from all facilities combined in the uranium fuel cycle, i.e., an ambient standard which would have to reflect AEC's findings as to the practicability of emission controls.

EPA can continue to have a major impact upon standards for facilities set by AEC through EPA's review of proposed standards, during which EPA can bring to bear its knowledge and perspective derived from its responsibility for setting ambient radiation standards.

The President expects that AEC and EPA continue to work together to carry out the responsibilities as outlined above.

E. PRESIDENT'S COMMISSION ON THE ACCIDENT AT THREE MILE ISLAND

Public Papers of the Presidents of the United States: Jimmy Carter, 1979

Remarks Announcing Actions in Response to the Commission's Report. December 7, 1979.

The purpose of this brief statement this afternoon is to outline to you and to the public, both in this country and in other nations of the world, my own assessment of the Kemeny report recommendations on the Three Mile Island accident. And I would like to add, of course, in the presentation some thoughts and actions of my own.

I have reviewed the report of the Commission, which I established to investigate the accident at the Three Mile Island nuclear power plant. The Commission, headed by Dr. John Kemeny, found very serious shortcomings in the way that both the Government and the utility industry regulate and manage nuclear power.

The steps that I am taking today will help to assure that nuclear power plants are operated safely. Safety, as it always has been and will remain, is my top priority.

As I've said before, in this country nuclear power is an energy source of last resort. By this I meant that as we reach our goals on conservation, on the direct use of coal, on development of solar power and synthetic fuels, and enhanced production of American oil and natural gas – as we reach those goals, then we can minimize our reliance on nuclear power.

Many of our foreign allies must place much greater reliance than we do on nuclear power, because they do not have the vast natural resources that give us so many alternatives. We must get on with the job of developing alternative energy resources, and we must also pass, in order to do this, the legislation that I have proposed to the Congress in making an effort, at every level of society, to conserve energy. To conserve energy and to develop energy resources in our country are the two basic answers for which we are seeking. But we cannot shut the door on nuclear power for the United States.

The recent events in Iran have shown us the clear, stark dangers that excessive dependence on imported oil holds for our Nation. We must make every effort to lead this country to energy security. Every domestic energy source, including nuclear power, is critical if we are to be free as a country from our present over-dependence on unstable and uncertain sources of high-priced foreign oil.

We do not have the luxury of abandoning nuclear power or imposing a lengthy moratorium on its further use. A nuclear power plant can displace 35,000 barrels of oil per day, or roughly 13 million barrels of oil per year. We must take every possible step to increase the safety of nuclear power production. I agree fully with the letter and the spirit and the intent of the Kemeny commission's recommendations, some of which are within my own power to implement, others of which rely on the Nuclear Regulatory Commission, or the NRC, or the utility industry itself.

To get the Government's own house in order, I will take several steps. First, I will send to the Congress a reorganization plan to strengthen the role of the Chairman of the NRC, to clarify assignment of authority and responsibility, and provide this person with the power to act on a daily basis as a chief executive officer with authority to put needed safety requirements in place and to implement better procedures. The Chairman must be able to select key personnel and to act on behalf of the Commission during any emergency.

Second, I intend to appoint a new Chairman of the Nuclear Regulatory Commission, someone from outside that agency, in the spirit of the Kemeny commission's recommendation. In the meantime, I've asked Commissioner Ahearne, now on the NRC, to serve as the Chairman. Mr. Ahearne will stress safety and the prompt implementation of the needed reforms.

In addition, I will establish an independent advisory committee to help keep me and the public of the United States informed of the progress of the NRC and the industry in achieving and in making clear the recommendations that nuclear power will be safer.

Third, I'm transferring responsibility to the Federal Emergency Management Agency, the FEMA, to head up all off site emergency activities and to complete a thorough review of emergency plans in all the States of our country with operating nuclear reactors by June 1980.

Fourth, I have directed the Nuclear Regulatory Commission and the other agencies of the Government to accelerate our program to place a resident Federal inspector at every reactor site.

Fifth, I'm asking all relevant Government agencies to implement virtually all of the other recommendations of the Kemeny commission – I believe there were 44 in all. A detailed fact sheet is being issued to the public, and a more extended briefing will be given to the press this afternoon.

With clear leadership and improved organization, the executive branch of Government and the NRC will be better able to act quickly on the crucial issues of improved training and standards, safety procedures, and the other Kemeny commission recommendations. But responsibility to make nuclear power safer does not stop with the Federal Government. In fact, the primary, day-by-day responsibility for safety rests with utility company management and with suppliers of nuclear equipment. There is no substitute for technically qualified and committed people working on the construction, the operation, and the inspection of nuclear powerplants.

Personal responsibility must be stressed. Some one person must always be designated as in charge, both at the corporate level and also at the power plant site. The industry owes it to the American people to strengthen its commitment to safety.

I call on the utilities to implement the following changes. First, building on the steps already taken, the industry must organize itself to develop enhanced standards for safe design, operation, and construction of plants. Second, the nuclear industry must work together to develop and to maintain in operation a comprehensive training, examination, and evaluation program for operators and for supervisors. This training program must pass muster with the NRC through accreditation of the training programs to be established. Third, control rooms in nuclear power plants must be modernized, standardized, and simplified as much as possible to permit better informed decision–making among regular operating hours and, of course, during emergencies.

I challenge our utility companies to bend every effort to improve the safety of nuclear power.

Finally, I would like to discuss how we manage this transition period during which the Kemeny recommendations are being implemented. There are a number of new nuclear plants now awaiting operating licenses or construction permits. Under law, the Nuclear Regulatory

Commission is an independent agency. Licensing decisions rest with the Nuclear Regulatory Commission, and, as the Kemeny commission noted, it has the authority to proceed with licensing these plants on a case-by-case basis, which may be used as circumstances surrounding a plant or its application dictate.

The NRC has indicated, however, that it will pause in issuing any new licenses and construction permits in order to devote its full attention to putting its own house in order and tightening up safety requirements. I endorse this approach which the NRC has adopted, but I urge the NRC to complete its work as quickly as possible and in no event later than 6 months from today. Once we've instituted the necessary reforms to assure safety, we must resume the licensing process promptly so that the new plants we need to reduce our dependence on foreign oil can be built and operated.

The steps I'm announcing today will help to ensure that our safety has the safety of nuclear plants. Nuclear power does have a future in the United States. It's an option that we must keep open. I will join with the utilities and their suppliers, the Nuclear Regulatory Commission, the executive departments and agencies of the Federal Government, and also the State and local governments to assure that the future is a safe one.

and now Dr. Frank Press, Stu Eizenstat, and John Deutch will be glad to answer your questions about these decisions and about nuclear power and the future of it in our country. Frank?

NOTE: The President spoke at 2:45 p.m. in Room 450 of the Old Executive Office Building.

Following the President's remarks, Frank Press, Director of the Office of Science and Technology Policy, Stuart E. Eizenstat, Assistant to the President for Domestic Affairs and Policy, and Under Secretary of Energy John .M. Deutch held a news conference on the announcements.

5. Low-Level Radioactive Waste

5. Low-Level Radioactive Waste
Contents

A. LOW-LEVEL RADIOACTIVE WASTE POLICY ACT OF 1985 (TITLE 1), AS AMENDED

Public Law 99–240   **99 Stat. 1842**

January 15, 1986

An Act

Low-Level Radioactive Waste Policy Amendments Act of 1985.

To amend the Low-Level Radioactive Waste Policy Act to improve procedures for the implementation of compacts providing for the establishment and operation of regional disposal facilities for low-level radioactive waste; to grant the consent of the Congress to certain interstate compacts on low-level radioactive waste; and for other purposes.[1]

State and local governments.

Be it enacted by the Senate and House of Representatives of the United States of America in Congress assembled,

Title I–Low-Level Radioactive Waste Policy Amendments Act of 1985

42 USC 2021b note.

Sec. 101. Short Title

This title may be cited as the "Low-Level Radioactive Waste Policy Amendments Act of 1985."

Sec. 102. Amendment to the Low-Level Radioactive Waste Policy Act

42 USC 2021b-2021d.
42 USC 2021b note.

The Low-Level Radioactive Waste Policy Act (42 USC 2021b et seq.) is amended by striking out sections 1, 2, 3, and 4 and inserting in lieu thereof the following:

42 USC 2021b note.

Sec. 1. Short Title

This Act may be cited as the "Low-Level Radioactive Waste Policy Act."

42 USC 2021b.

Sec. 2. Definitions

For purposes of this Act:

(1) Agreement State.–The term "agreement State" means a State that–

(A) has entered into an agreement with the Nuclear Regulatory Commission under section 274 of the Atomic Energy Act of 1954 (42 USC 2021); and

(B) has authority to regulate the disposal of low-level radioactive waste under such agreement.

(2) Allocation.–The term "allocation" means the assignment of a specific amount of low-level radioactive waste disposal capacity to a commercial nuclear power reactor for which access is required to be provided by sited States subject to the conditions specified under this Act.

(3) Commercial Nuclear Power Reactor.–The term "commercial nuclear power reactor" means any unit of a civilian light-water moderated utilization facility required to be licensed under section

---

[1]   P.L. 96–573, 94 Stat. 3347 (1980), "Low–Level Radioactive Waste Policy Act of 1980," was completely amended by P.L. 99–240.

103 or 104b. of the Atomic Energy Act of 1954 (42 USC 2133 or 2134(b)).

(4) Compact.–The term "compact" means a compact entered into by two or more States pursuant to this Act.

(5) Compact Commission.–The term "compact commission" means the regional commission, committee, or board established in a compact to administer such compact.

(6) Compact Region.–The term "compact region" means the area consisting of all States that are members of a compact.

(7) Disposal.–The term "disposal" means the permanent isolation of low-level radioactive waste pursuant to the requirements established by the Nuclear Regulatory Commission under applicable laws, or by an agreement State if such isolation occurs in such agreement State.

(8) Generate.–The term "generate", when used in relation to low-level radioactive waste, means to produce low-level radioactive waste.

(9) Low-level Radioactive Waste.–

(A) IN GENERAL–The term "low-level radioactive waste" means radioactive material that–

(i) is not high-level radioactive waste, spent nuclear fuel, or byproduct material (as defined in section 11e.(2) of the Atomic Energy Act of 1954 (42 USC 2014(e)(2))); and

(ii) the Nuclear Regulatory Commission, consistent with existing law and in accordance with paragraph (A), classifies as low-level radioactive waste.

(B) EXCLUSION–The term 'low-level radioactive waste' does not include byproduct material (as defined in paragraphs (3) and (4) of section 11e. of the Atomic Energy Act of 1954 (42 USC 2014(e)).[2]

(10) Non-sited Compact Region.–The term "non-sited compact region" means any compact region that is not a sited compact region.

(11) Regional Disposal Facility.–The term "regional disposal facility" means a non-Federal low-level radioactive waste disposal facility in operation on January 1, 1985, or subsequently established and operated under a compact.

(12) Secretary.–The term "Secretary" means the Secretary of Energy.

(13) Sited Compact Region.–

Nevada.
South Carolina.
Washington.

The term "sited compact region" means a compact region in which there is located one of the regional disposal facilities at Barnwell, in the State of South Carolina; Richland, in the State of Washington; or Beatty, in the State of Nevada.

(14) State.–The term "State" means any State of the United States, the District of Columbia, and the Commonwealth of Puerto Rico.

42 USC 2021c.

Sec. 3. Responsibilities for Disposal of Low-Level Radioactive Waste

Section 3(a)(1) State Responsibilities.–Each State shall be responsible for providing, either by itself or in cooperation with other States, for the disposal of–

---

[2]   Amended by P.L. 109–58, Title VI, § 651(e)(3)(B), 119 Stat. 808 (2005).
NOTE: For other requirements of P.L. 109–58, Title VI, § 651e., 119 Stat. 808 (2005), relating to the addition of paragraphs (3) and (4) to section 11e of the Atomic Energy Act of 1954, see footnote 8 of the Atomic Energy Act of 1954, as set forth in this NUREG.

(A) low-level radioactive waste generated within the State (other than by the Federal Government) that consists of or contains class A, B, or C radioactive waste as defined by section 61.55 of title 10, Code of Federal Regulations, as in effect on January 26, 1983;

Vessels.

(B) low-level radioactive waste described in subparagraph (A) that is generated by the Federal Government except such waste that is–

(i) owned or generated by the Department of Energy;

(ii) owned or generated by the United States Navy as a result of the decommissioning of vessels of the United States Navy; or

Research and development.

(iii) owned or generated as a result of any research, development, testing, or production of any atomic weapon; and

(C) low-level radioactive waste described in subparagraphs (A) and (B) that is generated outside of the State and accepted for disposal in accordance with section 5 or 6.

(2) No regional disposal facility may be required to accept for disposal any material–

(A) that is not low-level radioactive waste as defined by section 61.55 of title 10, Code of Federal Regulations, as in effect on January 26, 1983, or

(B) identified under the Formerly Utilized Sites Remedial Action Program.

Nothing in this paragraph shall be deemed to prohibit a State, subject to the provisions of its compact, or a compact region from accepting for disposal any material identified in subparagraph (A) or (B).

(b)(1) The Federal Government shall be responsible for the disposal of–

(A) low-level radioactive waste owned or generated by the Department of Energy;

Vessels.

(B) low-level radioactive waste owned or generated by the United States Navy as a result of the decommissioning of vessels of the United States Navy;

Health.
Research and development.

(C) low-level radioactive waste owned or generated by the Federal Government as a result of any research, development, testing, or production of any atomic weapon; and

(D) any other low-level radioactive waste with concentrations of radionuclides that exceed the limits established by the Commission for class C radioactive waste, as defined by section 61.55 of title 10, Code of Federal Regulations, as in effect on January 26, 1983.

42 USC 2011 note.
Safety.

(2) All radioactive waste designated a Federal responsibility pursuant to subparagraph (b)(1)(D) that results from activities licensed by the Nuclear Regulatory Commission under the Atomic Energy Act of 1954, as amended, shall be disposed of in a facility licensed by the Nuclear Regulatory Commission that the Commission determines is adequate to protect the public health and safety.

Report.

(3) Not later than 12 months after the date of enactment of this Act, the Secretary shall submit to the Congress a comprehensive report setting forth the recommendations of the Secretary for ensuring the safe disposal of all radioactive waste designated a Federal responsibility pursuant to subparagraph (b)(1)(D). Such report shall include–

(A) an identification of the radioactive waste involved, including the source of such waste, and the volume, concentration, and other relevant characteristics of such waste;

(B) an identification of the Federal and non-Federal options for disposal of such radioactive waste;

(C) a description of the actions proposed to ensure the safe disposal of such radioactive waste;

(D) a description of the projected costs of undertaking such actions;

(E) an identification of the options for ensuring that the beneficiaries of the activities resulting in the generation of such radioactive wastes bear all reasonable costs of disposing of such wastes; and

(F) an identification of any statutory authority required for disposal of such waste.

Prohibition.
Report.
(4) The Secretary may not dispose of any radioactive waste designated a Federal responsibility pursuant to paragraph (b)(1)(D) that becomes a Federal responsibility for the first time pursuant to such paragraph until ninety days after the report prepared pursuant to paragraph (3) has been submitted to the Congress.

42 USC 2021d.
Sec. 4. Regional Compacts for Disposal of Low-Level Radioactive Waste

(a) In General–

(1) Federal Policy.–It is the policy of the Federal Government that the responsibilities of the States under section 3 for the disposal of low-level radioactive waste can be most safely and effectively managed on a regional basis.

(2) Interstate Compacts.–To carry out the policy set forth in paragraph (1), the States may enter into such compacts as may be necessary to provide for the establishment and operation of regional disposal facilities for low-level radioactive waste.

(b) Applicability To Federal Activities.–

(1) In General.–

Prohibition.
(A) Activities of The Secretary.–Except as provided in subparagraph (B), no compact or act taken under a compact shall be applicable to the transportation, management, or disposal of any low-level radioactive waste designated in section 3(a)(1)(B) (i)-(iii).

(B) Federal Low-level Radioactive Waste Disposed of At Non-federal Facilities.–Low-level radioactive waste owned or generated by the Federal Government that is disposed of at a regional disposal facility or non-Federal disposal facility within a State that is not a member of a compact shall be subject to the same conditions, regulations, requirements, fees, taxes, and surcharges imposed by the compact commission, and by the State in which such facility is located, in the same manner and to the same extent as any low-level radioactive waste not generated by the Federal Government.

Prohibition.
(2) Federal Low-level Radioactive Waste Disposal Facilities.–Any low-level radioactive waste disposal facility established or operated exclusively for the disposal of low-level radioactive waste owned or generated by the Federal Government shall not be subject to any compact or any action taken under a compact.

Prohibition.

Regulations.
Transportation.

Health.
Pollution.
Safety.

Government
organization
and employees.

28 USC 2671
et seq.

Prohibition.

Prohibition.

Prohibition.

42 USC 2021e.

(3) Effect of Compacts On Federal Law.–Nothing contained in this Act or any compact may be construed to confer any new authority on any compact commission or State–

(A) to regulate the packaging, generation, treatment, storage, disposal, or transportation of low-level radioactive waste in a manner incompatible with the regulations of the Nuclear Regulatory Commission or inconsistent with the regulations of the Department of Transportation;

(B) to regulate health, safety, or environmental hazards from source material, byproduct material, or special nuclear material;

(C) to inspect the facilities of licensees of the Nuclear Regulatory Commission;

(D) to inspect security areas or operations at the site of the generation of any low-level radioactive waste by the Federal Government, or to inspect classified information related to such areas or operations; or

(E) to require indemnification pursuant to the provisions of chapter 171 of title 28, United States Code (commonly referred to as the Federal Tort Claims Act), or section 170 of the Atomic Energy Act of 1954 (42 USC 2210) (commonly referred to as the Price-Anderson Act), whichever is applicable.

(4) Federal Authority.–Except as expressly provided in this Act, nothing contained in this Act or any compact may be construed to limit the applicability of any Federal law or to diminish or otherwise impair the jurisdiction of any Federal agency, or to alter, amend, or otherwise affect any Federal law governing the judicial review of any action taken pursuant to any compact.

(5) State Authority Preserved.–

Except as expressly provided in this Act, nothing contained in this Act expands, diminishes, or otherwise affects State law.

(c) Restricted Use of Regional Disposal Facilities.–Any authority in a compact to restrict the use of the regional disposal facilities under the compact to the disposal of low-level radioactive waste generated within the compact region shall not take effect before each of the following occurs:

(1) January 1, 1986; and

(2) the Congress by law consents to the compact.

(d) Congressional Review.–Each compact shall provide that every 5 years after the compact has taken effect the Congress may by law withdraw its consent.

Sec. 5. Limited Availability of Certain Regional Disposal Facilities During Transition and Licensing Periods

(a) Availability of Disposal Capacity.–

(1) Pressurized Water and Boiling Water Reactors.–During the seven-year period beginning January 1, 1986, and ending December 31, 1992, subject to the provisions of subsections (b) through (g), each State in which there is located a regional disposal facility referred to in paragraphs (1) through (3) of subsection (b) shall make disposal capacity available for low-level radioactive waste generated by pressurized water and boiling water commercial nuclear power reactors in accordance with the allocations established in subsection (c).

(2) Other Sources of Low-level Radioactive Waste.–During the seven-year period beginning January 1, 1986 and ending December 31, 1992, subject to the provisions of subsections (b) through (g),

each State in which there is located a regional disposal facility referred to in paragraphs (1) through (3) of subsection (b) shall make disposal capacity available for low-level radioactive waste generated by any source not referred to in paragraph (1).

(3) Allocation of Disposal Capacity.–

(A) During the seven-year period beginning January 1, 1986 and ending December 31, 1992, low-level radioactive waste generated within a sited compact region shall be accorded priority under this section in the allocation of available disposal capacity at a regional disposal facility referred to in paragraphs (1) through (3) of subsection (b) and located in the sited compact region in which such waste is generated.

(B) Any State in which a regional disposal facility referred to in paragraphs (1) through (3) of subsection (b) is located may, subject to the provisions of its compact, prohibit the disposal at such facility of low-level radioactive waste generated outside of the compact region if the disposal of such waste in any given calendar year, together with all other low-level radioactive waste would result in that facility disposing of a total annual volume of low-level radioactive waste in excess of 100 per centum of the average annual volume for such facility designated in sub-section (b): *Provided, however,* That in the event that all three States in which regional disposal facilities referred to in paragraphs (1) through (3) of subsection (b) act to prohibit the disposal of low-level radioactive waste pursuant to this subparagraph, each such State shall, in accordance with any applicable procedures of its compact, permit, as necessary, the disposal of additional quantities of such waste in increments of 10 per centum of the average annual volume for each such facility designated in subsection (b).

Prohibition.

(C) Nothing in this paragraph shall require any disposal facility or State referred to in paragraphs (1) through (3) of subsection (b) to accept for disposal low-level radioactive waste in excess of the total amounts designated in subsection (b).

Prohibition.

(4) Cessation of Operation of Low-level Radioactive Waste Disposal Facility.–No provision of this section shall be construed to obligate any State referred to in paragraphs (1) through (3) of subsection (b) to accept low-level radioactive waste from any source in the event that the regional disposal facility located in such State ceases operations.

(b) Limitations.–The availability of disposal capacity for low-level radioactive waste from any source shall be subject to the following limitations:

(1) Barnwell, South Carolina.–The State of South Carolina, in accordance with the provisions of its compact, may limit the volume of low-level radioactive waste accepted for disposal at the regional disposal facility located in Barnwell, South Carolina to a total of 8,400,000 cubic feet of low-level radioactive waste during the 7-year period beginning January 1, 1986 and ending December 31, 1992 (as based on an average annual volume of 1,200,000 cubic feet of low-level radioactive waste).

(2) Richland, Washington.–The State of Washington, in accordance with the provisions of its compact, may limit the volume of low-level radioactive waste accepted for disposal at the regional disposal facility located at Richland, Washington to a total of

9,800,000 cubic feet of low-level radioactive waste during the 7-year period beginning January 1, 1986, and ending December 31, 1992 (as based on an average annual volume of 1,400,000 cubic feet of low-level radioactive waste).

(3) Beatty, Nevada.–The State of Nevada, in accordance with the provisions of its compact, may limit the volume of low-level radioactive waste accepted for disposal a the regional disposal facility located at Beatty, Nevada to a total of 1,400,000 cubic feet of low-level radioactive waste during the 7-year period beginning January 1, 1986, and ending December 31, 1992 (as based on an average annual volume of 200,000 cubic feet of low-level radioactive waste).

(c) Commercial Nuclear Power Reactor Allocations.–

(1) Amount.–Subject to the provisions of subsections (a) through (g) each commercial nuclear power reactor shall upon request receive an allocation of low-level radioactive waste disposal capacity (in cubic feet) at the facilities referred to in subsection (b) during the 4-year transition period beginning January 1, 1986 and ending December 31, 1989, and during the 3-year licensing period beginning January 1, 1990, and ending December 31, 1992, in an amount calculated by multiplying the appropriate number from the following table by the number of months remaining in the applicable period as determined under paragraph (2).

Reactor Type	4-year Licensing Period		3-year Licensing Period	
	In Sited Region	All Other Locations	In Sited Region	All Other Locations
PWR	1027	871	934	685
BWR	2300	1951	2091	1533

(2) Method of Calculation.–For purposes of calculating the aggregate amount of disposal capacity available to a commercial nuclear power reactor under this subsection, the number of months shall be computed beginning with the first month of the applicable period, or the sixteenth month after receipt of a full power operating license, whichever occurs later.

(3) Unused Allocations.–Any unused allocation under paragraph (1) received by a reactor during the transition period or the licensing period may be used at any time after such reactor receives its full power license or after the beginning of the pertinent period, whichever is later, but not in any event after December 31, 1992, or after commencement of operation of a regional disposal facility in the compact region or State in which such reactor is located, whichever occurs first.

(4) Transferability.–Any commercial nuclear power reactor in a State or compact region that is in compliance with the requirements of subsection (e) may assign any disposal capacity allocated to it under this subsection to any other person in each State or compact region. Such assignment may be for valuable consideration and shall be in writing, copies of which shall be filed at the affected compact commissions and States, along with the assignor's unconditional written waiver of the disposal capacity being assigned.

(5) Unusual Volumes.–

(A) The Secretary may, upon petition by the owner or operator of any commercial nuclear power reactor, allocate to such reactor disposal capacity in excess of the amount calculated under paragraph (1) if the Secretary finds and states in writing his reasons for so finding that making additional capacity available for such reactor through this paragraph is required to permit unusual or unexpected operating, maintenance, repair or safety activities.

Prohibition.

(B) The Secretary may not make allocations pursuant to subparagraph (A) that would result in the acceptance for disposal of more than 800,000 cubic feet of low-level radioactive waste or would result in the total of the allocations made pursuant to this subsection exceeding 11,900,000 cubic feet over the entire seven-year interim access period.

Prohibition.

(6) Limitation.–During the seven-year interim access period referred to in subsection (a), the disposal facilities referred to in subsection (b) shall not be required to accept more than 11,900,000 cubic feet of low-level radioactive waste generated by commercial nuclear power reactors.

Prohibition.

(d)(1) Surcharges.–The disposal of any low-level radioactive waste under this section (other than low-level radioactive waste generated in a sited compact region) may be charged a surcharge by the State in which the applicable regional disposal facility is located, addition to the fees and surcharges generally applicable for disposal of low-level radioactive waste in the regional disposal facility involved. Except as provided in sub-section (e)(2), such surcharges shall not exceed–

(A) in 1986 and 1987, $10 per cubic foot of low-level radioactive waste;

(B) in 1988 and 1989, $20 per cubic foot of low-level radioactive waste; and

(C) in 1990, 1991, and 1992, $40 per cubic foot of low-level radioactive waste.

(2) Milestone Incentives.–

(A) Escrow Account.–Twenty-five per centum of all surcharge fees received by a State pursuant to paragraph (1) during the seven-year period referred to in subsection (a) shall be transferred on a monthly basis to an escrow account held by the Secretary. The Secretary shall deposit all funds received in a special escrow account. The funds so deposited shall not be the property of the United States. The Secretary shall act as trustee for such funds and shall invest them in interest-bearing United States Government Securities with the highest available yield. Such funds shall be held by the Secretary until–

(i) paid or repaid in accordance with subparagraph (B) or (C); or

(ii) paid to the State collecting such fees in accordance with subparagraph (F).

(B) Payments.–

(i) July 1, 1986.–The twenty-five per centum of any amount collected by a State under paragraph (1) for low-level radioactive waste disposed of under this section during the period beginning on the date of enactment of the Low-Level Radioactive Waste Policy Amendments Act of 1985 and ending June 30, 1986, and transferred to the Secretary under subparagraph (A), shall be paid by the Secretary in accordance with subparagraph (D) if the milestone described in sub-section (e)(1)(A) is met by the State in which such waste originated.

(ii) January 1, 1988.–The twenty-five per centum of any amount collected by a State under paragraph (1) for low-level radioactive waste disposed of under this section during the period beginning July 1, 1986 and ending December 31, 1987, and transferred to the Secretary under subparagraph (A), shall be paid by the Secretary in accordance with subparagraph (D) if the milestone described in subsection (e)(1)(B) is met by the state in which such waste originated (or its compact region, where applicable).

(iii) January 1, 1990.–The twenty-five per centum of any amount collected by a State under paragraph (1) for low-level radioactive waste disposed of under this section during the period beginning January 1, 1988 and ending December 31, 1989, and transferred to the Secretary under subparagraph (A), shall be paid by the Secretary in accordance with subparagraph (D) if the milestone described in subsection (e)(1)(C) is met by the State in which such waste originated (or its compact region, where applicable).

(iv) The twenty-five per centum of any amount collected by a State under paragraph (1) for low-level radioactive waste disposed of under this section during the period beginning January 1, 1990 and ending December 31, 1992, and transferred to the Secretary under subparagraph (A), shall be paid by the Secretary in accordance with subparagraph (D) if, by January 1, 1993, the State in which such waste originated (or its compact region, where applicable) is able to provide for the disposal of all low-level radioactive waste generated within such State or compact region.

(C) Failure To Meet January 1, 1993 Deadline.–If, by January 1, 1993, a State (or, where applicable, a compact region) in which low-level radioactive waste is generated is unable to provide for the disposal of all such waste generated within such State or compact region–

(i) each State in which such waste is generated, upon the request of the generator or owner of the waste, shall take title to the waste, shall be obligated to take possession of the waste, and shall be liable for all damages directly or indirectly incurred by such generator or owner as a consequence of the failure of the State to take possession of the waste as soon after January 1, 1993 as the generator or owner notifies the State that the waste is available for shipment; or[3]

(ii) if such State elects not to take title to, take possession of, and assume liability for such waste, pursuant to clause (i), twenty-five per centum of any amount collected by a State under paragraph (1) for low-level radioactive waste disposed of under this section during the period beginning January 1, 1990 and ending December 31, 1992 shall be repaid, with interest, to each generator from whom such surcharge was

---

[3]   The United States Supreme Court struck down this provision, because it was unconditional (*N.Y. vs. United States* 112 S. Ct. 2408 (June 19, 1992)).

collected. Repayments made pursuant to this clause shall be made on a monthly basis, with the first such repayment beginning on February 1, 1993, in an amount equal to one thirty-sixth of the total amount required to be repaid pursuant to this clause, and shall continue until the State (or, where applicable, compact region) in which such low-level radioactive waste in generated is able to provide for the disposal of all such waste generated within such State or compact region or until January 1, 1996, whichever is earlier.

If a State in which low-level radioactive waste is generated elects to take title to, take possession of, and assume liability for such waste pursuant to clause (i), such State shall be paid such amounts as are designated in subparagraph (B)(iv). If a State (or, where applicable, a compact region) in which low-level radioactive waste is generated provides for the disposal of such waste at any time after January 1, 1993 and prior to January 1, 1996, such State (or, where applicable, compact region) shall be paid in accordance with subparagraph (D) a lump sum amount equal to twenty-five per centum of any amount collected by a State under paragraph (1): *Provided, however,* That such payment shall be adjusted to reflect the remaining number of months between January 1, 1993 and January 1, 1996 for which such State (or, where applicable, compact region) provides for the disposal of such waste. If a State (or, where applicable, a compact region) in which low-level radioactive waste is generated is unable to provide for the disposal of all such waste generated within such State or compact region by January 1, 1996, each State in which such waste is generated, upon the request of the generator or owner of the waste shall take title to the waste, be obligated to take possession of the waste, and shall be liable for all damages directly or indirectly incurred by such generator or owner as a consequence of the failure of the State to take possession of the waste as soon after January 1, 1996, as the generator or owner notifies the State that the waste is available for shipment.

(D) Recipients of Payments.–The payments described in subparagraphs (B) and (C) shall be paid within thirty days after the applicable date–

(i) if the State in which such waste originated is not a member of a compact region, to such State;

(ii) if the State in which such waste originated is a member of the compact region, to the compact commission serving such State.

(E) Uses of Payments.–

(i) Limitations.–Any amount paid under subparagraphs (B) or (C) may only be used to–

(I) establish low-level radioactive waste disposal facilities;

Reports.

(II) mitigate the impact of low-level radioactive waste disposal facilities on the host State;

(III) regulate low-level radioactive waste disposal facilities; or

(IV) ensure the decommissioning, closure, and care during the period of institutional control of low-level radioactive waste disposal facilities.

(ii) Reports.–

(I) Recipient.–Any State or compact commission receiving a payment under subparagraphs (B) or (C) shall, on December 31 of each year in which any such funds are expended, submit a report to the Department of Energy itemizing any such expenditures.

(II) Department of Energy.–Not later than six months after receiving the reports under subclause (I), the Secretary shall submit to the Congress a summary of all such reports that shall include an assessment of the compliance of each such State or compact commission with the requirements of clause (i).

(F) Payment To States.–Any amount collected by a State under paragraph (1) that is placed in escrow under subparagraph (A) and not paid to a State or compact commission under subparagraphs (B) and (C) or not repaid to a generator under subparagraph (C) shall be paid from such escrow account to such State collecting such payment under paragraph (1). Such payment shall be made not later than 30 days after a determination of ineligibility for a refund is made.

Prohibition.

(G) Penalty Surcharges.–No rebate shall be made under this subsection of any surcharge or penalty surcharge paid during a period of noncompliance with subsection (e)(1).

(e) Requirements For Access To Regional Disposal Facilities.–

(1) Requirements For Non-sited Compact Regions and Non-member States.–Each non-sited compact region, or State that is not a member of a compact region that does not have an operating disposal facility, shall comply with the following requirements:

(A) By July 1, 1986, each such non-member State shall ratify compact legislation or, by the enactment of legislation or the certification of the Governor, indicate its intent to develop a site for the location of a low-level radioactive waste disposal facility within such State.

(B) By January 1, 1988.–

(i) each non-sited compact region shall identify the State in which its low-level radioactive waste disposal facility is to be located, or shall have selected the developer for such facility and the site to be developed, and each compact region or the State in which its low-level radioactive waste disposal facility is to be located shall develop a siting plan for such facility providing detailed procedures and a schedule for establishing a facility location and preparing a facility license application and shall delegate authority to implement such plan;

(ii) each non-member State shall develop a siting plan providing detailed procedures and a schedule for establishing a facility location and preparing a facility license application for a low-level radioactive waste disposal facility and shall delegate authority to implement such plan; and

(iii) The siting plan required pursuant to this paragraph shall include a description of the optimum way to attain operation of the low-level radioactive waste disposal facility involved, within the time period specified in this Act. Such plan shall include a description of the objectives and a sequence of deadlines for all entities required to take action to implement such plan, including, to the extent practicable, an identification of the activities in which a delay in the start, or completion, of such activities will cause a delay in beginning facility operation. Such plan shall also identify, to the extent practicable, the process for (1) screening for broad siting areas; (2) identifying and evaluating specific candidate sites; and (3) characterizing the preferred site(s), completing all necessary environmental assessments, and preparing a license application for submission to the Nuclear Regulatory Commission or an Agreement State.

(C) By January 1, 1990.–

(i) a complete application (as determined by the Nuclear Regulatory Commission or the appropriate agency of an agreement State) shall be filed for a license to operate a low-level radioactive waste disposal facility within each non-sited compact region or within each non-member State; or

(ii) the Governor (or, for any State without a Governor, the chief executive officer) of any State that is not a member of a compact region in compliance with clause (i), or has not complied with such clause by its own actions, shall provide a written certification to the Nuclear Regulatory Commission, that such State will be capable of providing for, and will provide for, the storage, disposal, or management of any low-level radioactive waste generated within such State and requiring disposal after December 31, 1992, and include a description of the actions that will be taken to ensure that such capacity exists.

(D) By January 1, 1992, a complete application (as determined by the Nuclear Regulatory Commission or the appropriate agency of an agreement State) shall be filed for a license to operate a low-level radioactive waste disposal facility within each non-sited compact region or within each non-member State.

Federal Register, publication. Contracts.

(E) The Nuclear Regulatory Commission shall transmit any certification received under subparagraph (C) to the Congress and publish any such certification in the Federal Register.

(F) Any State may, subject to all applicable provisions, if any, of any applicable compact, enter into an agreement with the compact commission of a region in which a regional disposal facility is located to provide for the disposal of all low-level radioactive waste generated within such State, and, by virtue of such agreement, may, with the approval of the State in which the regional disposal facility is located, be deemed to be in compliance with subparagraphs (A), (B), (C), and (D).

(2) Penalties For Failure To Comply.–

(A) By July 1, 1986.–If any State fails to comply with subparagraph (1)(A)–

(i) any generator of low-level radioactive waste within such region or non-member State shall, for the period beginning July 1, 1986, and ending December 31, 1986, be charged 2 times the surcharge otherwise applicable under subsection (d); and

(ii) on or after January 1, 1987, any low-level radioactive waste generated within such region or non-member State may be denied access to the regional disposal facilities referred to in paragraphs (1) through (3) of subsection (b).

(B) By January 1, 1988.–If any non-sited compact region or non-member State fails to comply with paragraph (1)(B)–

(i) any generator of low-level radioactive waste within such region or non-member State shall–

(I) for the period beginning January 1, 1988, and ending June 30, 1988, be charged 2 times the surcharge otherwise applicable under subsection (d); and

(II) for the period beginning July 1, 1988, and ending December 31, 1988, be charged 4 times the surcharge otherwise applicable under subsection (d); and

(ii) on or after January 1, 1989, any low-level radioactive waste generated within such region or non-member State may be denied access to the regional disposal facilities referred to in paragraphs (1) through (3) of subsection (b).

(C) By January 1, 1990.–If any non-sited compact region or non-member State fails to comply with paragraph (1)(C), any low-level radioactive waste generated within such region or non-member State may be denied access to the regional disposal facilities referred to in paragraphs (1) through (3) of subsection (b).

(D) By January 1, 1992.–If any non-sited compact region or non-member State fails to comply with paragraph (1)(D), any generator of low-level radioactive waste within such region or non-member State shall, for the period beginning January 1, 1992 and ending upon the filing of the application described in paragraph (1)(D), be charged 3 times the surcharge otherwise applicable under subsection (d).

Prohibition.

(3) Denial of Access.–No denial or suspension of access to a regional disposal facility under paragraph (2) may be based on the source, class, or type of low-level radioactive waste.

Termination.

(4) Restoration of Suspended Access; Penalties For Failure To Comply.–Any access to a regional disposal facility that is suspended under paragraph (2) shall be restored after the non-sited compact region or non-member State involved complies with such requirement. Any payment of surcharge penalties pursuant to paragraph (2) for failure to comply with the requirements of subsection (e) shall be terminated after the non-sited compact region or non-member State involved complies with such requirements.

(f)(1) Administration.–Each State and compact commission in which a regional disposal facility referred to in paragraphs (1) through (3) of subsection (b) is located shall have authority–

(A) to monitor compliance with the limitations, allocations, and requirements established in this section; and

(B) to deny access to any non-Federal low-level radioactive waste disposal facilities within its borders to any low-level radioactive waste that–

(i) is in excess of the limitations or allocations established in this section; or

(ii) is not required to be accepted due to the failure of a compact region or State to comply with the requirements of subsection (e)(1).

(2) Availability of Information During Interim Access Period.–

Nevada.
South Carolina.
Washington.

(A) The States of South Carolina, Washington, and Nevada may require information from disposal facility operators, generators, intermediate handlers, and the Department of Energy that is reasonably necessary to monitor the availability of disposal capacity, the use and assignment of allocations and the applicability of surcharges.

Nevada.
South Carolina.
Washington.

(B) The States of South Carolina, Washington, and Nevada may, after written notice followed by a period of at least 30 days, deny access to disposal capacity to any generator or intermediate handler who fails to provide information under subparagraph (A).

(C) Proprietary Information.–

(i) Trade secrets, proprietary and other confidential information shall be made available to a State under this subsection upon request only if such State–

(I) consents in writing to restrict the dissemination of the information to those who are directly involved in monitoring under subparagraph (A) and who have a need to know;

(II) accepts liability for wrongful disclosure; and

(III) demonstrates that such information is essential to such monitoring.

(ii) The United States shall not be liable for the wrongful disclosure by any individual or State of any information provided to such individual or State under this subsection.

Commerce and
trade.
Government
organization
and employees.
Prohibition.

(iii) Whenever any individual or State has obtained possession of information under this subsection, the individual shall be subject to the same provisions of law with respect to the disclosure of such information as would apply to an officer or employee of the United States or of any department or agency thereof and the State shall be subject to the same provisions of law with respect to the disclosure of such information as would apply to the United States or any department or agency thereof. No State or State officer or employee who receives trade secrets, proprietary information, or other confidential information under this Act may be required to disclose such information under State law.

(g) Nondiscrimination.–Except as provided in subsections (b) through (e), low-level radioactive waste disposed of under this section shall be subject without discrimination to all applicable legal requirements of the compact region and State in which the disposal facility is located as if such low-level radioactive waste were generated within such compact region.

42 USC 2021f.
Defense and
national
security.
Health.
Safety.

Sec. 6. Emergency Access

(a) In General.–The Nuclear Regulatory Commission may grant emergency access to any regional disposal facility or non-Federal disposal facility within a State that is not a member of a compact for specific low-level radioactive waste, if necessary to eliminate an immediate and serious threat to the public health and safety or the common defense and security. The procedure for granting emergency access shall be as provided in this section.

(b) Request For Emergency Access.–Any generator of low-level radioactive waste, or any Governor (or, for any State without a Governor, the chief executive officer of the State) on behalf of any generator or generators located in his or her State, may request that the Nuclear Regulatory Commission grant emergency access to a regional disposal facility or a non-Federal disposal facility within a State that is not a member of a compact for specific low-level radioactive waste. Any such request shall contain any information and certifications the Nuclear Regulatory Commission may require.

Health.

(c) Determination of Nuclear Regulatory Commission.–

Defense and
national
security.
Safety.

(1) Required Determination.–Not later than 45 days after receiving a request under subsection (b), the Nuclear Regulatory Commission shall determine whether–

(A) emergency access is necessary because of an immediate and serious threat to the public health and safety or the common defense and security; and

(B) The threat cannot be mitigated by any alternative consistent with the public health and safety, including storage of low-level radioactive waste at the site of generation or in a storage facility obtaining access to a disposal facility by voluntary agreement, purchasing disposal capacity available for assignment pursuant to section 5(c) or ceasing activities that general low-level radioactive waste.

(2) Required Notification.–If the Nuclear Regulatory Commission makes the determinations required in paragraph (1) in the affirmative, it shall designate an appropriate non-Federal disposal facility or facilities, and notify the Governor (or chief executive officer) of the State in which such facility is located and the appropriate compact commission that emergency access is required. Such notification shall specifically describe the low-level radioactive waste as to source, physical and radiological characteristics, and the minimum volume and duration, not exceeding 180 days, necessary to alleviate the immediate threat to public health and safety or the common defense and security. The Nuclear Regulatory Commission shall also notify the Governor (or chief executive officer) of the State in which the low-level radioactive waste requiring emergency access was generated that emergency access has been granted and that, pursuant to subsection (e), no extension of emergency access may be granted absent diligent State action during the period of the initial grant.

Prohibition.

Defense and
national
security.
Health.
Safety.

(d) Temporary Emergency Access.–Upon determining that emergency access is necessary because of an immediate and serious threat to the public health and safety or the common defense and security, the Nuclear Regulatory Commission may at its discretion grant temporary emergency access, pending its determination whether the threat could be mitigated by any alternative consistent with the public health and safety. In granting access under this subsection, the Nuclear Regulatory Commission shall provide the same notification and information required under sub-section (c). Absent a determination that no alternative consistent with the public health and safety would mitigate the threat, access granted under this subsection shall expire 45 days after the granting of temporary emergency access under this subsection.

Defense and
national
security.
Health.
Safety.

(e) Extension of Emergency Access.–

The Nuclear Regulatory Commission may grant one extension of emergency access beyond the period provided in subsection (c), if it determines that emergency access continues to be necessary because of an immediate and serious threat to the public health and safety or the common defense and security that cannot be mitigated by any alternative consistent with the public health and safety, and that the generator of low-level radioactive waste granted emergency access and the State in which such low-level radioactive waste was generated have diligently though unsuccessfully acted during the period of the initial grant to eliminate the need for emergency access. Any extension granted under this subsection shall be for the minimum volume and duration the Nuclear Regulatory Commission finds necessary to eliminate the immediate threat to public health and safety or the common defense and security, and shall not in any event exceed 180 days.

(f) Reciprocal Access.–Any compact region or State not a member of a compact that provides emergency access to non-Federal disposal facilities within its borders shall be entitled to reciprocal access to any subsequently operating non-Federal disposal facility that serves the State or compact region in which low-level radioactive waste granted emergency access was generated. The compact commission or State having authority to approve importation of low-level radioactive waste to the disposal facility to which emergency access was granted shall designate for reciprocal access an equal volume of low-level radioactive waste having similar characteristics to that provided emergency access.

(g) Approval By Compact Commission.–Any grant of access under this section shall be submitted to the compact commission for the region in which the designated disposal facility is located for such approval as may be required under the terms of its compact. Any such compact commission shall act to approve emergency access not later than 15 days after receiving notification from the Nuclear Regulatory Commission, or reciprocal access not later than 15 days after receiving notification from the appropriate authority under subsection (f).

Prohibitions.

(h) Limitations.–No State shall be required to provide emergency or reciprocal access to any regional disposal facility within its borders for low-level radioactive waste not meeting criteria established by the license or license agreement of such facility, or in excess of the approved capacity of such facility, or to delay the closing of any such facility pursuant to plans established before receiving a request for emergency or reciprocal access. No State shall, during any 12-month period, be required to provide emergency or reciprocal access to any regional disposal facility within its borders for more than 20 percent of the total volume of low-level radioactive waste accepted for disposal at such facility during the previous calendar year.

(i) Volume Reduction and Surcharges.–Any low-level radioactive waste delivered for disposal under this section shall be reduced in volume to the maximum extent practicable and shall be subject to surcharges established in this Act.

(j) Deduction From Allocation.–Any volume of low-level radioactive waste granted emergency or reciprocal access under this section, if generated by any commercial nuclear power reactor, shall be deducted from the low-level radioactive waste volume allocable under section 5(c).

Prohibition.

(k) Agreement States.–Any agreement under section 274 of the Atomic Energy Act of 1954 (42 USC 2021) shall not be applicable to the determinations of the Nuclear Regulatory Commission under this section.

42 USC 2021g.

Sec. 7. Responsibilities of the Department of Energy

(a) Financial and Technical Assistance.–The Secretary shall, to the extent provided in appropriations Act, provide to those compact regions, host States, and nonmember States determined by the Secretary to require assistance for purposes of carrying out this Act–

Health.
Safety.
Science and
technology.
Transportation.

(1) continuing technical assistance to assist them in fulfilling their responsibilities under this Act. Such technical assistance shall include, but not be limited to, technical guidelines for site selection, alternative technologies for low-level radioactive waste disposal, volume reduction options, management techniques to reduce low-level waste generation, transportation practices for shipment of low-level wastes, health and safety considerations in the storage, shipment and disposal of low-level radioactive wastes, and establishment of a

computerized database to monitor the management of low-level radioactive wastes; and

(2) through the end of fiscal year 1993, financial assistance to assist them in fulfilling their responsibilities under this Act.

Science and technology. Transportation.

(b) Reports.–The Secretary shall prepare and submit to the Congress on an annual basis a report which (1) summarizes the progress of low-level waste disposal siting and licensing activities within each compact region, (2) reviews the available volume reduction technologies, their applications, effectiveness, and costs on a per unit volume basis, (3) reviews interim storage facility requirements, costs, and usage, (4) summarizes transportation requirements for such wastes on an inter- and intra-regional basis, (5) summarizes the data on the total amount of low-level waste shipped for disposal on a yearly basis, the proportion of such wastes subjected to volume reduction, the average volume reduction attained, and the proportion of wastes stored on an interim basis, and (6) projects the interim storage and final disposal volume requirements anticipated for the following year, on a regional basis.

42 USC 2021h.

Sec. 8. Alternative Disposal Methods

(a) Not later than 12 months after the date of enactment of the Low-Level Radioactive Waste Policy Amendments Act of 1985, the Nuclear Regulatory Commission shall, in consultation with the States and other interested persons, identify methods for the disposal of low-level radioactive waste other than shallow land burial, and establish and publish technical guidance regarding licensing of facilities that use such methods.

(b) Not later than 24 months after the date of enactment of the Low-Level Radioactive Waste Policy Amendments Act of 1985, the Commission shall, in consultation with the States and other interested persons, identify and publish all relevant technical information regarding the methods identified pursuant to subsection (a) that a State or compact must provide to the Commission in order to pursue such methods, together with the technical requirements that such facilities must meet, in the judgment of the Commission, if pursued as an alternative to shallow land burial. Such technical information and requirements shall include, but need not be limited to, site suitability, site design, facility operation, disposal site closure, and environmental monitoring, as necessary to meet the performance objectives established by the Commission for a licensed low-level radioactive waste disposal facility. The Commission shall specify and publish such requirements in a manner and form deemed appropriate by the Commission.

42 USC 2021i.

Sec. 9. Licensing Review and Approval

In order to ensure the timely development of new low-level radioactive waste disposal facilities, the Nuclear Regulatory Commission or, as appropriate, agreement States, shall consider an application for a disposal facility license in accordance with the laws applicable to such application, except that the Commission and the agreement state shall–

(1) not later than 12 months after the date of enactment of the Low-Level Radioactive Waste Policy Amendments Act of 1985, establish procedures and develop the technical capability for processing applications for such licenses;

(2) to the extent practicable, complete all activities associated with the review and processing of any application for such a license (except for public hearings) no later than 15 months after the date of receipt of such application; and

(3) to the extent practicable, consolidate all required technical and environmental reviews and public hearings.

42 USC 2021j.

Sec. 10. Radioactive Waste Below Regulatory Concern

(a) Not later than 6 months after the date of enactment of the Low-Level Radioactive Waste Policy Amendments Act of 1985, the Commission shall establish standards and procedures, pursuant to existing authority, and develop the technical capability for considering and acting upon petitions to exempt specific radioactive waste streams from regulation by the Commission due to the presence of radionuclides in such waste streams in sufficiently low concentrations or quantities as to be below regulatory concern.

(b) The standards and procedures established by the Commission pursuant to subsection (a) shall set forth all information required to be submitted to the Commission by licensees in support of such petitions, including, but not limited to–

(1) a detailed description of the waste materials, including their origin, chemical composition, physical state, volume, and mass; and

Health.
Safety.
Regulation.

(2) The concentration or contamination levels, half-lives, and identities of the radionuclides present.

Such standards and procedures shall provide that, upon receipt of a petition to exempt a specific radioactive waste stream from regulation by the Commission, the Commission shall determine in an expeditious manner whether the concentration or quantity of radionuclides present in such waste stream requires regulation by the Commission in order to protect the public health and safety. Where the Commission determines that regulation of a radioactive waste stream is not necessary to protect the public health and safety, the Commission shall take such steps as may be necessary, in an expeditious manner, to exempt the disposal of such radioactive waste from regulation by the Commission.

B. LOW-LEVEL RADIOACTIVE WASTE POLICY AMENDMENTS ACT OF 1985 (TITLE II, INCLUDES: NW, CENTRAL, SE, CENTRAL MIDWEST, MW, ROCKY MOUNTAIN, NE INTERSTATE COMPACTS)

Title II–Omnibus Low-Level Radioactive Waste Interstate Compact Consent Act

42 USC 2021d note.

Sec. 201. Short Title

This Title may be cited as the "Omnibus Low-Level Radioactive Waste Interstate Compact Consent Act."

Subtitle A–General Provisions

42 USC 2021d note.

Sec. 211. Congressional Finding

The Congress hereby finds that each of the compacts set forth in subtitle B is in furtherance of the Low-Level Radioactive Waste Policy Act.

42 USC 2021d note.

Sec. 212. Conditions of Consent to Compacts

The consent of the Congress to each of the compacts set forth in subtitle B–

Effective date.

(1) shall become effective on the date of the enactment of this Act;

(2) is granted subject to the provisions of the Low-Level Radioactive Waste Policy Act, as amended; and

(3) is granted only for so long as the regional commission, committee, or board established in the compact complies with all of the provisions of such Act.

42 USC 2021d note.

Sec. 213. Congressional Review

The Congress may alter, amend, or repeal this Act with respect to any compact set forth in subtitle B after the expiration of the 10-year period following the date of the enactment of this Act, and at such intervals thereafter as may be provided in such compact.

Subtitle B–Congressional Consent to Compacts

42 USC 2021d note.
Alaska.
Hawaii.
Idaho.
Montana.
Oregon.
Utah.
Washington.
Wyoming.

Sec. 221. Northwest Interstate Compact on Low-Level Radioactive Waste Management

The Consent of Congress is hereby given to the states of Alaska, Hawaii, Idaho, Montana, Oregon, Utah, Washington, and Wyoming to enter into the Northwest Interstate Compact on Low-level Radioactive Waste Management, and to each and every part and article thereof. Such compact reads substantially as follows:

Northwest Interstate Compact on Low-Level Radioactive Waste Management

Article I–Policy and Purpose

Health.
Safety.

The party states recognize that low-level radioactive wastes are generated by essential activities and services that benefit the citizens of the states. It is further recognized that the protection of the health and safety of the citizens of the party states and the most economical management of low-level radioactive wastes can be accomplished through cooperation of the states in minimizing the amount of handling and transportation required to dispose of such wastes and through the cooperation of the states in providing facilities that serve the region. It is the policy of the party states to undertake the necessary cooperation to protect the health and safety of the citizens of the party states and to provide for the most economical management of low-level radioactive wastes on a continuing basis. It is the purpose of this compact to provide the means for such a cooperative effort among the party states so that the protection of the citizens of the states and the maintenance of the viability of the states' economies will be enhanced while sharing the responsibilities of radioactive low-level waste management.

Article II–Definitions

As used in this compact:

(1) "Facility" means any site, location, structure, or property used or to be used for the storage, treatment, or disposal of low-level waste, excluding federal waste facilities;

(2) "Low-level waste" means waste material which contains radioactive nuclides emitting primarily beta or gamma radiation, or both, in concentrations or quantities which exceed applicable federal or state standards for unrestricted release. Low-level waste does not include waste containing more than ten (10) nanocuries of transuranic contaminants per gram of material, nor spent reactor fuel, nor material classified as either high-level waste or waste which is

unsuited for disposal by near-surface burial under any applicable federal regulations;

(3) "Generator" means any person, partnership, association, corporation, or any other entity whatsoever which, as a part of its activities, produces low-level radioactive waste;

(4) "Host state" means a state in which a facility is located.

Article III–Regulatory Practices

Transportation.

Each party state hereby agrees to adopt practices which will require low-level waste shipments originating within its borders and destined for a facility within another party state to conform to the applicable packaging and transportation requirements and regulations of the host state. Such practices shall include:

(1) Maintaining an inventory of all generators within the state that have shipped or expect to ship low-level waste to facilities in another party state;

(2) Periodic unannounced inspection of the premises of such generators and the waste management activities thereon;

(3) Authorization of the containers in which such waste may be shipped, and a requirement that generators use only that type of container authorized by the state;

(4) Assurance that inspections of the carriers which transport such waste are conducted by proper authorities, and appropriate enforcement action taken for violations;

Transportation.

(5) After receiving notification from a host state that a generator within the party state is in violation of applicable packaging or transportation standards, the party state will take appropriate action to assure that such violations do not recur. Such action may include inspection of every individual low-level waste shipment by that generator.

Each party state may impose fees upon generators and shippers to recover the cost of the inspections and other practices under this article. Nothing in this article shall be construed to limit any party state's authority to impose additional or more stringent standards on generators or carriers than those required under this article.

Article IV–Regional Facilities

(1) Facilities located in any party state, other than facilities established or maintained by individual low-level waste generators for the management of their own low-level waste, shall accept low-level waste generated in any party state if such waste has been packaged and transported according to applicable laws and regulations.

(2) No facility located in any party state may accept low-level waste generated outside of the region comprised of the party states, except as provided in article V.

(3) Until such time as paragraph (2) of article IV takes effect, facilities located in any party state may accept low-level waste generated outside of any of the party states only if such waste is accompanied by a certificate of compliance issued by an official of the state in which such waste shipment originated. Such certificate shall be in such form as may be required by the host state, and shall contain at least the following:

(A) The generator's name and address;

(B) A description of the contents of the low-level waste container.

Regulations.

(C) A statement that the low-level waste being shipped has been inspected by the official who issued the certificate or by his agent or by a representative of the United States Nuclear Regulatory Commission, and found to have been packaged in compliance with applicable Federal regulations and such additional requirements as may be imposed by the host state;

(D) A binding agreement by the state of origin to reimburse any party state for any liability or expense incurred as a result of an accidental release of such waste during shipment or after such waste reaches the facility.

Health.
Safety.

(4) Each party state shall cooperate with the other party states in determining the appropriate site of any facility that might be required within the region comprised of the party states, in order to maximize public health and safety while minimizing the use of any one (1) party state as the host of such facilities on a permanent basis. Each party state further agrees that decisions regarding low-level waste management facilities in their region will be reached through a good faith process which takes into account the burdens borne by each of the party states as well as the benefits each has received.

Hazardous
materials.
Idaho.
Oregon.
Prohibition.
Washington.

(5) The party states recognize that the issue of hazardous chemical waste management is similar in many respects to that of low-level waste management. Therefore, in consideration of the State of Washington allowing access to its low-level waste disposal facility by generators in other party states, party states such as Oregon and Idaho which host hazardous chemical waste disposal facilities will allow access to such facilities by generators within other party states. Nothing in this compact shall be construed to prevent any party state from limiting the nature and type of hazardous chemical or low-level wastes to be accepted at facilities within its borders of from ordering the closure of such facilities, so long as such action by a host state is applied equally to all generators within the region comprised of the party states.

(6) Any host state may establish a schedule of fees and requirements related to its facility, to assure that closure, perpetual care, and maintenance and contingency requirements are met, including adequate bonding.

Article V–Northwest Low–Level Waste Compact Committee

The governor of each party state shall designate one (1) official of that state as the person responsible for administration of this compact. The officials so designated shall together comprise the northwest low-level waste compact committee. The committee shall meet as required to consider matters arising under this compact. The parties shall inform the

Regulations.

committee of existing regulations concerning low-level waste management in their states, and shall afford all parties a reasonable opportunity to review and comment upon any proposed modification in such regulations. Notwithstanding any provision of article IV to the contrary, the committee may enter into arrangements with states, provinces, individual generators, or regional compact entities outside the region comprised of the party states for access to facilities on such terms and conditions as the committee may deem appropriate. However, it shall require a two-thirds (2/3) vote of all such members, including the affirmative vote of the member of any party state in which a facility affected by such arrangement is located, for the committee to enter into such arrangement.

Article VI–Eligible Parties and Effective Date

Alaska. Hawaii.
Idaho. Montana.
Oregon. Utah.
Washington.
Wyoming.

(1) Each of the following states is eligible to become a party to this compact: Alaska, Hawaii, Idaho, Montana, Oregon, Utah, Washington, and Wyoming. As to any eligible party, this compact shall become effective upon enactment into law by that party, but it shall not become initially effective until enacted into law by two (2) states. Any party state may withdraw from this compact by enacting a statute repealing its approval.

Effective date.
Wyoming.

(2) After the compact has initially taken effect pursuant to paragraph (1) of this article, any eligible party state may become a party to this compact by the execution of an executive order by the governor of the state. Any state which becomes a party in this manner shall cease to be a party upon the final adjournment of the next general or regular session of its legislature or July 1, 1983, whichever occurs first, unless the compact has by then been enacted as a statute by that state.

42 USC 2021b
note.
Effective date.

(3) Paragraph (2) of article IV of this compact shall take effect on July 1, 1983, if consent is given by Congress. As provided in Public Law 96-573, Congress may withdraw its consent to the compact after every five (5) year period.

Article VII–Severability

Provisions held invalid.

If any provision of this compact, or its application to any person or circumstances, is held to be invalid, all other provisions of this compact, and the application of all of its provisions to all other persons and circumstances, shall remain valid; and to this end the provisions of this compact are severable.

**42 USC 2021d
note.** Arkansas.
Iowa. Kansas.
Louisiana.
Minnesota.
Missouri.
Nebraska.
North Dakota.
Oklahoma.

Sec. 222. Central Interstate Low-Level Radioactive Waste Compact

The consent of Congress is hereby given to the states of Arkansas, Iowa, Kansas, Louisiana, Minnesota, Missouri, Nebraska, North Dakota, and Oklahoma to enter into the Central Interstate Low-Level Radioactive Waste Compact, and to each and every part and article thereof. Such compact reads substantially as follows:

Central Interstate Low-Level Radioactive Waste Compact

Article I–Policy and Purpose

42 USC 2021b
note.
Environmental
protection.
Health.
Safety.

The party states recognize that each state is responsible for the management of its non-federal low-level radioactive wastes. They also recognize that the Congress, by enacting the Low-Level Radioactive Waste Policy Act (Public Law 96-573) has authorized and encouraged states to enter into compacts for the efficient management of wastes. It is the policy of the party states to cooperate in the protection of the health, safety and welfare of their citizens and the environment and to provide for and encourage the economical management of low-level radioactive wastes. It is the purpose of this compact to provide the framework for such a cooperative effort; to promote the health, safety and welfare of the citizens and the environment of the region; to limit the number of facilities needed to effectively and efficiently manage low-level radioactive wastes and to encourage the reduction of the generation thereof; and to distribute the costs, benefits and obligations among the party states.

Article II–Definitions

As used in this compact, unless the context clearly requires a different construction:

a. "Commission" means the Central Interstate Low-Level Radioactive Waste Commission;

b. "disposal" means the isolation and final disposition of waste;

c. "extended care" means the care of a regional facility including necessary corrective measures subsequent to its active use for waste management until such time as the regional facility no longer poses a threat to the environment or public health;

d. "facility" means any site, location, structure or property used or to be used for the management of waste;

e. "generator" means any person who, in the course of or as incident to manufacturing, power generation, processing, medical diagnosis and treatment, biomedical research, other industrial or commercial activity, other research or mining in a party state, produces or processes waste. "Generator" does not include any person who receives waste generated outside the region for subsequent shipment to a regional facility;

f. "host state" means any party state in which a regional facility is situated or is being developed;

42 USC 2021b
note.

g. "low-level radioactive waste" or "waste" means, as defined in the Low-Level Radioactive Waste Policy Act (Public Law 96-573), radioactive waste not classified as: High-level radioactive waste, transuranic waste, spent nuclear fuel, or by-product material as defined in section 11e.(2) of the Atomic Energy Act of 1954, as amended through 1978.

42 USC 2014.

h. "management of waste" means the storage, treatment or disposal of waste;

i. "notification of each party state" means transmittal of written notice to the Governor, presiding officer of each legislative body and any other persons designated by the party state's Commission member to receive such notice;

j. "party state" means any state which is a signatory party to this compact;

k. "person" means any individual, corporation, business enterprise, or other legal entity, either public or private;

l. "region" means the area of the party states;

m. "regional facility" means a facility which is located within the region and which has been approved by the Commission for the benefit of the party States;

n. "site" means any property which is owned or leased by a generator and is contiguous to or divided only by a public or private way from the source of generation;

o. "state" means a state of the United States, the District of Columbia, the Commonwealth of Puerto Rico, the U.S. Virgin Islands or any other territorial possession of the United States;

p. "storage" means the holding of waste for treatment or disposal; and

q. "treatment" means any method, technique or process, including storage for radioactive decay, designed to change the physical, chemical or biological characteristics or composition of any waste in order to render such waste after for transport or management, amendable for recovery, convertible to another usable material, or reduced in volume.

Article III–Rights and Obligations

a. There shall be provided within the region one or more regional facilities which together provide sufficient capacity to manage all wastes generated within the region. It shall be the duty of regional facilities to accept compatible wastes generated in and from party states, and meeting the requirements of this Act, and each party state shall have the right to have the wastes generated within its borders managed at such facility.

Regulation.

b. To the extent authorized by Federal law and host State law, a host state shall regulate and license any regional facility within its borders and ensure the extended care of such facility.

c. Rates shall be charged to any user of the regional facility, set by the operator of a regional facility and shall be fair and reasonable and be subject to the approval of the host state. Such approval shall be based upon criteria established by the Commission.

d. A host state may establish fees which shall be charged to any user of a regional facility and which shall be in addition to the rates approved pursuant to section c. of this Article, for any regional facility within its borders. Such fees shall be reasonable and shall provide the host state with sufficient revenue to cover any costs associated with such facilities. If such fees have been reviewed and approved by the Commission and to the extent that such revenue is insufficient, all party states shall share the costs in a manner to be determined by the Commission.

Regulation.
Transportation.

e. To the extent authorized by Federal law, each party state is responsible for enforcing any applicable Federal and state laws and regulations pertaining to the packaging and transportation of waste generated within or passing through its borders and shall adopt practices that will ensure that waste shipments originating within its borders and destined for a regional facility will conform to applicable packaging and transportation laws and regulations.

f. Each party state has the right to rely on the good faith performance of each other party state.

g. Unless authorized by the Commission, it shall be unlawful after January 1, 1986, for any person:

1. to deposit at a regional facility, waste not generated within the region;

2. to accept, at a regional facility, waste not generated within the region;

Exports.

3. to export from the region, waste which is generated within the region; and

4. to transport waste from the site at which it is generated, except to a regional facility.

Article IV–The Commission

Central
Interstate Low-
Level
Radioactive
Waste
Commission,
establishment.

a. There is hereby established the Central Interstate Low-Level Radioactive Waste Commission. The Commission shall consist of one voting member from each party state to be appointed according to the laws of each state. The appointing authority of each party state shall notify the Commission in writing of the identity of its member and any alternates. An alternate may act on behalf of the member only in the absence of such member. Each state is responsible for the expenses of its member of the Commission.

b. Each Commission member shall be entitled to one vote. Unless otherwise provided herein, no action of the Commission shall be bonding unless a majority of the total membership casts its vote in the affirmative.

c. The Commission shall elect from among its membership a chairman. The Commission shall adopt and publish, in convenient form, by-laws and policies which are not inconsistent with this compact.

d. The Commission shall meet at least once a year and shall also meet upon the call of the chairman, by petition of a majority of the membership or upon the call of a host state member.

e. The Commission may initiate any proceedings or appear as an intervenor or party in interest before any court of law, or any Federal, state or local agency, board or Commission that has jurisdiction over any matter arising under or relating to the terms of the provisions of this compact. The Commission shall determine in which proceedings it shall intervene or otherwise appear and may arrange for such expert testimony, reports, evidence or other participation in such proceedings as may be necessary to represent its views.

f. The Commission may establish such committees as it deems necessary for the purpose of advising the Commission on any and all matters pertaining to the management of waste.

Contracts.

g. The Commission may employ and compensate a staff limited only to those persons necessary to carry out its duties and functions. The Commission may also contract with and designate any person to perform necessary functions to assist the Commission. Unless otherwise required by the acceptance of a Federal grant, the staff shall serve at the Commission's pleasure irrespective of the civil service, personnel or other merit laws of any of the party states or the Federal government and shall be compensated from funds of the Commission.

h. Funding for the Commission shall be as follows:

1. The Commission shall set and approve its first annual budget as soon as practicable after its initial meeting. Party states shall equally contribute to the Commission budget on an annual basis, an amount not to exceed $25,000 until surcharges are available for that purpose. Host states shall begin imposition of the surcharges provided for in this section as soon as practicable and shall remit to the Commission funds resulting from collection of such surcharges within 60 days of their receipt; and

2. Each state hosting a regional facility shall annually levy surcharges on all users of such facilities, based on the volume and characteristics of wastes received at such facilities, the total of which:

(A) Shall be sufficient to cover the annual budget of the Commission; and

(B) shall be paid to the Commission, provided, however, that each host state collecting such surcharges may retain a portion of the collection sufficient to cover the administrative costs of collection, and that the remainder be sufficient only to cover the approved annual budget of the Commission.

Audit.
Report.

i. The Commission shall keep accurate accounts of all receipts and disbursements. An independent certified public account shall annually audit all receipts and disbursements of Commission funds and submit an audit report to the Commission. Such audit report shall be made a part of the annual report of the Commission required by this Article.

Grants.

j. The Commission may accept for any of its purposes and functions any and all donations, grants of money, equipment, supplies, materials and services, conditional or otherwise from any person and may receive, utilize and dispose of same, attendant upon any donation or grant accepted pursuant to this section, together with the identity of the donor, grantor or lender, shall be detailed in the annual report of the Commission.

k. (1) Except as otherwise provided herein, nothing in this compact shall be construed to alter the incidence of liability of any kind for any act, omission, course of conduct, or on account of any casual or other relationships. Generators, transporters of waste, owners and operators of facilities shall be liable for their acts, omissions, conduct or relationships in accordance with all laws relating thereto.

(2) The Commission herein established is a legal entity separate and distinct from the party states and shall be so liable for its actions. Liabilities of the Commission shall not be deemed liabilities of the party states. Members of the Commission shall not be personally liable for actions taken by them in their official capacity.

l. Any person or party state aggrieved by a final decision of the Commission may obtain judicial review of such decisions in the United States District Court in the District wherein the Commission maintains its headquarters by filing in such court a petition for review within 60 days after the Commission's final decision. Proceedings thereafter shall be in accordance with the rules of procedure applicable in such court.

m. The Commission shall:

1. Receive and approve the application of a non-party state to become a party state in accordance with Article VII;

2. submit an annual report, and otherwise communicate with, the Governors and the presiding officers of the legislative bodies of the party states regarding the activities of the Commission;

3. hear and negotiate disputes which may arise between the party states regarding this compact;

4. require of and obtain from the party states, and non-party states seeking to become party states, data and information necessary to the implementation of Commission and party states' responsibilities;

5. approve the development and operation of regional facilities in accordance with Article V;

6. notwithstanding any other provision of this compact, have the authority to enter into agreements with any person for the importation of waste into the region and for the right of access to facilities outside the region for waste generated within the region. Such authorization to import or export waste requires the approval of the Commission, including the affirmative vote of any host state which may be affected;

7. revoke the membership of a party state in accordance with Articles V and VII;

8. require all party states and other persons to perform their duties and obligations arising under this compact by an appropriate action in any forum designated in section e. of Article IV; and

9. take such action as may be necessary to perform its duties and functions as provided in this compact.

Reports.

Prohibition.

Prohibitions.

Report.

Contracts.
Exports.
Imports.

Article V–Development and Operation of
Regional Facilities

a. Following the collection of sufficient data and information from the states, the Commission shall allow each party state the opportunity to volunteer as a host for a regional facility.

b. If no state volunteers or if no proposal identified by a volunteer state is deemed acceptable by the Commission, based on the criteria in section c. of this Article, then the Commission shall publicly seek applicants for the development and operation of regional facilities.

c. The Commission shall review and consider each applicant's proposal based upon the following criteria:

1. The capability of the applicant to obtain a license from the applicable authority;

2. the economic efficiency of each proposed regional facility, including the total estimated disposal and treatment costs per cubic foot of waste;

3. financial assurances;

4. Accessibility to all party states; and

5. Such other criteria as shall be determined by the Commission to be necessary for the selection of the best proposal, based on the health, safety and welfare of the citizens in the region and the party states.

d. The Commission shall make a preliminary selection of the proposal or proposals considered most likely to meet the criteria enumerated in section c. and the needs of the region.

e. Following notification of each party state of the results of the preliminary selection process, the Commission shall:

1. Authorize any person whose proposal has been selected to pursue licensure of the regional facility or facilities in accordance with the proposal originally submitted to the Commission or as modified with the approval of the Commission; and

2. require the appropriate state or states or the U.S. Nuclear Regulatory Commission to process all applications for permits and licenses required for the development and operation of any regional facility or facilities within a reasonable period from the time that a completed application is submitted.

f. The preliminary selection or selections made by the Commission pursuant to this Article shall become final and receive the Commission's approval as a regional facility upon the issuance of license by the licensing authority. If a proposed regional facility fails to become licensed, the Commission shall make another selection pursuant to the procedures identified in this Article.

g. The Commission may, by two-thirds affirmative vote of its membership, revoke the membership of any party state which, after notice and hearing, shall be found to have arbitrarily or capriciously denied or delayed the issuance of a license or permit to any person authorized by the Commission to apply for such license or permit. Revocation shall be in the same manner as provided for in section e. of Article VII.

Health.
Safety.

Article VI–Other Laws and Regulations

Prohibition.

a. Nothing in this compact shall be construed to:

1. Abrogate or limit the applicability of any act of Congress or diminish or otherwise impair the jurisdiction of any Federal agency expressly conferred thereon by the Congress;

2. prevent the application of any law which is not otherwise inconsistent with this compact;

3. prohibit or otherwise restrict the management and waste on the site where it is generated if such is otherwise lawful;

4. Affect any judicial or administrative proceeding pending on the effective date of this compact;

5. Alter the relations between, and the respective internal responsibilities of, the government of a party state and its subdivisions; and

Research and development. Prohibition.

6. Affect the generation or management of waste generated by the Federal government or federal research and development activities.

b. No party state shall pass or enforce any law or regulation which is inconsistent with this compact.

Regulations.

c. All laws and regulations or parts thereof of any party state which are inconsistent with this compact are hereby declared null and void for purposes of this compact. Any legal right, obligation, violation or penalty arising under such laws or regulations prior to enactment of this compact shall not be affected.

Prohibition. Regulations.

d. No law or regulation of a party state or of any subdivision or instrumentality thereof may be applied so as to restrict or make more costly or inconvenient access to any regional facility by the generators of another party state than for the generators of the state where the facility is situated.

Article VII–Eligible Parties, Withdrawal, Revocation, Entry Into Force, Termination

Arkansas.
Iowa.
Kansas.
Louisiana.
Minnesota.
Missouri.
Nebraska.
North Dakota.
Oklahoma.

a. This compact shall have as initially eligible parties the states of Arkansas, Iowa, Kansas, Louisiana, Minnesota, Missouri, Nebraska, North Dakota and Oklahoma. Such initial eligibility shall terminate on January 1, 1984.

b. Any state may petition the Commission for eligibility. A petitioning state shall become eligible for membership in the compact upon the unanimous approval of the Commission.

Prohibition.

c. An eligible state shall become a member of the compact and shall be bound by it after such state has enacted the compact into law. In no event shall the compact take effect in any state until it has been entered into force as provided for in section f. of this Article.

Effective date. Prohibition.

d. Any party state may withdraw from this compact by enacting a statute repeating the same. Unless permitted earlier by unanimous approval of the Commission, such withdrawal shall take effect five-years after the Governor of the withdrawing state has given notice in writing of such withdrawal to each Governor of the party states. No withdrawal shall affect any liability already incurred by or chargeable to a party state prior to the time of such withdrawal.

e. Any party state which fails to comply with the terms of this compact or fulfill its obligations hereunder may, after notice and hearing have its privileges suspended or its membership in the compact revoked

Effective date.

by the Commission. Revocation shall take effect one year from the date such party state receives written notice from the Commission of its action. The Commission may require such party state to pay to the Commission, for a period not to exceed five years from the date of notice of revocation, an amount determined by the Commission based on the anticipated fees which the generators of such party state would have paid to each regional facility and an amount equal to that which such party state would have contributed in accordance with section d. of Article III, in the event of insufficient revenues. The Commission shall use such funds to ensure the continued availability of safe and economical waste management facilities for all remaining party states. Such state shall also pay an amount equal to that which such party state would have contributed to the annual budget of the Commission if such party state would have remained a member of the compact. All legal rights established under this compact of any party state which has its membership revoked shall cease upon the effective date of revocation; however, any legal obligations of such party state arising prior to the effective date of revocation shall not cease until they have been fulfilled. Written notice of revocation of any state's membership in the company shall be transmitted immediately following the vote of the Commission, by the chairman, to the Governor of the affected party state, all other Governors of the party states and the Congress of the United States.

f. This compact shall become effective after enactment by a least three eligible states and after consent has been given to it by the Congress. The Congress shall have the opportunity to withdraw such consent every five-years. Failure of the Congress to withdraw its consent affirmatively shall have the effect of renewing consent for an additional five-year period. The consent given to this compact by the Congress shall extend to any future admittance of new party states under sections b. and c. of this Article and to the power to ban the exportation of waste pursuant to Article III.

Prohibition.

g. The withdrawal of a party state from this compact under section d. of this Article or the revocation of a state's membership in this compact under section 3. of this Article shall not affect the applicability of this compact to the remaining party states.

Termination.

h. This compact shall be terminated when all party states have withdrawn pursuant to section d. of this Article.

Article VIII–Penalties

a. Each party state, consistent with its own law, shall prescribe and enforce penalties against any person for violation of any provision of this compact.

Regulations.

b. Each party state acknowledges that the receipt by a regional facility of waste packaged or transported in violation of applicable laws and regulations can result in sanctions which may include suspension or revocation of the violator's right of access to the regional facility.

Article IX–Severability and Construction

The provisions of this compact shall be severable and if any phrase, clause, sentence or provision of this compact is declared by a court of competent jurisdiction to be contrary to the Constitution of any participating state or of the United States or the applicability thereof to any government, agency, person or circumstances is held invalid, the validity of the remainder of this compact and the applicability thereof to any government, agency, person or circumstance shall not be affected thereby. If any provision of this compact shall be held contrary to the Constitution of any state participating therein, the compact shall remain in full force and effect as to the state affected as to all severable matters. The provisions of this compact shall be liberally construed to give effect to the purpose thereof.

42 USC 2021d note.
Alabama.
Florida.
Georgia.
Mississippi.
North Carolina.
South Carolina.
Tennessee.
Virginia.

Sec. 223. Southeast Interstate Low-Level Radioactive Waste Management Compact

In accordance with section 4(a)(2) of the Low-Level Radioactive Waste Policy Act (42 USC 2021d(a)(2), the consent of the Congress is hereby given to the States of Alabama, Florida, Georgia, Mississippi, North Carolina, South Carolina, Tennessee, and Virginia to enter into the Southeast Interstate Low-Level Radioactive Waste Management Compact. Such compact is substantially as follows:

Southeast Interstate Low-Level Radioactive Waste Management Compact

Article I–Policy and Purpose

Research and development.

There is hereby created the Southeast Interstate Low-Level Radioactive Waste Management Compact. The party states recognize and declare that each state is responsible for providing for the availability of capacity either within or outside the State for disposal of low-level radioactive waste generated within its borders, except for waste generated as a result of defense activities of the federal government or federal research and development activities. They also recognize that the management of low-level radioactive waste is handled most efficiently on a regional basis. The party states further recognize that the Congress of the United States, by enacting the Low-Level Radioactive Waste Policy Act (Public Law 96-573), has provided for encouraged the development of low level radioactive waste compacts as a tool for disposal of such waste. The party states recognize that the safe and efficient management of low-level radioactive waste generated within the region requires that sufficient capacity to dispose of such waste be properly provided.

42 USC 2021b note.

It is the policy of the party states to: enter into a regional low-level radioactive waste management compact for the purpose of providing the instrument and framework for a cooperative effort; provide sufficient facilities for the proper management of low-level radioactive waste generated in the region; promote the health and safety of the region; limit the number of facilities required to effectively and efficiently manage low-level radioactive waste generated in the region; encourage the reduction of the amounts of low-level waste generated in the region; distribute the costs, benefits, and obligations of successful low-level radioactive waste management equitably among the party states; and ensure the ecological and economical management of low-level radioactive wastes.

Regulations.

Implicit in the Congressional consent to this compact is the expectation by Congress and the party states that the appropriate federal agencies will actively assist the Compact Commission and the individual party states to this compact by:

　　1. expeditious enforcement of federal rules, regulations, and laws;

　　2. imposing sanctions against those found to be in violation of federal rules, regulations, and laws;

3. timely inspection of their licensees to determine their capability to adhere to such rules, regulations, and laws;

4. timely provision of technical assistance to this compact in carrying out their obligations under the Low-Level Radioactive Waste Policy Act, as amended.

42 USC 2021b note.

Article II–Definitions

As used in this compact, unless the context clearly requires a different construction:

1. "Commission" or "Compact Commission" means the Southeast Interstate Low-Level Radioactive Waste Management Commission.

2. "Facility" means a parcel of land, together with the structure, equipment, and improvements thereon or appurtenant thereto, which is used or is being developed for the treatment, storage, or disposal of low-level radioactive waste.

3. "Generator" means any person who produces or processes low-level radioactive waste in the course of, or as an incident to, manufacturing, power generation, processing, medical diagnosis and treatment, research, or other industrial or commercial activity. This does not include persons who provide a service to generators by arranging for the collection, transportation, storage, or disposal of wastes with respect to such waste generated outside the region.

4. "High-level waste" means irradiated reactor fuel, liquid wastes from reprocessing irradiated reactor fuel, and solids into which such liquid wastes have been converted, and other high-level radioactive waste as defined by the U.S. Nuclear Regulatory Commission.

5. "Host state" means any state in which a regional facility is situated or is being developed.

42 USC 2014.

6. "Low-level radioactive waste" or "waste" means radioactive waste not classified as high-level radioactive waste, transuranic waste, spent nuclear fuel, or by-product material as defined in section 11e(2) of the Atomic Energy Act of 1954, or as may be further defined by Federal law or regulation.

7. "Party state" means any state which is a signatory party to this compact.

8. "Person" means any individual, corporation, business enterprise, or other legal entity (either public or private).

9. "Region ' means the collective party states.

10. 'Regional facility" means (1) a facility as defined in this article which has been designated, authorized, accepted, or approved by the Commission to receive waste or (2) the disposal facility in Barnwell County, South Carolina, owned by the State of South Carolina and as licensed for the burial of low-level radioactive waste on July 1, 1982, but in no event shall this disposal facility serve as a regional facility beyond December 31, 1992.

11. "State" means a state of the United States, the District of Columbia, the Commonwealth of Puerto Rico, the Virgin Islands, or any other territorial possession of the United States.

42 USC 2021.

12. "Transuranic wastes" means waste material containing transuranic elements with contamination levels as determined by the regulations of (1) the U.S. Nuclear Regulatory Commission or (2) any host state, if it is an agreement state under section 274 of the Atomic Energy Act of 1954.

13. "Waste management" means the storage, treatment, or disposal of waste.

Article III–Rights and Obligations

Prohibition.

The rights granted to the party states by this compact are additional to the rights enjoyed by sovereign states, and nothing in this compact shall be construed to infringe upon, limit, or abridge those rights.

(A) Subject to any license issued by the U.S. Nuclear Regulatory Commission or a host state, each party state shall have the right to have all wastes generated within its borders stored, treated, or disposed of, as applicable, at regional facilities and, additionally, shall have the right of access to facilities made available to the region through agreements entered into by the Commission pursuant to Article 4(e)(9). The right of access by a generator within a party state to any regional facility is limited by its adherence to applicable state and federal law and regulation.

(B) If no operating regional facility is located within the borders of a party state and the waste generated within its borders must therefore be stored, treated, or disposed of at a regional facility in another party state, the party state without such facilities may be required by the host state or states to establish a mechanism which provides compensation for access to the regional facility according to terms and conditions established by the host state or states and approved by a two-thirds vote of the Commission.

(C) Each party state must establish the capability to regulate, license, and ensure the maintenance and extended care of any facility within its borders. Host states are responsible for the availability, the subsequent post-closure observation and maintenance, and the extended institutional control of their regional facilities in accordance with the provisions of Article 5, section (b).

Regulations.
Transportation.

(D) Each party state must establish the capability to enforce any applicable federal or state laws and regulations pertaining to the packaging and transportation of waste generated within or passing through its borders.

(E) Each party state must provide to the Commission on an annual basis any data and information necessary to the implementation of the Commission's responsibilities. Each party state shall establish the capability to obtain any data and information necessary to meet its obligation.

(F) Each party state must, to the extent authorized by federal law, require generators within its borders to use the best available waste management technologies and practices to minimize the volumes of waste requiring disposal.

Article IV–The Commission

(A) There is hereby created the Southeast Interstate Low-Level Radioactive Waste Management Commission ("Commission" or "Compact Commission"). The Commission shall consist of two voting members from each party state to be appointed according to the laws of each state. The appointing authorities of each state must notify the Commission in writing of the identity of its members and any alternates. An alternate may act on behalf of the member only in the member's absence.

(B) Each commission member is entitled to one vote. No action of the Commission shall be binding unless a majority of the total membership cast their vote in the affirmative, or unless a greater than majority vote is specifically required by any other provision of this compact.

(C) The Commission must elect from among its members a presiding officer. The Commission shall adopt and publish, in convenient form, bylaws which are consistent with this compact.

(D) The Commission must meet at least once a year and also meet upon the call of the presiding officer, by petition of a majority of the party states, or upon the call of a host state. All meetings of the Commission must be open to the public.

(E) The Commission has the following duties and powers:

1. To receive and approve the application of a nonparty state to become an eligible state in accordance with the provisions of Article 7(b).

2. To receive and approve the application of a nonparty state to become an eligible state in accordance with the provisions of Article 7(c).

3. To submit an annual report and other communications to the Governors and to the presiding officer of each body of the legislature of the party states regarding the activities of the Commission.

4. To develop and use procedures for determining, consistent with consideration for public health and safety, the type and number of regional facilities which are presently necessary and which are projected to be necessary to manage waste generated within the region.

5. To provide the party states with reference guidelines for establishing the criteria and procedures for evaluating alternative locations for emergency or permanent regional facilities.

6. To develop and adopt, within one year after the Commission is constituted as provided in Article 7(d) procedures and criteria for identifying a party state as a host state for a regional facility as determined pursuant to the requirements of this article. In accordance with these procedures and criteria, the Commission shall identify a host state for the development of a second regional disposal facility within three years after the Commission is constituted as provided for in Article 7(d), and shall seek to ensure that such facility is licensed and ready to operate as soon as required but in no event later than 1991.

In developing criteria, the Commission must consider the following; the health, safety, and welfare of the citizens of the party states; the existence of regional facilities within each party state; the minimization of waste transportation; the volumes and types of wastes generated within each party state; and the environmental, economic, and ecological impacts on the air, land, and water resources of the party states.

The Commission shall conduct such hearings, require such reports, studies, evidence, and testimony, and do what is required by its approved procedures in order to identify a party state as a host state for a needed facility.

7. In accordance with the procedures and criteria developed pursuant to section (e)(6) of this Article, to designate, by a two-thirds vote, a host state for the establishment of a needed regional facility. The Commission shall not exercise this authority unless the party states have failed to voluntarily pursue the development of such facility. The Commission shall have the authority to revoke the membership of a party state that willfully creates barriers to the siting of a needed regional facility.

8. To require of and obtain from party states, eligible states seeking to become party states, and nonparty states seeking to become eligible states, data and information necessary to the implementation of Commission responsibilities.

9. Notwithstanding any other provision of this compact, to enter into agreements with any person, state, or similar regional body or group of states for the importation of waste into the region and for the right of access to facilities outside the region for waste generated within the region. The authorization to import requires a two-thirds majority vote of the Commission, including an affirmative vote of both representatives of a host state in which any affected regional facility is located. This shall be done only after an assessment of the affected facility's capability to handle such wastes.

10. To act or appear on behalf of any party state or states, only upon written request of both members of the Commission for such state or states as an intervenor or party in interest before Congress, state legislatures, any court of law, or any federal, state, or local agency, board, or commission which has jurisdiction over the management of wastes. The authority to act, intervene, or otherwise appear shall be exercised by the Commission, only after approval by a majority vote of the Commission.

11. To revoke the membership of a party state in accordance with Article 7(f).

F. The Commission may establish any advisory committees as it deems necessary for the purpose of advising the Commission on any matters pertaining to the management of low-level radioactive waste.

G. The Commission may appoint or contract for and compensate such limited staff necessary to carry out its duties and functions. The staff shall serve at the commission's pleasure irrespective of the civil service, personnel, or other merit laws of any of the party states or the federal government and shall be compensated from funds of the Commission. In selecting any staff, the Commission shall assure that the staff has adequate experience and formal training to carry out such functions as may be assigned to it by the Commission. If the Commission has a headquarters it shall be in a party state.

H. Funding for the Commission must be provided as follows:

1. Each eligible state, upon becoming a party state, shall pay twenty-five thousand dollars to the Commission which shall be used for costs of the Commission's services.

2. Each state hosting a regional disposal facility shall annually levy special fees or surcharges on all users of such facility, based

upon the volume of wastes disposed of at such facilities, the total of which:

a. must be sufficient to cover the annual budget of the Commission;

b. must represent the financial commitments of all party states to the Commission;

c. must be paid to the Commission;

Provided, however, That each host state collecting such fees or surcharges may retain a portion of the collection sufficient to cover its administrative costs of collection and that the remainder be sufficient only to cover the approved annual budgets of the Commission.

3. The Commission must set and approve its first annual budget as soon as practicable after its initial meeting. Host states for disposal facilities must begin imposition of the special fees and surcharges provided for in this section as soon as practicable after becoming party states and must remit to the Commission funds resulting from collection of such special fees and surcharges within sixty days of their receipt.

Audit report.

I. The Commission must keep accurate accounts of all receipts and disbursements. An independent certified public accountant shall annually audit all receipts and disbursements of Commission funds and submit an audit report to the Commission. The audit report shall be made a part of the annual report of the Commission required by Article 4(e)(3).

Grants.

J. The Commission may accept for any of its purposes and functions any and all donations, grants of money, equipment, supplies, materials, and services (conditional or otherwise) from any state, or the United States, or any subdivision or agency thereof, or interstate agency, or from

Report.

any institution, person, firm, or corporation, and may receive, utilize, and dispose of the same. The nature, amount, and condition, if any, attendant upon any donation or grant accepted pursuant to this section, together with the identity of the donor, grantor, or lender shall be detailed in the annual report to the Commission.

K. The Commission is not responsible for any costs associated with:

(1) the creation of any facility,

(2) the operation of any facility,

(3) the stabilization and closure of any facility,

(4) the post-closure observation and maintenance of any facility, or

(5) the extended institutional control, after post-closure observation and maintenance of any facility.

L. As of January 1, 1986, the management of wastes at regional facilities is restricted to wastes generated within the region, and to wastes generated within nonparty states when authorized by the Commission pursuant to the provisions of this compact. After January 1, 1986, the Commission may prohibit the exportation of waste from the region for the purposes of management.

Exports.
Prohibition.

M. 1. The Commission herein established is a legal entity separate and distinct from the party states capable of acting in its own behalf and is liable for its actions. Liabilities of the Commission shall not be deemed liabilities of the party states. Members of the Commission shall not personally be liable for action taken by them in their official capacity.

Prohibition.

2. Except as specifically provided in this compact, nothing in this compact shall be construed to alter the incidence of liability of any kind for any act, omission, course of conduct, or on account of any casual or other relationships. Generators and transporters of wastes and owners and

operators of sites shall be liable for their acts, omissions, conduct, or relationships in accordance with all laws relating thereto.

Article V–Development and Operation of Facilities

A. Any party state which becomes a host state in which a regional facility is operated shall not be designated by the Compact Commission as a host state for an additional regional facility until each party state has fulfilled its obligation, as determined by the Commission, to have a regional facility operated within its borders.

Health.
Safety.

B. A host state desiring to close a regional facility located within its borders may do so only after notifying the Commission in writing of its intention to do so and the reasons therefor. Such notification shall be given to the Commission at least four years prior to the intended date of closure.

Notwithstanding the four-year notice requirement herein provided, a host state is not prevented from closing its facility or establishing conditions of its use and operations as necessary for the protection of the health and safety of its citizens. A host state may terminate or limit access to its regional facility if it determines that Congress has materially altered the conditions of this compact.

C. Each party state designated as a host state for a regional facility shall take appropriate steps to ensure that an application for a license to construct and operate a facility of the designated type is filed with and issued by the appropriate authority.

Prohibition.

D. No party state shall have any form of arbitrary prohibition on the treatment, storage, or disposal of low-level radioactive waste within its borders.

E.[4] No party state shall be required to operate a regional facility for longer than a 20-year period, or to dispose of more than 32,000,000 cubic feet of low-level radioactive waste, whichever first occurs.

Article VI–Other Laws and Regulations

Prohibition.

A. Nothing in this compact shall be construed to:

(1) Abrogate or limit the applicability of any act of Congress or diminish or otherwise impair the jurisdiction of any federal agency expressly conferred thereon by the Congress.

42 USC 2021.

(2) Abrogate or limit the regulatory responsibility and authority of the U.S. Nuclear Regulatory Commission or of an agreement state under section 274 of the Atomic Energy Act of 1954 in which a regional facility is located.

(3) Make inapplicable to any person or circumstance any other law of a party state which is not inconsistent with this compact.

(4) Make unlawful the continued development and operation of any facility already licensed for development or operation on the date this compact becomes effective, except that any such facility shall comply with Article 3, Article 4, and Article 5 and shall be subject to any action lawfully taken pursuant thereto.

---

[4]   Added by P.L. 101–171, 103 Stat. 1289 (1989).

Prohibition.

(5) Prohibit any storage or treatment of waste by the generator on its own premises.

(6) Affect any judicial or administrative proceeding pending on the effective date of this compact.

(7) Alter the relations between, and the respective internal responsibilities of, the government of a party state and its subdivisions.

42 USC 2021b
note.
Research and
development.

(8) Affect the generation, treatment, storage, or disposal of waste generated by the atomic energy defense activities of the Secretary of the United States Department of Energy or federal research and development activities as defined in Public Law 96-573.

(9) Affect the rights and powers of any party state and its political subdivisions to regulate and license any facility within its borders or to affect the rights and powers of any party state and its political subdivisions to tax or impose fees on the waste managed at any facility within its borders.

Prohibition.
Regulation.

B. No party shall pass any law or adopt any regulation which is inconsistent with this compact. To do so may jeopardize the membership status of the party state.

Prohibition.
Regulation.

C. Upon formation of the compact no law or regulation of a party state or of any subdivision or instrumentality thereof may be applied so as to restrict or make more inconvenient access to any regional facility by the generators of another party state than for the generators of the state where the facility is situated.

D. Restrictions of waste management of regional facilities pursuant to Article 4 shall be enforceable as a matter of state law.

Article VII–Eligible Parties; Withdrawal; Revocation; Entry Into Force; Termination

Alabama.
Florida.
Georgia.
Mississippi.
North Carolina.
South Carolina.
Tennessee.
Virginia.

A. This compact shall have as initially eligible parties the States of Alabama, Florida, Georgia, Mississippi, North Carolina, South Carolina, Tennessee, and Virginia.

B. Any state not expressly declared eligible to become a party state to this compact in section (A) of this Article may petition the Commission, once constituted, to be declared eligible. The Commission may establish such conditions as it deems necessary and appropriate to be met by a state wishing to become eligible to become a party state to this compact pursuant to such provisions of this section. Upon satisfactorily meeting the conditions and upon the affirmative vote of two-thirds of the Commission, including the affirmative vote of both representatives of a host state in which any affected regional facility is located, the petitioning state shall be eligible to become a party state to this compact and may become a party state in the manner as those states declared eligible in section (a) of this Article.

C. Each state eligible to become a party state to this compact shall be declared a party state upon enactment of this compact into law by the state and upon payment of the fees required by Article 4(H)(1). The Commission is the judge of the qualifications of the party states and of its members and of their compliance with the conditions and requirements of this compact and the laws of the party states relating to the enactment of this compact.

D. 1. The first three states eligible to become party states to this compact which enact this compact into law and appropriate the fees required by Article 4(H)(1) shall immediately, upon the appointment of their Commission members, constitute themselves as the Southeast Low-Level Radioactive Waste Management Commission; shall cause legislation to be introduced in Congress which grants the consent of Congress to this compact; and shall do those things necessary to organize the commission and implement the provisions of this compact.

2. All succeeding states eligible to become party states to this compact shall be declared party states pursuant to the provisions of section (C) of this Article.

Effective date.

3. The consent of Congress shall be required for the full implementation of this compact. The provisions of Article 5, section (D) shall not become effective until the effective date of the import ban authorized by Article 4, section (L) as approved by Congress. Congress may by law withdraw its consent only every five years.

Prohibition.

E. No state which holds membership in any other regional compact for the management of low-level radioactive waste may be considered by the Compact Commission for eligible state status or party state status.

F. Any party state which fails to comply with the provisions of this compact or to fulfill the obligations incurred by becoming a party state to this compact may be subject to sanctions by the Commission, including suspension of its rights under this compact and revocation of its status as a party state. Any sanction shall be imposed only upon the affirmative vote of at least two-thirds of the Commission members. Revocation of party state status may take effect on the date of the meeting at which the Commission approves the resolution imposing such sanction, but in no event shall revocation take effect later than ninety days from the date of such meeting. Rights and obligations incurred by being declared a party state to this compact shall continue until the effective date of the sanction imposed or as provided in the resolution of the Commission imposing the sanction.

The Commission must, as soon as practicable after the meeting at which a resolution revoking status as a party state is approved, provide written notice of the action, along with a copy of the resolution, to the Governors, the Presidents of the Senates, and the Speakers of the Houses of Representatives of the party states, as well as chairmen of the appropriate committees of Congress.

G. Subject to the provisions of Article 7, section H., any party state may withdraw from the compact by enacting a law repealing the compact, provided that if a regional facility is located within such state, such regional facility shall remain available to the region for four years after the date the Commission receives verification in writing from the Governor of such party state of the rescission of the Compact. The Commission, upon receipt of the verification, shall as soon as practicable provide copies of such verification to the Governor, the presidents of the Senates, and the Speakers of the Houses of Representatives of the party states as well as the chairmen of the appropriate committees of the Congress.[5]

H. The right of a party state to withdraw pursuant to section G. shall terminate thirty days following the commencement of operation of the second host state disposal facility. Thereafter a party state may withdraw only with the unanimous approval of the Commission and with the consent of Congress.

---

[5]   Amended by P.L. 101–171, 103 Stat. 1290 (1989).

South Carolina.

For purposes of this section, the low-level radioactive waste disposal facility located in Barnwell County, South Carolina shall be considered the first host state disposal facility.[6]

I.[7] This compact may be terminated only by the affirmative action of the Congress or by rescission of all laws enacting the compact in each party state.

Article VIII–Penalties

A. Each party state, consistently with its own law, shall prescribe and enforce penalties against any person not an official of another state for violation of any provisions of this compact.

Regulation.

B. Each party state acknowledges that the receipt by a host state of waste packaged or transported in violation of applicable laws and regulations can result in the imposition of sanctions by the hose state which may include suspension or revocation of the violator's right of access to the facility in the host state.

Article IX–Severability and Construction

Provisions held invalid.

The provisions of this compact shall be severable and if any phrase, clause, sentence, or provision of this compact is declared by a court of competent jurisdiction to be contrary to the Constitution, of any participating state or of the United States, or the applicability thereof to any government, agency, person, or circumstance is held invalid, the validity of the remainder of this compact and the applicability thereof to any other government, agency, person, or circumstance shall not be affected thereby. If any provision of this compact shall be held contrary to the Constitution of any State participating therein, the compact shall remain in full force and effect as to the state affected as to all severable matters. The provisions of this compact shall be liberally construed to give effect to the purposes thereof.

Sec. 224. Central Midwest Interstate Low-Level Radioactive Waste Compact

42 USC 2021d note.
Illinois.
Kentucky.

In accordance with section 4(a)(2) of the Low-Level Radioactive Waste Policy Act (42 USC 2021d(a)(2)), the consent of the Congress hereby is given to the States of Illinois and Kentucky to enter into the Central Midwest Interstate Low-Level Radioactive Waste Compact. Such compact is substantially as follows:

Central Midwest Interstate Low-Level Radioactive Waste Compact

Article I–Policy and Purpose

There is created the Central Midwest Interstate Low-Level Radioactive Waste Compact.

42 USC 2021b note.

The states party to this compact recognize that the Congress of the United States, by enacting the Low-Level Radioactive Waste Policy Act (42 USC 2021), has provided for and encouraged the development of low-level radioactive waste compacts as a tool for managing such waste. The party states also recognize that the management of low-level

---

[6]   Amended by P.L. 101–171, 103 Stat. 1290 (1989).
[7]   Added by P.L. 101–171, 103 Stat. 1290 (1989).

radioactive waste is handled most efficiently on a regional basis; and, that the safe and efficient management of low-level radioactive waste generated within the region requires that sufficient capacity to manage such waste be properly provided.[8]

a) It is the policy of the party states to enter into a regional low-level radioactive waste management compact for the purpose of:

1) providing the instrument and the framework for a cooperative effort;

2) providing sufficient facilities for the proper management of low-level radioactive waste generated in the region;

3) protecting the health and safety of the citizens of the region;

4) limiting the number of facilities required to manage low-level radioactive waste generated in the region effectively and efficiency;

5) promoting the volume and source reduction of low-level radioactive waste generated in the region;

6) distributing the costs, benefits and obligations of successful low-level radioactive waste management equitably among the party states and among generators and other persons who use regional facilities to manage their waste;

7) ensuring the ecological and economical management of low-level radioactive waste, including the prohibition of shallow-land burial of waste; and

8) promoting the use of above-ground facilities and other disposal technologies providing greater and safer confinement of low-level radioactive waste than shallow-land burial facilities.

b) Implicit in the Congressional consent to this compact is the expectation by the Congress and the party states that the appropriate federal agencies will actively assist the Compact Commission and the individual party states to this compact by:

1) expeditious enforcement of federal rules, regulations and laws;

2) imposition of sanctions against those found to be in violation of federal rules, regulations and laws; and

3) timely inspection of their licensees to determine their compliance with these rules, regulations and laws.

Article II–Definitions

As used in this compact, unless the context clearly requires a different construction:

a) "Commission" means the Central Midwest Interstate Low-Level Radioactive Waste Commission.

b) "Decommissioning" means the measures taken at the end of a facility's operating life to assure the continued protection of the public from any residual radioactivity or other potential hazards present at a facility.

c) "Disposal" means the isolation of waste from the biosphere in a permanent facility designed for that purpose.

d) "Eligible" state means either the State of Illinois or the Commonwealth of Kentucky.

e) "Extended care" means the continued observation of a facility after closure for the purpose of detecting a need for maintenance, ensuring environmental safety, and determining compliance with applicable

Health.

Safety.

Regulations.

---

[8]   Amended by P.L. 103–439, 108 Stat. 4607 (1994).

licensure and regulatory requirements and includes undertaking any action or clean-up necessary to protect public health and the environment from radioactive releases from a regional facility.

f) "Facility" means a parcel of land or site, together with the structures, equipment and improvements on or appurtenant to the land or site, which is used or is being developed for the treatment, storage or disposal of low-level radioactive waste.

g) "Generator" means a person who produces or possesses low-level radioactive waste in the course of or incident to manufacturing, power generation, processing medical diagnosis and treatment, research, or other industrial or commercial activity and who, to the extent required by law, is licensed by the U. S. Nuclear Regulatory Commission or a party state, to produce or possess such waste.

h) "Host state" means any party state that is designated by the Commission to host a regional facility, provided that a party state with a total volume of waste recorded on low-level radioactive waste manifests for any year that is less than 10 percent of the total volume recorded on such manifests for the region during the same year shall not be designated a host state.

i) "Institutional control" means those activities carried out by the host state to physically control access to the disposal site following transfer of control of the disposal site from the disposal site operator to the state or federal government. These activities must include, but need not be limited to, environmental monitoring, periodic surveillance, minor custodial care, and other necessary activities at the site as determined by the host state, and administration of funds to cover the costs for these activities. The period of institutional control will be determined by the host state, but institutional control may not be relied upon for more than 100 years following transfer of control of the disposal site to the state or federal government.

j) "Long-term liability" means the financial obligation to compensate any person for medical and other expenses incurred from damages to human health, personal injuries suffered from damages to human health and damages or losses to real or personal property, and to provide for the costs for accomplishing any necessary corrective action or clean-up on real or personal property caused by radioactive releases from a regional facility.

42 USC 2014.

k) "Low-level radioactive waste" or "waste" means radioactive waste not classified as (1) high-level radioactive waste, (2) transuranic waste, (3) spent nuclear fuel, or (4) by-product material as defined in section 11e.(2) of the Atomic Energy Act of 1954. This definition shall apply notwithstanding any declaration by the federal government, a state or any regulatory agency that any radioactive material is exempt from any regulatory control.[9]

l) "Management plan" means the plan adopted by the Commission for the storage, transportation, treatment and disposal of waste within the region.

m) "Manifest" means a shipping document identifying the generator of waste, the volume of waste, the quantity of radionuclides in the shipment, and such other information as may be required by the appropriate regulatory agency.

[9]   Amended by P.L. 103–439, 108 Stat. 4608 (1994).

n) "Party state" means any eligible state which enacts the compact into law and pays the membership fee.

o) "Person" means any individual, corporation, business enterprise or other legal entity, either public or private, and any legal successor, representative, agent or agency of that individual, corporation, business enterprise, or legal entity.

p) "Region" means the geographical area of the party states.

q) "Regional facility" means any facility as defined in Article II(f) that is (1) located within the region, and (2) established by a party state pursuant to designation of that state as a host state by the Commission.[10]

r) "Shallow-land burial" means a land disposal facility in which radioactive waste is disposed of in or within the upper 30 meters of the earth's surface; however, this definition shall not include an enclosed, engineered, strongly structurally enforced and solidified bunker that extends below the earth's surface.

s) "Site" means the geographic location of a facility.

t) "Source reduction" means those administrative practices that reduce the radionuclide levels in low-level radioactive waste or that prevent the generation of additional low-level radioactive waste.

u) "State" means a state of the United States, the District of Columbia, the Commonwealth of Puerto Rico, the Virgin Islands or any other territorial possession of the United States.

v) "Storage" means the temporary holding of waste for treatment or disposal.

w) "Treatment" means any method, technique or process, including storage for radioactive decay, designed to change the physical, chemical or biological characteristics or composition of any waste in order to render the waste safer for transport or management, amenable to recovery, convertible to another usable material or reduced in volume.

x) "Volume reduction" means those methods including, but not limited to, biological, chemical, mechanical and thermal methods used to reduce the amount of space that waste materials occupy and to put them into a form suitable for storage or disposal.

y) "Waste management" means the source and volume reduction, storage, transportation, treatment or disposal of waste.

Article III–The Commission

Central Midwest Interstate Low-Level Radioactive Waste Commission establishment.

a) There is created the Central Midwest Interstate Low-Level Radioactive Waste Commission. Upon the eligible states becoming party states, the Commission shall consist of two voting Commissioners from each state eligible to be designated a host state under Article VI(b), one voting Commissioner from any other party state, and for each regional facility, one non-voting Commissioner who is an elected official of local government and a resident of the county where that regional facility is located. The Governor of each party state shall notify the Commission in writing of its Commissioners and any alternates.[11]

b) Each voting Commissioner is entitled to one vote. No action of the Commission is binding unless a majority of the voting membership casts its vote in the affirmative. In addition, no agreement by the Commission under Article III(i)(1), Article III(i)(2), or Article III(i)(3) is valid unless

---

[10]   Amended by P.L. 103–439, 108 Stat. 4608 (1994).
[11]   Amended by P.L. 103–439, 108 Stat. 4608 (1994).

all voting Commissioners from the party state in which the facility where waste would be sent cast their votes in the affirmative.[12]

Public information.

c) The Commission shall elect annually from among its members a chairperson. The Commission shall adopt and publish, in convenient form, by-laws and policies that are not inconsistent with this compact, including procedures that conform with the provisions of the Federal Administrative Procedure Act (5 USC ss. 500 to 559) to the greatest extent practicable in regard to notice, conduct and recording of meetings; access by the public to records; provision of information to the public; conduct of adjudicatory hearings; and issuance of decisions.

d) The Commission shall meet at least once annually and shall also meet upon the call of any voting Commissioner.[13]

e) All meetings of the Commission and its designated committees shall be open to the public with reasonable advance notice. The Commission may, by majority vote, close a meeting to the public for the purpose of considering sensitive personnel or legal strategy matters. However, all Commission actions and decisions shall be made in open meetings and appropriately recorded. A roll call may be required upon request of any voting Commissioner.[14]

f) The Commission may establish advisory committees for the purpose of advising the Commission on any matters pertaining to waste management, waste generation and source and volume reduction.

g) The Office of the Commission shall be in Illinois. The Commission may appoint or contract for and compensate such staff necessary to carry out its duties and functions. The staff shall serve at the Commission's pleasure with the exception that staff hired as the result of securing federal funds shall be hired and governed under applicable federal statutes and regulations. In selecting any staff, the Commission shall assure that the staff has adequate experience and formal training to carry out the functions assigned to it by the Commission.[15]

Public inspection. Records.

h) All files, records and data of the Commission shall be open to reasonable public inspection and may be copied upon payment of reasonable fees to be established where appropriate by the Commission, except for information privileged against introduction in judicial proceedings. Such fees may be waived or shall be reduced substantially for not-for-profit organizations.

i) The Commission may:

Contracts. Prohibitions.

1) Enter into an agreement with any person to allow waste from outside the region to be disposed of at facilities in the region. However, no such agreement shall be effective unless and until ratified by a law enacted by the party state to which the waste would be sent for disposal.

2) Enter into an agreement with any person to allow waste described in Article VII(a)(6) to be treated, stored, or disposed of at regional facilities. However, no such agreement shall be effective unless and until ratified by a law enacted by the host state of the regional facility to which the waste would be sent for treatment, storage, or disposal.

---

[12]   Amended by P.L. 103–439, 108 Stat. 4608 (1994).
[13]   Amended by P.L. 103–439, 108 Stat. 4608 (1994).
[14]   Amended by P.L. 103–439, 108 Stat. 4608 (1994).
[15]   Amended by P.L. 103–439, 108 Stat. 4608 (1994).

Reports.

3) Enter into an agreement with any person to allow waste from outside the region to be treated or stored at facilities in the region.

However, any such agreement shall be revoked as a matter of law if, within one year of the effective date of the agreement, a law is enacted ordering such revocation by the party state to which the waste would be sent for treatment or storage.

Prohibition.

4) Approve, or enter into an agreement with any person for, the export of waste from the region.

5) Approve the disposal of waste generated within the region at a facility in the region other than a regional facility, subject to the limitations of Articles V(f) and VII(a)(6).

6) Require that waste generated within the region be treated or stored at available regional facilities, subject to the limitations of Articles V(f), VII(a)(3) and VII(a)(6).

7) Appear as an intervenor or party in interest before any court of law or any federal, state or local agency, board or commission in any matter related to waste management. In order to represent its views, the Commission may arrange for any expert testimony, reports, evidence or other participation.

8) Review the emergency closure of a regional facility, determine the appropriateness of that closure, and take whatever actions are necessary to ensure that the interests of the region are protected, provided that a party state with a total volume of waste recorded on low-level radioactive waste manifests for any year that is less than 10 percent of the total volume recorded on such manifests for the region during the same year shall not be designated a host state or be required to store the region's waste. In determining the 10 percent exclusion, there shall not be included waste recorded on low-level radioactive waste manifests by a person whose principal business is providing a service by arranging for the collection, transportation, treatment, storage or disposal of such waste.

9) Take any action which is appropriate and necessary to perform its duties and functions as provided in this compact.

10) Suspend the privileges or revoke the membership of a party state.[16]

j) The Commission shall:

Report.

1) Submit within 10 days of its execution to the governor and the appropriate officers of the legislative body of the party state in which any affected facility is located a copy of any agreement entered into by the Commission under Article III(i)(1), Article III(i)(2) or Article III(i)(3).

2) Submit an annual report to, and otherwise communicate with, the governors and the appropriate officers of the legislative bodies of the party states regarding the activities of the Commission. The annual report shall include a description of the status of the activities taken pursuant to any agreement entered into by the Commission under Article III(i)(1), Article III(i)(2) or Article III(i)(3) and any violation of any provision thereof, and a description of the source, volume, activity, and current status of any waste from outside the region or waste described under Article VII(a)(6) that was treated, stored, or disposed of in the region in the previous year.

---

[16]   Amended by P.L. 103–439, 108 Stat. 4609 (1994).

3) Hear, negotiate, and, as necessary, resolve by final decision disputes which may arise between the party states regarding this compact.

4) Adopt and amend, as appropriate, a regional management plan that plans for the establishment of needed regional facilities.

5) Adopt an annual budget.[17]

k) Funding of the budget of the Commission shall be provided as follows:

1) Each state, upon becoming a party state, shall pay $50,000 to the Commission which shall be used for the administrative costs of the Commission.

2) Each state hosting a regional facility shall levy surcharges on each user of the regional facility based upon its portion of the total volume and characteristics of wastes managed at that facility. The surcharges collected at all regional facilities shall:

A) be sufficient to cover the annual budget of the Commission; and

B) be paid to the Commission, provided, however, that each host state collecting surcharges may retain a portion of the collection sufficient to cover its administrative costs of collection.

l) The Commission shall keep accurate accounts of all receipts and disbursements. The Commission shall contract with an independent certified public accountant to annually audit all receipts and disbursements of Commission funds and to submit an audit report to the Commission. The audit report shall be made a part of the annual report of the Commission required by this Article.

m) The Commission may accept for any of its purposes and functions and may utilize and dispose of any donations, grants of money, equipment, supplies, materials and services from any state or the United States (or any subdivision or agency thereof), or interstate agency, or from any institution, person, firm or corporation. The nature, amount and condition, if any, attendant upon any donation or grant accepted or received by the Commission together with the identity of the donor, grantor, or lender, shall be detailed in the annual report of the Commission. The Commission shall establish guidelines for the acceptance of donations, grants, equipment, supplies, materials and services and shall review such guidelines annually.

n) The Commission is not liable for any costs associated with any of the following:

1) the licensing and construction of any facility;

2) the operation of any facility;

3) the stabilization and closure of any facility;

4) the extended care of any facility;

5) the institutional control, after extended care of any facility; or

6) the transportation of waste to any facility.

o) The Commission is a legal entity separate and distinct from the party states and is liable for its actions as a separate and distinct legal entity. Commissioners are not personally liable for actions taken by them in their official capacity.[18]

Audit.
Contracts.
Reports.

Grants.
Report.

Transportation.

---

[17]   Amended by P.L. 103–439, 108 Stat. 4609 (1994).
[18]   Amended by P.L. 103–439, 108 Stat. 4610 (1994).

p) Except as provided under Article III(n), Article III(o), Article VI(p) and Article VI(q), nothing in this compact alters liability for any action, omission, course of conduct or liability resulting from any causal or other relationships.[19]

q) Any person aggrieved by a final decision of the Commission, which adversely affects the legal rights, duties or privileges of such person, may petition a court of competent jurisdiction, within 60 days after the Commission's final decision, to obtain judicial review of said final decision.

Article IV–Regional Management Plan

The Commission shall adopt a regional management plan designed to ensure the safe and efficient management of waste generated within the region. In adopting a regional waste management plan the Commission shall:

Health.
Safety.

a) Adopt procedures for determining, consistent with considerations of public health and safety, the type and number of regional facilities which are presently necessary and which are projected to be necessary to manage waste generated within the region.

b) Develop and adopt policies promoting source and volume reduction of waste generated within the region.

c) Develop alternative means for the treatment, storage and disposal of waste, other than shallow-land burial or underground injection well.

d) Prepare a draft regional management plan that shall be made available in a convenient form to the public for comment. The Commission shall conduct one or more public hearings in each party state prior to the adoption of the regional management plan. The regional management plan shall include the Commission's response to public and party state comment.

Article V–Rights and Obligations of Party States

a) Each party state shall act in good faith in the performance of acts and courses of conduct which are intended to ensure the provision of facilities for regional availability and usage in a manner consistent with this compact.

b) Other than the provisions of Article V(f) and VII(a)(6), each party state has the right to have all wastes generated within borders managed at regional facilities. This right shall be subject to the provisions of this Compact. All party states have an equal right of access to any facility outside the region made available to the region by any agreement entered into by the Commission pursuant to Article III(i)(4).[20]

Exports.

c) Party states or generators may negotiate for the right of access to a facility outside the region and may export waste outside the region subject to Commission approval under Article III(i)(4).[21]

---

[19]   Amended by P.L. 103–439, 108 Stat. 4610 (1994).
[20]   Amended by P.L. 103–439, 108 Stat. 4610 (1994).
[21]   Amended by P.L. 103–439, 108 Stat. 4610 (1994).

Contracts.
Prohibition.
Regulations.
Transportation.

d) To the extent permitted by federal law, each party state may enforce any applicable federal and state laws, regulations and rules pertaining to the packaging and transportation of waste generated within or passing through its borders. Nothing in this Section shall be construed to require a party state to enter into any agreement with the U. S. Nuclear Regulatory Commission.

e) Each party state shall provide to the Commission any data and information the Commission requires to implement its responsibilities. Each party state shall establish the capability to obtain any data and information required by the Commission.

Kentucky.
Prohibition.

f) Waste originating from the Maxey Flats nuclear waste disposal site in Fleming County, Kentucky shall not be shipped to any facility in Illinois for storage, treatment or disposal. Disposition of these wastes shall be the sole responsibility of the Commonwealth of Kentucky and such waste shall not be subject to the provisions of Articles IX(b)(3) and (4) of this compact.[22]

Article VI–Development and Operation of Facilities

a) Any party state may volunteer to become a host state, and the Commission may designate that state as a host state.

b) If all regional facilities required by the regional management plan are not developed pursuant to Article VI(a), or upon notification that an existing regional facility will be closed, the Commission may designate a party state as a host state. A party state shall not be designated as a host state for any regional facility under this Article VI(b) unless that state's total volume of waste recorded on low-level radioactive waste manifests for any year is more than 10% of the total volume recorded on such manifests for the region during the same year. In determining the 10% exclusion, there shall not be included waste recorded on low-level radioactive waste manifests by a person whose principal business is providing a service by arranging for the collection, transportation, treatment, storage or disposal of such waste, or waste described in Article VII(a)(6).[23]

Prohibition.

c) Each party state designated as host state is responsible for determining possible facility locations within its borders. The selection of a facility site shall not conflict with applicable federal and host state laws, regulations and rules not inconsistent with this compact and shall be based on factors including, but not limited to, geological, environmental, engineering and economic viability of possible facility locations.[24]

d) Any party state designated as a host state may request the Commission to relieve that state of the responsibility to serve as a host state. The Commission may relieve a party state of this responsibility upon a showing by the requesting party state that no feasible potential regional facility site of the type it is designated to host exists within its borders or for other good cause shown and consistent with the purposes of this Compact.[25]

---

[22]   Amended by P.L. 103–439, 108 Stat. 4610 (1994).
[23]   Amended by P.L. 103–439, 108 Stat. 4610 (1994).
[24]   Repealed by P.L. 103–439, 108 Stat. 4611 (1994), and redesignated section d to section c.
[25]   Amended by P.L. 103–439, 108 Stat. 4611 (1994), redesignating section e to section d.

e) After a state is designated a host state by the Commission, it is responsible for the timely development and operation of a regional facility.[26]

f) To the extent permitted by federal and state law, a host state shall regulate and license any facility within its borders and ensure the extended care of that facility.[27]

g) The Commission may designate a party state as a host state while a regional facility is in operation if the Commission determines that an additional regional facility is or may be required to meet the needs of the region.[28]

h) Designation of a host state is for a period of 20 years or the life of the regional facility which is established under that designation, whichever is shorter. Upon request of a host state, the Commission may modify the period of its designation.[29]

i) A host state may establish a fee system for any regional facility within its borders. The fee system shall be reasonable and equitable. This fee system shall provide the host state with sufficient revenue to cover any costs including, but not limited to, the planning, siting, licensing, operation, pre-closure corrective action or clean-up, monitoring, inspection, decommissioning, extended care and long-term liability, associated with such facilities. This fee system may provide for payment to units of local government affected by a regional facility for costs incurred in connection with such facility. This fee system may also include reasonable revenue beyond the costs incurred for the host state, subject to approval by the Commission. The fee system shall include incentives for source or volume reduction and may be based on the hazard of the waste. A host state shall submit an annual financial audit of the operation of the regional facility to the Commission.[30]

42 USC 10101
note.
Health.
Safety.

j) A host state shall ensure that a regional facility located within its borders which is permanently closed is properly decommissioned.
A host state shall also provide for the extended care of a closed or decommissioned regional facility within its borders so that the public health and safety of the state and region are ensured, unless, pursuant to the federal Nuclear Waste Policy Act of 1982, the federal government has assumed title and custody of the regional facility and the federal government thereby has assumed responsibility to provide for the extended care of such facility.[31]

Environmental
protection.
Health.
Prohibition.
Safety.

k) A host state intending to close a regional facility located within its borders shall notify the Commission in writing of its intention and the reasons. Notification shall be given to the Commission at least five years prior to the intended date of closure. This Section shall not prevent an emergency closing of a regional facility by a host state to protect its air, land and water resources and the health and safety of its citizens. However, a host state which has an emergency closing of a regional facility shall notify the Commission in writing within 3 working days of its action and shall, within 30 working days of its action, demonstrate justification for the closing.[32]

---

[26]   Amended by P.L. 103–439, 108 Stat. 4612 (1994), redesignating section f to e.
[27]   Amended by P.L. 103–439, 108 Stat. 4612 (1994), redesignating section g to f.
[28]   Amended by P.L. 103–439, 108 Stat. 4612 (1994), redesignating section h to g.
[29]   Amended by P.L. 103–439, 108 Stat. 4612 (1994), redesignating section i to h.
[30]   Amended by P.L. 103–439, 108 Stat. 4612 (1994), redesignating section j to i.
[31]   Amended by P.L. 103–439, 108 Stat. 4612 (1994), redesignating section k to j.
[32]   Amended by P.L. 103–439, 108 Stat. 4611 (1994), and redesignated it as section k.

Prohibition.
Transportation.

l) If a regional facility closes before an additional or new facility becomes operational, waste generated within the region may be shipped temporarily to any location agreed on by the Commission until a regional facility is operational, provided that the region's waste shall not be stored in a party state with a total volume of waste recorded on low-level radioactive waste manifests for any year which is less than 10% of the total volume recorded on the manifests for the region during the same year. In determining the 10% exclusion, there shall not be included waste recorded on low-level radioactive waste manifests by a person whose principal business is providing a service by arranging for the collection, transportation, treatment, storage or disposal of such waste, or waste described in Article VII(a)(6).[33]

m) A party state which is designated as a host state by the Commission and fails to fulfill its obligations as a host state may have its privileges under the compact suspended or membership in the compact revoked by the Commission.[34]

n) The host state shall create an "Extended Care and Long-Term Liability Fund" and shall allocate sufficient fee revenues, received pursuant to Article VI(i), to provide for the costs of:

1) decommissioning and other procedures required for the proper closure of a regional facility;

2) monitoring, inspection and other procedures required for the proper extended care of a regional facility;

Environmental.
Health.
Protection.
Contracts.
Gifts and
property.
Health.
Insurance.

3) undertaking any corrective action or clean-up necessary to protect human health and the environment from radioactive releases from a regional facility; and

4) compensating any person for medical and other expenses incurred from damages to human health, personal injuries suffered from damages to human health and damages or losses to real or personal property, and accomplishing any necessary corrective action or clean-up on real or personal property caused by radioactive releases from a regional facility; the host state may allocate monies in this Fund in amounts as it deems appropriate to purchase insurance or to make other similar financial protection arrangements consistent with the purposes of this Fund; this Article VI(n) shall in no manner limit the financial responsibilities of the site operator under Article VI(o), the party states under Article VI(p), or any person who sends waste to a regional facility, under Article VI(q).[35]

Health.
Real property.
Insurance.

o) The operator of a regional facility shall purchase an amount of property and third-party liability insurance deemed appropriate by the host state, pay the necessary periodic premiums at all times and make periodic payments to the Extended Care and Long-Term Liability Fund as set forth in Article VI(n) for such amounts as the host state reasonably determines is necessary to provide for future premiums to continue such insurance coverage, in order to pay the costs of compensating any person for medical and other expenses incurred from damages to human health, personal injuries suffered from damages to human health and damages or losses to real or personal property, and accomplishing any necessary corrective action or clean-up on real or personal property caused by radioactive releases from a regional facility. In the event of such costs

---

[33]   Amended by P.L. 103–439, 108 Stat. 4611 (1994), and redesignated it as section l.
[34]   Amended by P.L. 103–439, 108 Stat. 4612 (1994), redesignating section n as section m.
[35]   Amended by P.L. 103–439, 108 Stat. 4611 (1994), and redesignated it as section n.

resulting from radioactive releases from a regional facility, the host state should, to the maximum extent possible, seek to obtain monies from such insurance prior to using monies from the Extended Care and Long-Term Liability Fund.[36]

Contracts.
Prohibition.

p) All party states shall be liable for the cost of extended care and long-term liability in excess of monies available from the Extended Care and Long-Term Liability Fund, as set forth in Article VI(n) and from the property and third-party liability insurance as set forth in Article VI(o). A party state may meet such liability for costs by levying surcharges upon generators located in the party state. The extent of such liability shall be based on the proportionate share of the total volume of waste placed in the regional facility by generators located in each such party state. Such liability shall be joint and several among the party states with a right of contribution between the party states. However, this Section shall not apply to a party state with a total volume of waste recorded on low-level radioactive waste manifests for any year that is less than 10% of the total volume recorded on such manifests for the region during the same year.[37]

q)[38] Any person who sends waste from outside the region or waste described in Article VII(a)(6) for treatment, storage or disposal at a regional facility shall be liable for the cost of extended care and long-term liability of that regional facility in excess of the monies available from the Extended Care and Long-Term Liability Fund as set forth in Article VI(n) and from the property and third-party liability insurance as set forth in Article VI(o). The extent of the liability for the person shall be based on the proportionate share of the total volume of waste sent by that person to the regional facility.

Article VII–Other Laws and Regulations

Prohibitions.

a) Nothing in this compact:

1) abrogates or limits the applicability of any act of Congress or diminishes or otherwise impairs the jurisdiction of any federal agency expressly conferred thereon by the Congress;

2) prevents the enforcement of any other law of a party state which is not inconsistent with this compact;

3) prohibits any storage or treatment of waste by the generator on its own premises;

4) affects any administrative or judicial proceeding pending on the effective date of this compact;

5) alters the relations between the respective internal responsibility of the government of a party state and its subdivisions;

Research and
development.

6) establishes any right to the treatment, storage or disposal at any facility in the region or provides any authority to prohibit export from the region of waste that is owned or generated by the United States Department of Energy, owned or generated by the United States Navy as a result of the decommissioning of vessels of the United States Navy, or owned or generated as the result of any research, development, testing or production of any atomic weapon; or.[39]

---

[36]   Amended by P.L. 103–439, 108 Stat. 4612 (1994), and redesignated it as section o.
[37]   Amended by P.L. 103–439, 108 Stat. 4612 (1994), and redesignated it as section p.
[38]   Added by P.L. 103–439, 108 Stat. 4612 (1994).
[39]   Amended by P.L. 103–439, 108 Stat. 4613 (1994).

Taxes.
Transportation.

7) affects the rights and powers of any party state or its political subdivisions, to the extent not inconsistent with this compact, to regulate and license any facility or the transportation of waste within its borders or affects the rights and powers of any state or its political subdivisions to tax or impose fees on the waste managed at any facility within its borders;

Contracts.

8) requires a party state to enter into an agreement with the U. S. Nuclear Regulatory Commission; or

9) alters or limits liability of transporters of waste and owners and operators of sites for their acts, omissions, conduct or relationships in accordance with applicable laws.

b) For purposes of this compact, all state laws or parts of laws in conflict with this compact are hereby superseded to the extent of the conflict.

Prohibition.
Regulations.

c) No law, rule, regulation, fee or surcharge of a party state, or of any of its subdivisions or instrumentalities, may be applied in a manner which discriminates against the generators of another party state.

Prohibition.

d) No person who provides a service by arranging for collection, transportation, treatment, storage or disposal of waste from outside the region shall be allowed to dispose of any waste, regardless of origin, in the region unless specifically permitted under an agreement entered into by the Commission in accordance with the requirements of Article III(i)(1).[40]

Article VIII–Eligible Parties, Withdrawal, Revocation, Entry into Force, Termination

Illinois.
Kentucky.

a) Eligible parties to this compact are the State of Illinois and Commonwealth of Kentucky. Eligibility terminates on April 15, 1985.

b) An eligible state becomes a party state when the state enacts the compact into law and pays the membership fee required in Article III(k)(1).

c) The Commission is formed upon the appointment of the Commissioners and the tender of the membership fee payable to the Commission by the eligible states. The Governor of Illinois shall convene the initial meeting of the Commission. The Commission shall cause legislation to be introduced in the Congress which grants the consent of the Congress to this compact, and shall take action necessary to organize the Commission and implement the provisions of this compact.[41]

d) Other than the special circumstances for withdrawal in section (f) of this Article, either party state may withdraw from this compact at any time by repealing the authorizing legislation, but no withdrawal may take effect until 5 years after the governor of the withdrawing state gives notice in writing of the withdrawal to the Commission and to the governor of the other state. Withdrawal does not affect any liability already incurred by or chargeable to a party state prior to the time of such withdrawal. Any host state which grants a disposal permit for waste generated in a withdrawing state shall void the permit when the withdrawal of that state is effective.

---

[40]   Amended by P.L. 103–439, 108 Stat. 4613 (1994).
[41]   Amended by P.L. 103–439, 108 Stat. 4613 (1994).

Effective date.

Exports.
Prohibition.

Effective date.

Regulation.

e) This compact becomes effective July 1, 1984, or at any date subsequent to July 1, 1984, upon enactment by the eligible states. However, Article IX(b) shall not take effect until the Congress has by law consented to this compact. The Congress shall have an opportunity to withdraw such consent every 5 years. Failure of the Congress affirmatively to withdraw its consent has the effect of renewing consent for an additional 5 year period. The consent given to this compact by the Congress shall extend to the power of the region to ban the shipment of waste into the region pursuant to Article III(i)(1) and to prohibit exportation of waste generated within the region under Article III(i)(4).[42]

f) A state which has been designated a host state may withdraw from the compact. The option to withdraw must be exercised within 90 days of the date the governor of the designated state receives written notice of the designation. Withdrawal becomes effective immediately after notice is given in the following manner. The governor of the withdrawing state shall give notice in writing to the Commission and to the governor of each party state. A state which withdraws from the compact under this section forfeits any funds already paid pursuant to this compact. A designated host state which withdraws from the compact after 90 days and prior to fulfilling its obligations shall be assessed a sum the Commission determines to be necessary to cover the costs borne by the Commission and remaining party states as a result of that withdrawal.

Article IX–Penalties

a) Each party state shall prescribe and enforce penalties against any person who is not an official of another state for violation of any provision of this compact.

b) Unless authorized by the Commission pursuant to Article III(i), or otherwise provided in this Compact, after January 1, 1986 it is a violation of this Compact:

1) for any person to deposit at a facility in the region waste from outside the region;

2) for any facility in the region to accept waste from outside the region;

3) for any person to export from the region waste that is generated within the region;

4) for any person to dispose of waste at a facility other than a regional facility;

5) for any person to deposit at a regional facility waste described in Article VII(a)(6); or

6) for any regional facility to accept waste described in Article VII(a)(6).[43]

c)[44] It is a violation of this compact for any person to treat or store waste at a facility other than a regional facility if such treatment or storage is prohibited by the Commission under Article III(i)(6).

d) Each party state acknowledges that the receipt by a host state of waste packaged or transported in violation of applicable laws, rules or

---

[42]   Amended by P.L. 103–439, 108 Stat. 4613 (1994).
[43]   Amended by P.L. 103–439, 108 Stat. 4613 (1994).
[44]   Added by P.L. 103–439, 108 Stat. 4614 (1994), and redesignated sections c and d as sections d and e, respectively.

regulations may result in the imposition of sanctions by the host state which may include suspension or revocation of the violator's right of access to the facility in the host state.

e) Each party state has the right to seek legal recourse against any party state which acts in violation of this compact.

Article X–Severability and Construction

Provisions held invalid.

The provisions of this compact shall be severable and if any phrase, clause, sentence or provision of this compact is declared by a court of competent jurisdiction to be contrary to the Constitution of any participating state or the United States, or if the applicability thereof to any government, agency, person or circumstance is held invalid, the validity of the remainder of this compact and the applicability thereof to any government, agency, person or circumstance shall not be affected thereby. If any provision of this compact shall be held contrary to the Constitution of any state participating therein, the compact shall remain in full force and effect as to the state affected as to all severable matters.

42 USC 2021d note.

Iowa. Indiana.
Michigan.
Minnesota.
Missouri. Ohio.
Wisconsin.

Sec. 225. Midwest Interstate Low-Level Radio Active Waste Management Compact

The consent of Congress is hereby given to the States of Iowa, Indiana, Michigan, Minnesota, Missouri, Ohio, and Wisconsin to enter into the Midwest Interstate Compact on Low-Level Radioactive Waste Management. Such compact is as follows:

Midwest Interstate Low-Level Radio Active Waste Management Compact

Article I–Policy and Purpose

There is created the Midwest Interstate Low-Level Radioactive Waste Compact.

Research and development.

The states party to this compact recognize that the Congress of the United States, by enacting the Low-Level Radioactive Waste Policy Act (42 USC 2021b to 2021d), has provided for and encouraged the development of low-level radioactive waste compacts as a tool for managing such waste. The party states acknowledge that the Congress has declared that each state is responsible for providing for the availability of capacity either within or outside the state for the disposal of low-level radioactive waste generated within its borders, except for waste generated as a result of certain defense activities of the federal government or federal research and development activities. The party states also recognize that the management of low-level radioactive waste is handled most efficiently on a regional basis; and, that the safe and efficient management of low-level radioactive waste generated within the region requires that sufficient capacity to manage such waste be properly provided.

a. It is the policy of the party states to enter into a regional low-level radioactive waste management compact for the purpose of:

1. Providing the instrument and framework for a cooperative effort;

2. Providing sufficient facilities for the proper management of low-level radioactive waste generated in the region;

Health.

Safety.

3. Protecting the health and safety of the citizens of the region;

4. Limiting the number of facilities required to effectively and efficiently manage low-level radioactive waste generated in the region;

5. Encouraging the reduction of the amounts of low-level radioactive waste generated in the region;

6. Distributing the costs, benefits and obligations of successful low-level radioactive waste management equitably among the party states, and among generators and other persons who use regional facilities to manage their waste; and

7. Ensuring the ecological and economical management of low-level radioactive wastes.

Regulations.

b. Implicit in the Congressional consent to this compact is the expectation by the Congress and the party states that the appropriate federal agencies will actively assist the Compact Commission and the individual party states to this compact by:

1. Expeditious enforcement of federal rules, regulations and laws;

2. Imposition of sanctions against those found to be in violation of federal rules, regulations and laws; and

3. Timely inspection of their licensees to determine their compliance with these rules, regulations and laws.

Article II–Definitions

As used in this compact, unless the context clearly requires a different construction:

a. "Care" means the continued observation of a facility after closure for the purposes of detecting a need for maintenance, ensuring environmental safety, and determining compliance with applicable licensing and regulatory requirements and including the correction of problems which are detected as a result of that observation.

b. "Commission" means the Midwest Interstate Low-Level Radioactive Waste Commission.

c. "Decommissioning" means the measures taken at the end of a facility's operating life to assure the continued protection of the public from any residual radioactivity or other potential hazards present at a facility.

d. "Disposal" means the isolation of waste from the biosphere in a permanent facility designed for that purpose.

e. "Eligible state" means a state qualified to be a party state to this compact as provided in Article VIII.

f. "Facility" means a parcel of land or site, together with the structures, equipment and improvements on or appurtenant to the land or site, which issued or is being developed for the treatment, storage or disposal of low-level radioactive waste.

g. "Generator" means any person who produces or possesses low-level radioactive waste in the course of or incident to manufacturing, power generation, processing, medical diagnosis and treatment, research, or other industrial or commercial activity and who, to the extent required by law, is licensed by the U. S. Nuclear Regulatory Commission or a party state, to produce or possess such waste. Generator does not include a person who provides a service by arranging for the collection, transportation, treatment, storage or disposal of wastes generated outside the region.

h. "Host state" means any state which is designated by the Commission to host a regional facility.

i. "Low-Level radioactive waste" or "waste" means radioactive waste not classified as high-level radioactive waste, transuranic waste, spent nuclear fuel or by-product material as defined in section 11(e)(2) of the Atomic Energy Act of 1954 (42 USC 2014).

j. "Management plan" means the plan adopted by the Commission for the storage, transportation, treatment and disposal of waste within the region.

k. "Party state" means any eligible state which enacts the compact into law.

l. "Person" means any individual, corporation, business enterprise or other legal entity either public or private and any legal successor, representative, agent or agency of that individual, corporation, business enterprise or legal entity.

m. "Region" means the area of the party states.

n. "Regional facility" means a facility which is located within the region and which is established by a party state pursuant to designation of that state as a host state by the Commission.

o. "Site" means the geographic location of a facility.

p. "State" means a state of the United States, the District of Columbia, the commonwealth of Puerto Rico, the Virgin Islands, or any other territorial possession of the United States.

q. "Storage" means the temporary holding of waste for treatment or disposal.

r. "Treatment" means any method, technique or process, including storage for radioactive decay, designed to change the physical, chemical or biological characteristics or composition of any waste in order to render the waste safer for transport or management, amenable to recovery, convertible to another usable material, or reduced in volume.

s. "Waste management" means the storage, transportation, treatment, or disposal of waste.

Article III–The Commission

Midwest Interstate Low-Level Radioactive Waste Commission, establishment. Prohibition.

a. There is hereby created the Midwest Interstate Low-Level Radioactive Waste Commission. The Commission consists of one voting member from each party state. The Governor of each party state shall notify the Commission in writing of its member and any alternates. An alternate may act on behalf of the member only in that member's absence. The method for selection and the expenses of each Commission member shall be the responsibility of the member's respective state.

b. Each Commission member is entitled to one vote. No action of the Commission is binding unless a majority of the total membership cast their vote in the affirmative.

c. The Commission shall elect annually from among its members a chairperson. The Commission shall adopt and publish, in convenient form, bylaws, and policies which are not inconsistent with this compact, including procedures which substantially conform with the provisions of federal law on administrative procedure compiled at 5 USC 500 to 559 in regard to notice, conduct and recording of meetings; access by the public to records; provision of information to the public; conduct of adjudicatory hearings; and issuance of decisions.

d. The Commission shall meet at least once annually and shall also meet upon the call of the chairperson or a Commission member.

e. All meetings of the commission shall be open to the public with reasonable advance notice. The Commission may, by majority vote, close a meeting to the public for the purpose of considering sensitive personnel or legal strategy matters. However, all Commission actions and decisions shall be made in open meetings and appropriately recorded.

f. The Commission may establish advisory committees for the purpose of advising the Commission on any matters pertaining to waste management.

Contracts.

g. The office of the Commission shall be in a party state. The Commission may appoint or contract for and compensate such limited staff necessary to carry out its duties and functions. The staff shall serve at the Commission's pleasure with the exception that staff hired as the result of securing federal funds shall be hired and governed under applicable federal statutes and regulations. In selecting any staff, the Commission shall assure that the staff has adequate experience and formal training to carry out the functions assigned to it by the Commission.

h. The Commission may:

Contracts.

1. Enter into an agreement with any person, state, or group of states for the right to use regional facilities for waste generated outside of the region and for the right to use facilities outside the region for waste generated within the region. The right of any person to use a regional facility for waste generated outside of the region requires an affirmative vote of a majority of the Commission, including the affirmative vote of the member of the host state in which any affected regional facility is located.

2. Approve the disposal of waste generated within the region at a facility other than a regional facility.

Reports.

3. Appear as an intervenor or party in interest before any court of law or any federal, state or local agency, board or commission in any matter related to waste management. In order to represent its views, the Commission may arrange for any expert testimony, reports, evidence or other participation.

4. Review the emergency closure of a regional facility, determine the appropriateness of that closure, and take whatever actions are necessary to ensure that the interests of the region are protected.

5. Take any action which is appropriate and necessary to perform its duties and functions as provided in this compact.

6. Suspend the privileges or revoke the membership of a party state by a two-thirds vote of the membership in accordance with Article VIII.

i. The Commission shall:

1. Receive and act on the petition of a nonparty state to become an eligible state.

Report.

2. Submit an annual report to, and otherwise communicate with, the governors and the appropriate officers of the legislative bodies of the party states regarding the activities of the Commission.

3. Hear, negotiate, and, as necessary, resolve by final decision disputes which may arise between the party states regarding this compact.

4. Adopt and amend, by a two-thirds vote of the membership, in accordance with the procedures and criteria developed pursuant to Article IV, a regional management plan which designates host states for the establishment of needed regional facilities.

5. Adopt an annual budget.

j. Funding of the budget of the Commission shall be provided as follows:

1. Each state, upon becoming a party state, shall pay $50,000 or $1,000 per cubic meter of waste shipped from that state in 1980, whichever is lower, to the Commission which shall be used for the administrative costs of the Commission;

2. Each state hosting a regional facility shall levy surcharges on all users of the regional facility based upon its portion of the total volume and characteristics of wastes managed at that facility. The surcharges collected at all regional facilities shall:

(a) Be sufficient to cover the annual budget of the Commission; and

(b) Represent the financial commitments of all party states to the Commission; and

(c) Be paid to the Commission, provided, however, that each host state collecting surcharges may retain a portion of the collection sufficient to cover its administrative costs of collection, and that the remainder be sufficient only to cover the approved annual budget of the Commission.

Audit.
Contracts.
Report.

k. The Commission shall keep accurate accounts of all receipts and disbursements. The Commission shall contract with an independent certified public accountant to annually audit all receipts and disbursements of Commission funds, and to submit an audit report to the Commission. The audit report shall be made a part of the annual report of the Commission required by this Article.

Grants.

l. The Commission may accept for any of its purposes and functions and may utilize and dispose of any donations, grants of money, equipment, supplies, materials and services from any state or the United States (or any subdivision or agency thereof), or interstate agency, or from any institution, person, firm or corporation. The nature, amount and condition, if any, attendant upon any donation or grant accepted or received by the Commission together with the identity of the donor, grantor or lender, shall be detailed in the annual report of the Commission.

Report.

m. The Commission is not liable for any costs associated with any of the following:

1. The licensing and construction of any facility.

2. The operation of any facility,

3. The stabilization and closure of any facility,

4. The care of any facility,

5. The extended institutional control, after care of any facility, or

Transportation.

6. The transportation of waste to any facility.

n. 1. The Commission is a legal entity separate and distinct from the party states and is liable for its actions as a separate and distinct legal entity. Liabilities of the Commission are not liabilities of the party states. Members of the Commission are not personally liable for actions taken by them in their official capacity.

Prohibition.

2. Except as provided under sections m. and n.1. of this article, nothing in this compact alters liability for any act, omission, course of conduct or liability resulting from any causal or other relationships.

o. Any person aggrieved by a final decision of the Commission may obtain judicial review of such decision in any court of competent jurisdiction by filing in such court a petition for review within 60 days after the Commission's final decision.

Article IV–Regional Management Plan

The Commission shall adopt a regional management plan designed to ensure the safe and efficient management of waste generated within the region. In adopting a regional waste management plan the Commission shall:

Health.
Safety.

a. Adopt procedures for determining, consistent with considerations for public health and safety, the type and number of regional facilities which are presently necessary and which are projected to be necessary to manage waste generated within the region.

b. Develop and consider policies promoting source reduction of waste generated within the region.

c. Develop and adopt procedures and criteria for identifying a party state as a host state for a regional facility. In developing these criteria, the Commission shall consider all the following:

Health.
Safety.
Transportation.

1. The health, safety, and welfare of the citizens of the party states.

2. The existence of regional facilities within each party state.

3. The minimization of waste transportation.

4. The volumes and types of wastes generated within each party state.

5. The environmental, economic, and ecological impacts on the air, land and water resources of the party states.

Reports.
Studies.

d. Conduct such hearings, and obtain such reports, studies, evidence and testimony required by its approved procedures prior to identifying a party state as a host state for a needed regional facility.

e. Prepare a draft management plan, including procedures, criteria and host states, including alternatives, which shall be made available in a convenient form to the public for comment. Upon the request of a party state, the Commission shall conduct a public hearing in that state prior to the adoption of the management plan. The management plan shall include the Commission's response to public and party state comment.

Article V–Rights and Obligations of Party States

a. Each party state shall act in good faith in the performance of acts and courses of conduct which are intended to ensure the provision of facilities for regional availability and usage in a manner consistent with this compact.

b. Each party state has the right to have all wastes generated within its borders managed at regional facilities subject to the provisions contained in Article IX.c. All party states have an equal right of access to any facility made available to the region by any agreement entered into by the Commission pursuant to Article III.

Exports.

c. Party states or generators may negotiate for the right of access to a facility outside the region and may export waste outside the region subject to Commission approval under Article III.

Prohibition.
Regulations.
Transportation.

d. To the extent permitted by federal law, each party state may enforce any applicable federal and state laws, regulations and rules pertaining to the packaging and transportation of waste generated within or passing through its borders. Nothing in this section shall be construed to require a party state to enter into any agreement with the U. S. Nuclear Regulatory Commission.

e. Each party state shall provide to the Commission any data and information the Commission requires to implement its responsibilities.

Each party state shall establish the capability to obtain any data and information required by the Commission.

Article VI–Development and Operation of Facilities

a. Any party state may volunteer to become a host state, and the Commission may designate that state as a host state upon a two-thirds vote of its members.

b. If all regional facilities required by the regional management plan are not developed pursuant to section a., or upon notification that an existing regional facility will be closed, the Commission may designate a host state.

Prohibition.

c. Each party state designated as a host state is responsible for determining possible facility locations within its borders. The selection of a facility site shall not conflict with applicable federal and host state laws, regulations and rules not inconsistent with this compact and shall be based on factors including but not limited to geological, environmental and economic viability of possible facility locations.

d. Any party state designated as a host state may request the Commission to relieve that state of the responsibility to serve as a host state. The Commission may relieve a party state of this responsibility only upon a showing by the requesting party state that no feasible potential regional facility site of the type it is designated to host exists within its borders.

e. After a state is designated a host state by the Commission, it is responsible for the timely development and operation of a regional facility.

f. To the extent permitted by federal and state law, a host state shall regulate and license any facility within its borders and ensure the extended care of that facility.

g. The Commission may designate a party state as a host state while a regional facility is in operation if the Commission determines that an additional regional facility is or may be required to meet the needs of the region. The Commission shall make this designation following the procedures established under Article IV.

h. Designation of a host state is for a period of 20 years or the life of the regional facility which is established under that designation, whichever is longer. Upon request of a host state, the Commission may modify the period of its designation.

i. A host state may establish a fee system for any regional facility within its borders. The fee system shall be reasonable and equitable. This fee system shall provide the host state with sufficient revenue to cover any cost, including but not limited to the planning, siting, licensure, operation, decommissioning, extended care and long-term liability, associated with such facilities. This fee system may also include reasonable revenue beyond costs incurred for the host state, subject to

Audit.

approval by the Commission A host shall submit an annual financial audit of the operation of the regional facility to the Commission. The fee system may include incentives for source reduction and may be based on the hazard of the waste as well as the volume.

Health.
Safety.

j. A host state shall ensure that a regional facility located within its borders which is permanently closed is properly decommissioned. A host state shall also provide for the care of a closed or decommissioned regional facility within its borders so that the public health and safety of the state and region are ensured.

Prohibitions.

k. A host state intending to close a regional facility located within its borders shall notify the Commission in writing of its intention and the reasons. Notification shall be given to the Commission at least five years prior to the intended date of closure. This section shall not prevent an emergency closing of a regional facility by a host state to protect its air, land and water resources and the health and safety of its citizens. However, a host state which has an emergency closing of a regional facility shall notify the Commission in writing within three working days of its action and shall, within 30 working days of its action, demonstrate justification for the closing.

l. If a regional facility closes before an additional or new facility becomes operational, waste generated within the region may be shipped temporarily to any location agreed on by the Commission until a regional facility is operational.

m. A party state which is designated as a host state by the Commission and fails to fulfill its obligations as a host state may have its privileges under the compact suspended or membership in the compact revoked by the Commission.

Article VII–Other Laws and Regulations

Prohibitions.

a. Nothing in this compact:

1. Abrogates or limits the applicability of any act of Congress or diminishes or otherwise impairs the jurisdiction of any federal agency expressly conferred thereon by the Congress;

2. Prevents the enforcement of any other law of a party state which is not inconsistent with this compact;

3. Prohibits any storage or treatment of waste by the generator on its own premises;

4. Affects any administrative or judicial proceedings pending on the effective date of this compact;

5. Alters the relations between and the respective internal responsibility of the government of a party state and its subdivisions;

Research and development.

6. Affects the generation, treatment, storage or disposal of waste generated by the atomic energy defense activities of the Secretary of the U. S. Department of Energy or successor agencies or federal research and development activities as described in section 31 of the Atomic Energy Act of 1954 (42 USC 2051); or

Taxes.
Transportation.

7. Affects the rights and powers of any party state or its political subdivisions to the extent not inconsistent with this compact, to regulate and license any facility or the transportation of waste within its borders or affects the rights and powers of any party state and its political subdivisions to tax or impose fees on the waste managed at any facility within its borders.

Contracts.

8. Requires a party state to enter into any agreement with the U. S. Nuclear Regulatory Commission.

9. Alters or limits liability of transporters of waste, owners and operators of sites for their acts, omissions, conduct or relationships in accordance with applicable laws.

b. For purposes of this compact, all state laws or parts of laws in conflict with this compact are hereby superseded to the extent of the conflict.

Prohibition.
Regulations.

c. No law, rule or regulation of a party state or of any of its subdivisions or instrumentalities may be applied in a manner which discriminates against the generators of another party state.

Article VIII–Eligible Parties, Withdrawal, Revocation, Entry into Force, Termination

Delaware.
Illinois.
Indiana.
Iowa.
Kansas.
Kentucky.
Maryland.
Michigan.
Minnesota.
Missouri.
Nebraska.
North Dakota.
Ohio.
South Dakota.
Virginia.
Wisconsin.

a. Eligible parties to this compact are the states of Delaware, Illinois, Indiana, Iowa, Kansas, Kentucky, Maryland, Michigan, Minnesota, Missouri, Nebraska, North Dakota, Ohio, South Dakota, Virginia and Wisconsin. Eligibility terminates on July 1, 1984.

b. Any state not eligible for membership in the compact may petition the Commission for eligibility. The Commission may establish appropriate eligibility requirements. These requirements may include but are not limited to, an eligibility fee or designation as a host state. A petitioning state becomes eligible for membership in the compact upon the approval of the Commission, including the affirmative vote of all host states. Any state becoming eligible upon the approval of the Commission becomes a member of the compact in the same manner as any state eligible for membership at the time this compact enters into force.

c. An eligible state becomes a party state when the state enacts the compact into law and pays the membership fee required in Article IIIj.1.

d. The Commission is formed upon the appointment of Commission members and the tender of the membership fee payable to the Commission by three party states. The Governor of the first state to enact this compact shall convene the initial meeting of the Commission. The Commission shall cause legislation to be introduced in the Congress which grants the consent of the Congress to this compact, and shall take action necessary to organize the Commission and implement the provision of this compact.

Prohibition.

e. Any party state may withdraw from this compact by repealing the authorizing legislation but no withdrawal may take effect until five years after the governor of the withdrawing state gives notice in writing of the withdrawal to the Commission and to the governor of each party state. Withdrawal does not affect any liability already incurred by or chargeable to a party state prior to the time of such withdrawal. Any host state which grants a disposal permit for waste generated in a withdrawing state shall void the permit when the withdrawal of that state is effective.

f. Any party state which fails to comply with the terms of this compact or fails to fulfill its obligations may have its privileges suspended or its membership in the compact revoked by the Commission in accordance with Article III h.6. Revocation takes effect one year from the date the affected party state receives written notice from the Commission of its action. All legal rights of the affected party state established under this compact cease upon the effective date of revocation but any legal obligations of that party state arising prior to revocation continue until they are fulfilled. The chairperson of the Commission shall transmit written notice of a revocation of a party states's membership in the compact immediately following the vote of the Commission to the governor of the affected party state, all other governors of the party states and the Congress of the United States.

Effective date.

g. This compact becomes effective upon enactment by at least three eligible states and consent to this compact by Congress. The Congress shall have an opportunity to withdraw such consent every five years. Failure of the Congress to affirmatively withdraw its consent has the effect of renewing consent for an additional five year period. The consent given to this compact by the Congress shall extend to any future admittance of new party states under sections b. and c. of this article and to the power of the Commission to ban the shipment of waste from the region pursuant to Article III.

h. The withdrawal of a party state from this compact under section e. of this article or the suspension or revocation of a state's membership in this compact under section f. of this article does not affect the applicability of this compact to the remaining party states.

i. A state which has been designated by the Commission to be a host state has 90 days from receipt by the Governor of written notice of designation to withdraw from the compact without any right to receive refund of any funds already paid pursuant to this compact, and without any further payment. Withdrawal becomes effective immediately upon notice as provided in section e. of this article. A designated host state which withdraws from the compact after 90 days and prior to fulfilling its obligations shall be assessed a sum the Commission determines to be necessary to cover the costs borne by the Commission and remaining party states as a result of that withdrawal.

Article IX–Penalties

a. Each party state shall prescribe and enforce penalties against any person who is not an official of another state for violation of any provision of this compact.

b. Unless otherwise authorized by the Commission pursuant to Article III h. After January 1, 1986, it is a violation of this compact:

1. For any person to deposit at a regional facility waste not generated within the region;

2. For any regional facility to accept waste not generated within the region;

Exports.

3. For any person to export from the region waste which is generated within the region; or

4. For any person to dispose of waste at a facility other than a regional facility.

Regulations.

c. Each party state acknowledges that the receipt by a host state of waste packaged or transported in violation of applicable laws, rules and regulations may result in the imposition of sanctions by the host state which may include suspension or revocation of the violator's right of access to the facility in the host state.

d. Each party state has the right to seek legal recourse against any party state which acts in violation of this compact.

Article X–Severability and Construction

Provisions held invalid.

The provisions of this compact shall be severable and if any phrase, clause, sentence or provision of this compact is declared by a court of competent jurisdiction to be contrary to the Constitution of any participating state or of the United States or the applicability thereof to any government, agency, person or circumstance is held invalid, the validity of the remainder of this compact and the applicability thereof to any government, agency, person or circumstance shall not be affected thereby. If any provision of this compact shall be held contrary to the Constitution of any state participating therein, the compact shall remain in full force and effect as to the state affected as to all severable matters.

42 USC 2021d
note.
Arizona.
Colorado.
Nevada. New
Mexico. Utah.
Wyoming.

Sec. 226. Rocky Mountain Low-Level Radioactive Waste Compact

In accordance with section 4(a)(2) of the Low-Level Radioactive Waste Policy Act (42 USC 2021d(a)(2)), the consent of the Congress hereby is given to the States of Arizona, Colorado, Nevada, New Mexico, Utah, and Wyoming to enter into the Rocky Mountain Interstate Low-Level Radioactive Waste Compact. Such compact is substantially as follows:

Rocky Mountain Low-Level Radioactive Waste Compact

Article I–Findings and Purpose

Research and
development.

42 USC 2021b
note.

(a) The party states agree that each state is responsible for providing for the management of low-level radioactive waste generated within its borders, except for waste generated as a result of defense activities of the federal government or federal research and development activities. Moreover, the party states find that the United States Congress, by enacting the "Low-Level Radioactive Waste Policy Act" (P.L. 96-573), has encouraged the use of interstate compacts to provide for the establishment and operation of facilities for regional management of low-level radioactive waste.

Health.
Safety.

(b) It is the purpose of the party states, by entering into an interstate compact, to establish the means for cooperative effort in managing low-level radioactive waste; to ensure the availability and economic viability of sufficient facilities for the proper and efficient management of low-level radioactive waste generated within the region while preventing unnecessary and uneconomic proliferation of such facilities; to encourage reduction of the volume of low-level radioactive waste requiring disposal within the region; to restrict management within the region of low-level radioactive waste generated outside the region; to distribute the costs, benefits and obligations of low-level radioactive waste management equitably among the party states; and by these means to promote the health, safety and welfare of the residents within the region.

Article II–Definitions

As used in this compact, unless the context clearly indicates otherwise:

(a) "Board" means the Rocky Mountain low-level radioactive waste board;

(b) "Carrier" means a person who transports low-level waste;

(c) "Disposal" means the isolation of waste from the biosphere, with no intention of retrieval, such as by land burial;

(d) "Facility" means any property, equipment or structure used or to be used for the management of low-level waste;

(e) "Generate" means to produce low-level waste;

(f) "Host state" means a party state in which a regional facility is located or being developed;

(g) "Low-level waste" or "waste" means radioactive waste other than:

(i) Waste generated as a result of defense activities of the federal government or federal research and development activities;

(ii) High-level waste such as irradiated reactor fuel, liquid waste from reprocessing irradiated reactor fuel, or solids into which any such liquid waste has been converted;

(iii) Waste material containing transuranic elements with contamination levels greater than ten (10) nanocuries per gram of waste material;

(iv) By-product material as defined in section 11e.(2) of the "Atomic Energy Act of 1954," as amended November 8, 1978; or

(v) Wastes from mining, milling, smelting or similar processing of ores and mineral-bearing material primarily for minerals other than radium.

(h) "Management" means collection, consolidation, storage, treatment, incineration or disposal;

(i) "Operator" means a person who operates a regional facility;

(j) "Person" means an individual, corporation, partnership or other legal entity, whether public or private;

(k) "Region" means the combined geographic area within the boundaries of the party states; and

(l) "Regional facility" means a facility within any party state which either:

(i) has been approved as a regional facility by the board; or

Nevada.

(ii) is the low-level waste facility in existence on January 1, 1982, at Beatty, Nevada.

Article III–Rights, Responsibilities, and Obligations

Nevada.

(a) There shall be regional facilities sufficient to manage the low-level waste generated within the region. At least one (1) regional facility shall be open and operating in a party state other than Nevada within six (6) years after this compact becomes law in Nevada and in one (1) other state.

Health.
Safety.

(b) Low-level waste generated within the region shall be managed at regional facilities without discrimination among the party states; provided, however, that a host state may close a regional facility when necessary for public health or safety.

(c) Each party state which, according to reasonable projections made by the board, is expected to generate twenty percent (20%) or more in cubic feet except as otherwise determined by the board of the low-level waste generated within the region has an obligation to become a host state in compliance with subsection (d) of this article.

(d) A host state, or a party state seeking to fulfill its obligation to become a host state, shall:

(i) Cause a regional facility to be developed on a timely basis as determined by the board, and secure the approval of such regional facility by the board as provided in Article IV before allowing site preparation or physical construction to begin;

Health.
Safety.

(ii) Ensure by its own law, consistent with any applicable federal law, the protection and preservation of public health and safety in the siting, design, development, licensure or other regulation, operation, closure, decommissioning and long-term care of the regional facilities within the state;

(iii) Subject to the approval of the board, ensure that charges for management of low-level waste at the regional facilities within the state are reasonable;

(iv) Solicit comments from each other party state and the board regarding siting, design, development, licensure or other regulation, operation, closure, decommissioning and long-term care of the regional facilities within the state and respond in writing to such comments;

Report.

(v) Submit an annual report to the board which contains projections of the anticipated future capacity and availability of the regional facilities within the state, together with other information required by the board; and

Report.

(vi) Notify the board immediately if any exigency arises requiring the possible temporary or permanent closure of a regional facility within the state at a time earlier than was projected in the state's most recent annual report to the board.

Nevada.

(e) Once a party state has served as a host state, it shall not be obligated to serve again until each other party state having an obligation under subsection (c) of this article has fulfilled that obligation. Nevada, already being a host state, shall not be obligated to serve again as a host state until every other party state has so served.

(f) Each party state:

Regulations.
Transportation.

(i) Agrees to adopt and enforce procedures requiring low-level waste shipments originating within its borders and destined for a regional facility to conform to packaging and transportation requirements and regulations. Such procedures shall include but are not limited to:

(A) Periodic inspection of packaging and shipping practices;

(B) Periodic inspections of waste containers while in the custody of carriers; and

(C) Appropriate enforcement actions with respect to violations.

Regulations.
Transportation.

(ii) Agrees that after receiving notification from a host state that a person in the party state has violated packaging, shipping or transportation requirements or regulations, it shall take appropriate action to ensure that violations do not recur. Appropriate action may include but is not limited to the requirement that a bond be posted by the violator to pay the cost of repackaging at the regional facility and the requirement that future shipments be inspected;

(iii) May impose fees to recover the cost of the practices provided for in paragraph (i) and (ii) of this subsection;

(iv) Shall maintain an inventory of all generators within the state that may have low-level waste to be managed at a regional facility; and

Regulations.

(v) May impose requirements or regulations more stringent than those required by this subsection.

Article IV–Board Approval of Regional Facilities

(a) Within ninety (90) days after being requested to do so by a party state, the board shall approve or disapprove a regional facility to be located within that state.

(b) A regional facility shall be approved by the board if and only if the board determines that:

(i) There will be, for the foreseeable future, sufficient demand to render operation of the proposed facility economically feasible without endangering the economic feasibility of operation of any other regional facility; and

(ii) The facility will have sufficient capacity to serve the needs of the region for a reasonable period of years.

Article V–Surcharges

(a) The board shall impose a "compact surcharge" per unit of waste received at any regional facility. The surcharge shall be adequate to pay the costs and expenses of the board in the conduct of its authorized activities and may be increased or decreased as the board deems necessary.

(b) A host state may impose a "state surcharge" per unit of waste received at any regional facility within the state. The host state may fix and change the amount of the state surcharge subject to approval by the board. Money received from the state surcharge may be used by the host state for any purpose authorized by its own law, including but not limited to costs of licensure and regulatory activities related to the regional facility, reserves for decommissioning and long-term care of the regional facility and local impact assistance.

Article VI–The Board

Prohibition.
Rocky
Mountain
Low-Level
Radioactive
Waste Board,
establishment.

(a) The "Rocky Mountain low-level radioactive waste board", which shall not be an agency or instrumentality of any party state, is created.

(b) The board shall consist of one (1) member from each party state. The governor shall determine how and for what term its member shall be appointed, and how and for what term any alternate may be appointed to perform that member's duties on the board in the member's absence.

(c) Each party state is entitled to one (1) vote. A majority of the board constitutes a quorum. Unless otherwise provided in this compact, a majority of the total number of votes on the board is necessary for the board to take any action.

(d) The board shall meet at least once a year and otherwise as its business requires. Meetings of the board may be held in any place within the region deemed by the board to be reasonably convenient for the attendance of persons required or entitled to attend and where adequate accommodations may be found. Reasonable public notice and opportunity for comment shall be given with respect to any meeting; provided, however, that nothing in this subsection shall preclude the board from meeting in executive session when seeking legal advice from its attorneys or when discussing the employment, discipline or termination of any of its employees.

(e) The board shall pay necessary travel and reasonable per diem expenses of its members, alternates, and advisory committee members.

(f) The board shall organize itself for the efficient conduct of its business. It shall adopt and publish rules consistent with this compact regarding its organization and procedures. In special circumstances the board, with unanimous consent of its members, may take actions by telephone; provided, however, that any action taken by telephone shall be confirmed in writing by each member within thirty (30) days. Any action taken by telephone shall be noted in the minutes of the board.

(g) The board may use for its purposes the services of any personnel or other resources which may be offered by any party state.

(h) The board may establish its offices in space provided for that purpose by any of the party states, or, if space is not provided or is deemed inadequate, in any space within the region selected by the board.

Contracts.

(i) Consistent with available funds, the board may contract for necessary personnel services to carry out its duties. Staff shall be employed without regard for the personnel, civil service, or merit system laws of any of the party states and shall serve at the pleasure of the board. The board may provide appropriate employee benefit programs for its staff.

(j) The board shall establish a fiscal year which conforms to the extent practicable to the fiscal years of the party states.

Audit.
Report.

(k) The board shall keep an accurate account of all receipts and disbursements. An annual audit of the books of the board shall be conducted by an independent certified public accountant, and the audit report shall be made a part of the annual report of the board.

Report.

(l) The board shall prepare and include in the annual report a budget showing anticipated receipts and disbursements for the ensuing year.

(m) Upon legislative enactment of this compact, each party state shall consider the need to appropriate seventy thousand dollars ($70,000.00) to the board to support its activities prior to the collection of sufficient funds through the compact surcharge imposed pursuant to subsection (a) of article V of this compact.

Grants.
Report.

(n) The board may accept any donations, grants, equipment, supplies, materials or services, conditional or otherwise, from any source. The nature, amount and condition, if any, attendant upon any donation, grant or other resources accepted pursuant to this subsection, together with the identity of the donor or grantor, shall be detailed in the annual report of the board.

(o) In addition to the powers and duties conferred upon the board pursuant to other provisions of this compact, the board:

Report.

(i) Shall submit communications to the governors and to the presiding officers of the legislatures of the party states regarding the activities of the board, including an annual report to be submitted by December 15;

(ii) May assemble and make available to the governments of the party states and to the public through its members information concerning low-level waste management needs, technologies and problems;

(iii) Shall keep a current inventory of all generators within the region, based upon information provided by the party states;

(iv) Shall keep a current inventory of all regional facilities, including information on the size, capacity, location, specific wastes capable of being managed and the projected useful life of each regional facility;

(v) May keep a current inventory of all low-level waste facilities in the region, based upon information provided by the party states;

(vi) Shall ascertain on a continuing basis the needs for regional facilities and capacity to manage each of the various classes of low-level waste;

(vii) May develop a regional low-level waste management plan;

(viii) May establish such advisory committees as it deems necessary for the purpose of advising the board on matters pertaining to the management of low-level waste;

Contracts.
Prohibition.

(ix) May contract as it deems appropriate to accomplish its duties and effectuate its powers, subject to its projected available resources; but no contract made by the board shall bind any party state;

(x) Shall make suggestions to appropriate officials of the party states to ensure that adequate emergency response programs are available for dealing with any exigency that might arise with respect to low-level waste transportation or management;

(xi) Shall prepare contingency plans, with the cooperation and approval of the hose state, for management of low-level waste in the event any regional facility should be closed;

Records.

(xii) May examine all records of operators of regional facilities pertaining to operating costs, profits or the assessment or collection of any charge, fee or surcharge;

(xiii) Shall have the power to sue; and

(xiv) When authorized by unanimous vote of its members, may intervene as of right in any administrative or judicial proceeding involving low-level waste.

Article VII–Prohibited Acts and Penalties

(a) It shall be unlawful for any person to dispose of low-level waste within the region, except at a regional facility; provided, however, that a generator who, prior to January 1, 1982, had been disposing of only his own waste on his own property may, subject to applicable federal and state law, continue to do so.

Exports.

(b) After January 1, 1986, it shall be unlawful for any person to export low-level waste which was generated within the region outside the region unless authorized to do so by the board. In determining whether to grant such authorization, the factors to be considered by the board shall include, but not be limited to, the following:

(i) The economic impact of the export of the waste on the regional facilities;

(ii) The economic impact on the generator of refusing to permit the export of the waste; and

(iii) The availability of a regional facility appropriate for the disposal of the waste involved.

(c) After January 1, 1986, it shall be unlawful for any person to manage any low-level waste within the region unless the waste was generated within the region or unless authorized to do so both by the board and by the state in which said management takes place. In determining whether to grant such authorization, the factors to be considered by the board shall include, but not be limited to, the following:

Imports.

(i) the impact of importing waste on the available capacity and projected life of the regional facilities;

(ii) the economic impact on the regional facilities; and

(iii) the availability of a regional facility appropriate for the disposal of the type of waste involved.

(d) It shall be unlawful for any person to manage at a regional facility any radioactive waste other than low-level waste as defined in this compact, unless authorized to do so both by the board and the host state. In determining whether to grant such authorization, the factors to be considered by the board shall include, but not be limited to, the following:

(i) the impact of allowing such management on the available capacity and projected life of the regional facilities;

(ii) the availability of a facility appropriate for the disposal of the type of waste involved;

(iii) the existence of transuranic elements in the waste; and

(iv) the economic impact on the regional facilities.

(e) Any person who violates subsection (a) or (b) of this article shall be liable to the board for a civil penalty not to exceed ten (10) times the charges which would have been charged for disposal of the waste at a regional facility.

(f) Any person who violates subsection (c) or (d) of this article shall be liable to the board for a civil penalty not to exceed ten (10) times the charges which were charged for management of the waste at a regional facility.

(g) The civil penalties provided for in subsections (e) and (f) of this article may be enforced and collected in any court of general jurisdiction within the region where necessary jurisdiction is obtained by an appropriate proceeding commenced on behalf of the board by the attorney general of the party state wherein the proceeding is brought or by other counsel authorized by the board. In any such proceeding, the board, if it prevails, is entitled to recover reasonable attorney's fees as part of its costs.

(h) Out of any civil penalty collected for a violation of subsection (a) or (b) of this article, the board shall pay to the appropriate operator a sum sufficient in the judgment of the board to compensate the operator for any loss of revenue attributable to the violation. Such compensation may be subject to state and compact surcharges as if received in the normal course of the operator's business. The remainder of the civil penalty collected shall be allocated by the board. In making such allocation, the board shall give first priority to the needs of the long-term care funds in the region.

(i) Any civil penalty collected for a violation of subsection (c) or (d) of this article shall be allocated by the board. In making such allocation, the board shall give first priority to the needs of the long-term care funds in the region.

(j) Violations of subsection (a), (b), (c), or (d) of this article may be enjoined by any court of general jurisdiction within the region where necessary jurisdiction is obtained in any appropriate proceeding commenced on behalf of the board by the attorney general of the party state wherein the proceeding is brought or by other counsel authorized by the board. In any such proceeding, the board, if it prevails, is entitled to recover reasonable attorney's fees as part of its costs.

Prohibition.

(k) No state attorney general shall be required to bring any proceeding under any subsection of this article, except upon his consent.

Article VIII–Eligibility, Entry Into Effect, Congressional Consent, Withdrawal, Exclusion

Arizona.
Colorado.
Nevada.
New Mexico.
Utah.
Wyoming.

(a) Arizona, Colorado, Nevada, New Mexico, Utah, and Wyoming are eligible to become parties to this compact. Any other state may be made eligible by unanimous consent of the board.

(b) An eligible state may become a party state by legislative enactment of this compact or by executive order of its governor this adopting compact; provided, however, a state becoming a party by executive order shall cease to be a party state upon adjournment of the first general session of its legislature convened thereafter, unless before such adjournment the legislature shall have enacted this compact.

Effective date.

(c) This compact shall take effect when it has been enacted by the legislatures of two (2) eligible states. However, subsections (b) and (c) of article VII shall not take effect until Congress has by law consented to this compact. Every five (5) years after such consent has been given, Congress may by law withdraw its consent.

Nevada.

(d) A state which has become a party state by legislative enactment may withdraw by legislation repealing its enactment of this compact; but not such repeal shall take effect until two (2) years after enactment of the repealing legislation. If the withdrawing state is a host state, any regional facility in that state shall remain available to receive low-level waste generated within the region until five (5) years after the effective date of the withdrawal; provided, however, this provision shall not apply to the existing facility in Beatty, Nevada.

(e) A party state may be excluded from this compact by a two-thirds (2/3) vote of the members representing the other party states, acting in a meeting, on the ground that the state to be excluded has failed to carry out its obligation under this compact. Such an exclusion may be terminated upon a two-thirds (2/3) vote of the members acting in a meeting.

Article IX–Construction and Severability

(a) The provisions of this compact shall be broadly construed to carry out the purposes of the compact.

Prohibition.

(b) Nothing in this compact shall be construed to affect any judicial proceeding pending on the effective date of this compact.

Provisions held invalid.

(c) If any part or application of this compact is held invalid, the remainder, or its application to other situations or persons, shall not be affected.

42 USC 2021d note.

42 USC 2021d.
Connecticut.
Delaware.
Maryland.
New Jersey.

Sec. 227. Northeast Interstate Low-Level Radioactive Waste Management Compact

In accordance with   section 4(a)(2) of the Low-Level Radioactive Waste Policy Act, the consent of the Congress is hereby given to the States of Connecticut, New Jersey, Delaware, and Maryland to enter into the Northeast Interstate Low-Level Radioactive Waste Management Compact. Such compact is substantially as follows:

Northeast Interstate Low-Level Radioactive Waste Management Compact

Article I–Policy and Purpose

Research and development.

There is hereby created the Northeast Interstate Low-Level Radioactive Waste Management Compact. The party states recognize that the Congress has declared that each state is responsible for providing for the availability of capacity, either within or outside its borders, for disposal of low-level radioactive waste generated within its borders, except for waste generated as a result of atomic energy defense activities of the federal government, as defined in the Low-Level Radioactive Waste Policy Act (P. L. 96-573, "The Act"), or federal research and development activities. They also recognize that the management of low-level radioactive waste is handled most efficiently on a regional basis. The party states further recognize that the Congress of the United States, by enacting the Act has provided for and encouraged the development of regional low-level radioactive waste compacts to manage such waste. The party states recognize that the long-term, safe and efficient management of low-level radioactive waste generated within the region

42 USC 2021d note.

requires that sufficient capacity to manage such waste be properly provided.

Health.
Safety.

In order to promote the health and safety of the region, it is the policy of the party states to: enter into a regional low-level radioactive waste management compact as a means of facilitating an interstate cooperative effort, provide for proper transportation of low-level waste generated in the region, minimize the number of facilities required to effectively and efficiently manage low-level radioactive waste generated in the region, encourage the reduction of the amounts of low-level waste generated in the region, distribute the costs, benefits, and obligations of proper low-level radioactive waste management equitably among the party states, and ensure the environmentally sound and economical management of low-level radioactive waste.

Article II–Definitions

As used in this compact, unless the context clearly requires a different construction:

a. "commission" means the Northeast Interstate Low-Level Radioactive Waste Commission established pursuant to Article IV of this compact;

b. "custodial agency" means the agency the government designated to act on behalf of the government owner of the regional facility;

c. "disposal" means the isolation of low-level radioactive waste from the biosphere inhabited by man and his food chains;

d. "facility" means a parcel of land, together with the structures, equipment and improvements thereon or appurtenant thereto, which is used or is being developed for the treatment, storage or disposal of low-level waste, but shall not include on-site treatment or storage by a generator;

e. "generator" means a person who produces or processes low-level waste, but does not include persons who only provide a service by arranging for the collection, transportation, treatment, storage or disposal of wastes generated outside the region;

f. "high-level waste" means 1) the highly radioactive material resulting from the reprocessing of spent nuclear fuel, including liquid waste produced directly in reprocessing and any solid material derived from such liquid waste that contains fission products in sufficient concentration; and 2) any other highly radioactive material determined by the federal government as requiring permanent isolation;

g. "host state" means a party state in which a regional facility is located or being developed;

h. "institutional control" means the continued observation, monitoring, and care of the regional facility following transfer of control of the regional facility from the operator to the custodial agency;

i. "low-level waste" means radioactive waste that 1) is neither high-level waste nor transuranic waste, nor spent nuclear fuel, nor by-product material as defined in section 11e (2) of the Atomic Energy Act of 1954 as amended; and 2) is classified by the federal government as low-level waste, consistent with existing law; but does not include waste generated as a result of atomic energy defense activities of the federal government, as defined in P. L. 96-573, or federal research and development activities;

j. "party state" means any state which is a signatory party in good standing to this compact;

k. "person" means an individual, corporation, business enterprise or other legal entity, either public or private and their legal successors;

l. "post-closure observation and maintenance" means the continued monitoring of a closed regional facility to ensure the integrity and environmental safety of the site through compliance with applicable licensing and regulatory requirements; prevention of unwarranted intrusion, and correction of problems;

m. "region" means the entire area of the party states;

n. "regional facility" means a facility as defined in this section which has been designated or accepted by the Commission;

o. "state" means a state of the United States, the District of Columbia, the Commonwealth of Puerto Rico, the Virgin Islands or any other territory subject to the laws of the United States;

p. "storage" means the holding of waste for treatment or disposal;

q. "transuranic waste" means waste material containing radionuclides with an atomic number greater than 92 which are excluded from shallow land burial by the federal government;

r. "treatment" means any method, technique or process, including storage for decay, designed to change the physical, chemical or biological characteristics or composition of any waste in order to render such waste safer for transport or disposal, amenable for recovery, convertible to another usable material or reduced in volume;

s. "waste" means low-level radioactive waste as defined in this section;

t. "waste management" means the storage, treatment, transportation, and disposal, where applicable, of waste.

Article III–Rights and Obligations

a. There shall be provided within the region one or more regional facilities which, together with such other facilities as may be made available to the region, will provide sufficient capacity to manage all wastes generated within the region.

Exports.
Prohibition.
1. Regional facilities shall be entitled to waste generated within the region, unless otherwise provided by the Commission. To the extent regional facilities are available, no waste generated within a party state shall be exported to facilities outside the region unless such exportation is approved by the Commission and the affected host state(s).

Prohibition.
2. After January 1, 1986, no person shall deposit at a regional facility waste generated outside the region, and further, no regional facility shall accept waste generated outside the region, unless approved by the Commission and the affected host state(s).

b. The rights, responsibilities and obligations of each party state to this compact are as follows:

1. Each party state shall have the right to have all wastes generated within its borders managed at regional facilities, and shall have the right of access to facilities made available to the region through agreements entered into by the Commission pursuant to Article IV(i)(11). The right of access by a generator within a party state to any regional facility is limited by the generator's adherence to applicable state and federal laws and regulations and the provisions of this compact.

Regulations.
Transportation.
2. To the extent not prohibited by federal law, each party state shall institute procedures which will require shipments of low-level waste generated within or passing through its borders to be consistent

with applicable federal packaging and transportation regulations and applicable host state packaging and transportation regulations for management of low-level waste; provided, however, that these practices shall not impose unreasonable, burdensome impediments to the management of low-level waste in the region. Upon notification by a host state that a generator, shipper, or carrier within the party state is in violation of applicable packaging or transportation regulations, the party state shall take appropriate action to ensure that such violations do not recur.

3. Each party state may impose reasonable fees upon generators, shippers, or carriers to recover the cost of inspections and other practices under this compact.

4. Each party state shall encourage generators within its borders to minimize the volumes of waste requiring disposal.

5. Each party state has the right to rely on the good faith performance by every other party state of acts which ensure the provision of facilities for regional availability and their use in a manner consistent with this compact.

6. Each party state shall provide to the Commission any data and information necessary for the implementation of the Commission's responsibilities, and shall establish the capability to obtain any data and information necessary to meet its obligation as herein defined.

7. Each party state shall have the capability to host a regional facility in a timely manner and to ensure the post-closure observation and maintenance, and institutional control of any regional facility within its borders.

8. No non-host party state shall be liable for any injury to persons or property resulting from the operation of a regional facility or the transportation of waste to a regional facility; however, if the host state itself is the operator of the regional facility, its liability shall be that of any private operator.

c. The rights, responsibilities and obligations of a host state are as follows:

1. To the extent not prohibited by federal law, a host state shall ensure the timely development and the safe operation, closure, post-closure observation and maintenance, and institutional control of any regional facility within its borders.

2. In accordance with procedures established in Articles V and IX, the host state shall provide for the establishment of a reasonable

structure of fees sufficient to cover all costs related to the development, operation, closure, post-closure observation and maintenance, and maintenance, and institutional control of a regional facility. It may also establish surcharges to cover the regulatory costs, incentives, and compensation associated with a regional facility; provided, however, that without the express approval of the Commission, no distinction in fees or surcharges shall be made between persons of the several states party to this compact.

3. To the extent not prohibited by federal law, a host state may establish requirements and regulations pertaining to the management of waste at a regional facility; provided, however, that such requirements shall not impose unreasonable impediments to the management of low-level waste within the region. Nor may a host state or a subdivision impose such restrictive requirements on the siting or operation of a regional facility that, along or as a whole, they

serve as unreasonable barriers or prohibitions to the siting or operation of such a facility.

4. Each host state shall submit to the Commission annually a report concerning each operating regional facility within its borders. The report shall contain projections of the anticipated future capacity and availability of the regional facility, a financial audit of its operations, and other information as may be required by the Commission; and in the case of regional facilities in institutional control or otherwise no longer operating, the host states shall furnish such information as may be required on the facilities still subject to their jurisdiction.

5. A host state shall notify the Commission immediately if any exigency arises which requires the permanent, temporary, or possible closure of any regional facility located therein at a time earlier than projected in its most recent annual report to the Commission. The Commission may conduct studies, hold hearings, or take such other measures to ensure that the actions taken are necessary and compatible with the obligations of the host state under this compact.

Article IV–The Commission

a. There is hereby created the Northeast Interstate Low-Level Radioactive Waste Commission. The Commission shall consist of one member from each party state to be appointed by the Governor according to procedures of each party state, except that a host state shall have two members during the period that it has an operating regional Governor shall notify the Commission in writing of the identity of the facility. The member and one alternate, who may act on behalf of the member only in the member's absence.

b. Each Commission member shall be entitled to one vote. No action of the Commission shall be binding unless a majority of the total membership cast their vote in the affirmative.

c. The Commission shall elect annually from among its members a presiding officer and such other officers as it deems appropriate. The Commission shall adopt and publish, in convenient form, such rules and regulations as are necessary for due process in the performance of its duties and powers under this compact.

d. The Commission shall meet at least once a year and shall also meet upon the call of the presiding officer, or upon the call of a party state member.

e. All meetings of the Commission shall be open to the public with reasonable prior public notice. The Commission may, by majority vote, close a meeting to the public for the purpose of considering sensitive personnel or legal matters. All Commission actions and decisions shall be made in open meetings and appropriately recorded. A roll call vote may be required upon request of any party state or the presiding officer.

f. The Commission may establish such committees as it deems necessary.

g. The commission may appoint, contract for, and compensate such limited staff as it determines necessary to carry out its duties and functions. The staff shall serve at the Commission's pleasure irrespective of the civil service, personnel or other merit laws of any of the party states or the federal government and shall be compensated from funds of the Commission.

h. The Commission shall adopt an annual budget for its operations.

Margin notes:

Report.

Audit.

Report.

Studies.

Northeast Interstate Low-Level Radioactive Waste Commission, establishment.

Prohibition.

Regulations.

Contracts.

i. The Commission shall have the following duties and powers:

1. The Commission shall receive and act on the application of a non-party state to become an eligible state in accordance with Article VII(e).

2. The Commission shall receive and act on the application of an eligible state to become a party state in accordance with Article VII(b).

Report.

3. The commission shall submit an annual report to and otherwise communicate with the governors and the presiding officer of each body of the legislature of the party states regarding the activities of the Commission.

4. Upon request of party states, the Commission shall meditate disputes which arise between the party states regarding this compact.

5. The Commission shall develop, adopt and maintain a regional management plan to ensure safe and effective management of waste within the region, pursuant to Article V.

Report.
Studies.

6. The Commission may conduct such legislative or adjudicatory hearings, and require such reports, studies, evidence and testimony as are necessary to perform its duties and functions.

Regulation.

7. The Commission shall establish by regulation, after public notice and opportunity for comment, such procedural regulations as deemed necessary to ensure efficient operation, the orderly gathering of information, and the protection of the rights of due process of affected persons.

8. In accordance with the procedures and criteria set forth in Article V, the Commission shall accept a host state's proposed facility as a regional facility.

9. In accordance with the procedures and criteria set forth in Article V, the Commission may designate, by a two-thirds vote, host states for the establishment of needed regional facilities. The Commission shall not exercise this authority unless the party states have failed to voluntarily pursue the development of such facilities.

Prohibition.

10. The commission may require of and obtain from party states, eligible states seeking to become party states, and non-party states seeking to become eligible states, data and information necessary for the implementation of Commission responsibilities.

Contracts.
Imports.

11. The Commission may enter into agreements with any person, state, regional body, or group of states for the importation of waste into the region and for the right of access to facilities outside the region for waste generated within the region. Such authorization to import requires a two-thirds majority vote of the Commission, including an affirmative vote of the representatives of the host state in which any affected regional facility is located. This shall be done only after the Commission and the host state have made an assessment of the affected facilities' capability to handle such wastes and of relevant environmental, economic, and public health factors, as defined by the appropriate regulatory authorities.

Exports.

12. The Commission may, upon petition, grant an individual generator or group of generators in the region the right to export wastes to a facility located outside the region. Such grant of right shall be for a period of time and amount of waste and on such other terms and conditions as determined by the Commission and approved by the affected host states.

Report.

13. The Commission may appear as an intervenor or party in interest before any court of law, federal, state or local agency, board

or commission that has jurisdiction over the management of wastes. Such authority to intervene or otherwise appear shall be exercised only after a two-thirds vote of the Commission.

views, the Commission may arrange for any expert testimony, reports, evidence or other participation as it deems necessary.

14. The Commission may impose sanctions, including but not limited to, fines, suspension of privileges and revocation of the membership of a party state in accordance with Article VII. The Commission shall have the authority to revoke, in accordance with Article VII(g), the membership of a party state that creates unreasonable barriers to the siting of a needed regional facility or refuses to accept host state responsibilities upon designation by the Commission.

Regulation.

15. The Commission shall establish by regulation criteria for and shall review the fee and surcharge systems in accordance with Articles V and IX.

16. The Commission shall review the capability of party states to ensure the siting, operation, post-closure observation and maintenance, and institutional control of any facility within its borders.

17. The Commission shall review the compact legislation every five years prior to federal congressional review provided for in the Act, and may recommend legislative action.

Regulations.

18. The Commission has the authority to develop and provide to party states such rules, regulations and guidelines as it deems appropriate for the efficient, consistent, fair and reasonable implementation of the compact.

j. There is hereby established a Commission operating account. The Commission is authorized to expend monies from such account for the expenses of any staff and consultants designated under section (g) of this Article and for official Commission business. Financial support of the Commission account shall be provided as follows:

1. Each eligible state, upon becoming a party state, shall pay $70,000 to the Commission, which shall be used for administrative cost of the Commission.

2. The commission shall impose a "commission surcharge" per unit of waste received at any regional facility as provided in Article V.

3. Until such time as at least one regional facility is in operation and accepting waste for management, or to the extent that revenues under paragraphs (1) and (2) of this section are unavailable or insufficient to cover the approved annual budget of the Commission, each party state shall pay an apportioned amount of the difference between the funds available and the total budget in accordance with the following formula:

(a) 20 percent in equal shares;

(b) 30 percent in the proportion that the population of the party state bears to the total population of all party states, according to the most recent U. S. census;

(c) 50 percent in the proportion that the waste generated for management in each party state bears to the total waste generated for management in the region for the most recent calendar year in which reliable data are available, as determined by the Commission.

Audit.
Report.

k. The Commission shall keep accurate accounts of all receipts and disbursements. An independent certified public accountant shall annually audit all receipts and disbursements of Commission accounts and funds and submit an audit report to the Commission. Such audit report shall be made a part of the annual report of the Commission required by Article IV(i)(3).

Grants.
Loans.

Report.

l. The Commission may accept, receive, utilize and dispose for any of this purposes and functions any and all donations, loans, grants of money, equipment, supplies, materials and services (conditional or otherwise) from any state or the United States or any subdivision or agency thereof, or interstate agency, or from any institution, person, firm or corporation. The nature, amount and condition, if any, attendant upon any donation, loans, or grant accepted pursuant to this paragraph, together with the identity of the donor, grantor, or lender, shall be detailed in the annual report of the Commission. The Commission shall by rule establish guidelines for the acceptance of donations, loans, grants of money, equipment, supplies, materials and services. This shall provide that no donor, grantor or lender may derive unfair or unreasonable advantage in any proceeding before the Commission.

Prohibitions.

m. The Commission herein established is a body corporate and politic, separate and distinct from the party states and shall be so liable for its own actions. Liabilities of the Commission shall not be deemed liabilities of the party states, nor shall members of the Commission be personally liable for action taken by them in their official capacity.

1. The Commission shall not be responsible for any costs or expenses associated with the creation, operation, closure, post-closure observation and maintenance, and institutional control of any regional facility, or any associated regulatory activities of the party states.

2. Except as otherwise provided herein, this compact shall not be construed to alter the incidence of liability of any kind for any act, omission, or course of conduct. Generators, shippers and carriers of wastes, and owners and operators of sites shall be liable for their acts, omissions, conduct, or relationships in accordance with all laws relating thereto.

Courts, U.S.
District of
Columbia.

n. The United States district courts in the District of Columbia shall have original jurisdiction of all actions brought by or against the Commission. Any such action initiated in a state court shall be removed to the designated United States district court in the manner provided by Act of June 25, 1948 as amended (28 USC 14446). This section shall not

Prohibition.

alter the jurisdiction of the United States Court of Appeals for the District of Columbia Circuit to review the final administrative decisions of the Commission as set forth in the paragraph below.

Courts, U.S.

o. The United States Court of Appeals for the District of Columbia Circuit shall have jurisdiction to review the final administrative decisions of the Commission.

1. Any person aggrieved by a final administrative decision may obtain review of the decision by filing a petition for review within 60 days after the Commission's final decision.

2. In the event that review is sought of the Commission's decision relative to the designation of a host state, the Court of Appeals shall accord the   matter an expedited review, and, if the Court does not rule within 90 days after a petition for review has been filed, the Commission's decision shall be deemed to be affirmed.

Prohibition.

3. The courts shall not substitute their judgement for that of the Commission as to the decisions of policy or weight of the evidence on questions of fact. The Court may affirm the decision of the Commission or remand the case for further proceedings if it finds that the petitioners has been aggrieved because the finding, inferences, conclusions or decisions of the Commission are:

a. in violation of the Constitution of the United States;

b. in excess of the authority granted to the Commission by this compact;

c. made upon unlawful procedure to the detriment of any person;

d. Arbitrary or capricious or characterized by abuse of discretion or clearly unwarranted exercise of discretion.

Regulations.

4. The Commission shall be deemed to be acting in a legislative capacity except in those instances where it decides, pursuant to its rules and regulations, that its determinations are adjudicatory in nature.

Article V–Host State Selection and Development and Operation of Regional Facilities

a. The Commission shall develop, adopt, maintain, and implement a regional management plan to ensure the safe and efficient management of waste within the region. The plan shall include the following:

1. A current inventory of all generators within the region;

2. A current inventory of all facilities within the region, including information on the size, capacity, location, specific waste being handled, and projected useful life of each facility;

Health.
Safety.

3. consistent with considerations for public health and safety as defined by appropriate regulatory authorities, a determination of the type and number of regional facilities which are presently necessary and projected to be necessary to manage waste generated within the region;

4. reference guidelines, as defined by appropriate regulatory authorities, for the party states for establishing the criteria and procedures to evaluate locations for regional facilities.

b. The Commission shall develop and adopt criteria and procedures for reviewing a party state which volunteers to host a regional facility within its borders. These criteria shall be developed with public notice and shall include the following factors: the capability of the volunteering party state to host a regional facility in a timely manner and to ensure its post-closure observation and maintenance, and institutional control; and the anticipated economic feasibility of the proposed facility.

1. Any party state may volunteer to host a regional facility within its borders. The Commission may set terms and conditions to encourage a party state to volunteer to be the first host state.

2. Consistent with the review required above, the Commission shall, upon a two-thirds affirmative vote, designate a volunteering party state to serve as a host state.

c. If all regional facilities required by the regional management plan are not developed pursuant to section (b), or upon notification that an existing facility will be closed, or upon determination that an additional regional facility is or may be required, the Commission shall convene to consider designation of a host state.

1. The Commission shall develop and adopt procedures for designating a party state to be a host state for a regional facility. The Commission shall base its decision on the following criteria:

a. the health, safety and welfare of citizens of the party states as defined by the appropriate regulatory authorities;

b. the environmental, economic, and social effects of a regional facility on the party states;

The Commission shall also base its decision on the following criteria:

c. economic benefits and costs;

d. the volumes and types of waste generated within each party state;

e. the minimization of waste transportation; and

f. the existence of regional facilities within the party states.

2. Following its established criteria and procedures, the Commission shall designate by a two-thirds affirmative vote a party state to serve as a host state. A current host state shall have the right of first refusal for a succeeding regional facility.

3. The Commission shall conduct such hearings and studies, and take such evidence and testimony as is required by its approved procedures prior to designating a host state. Public hearings shall be held upon request in each candidate host state prior to final evaluation and selection.

4. A party state which has been designated as a host state by the Commission and which fails to fulfill its obligations as a host state may have its privileges under the compact suspended or membership in the compact revoked by the Commission.

d. Each host state shall be responsible for the timely identification of a site and the time development and operation of a regional facility. The proposed facility shall meet geologic, environmental and economic criteria which shall not conflict with applicable federal and host state laws and regulations.

1. To the extent not prohibited by federal law, a host state may regulate and license any facility within its borders.

2. To the extent not prohibited by federal law, a host state shall ensure the safe operation, closure, post-closure observation and maintenance, and institutional control of a facility, including adequate financial assurances by the operator and adequate emergency response procedures. It shall periodically review and report to the Commission on the status of the post-closure and institutional control funds and the remaining useful life of the facility.

3. A host state shall solicit comments from each party state and the Commission regarding the siting, operation, financial assurances, closure, post-closure observation and maintenance, and institutional control of a regional facility.

e. A host state intending to close a regional facility within its borders shall notify the Commission in writing of its intention and reasons therefore.

1. Except as otherwise provided, such notification shall be given to the Commission at least five years prior to the scheduled date of closure.

2. A host state may close a regional facility within its borders in the event of an emergency of if a condition exists which constitutes a substantial threat to public health and safety. A host state shall notify

the Commission in writing within three days of its action and shall within 30 working days, show justification for the closing.

3. In the event that a regional facility closes before an additional or new facility becomes operational, the Commission shall make interim arrangements for the storage or disposal of waste generated within the region until such time that a new regional facility is operational.

f. Fees and surcharges shall be imposed equitably upon all users of a regional facility, based upon criteria established by the Commission.

Regulations.

1. A host state shall, according to its lawful administrative procedures, approve fee schedules to be charged to all users of the regional facility within its borders. Except as provided herein, such fee schedules shall be established by the operator of a regional facility, under applicable state regulations, and shall be reasonable and sufficient to cover all costs related to the development, operation, closure, post-closure observation and maintenance, institutional control of the regional facility. The host state shall determine a schedule for contributions to the post-closure observation and maintenance, and institutional control funds. Such fee schedules shall not be approved unless the Commission has been given reasonable opportunity to review and make recommendations on the proposed fee schedules.

Regulation.

2. A host state may, according to its lawful administrative procedures impose a state surcharge per unit of waste received at any regional facility within its borders. The state surcharge shall be in addition to the fees charged for waste management. The surcharge shall be sufficient to cover all reasonable costs associated with administration and regulation of the facility. The surcharge shall not be established unless the Commission has been provided reasonable opportunity to review and make recommendations on the proposed state surcharge.

3. The Commission shall impose a commission surcharge per unit of waste received at any regional facility. The total monies collected shall be adequate to pay the cost and expenses of the Commission and shall be remitted to the Commission on a timely basis as determined by the Commission. The surcharge may be increased or decreased as the Commission deems necessary.

4. Nothing herein shall be construed to limit the ability of the host state, or the political subdivision in which the regional facility is situated, to impose surcharges for purposes including, but not limited to, host community compensation and host community development incentives. Such surcharges shall be reasonable and shall not be imposed unless the Commission has been provided reasonable opportunity to review and make recommendations on the proposed surcharge. Such surcharge may be recovered through the approved fee and surcharge schedules provided for in this section.

Article VI–Other Laws and Regulations

42 USC 2021.
Prohibition.

a. Nothing in this compact shall be construed to abrogate or limit the regulatory responsibility or authority of the U.S. Nuclear Regulatory Commission or of an Agreement State under section 274 of the Atomic Energy Act of 1954, as amended.

b. The laws or portions of those laws of a party state that are not inconsistent with this compact remain in full force.

Prohibition.

Prohibition.

Prohibition.

42 USC 2021b
note.
Research and
development.

Taxes.
Transportation.

Prohibition.

Prohibition.

Prohibition.

Connecticut.
Delaware.
Maine.
Maryland.
Massachusetts.
New
Hampshire.
New Jersey.
New York.
Pennsylvania.
Rhode Island.
Vermont.

c. Nothing in this compact shall make unlawful the continued development and operation of any facility already licensed for development or operation on the date this compact becomes effective.

d. No judicial or administrative proceeding pending on the effective date of the compact shall be affected by the compact.

e. Except as provided for in Article III(b)(2) and (c)(3), this compact shall not affect the relations between and the respective internal responsibilities of the government of a party state and its subdivisions.

f. The generation, treatment, storage, transportation, or disposal of waste generated by the atomic energy defense activities of the federal government, as defined in P.L. 96-573, or federal research and development activities are not affected by this compact.

g. To the extent that the rights and powers of any state or political subdivision to license and regulate any facility within its borders and to impose taxes, fees, and surcharges on the waste managed at that regional facility do not operate as an unreasonable impediment to the transportation, treatment or disposal of waste, such rights and powers shall not be diminished by this compact.

h. No party state shall enact any law or regulation or attempt to enforce any measure which is inconsistent with this compact. Such measures may provide the basis for the Commission to suspend or terminate a party state's membership and privileges under this compact.

i. All laws and regulations, or parts thereof of any party state or subdivision or instrumentality thereof which are inconsistent with this compact are hereby repealed and declared null and void. Any legal right, obligation, violation or penalty arising under such laws or regulations prior to the enactment of this compact, or not in conflict with it, shall not be affected.

j. Subject to Article III(c)(2), no law or regulation of a party state or subdivision or instrumentality thereof may be applied so as to restrict or make more costly or inconvenient access to any regional facility by the generators of another party state than for the generators of the state where the facility is situated.

k. No law, ordinance, or regulation of any party state or any subdivision or instrumentality thereof shall prohibit, suspend, or unreasonably delay, limit or restrict the operation of a siting or licensing agency in the designation, siting, or licensing of a regional facility. Any such provision in existence at the time of ratification of this compact is hereby repealed.

Article VII–Eligible Parties, Withdrawal, Revocation, Entry Into Force, Termination

a. The initially eligible parties to this compact shall be the eleven states of Connecticut, Delaware, Maine, Maryland, Massachusetts, New Hampshire, New Jersey, New York, Pennsylvania, Rhode Island, and Vermont. Initial eligibility will expire June 30, 1984.

b. Each state eligible to become a party state to this compact shall be declared a party state upon enactment of this compact into law by the state, repeal of all statutes or statutory provisions that pose unreasonable impediments to the capability of the state to host a regional facility in a timely manner, and upon payment of the fees required by Article IV(j)(1). An eligible state may become a party to this compact by an executive order by the governor of the state and upon payment of the fees required by Article IV(j)(1). However, any state which becomes a party state by executive order shall cease to be a party state upon the final adjournment of the next general or regular session of its legislature, unless this compact has by then been enacted as a statute by the state and all statutes and statutory provisions that conflict with the compact have been repealed.

Effective date.

c. The compact shall become effective in a party state upon enactment by that state. It shall not become initially effective in the region until enacted into law by three party states and consent given to it by the Congress.

d. The first three states eligible to become party states to this compact which adopt this compact into law as required in Article VII(b) shall immediately, upon the appointment of their Commission members, constitute themselves as the Northeast Interstate Low-Level Radioactive Waste Commission. They shall cause legislation to be introduced in the Congress which grants the consent of the Congress to this compact, and shall do those things necessary to organize the Commission and implement the provisions of this compact.

1. The Commission shall be the judge of the qualifications of the party states and of its members and of their compliance with the conditions and requirements of this compact and of the laws of the party states relating to the enactment of this compact.

2. All succeeding states eligible to become party states to this compact shall be declared party states pursuant to the provisions of section (b) of this Article.

e. Any state not expressly declared eligible to become a party state to this compact in section (a) of this Article may petition the Commission to be declared eligible. The Commission may establish such conditions as it deems necessary and appropriate to be met by a state requesting eligibility as a party state to this compact pursuant to the provisions of this section, including a public hearing on the application. Upon satisfactorily meeting such conditions and upon the affirmative vote of two-thirds of the Commission, including the affirmative vote of the representatives of the host states in which any affected regional facility is located, the petitioning state shall be eligible to become a party state to this compact and may become a party state in the same manner as those states declared eligible in section (a) of this Article.

Prohibition.

f. No state holding membership in any other regional compact for the management of low-level radioactive waste may become a member of this compact.

g. Any party state which fails to comply with the provisions of this compact or to fulfill its obligations hereunder may have its privileges suspended or, upon a two-thirds vote of the Commission, after full opportunity for hearing and comment, have its membership in the compact revoked. Revocation shall take effect one year from the date the affected party state receives written notice from the Commission of its action. All legal rights of the affected party state established under this compact shall cease upon the effective date of revocation, except that any legal obligations of that party state arising prior to revocation will not cease until they have been fulfilled. As soon as practicable after a Commission decision suspending or revoking party state status, the Commission shall provide written notice of the action and a copy of the resolution to the governors and the presiding officer of each body of the state legislatures of the party states, and to chairmen of the appropriate committees of the Congress.

h. Any party state may withdraw from this compact by repealing its authorization legislation, and all legal rights under this compact of the party state cease upon repeal. However, no such withdrawal shall take effect until five years after the Governor of the withdrawing state has given notice in writing of such withdrawal to the Commission and to the governor of each party state. No withdrawal shall affect any liability already incurred by or chargeable to a party state prior to that time.

1. Upon receipt of the notification, the Commission shall, as soon as practicable, provide copies to the governors and the presiding officer of each body of the state legislatures of the party states, and to the chairmen of the appropriate committees of the Congress.

2. A regional facility in a withdrawing state shall remain available to the region for five years after the date the Commission receives written notification of the intent to withdraw or until the prescheduled dates of closure, whichever occurs first.

i. This compact may be terminated only by the affirmative action of the Congress or by the repeal of all laws enacting the compact in each party state. The Congress may by law withdraw its consent every five years after the compact takes effect.

1. The consent given to this compact by the Congress shall extend to any future admittance of new party states under sections (b) and (e) of this Article.

2. The withdrawal of a party state from this compact under section (h) or the revocation of a state's membership in this compact under section (g) of this Article shall not affect the applicability of the compact to the remaining party states.

Article VIII–Penalties

a. Each party state, consistent with federal and host state regulations and laws, shall enforce penalties against any person not acting as an official of a party state for violation of this compact in the party state. Each party state acknowledges that the shipment to a host state of waste packaged or transported in violation of applicable laws and regulations can result in the imposition of sanctions by the host state. These sanctions may include, but are not limited to, suspension or revocation of the violator's rights of access to the facility in the host state.

Regulations.   b. Without the express approval of the Commission, it shall be unlawful for any person to dispose of any low-level waste within the region except at a regional facility; provided, however, that this restriction shall not apply to waste which is permitted by applicable federal or state regulations to be discarded without regard to its radioactivity.

c. Unless specifically approved by the Commission and affected host state(s) pursuant to Article IV, it shall be a violation of this compact for: 1) any person to deposit at a regional facility waste not generated within the region; 2) any regional facility to accept waste not generated within the region; and 3) any person to export from the region waste generated within the region.

d. Primary responsibility for enforcing provisions of the law will rest with the affected state or states. The Commission, upon a two-thirds vote of its members, may bring action to seek enforcement or appropriate remedies against violators of the provisions and regulations for this compact as provided for in Article IV.

Article IX–Compensation Provisions

a. The responsibility for ensuring compensation and clean-up during the operational and post-closure periods rests with the host state, as set forth herein.

1. The host state shall ensure the availability of funds and procedures for compensation of injured persons, including facility employees, and property damage (except any possible claims for diminution of property values) due to the existence and operation of a regional facility, and for clean-up and restoration of the facility and surrounding areas.

2. The state may satisfy this obligation by requiring bonds, insurance, compensation funds, or any other means or combination of means, imposed either on the facility operator or assumed by the state itself, or both. Nothing in this article alters the liability of any person or governmental entity under applicable state and federal laws.

Prohibition.

b. The Commission shall provide a means of compensation for persons injured or property damaged during the institutional control period due to the radioactive and waste management nature of the regional facility. This responsibility may be met by a special fund, insurance, or other means.

1. The Commission is authorized, at its discretion, to impose a waste management surcharge, to be collected by the operator or owner of the regional facility; to establish a separate insurance entity, formed by but separate from the Commission itself, but under such terms and conditions as it decides, and exempt from state insurance regulation; to contract with this company or other entity for coverage; or to take any other measures, or combination of measures, to implement the goals of this section.

Contracts.
Insurance.

2. The existence of this fund or other means of compensation shall not imply liability by the Commission, the non-host party states, or any of their officials and staff, which are exempted from liability by other provisions of this compact. Claims or suits for compensation shall be directed against the fund, the insurance company, or other entity, unless the Commission, by regulation, directs otherwise.

Regulations.

c. Notwithstanding any other provisions, the Commission fund, insurance, or other means of compensation shall also be available for third party relief during the operational and post-closure periods, as the Commission may direct, but only to the extent that no other funds, insurance, tort compensation, or other means are available from the host state or other entities, under section a. of this Article or otherwise; provided, that this Commission contribution shall not apply to clean-up or restoration of the regional facility and its environs during the operational and post-closure period.

d. The liability of the Commission's fund, insurance entity, or any other means of compensation shall be limited to the amount currently contained therein; provided that the Commission may set some lower limit to ensure the integrity and availability of the fund or other entity for liability.

Article X–Severability and Construction

Provisions
held invalid.

The provisions of this compact shall be severable, and if any phrase, clause, sentence or provision of this compact is declared by a federal court of competent jurisdiction to be contrary to the Constitution of the

United States or the applicability thereof to any government, agency, person or circumstance is held invalid, the validity of the remainder of this compact and the applicability thereof to any other government, agency, person or circumstance shall not be affected thereby. The provisions of this compact shall be liberally construed to give effect to the purpose thereof.

C. APPALACHIAN STATES LOW-LEVEL RADIOACTIVE WASTE COMPACT CONSENT ACT

Public Law 100–319   **102 Stat. 471**

May 19, 1988

An Act

Appalachian States Low-Level Radioactive Waste Compact Consent Act.

To grant the consent of the Congress to the Appalachian States Low-Level Radioactive Waste Compact.

Be it enacted by the Senate and House of Representatives of the United States of America in Congress assembled,

42 USC 2021d note.

Sec. 1. Short Title

This Act may be cited as the "Appalachian States Low-Level Radioactive Waste Compact Consent Act."

Sec. 2. Congressional Finding

The Congress finds that the compact set forth in section 5 is in furtherance of the Low-Level Radioactive Waste Policy Act.

42 USC 2021d note.

Sec. 3. Conditions of Consent to Compact

The consent of the Congress to the compact set forth in section 5–

(1) shall become effective on the date of the enactment of this Act,

(2) is granted subject to the provisions of the Low-Level Radioactive Waste Policy Act, and

(3) is granted only for so long as the Appalachian States Low-Level Radioactive Waste Commission, advisory committees, and regional boards established in the compact comply with all the provisions of such Act.

42 USC 2021d note.

Sec. 4. Congressional Review

The Congress may alter, amend, or repeal this Act with respect to the compact set forth in section 5 after the expiration of the 10-year period following the date of the enactment of this Act, and at such intervals thereafter as may be provided for in such compact.

42 USC 2021d note.
Delaware.
Maryland.
Pennsylvania.
West Virginia.
Waste disposal.

Sec. 5. Appalachian States Low-Level Radioactive Waste Compact

In accordance with section 4(a)(2) of the Low-Level Radioactive Waste Policy Act (42 U.S.C. 2021d(A)(2)), the consent of Congress is given to the States of Pennsylvania, West Virginia, and any eligible States as defined in Article 5(A) of the Appalachian States Low-Level Radioactive Waste Compact to enter into such compact. Such compact is substantially as follows:

Appalachian States Low-Level Radioactive Waste Compact

Preamble

Whereas, The United States Congress, by enacting the Low-Level Radioactive Waste Policy Act (42 USC sections2021b-2021d) has encouraged the use of interstate compacts to provide for the establishment and operation of facilities for regional management of low-level radioactive waste;

Whereas, Under section 4(a)(1)(A) of the Low-Level Radioactive Waste Policy Act (42 USC section 2021d(a)(1)(A)), each state is responsible for providing for the capacity for disposal of low-level radioactive waste generated within its borders;

Whereas, To promote the health, safety and welfare of residents within, the Commonwealth of Pennsylvania and other eligible states as defined in Article 5(A) of this compact shall enter into a compact for the regional management and disposal of low-level radioactive waste.

Now, therefore, the Commonwealth of Pennsylvania and the state of West Virginia and other eligible states hereby agree to enter into the Appalachian States Low-Level Radioactive Waste Compact.

Article I–Definitions

As used in this compact, unless the context clearly indicates otherwise:

(a) "Broker" means any intermediate person who handles, treats, processes, stores, packages, ships or otherwise has responsibility for or possesses low-level waste obtained from a generator.

(b) "Carrier" means a person who transports low-level waste to a regional facility.

(c) "Commission" means the Appalachian States Low-Level Radioactive Waste Commission.

(d) "Disposal" means the isolation of low-level waste from the biosphere.

(e) "Facility" means any real personal property within the region, and improvements thereof or thereon, and any and all plant structures, machinery and equipment acquired, constructed, operated or maintained for the management or disposal of low-level waste.

(f) "Generate" means to produce low-level waste requiring disposal.

(g) "Generator" means a person whose activity results in the production of low-level waste requiring disposal.

(h) "Hazardous life" means the time required for radioactive materials to decay to safe levels, as defined by the time period for the concentration of radioactive materials within a given container or package to decay to maximum permissible concentrations as defined by Federal law or by standards to be set by a host state, whichever is more restrictive.

(i) "Host state" means Pennsylvania or other party state so designated by the Commission in accordance with Article 3 of this compact.

(j) "Institutional control period" means the time of the continued observation, monitoring and care of the regional facility following transfer of control from the operator to the custodial agency.

(k) "Low-level waste" means radioactive waste that:

(1) is neither high-level waste or transuranic waste, nor spent nuclear fuel, nor by-product material as defined in section 11(e)(2) of the Atomic Energy Act of 1954 as amended; and

(2) is classified by the Federal Government as low-level waste, consistent with existing law; but does not include waste generated as a result of atomic energy defense activities of the Federal Government, as defined in Public Law 96-573, or Federal research and development activities.

(l) "Management" means the reproduction, collection, consolidation, storage, packaging or treatment of low-level waste.

(m) "Operator" means a person who operates a regional facility.

(n) "Party state" means any state that has become a party in accordance with Article 5 of this compact.

(o) "Person" means an individual, corporation, partnership or other legal entity, whether public or private.

(p) "Region" means the combined geographic area within the boundaries of the party states.

(q) "Regional facility" means a facility within any party state which has been approved by the Commission for the disposal of low-level waste.

(r) "Shallow land burial" means the disposal of low-level radioactive waste directly in subsurface trenches without additional confinement in engineered structures or by proper packaging in containers as determined by the law of the host state.

(s) "Transuranic waste" means low-level waste containing radionuclides with an atomic number greater than 92 which are excluded from shallow-land burial by the Federal Government.

Establishment.

(A) Creation and Organization.

(1) Creation–There is hereby created the Appalachian States Low-Level Radioactive Waste Commission. The Commission is hereby created as a body corporate and politic, with succession for the duration of this compact, as an agency and instrumentality of the governments of the respective signatory parties, but separate and distinct from the respective signatory party states. The Commission shall have central offices located in Pennsylvania.

(2) Commission Membership–The Commission shall consist of two voting members from each party state to be appointed according to the laws of each party state and two additional voting members from each host state to be appointed according to the laws of each host state. Upon selection of the site of the regional facility, an additional voting member shall be appointed to the Commission who shall be a resident of the county or municipality where the facility is to be located. The appointing authority of each party state shall notify the Commission in writing of the identities of the members and of any alternates. An alternate may vote and act in the member's absence. No member shall have a financial interest in any industry which generates low-level radioactive waste, any low-level radioactive waste regional facility or any related industry for the duration of the member's term. No more than one-half the members and alternates from any party state shall have been employed by or be employed by a low-level waste generator or related industry upon appointment to or during their tenure of office; provided, that no member shall have been employed by or be employed by a regional facility operator. No member or alternate from any party state shall accept employment from any regional facility operator or brokers for at least three years after leaving office.

(3) Compensation–Members of the Commission and alternates shall serve without compensation from the Commission but may be reimbursed for necessary expenses incurred in and incident to the performance of their duties.

(4) Voting Power–Each Commission member is entitled to one vote. Unless otherwise provided in this compact, affirmative votes by a majority of a host state's members are necessary for the Commission to take any action related to the regional facility and

the disposal and management of low-level waste within that host state.

(5) Organization and –

(a) The Commission shall provide for its own organization and procedures and shall adopt by-laws not inconsistent with this compact and any rules and regulations necessary to implement this compact. It shall meet at least once a year in the county selected to host a regional facility and shall elect a chairman and vice chairman from among its members. In the absence of the chairman, the vice chairman shall serve.

Public
information.

(b) All meetings of the Commission shall be open to the public with at least 14 days' advance notice, except that the chairman may convene an emergency meeting with less advance notice. Each municipality and county selected to host a regional facility shall be specifically notified in advance of all Commission meetings. All meetings of the Commission shall be conducted in a manner that substantially conforms to the Administrative Procedure Act (5 U.S.C. Chapter 5, Subchapter II, and Chapter 7). The Commission may, by a two-thirds vote, including approval of a majority of each host state's Commission members, hold an Executive Session closed to the public for the purpose of: considering or discussing legally privileged or proprietary information; to consider dismissal, disciplining of or hearing complaints or charges brought against an employee or other public agents unless such person requests such public hearing; or to consult with its attorney regarding information or strategy in connection with specific litigation. The reason for the Executive Session must be announced at least 14 days prior to the Executive Session, except that the chairman may convene an emergency meeting with less advance notice, in which case the reason for the Executive Session must be announced at the open meeting immediately subsequent to the Executive Session. All action taken in violation of this open meeting shall be null and void.

Records.
Public
Information.

(c) Detailed written minutes shall be kept of all meetings of the Commission. All decisions, files, records and data of the Commission, except for information privileged against introduction in judicial proceedings, personnel records and minutes of a properly convened Executive Session, shall be open to public inspection subject to a procedure that substantially conforms to the Freedom of Information Act (Public Law 89-554, 5 USC section552) and applicable Pennsylvania law and may be copied upon request and payment of fees which shall be no higher than necessary to recover copying costs.

(d) The Commission shall select an appropriate staff, including an Executive Director, to carry out the duties and functions assigned by the Commission. Notwithstanding any other provision of law, the Commission may hire and/or retain its own legal counsel.

(e) Any person aggrieved by a final decision of the Commission which adversely affects the legal rights, duties or privileges of such person may petition a court of competent jurisdiction, within 60 days after the Commission's final decision, to obtain judicial review of said final decisions.

(f) Liabilities of the Commission shall not be deemed liabilities of the party states. Members of the Commission shall not be personally liable for actions taken in their official capacity.

(B) Powers and Duties.

The Commission:

Regulations.
Research and
development.

(a) Shall conduct research and establish regulations to promote a reasonable reduction of volume and curie content of low-level wastes generated in the region. The regulations shall be reviewed and, if necessary, revised by the Commission at least annually.

(b) Shall ensure, to the extent authorized by Federal law, that low-level wastes are safely disposed of within the region except that the Commission shall have no power or authority to license, regulate or otherwise develop a regional facility, such powers and authority being reserved for the host state(s) as permitted under the law.

(c) Shall designate as "host states" any party state which generates 25 percent or more of Pennsylvania's volume or total curie content of low-level waste generated based on a comparison of averages over three successive years, as determined by the Commission. This determination shall be based on volume or total curie content, whichever is greater.

(d) Shall ensure, to the extent authorized by Federal law, that low-level waste packages brought into the regional facility for disposal conform to applicable state and Federal regulations. Low-level waste brokers or generators who violate these regulations will be subject to a fine or other penalty imposed by the Commission, including restricted access to a regional facility. The Commission may impose such fines and/or penalties in addition to any other penalty levied by the party states pursuant to Article 4(D).

(e) Shall establish such advisory committees as it deems necessary for the purpose of advising the Commission on matters pertaining to the management and disposal of low-level waste.

Contracts.

(f) May contract to accomplish its duties and effectuate its powers subject to projected available resources. No contract made by the Commission shall bind a party state.

(g) Shall prepare contingency plans for management and disposal of low-level waste in the event any regional facility should be closed or otherwise unavailable.

Records.

(h) Shall examine all records of operators of regional facilities pertaining to operating costs, profits or the assessment or collection of any charge, fee or surcharge and may make recommendations to the host state(s) which shall review the recommendations in accordance with its (their) own sovereign laws.

(i) Shall have the power to sue and be sued subject to Article (2)(A)(5)(e) and may seek to intervene in any administrative or judicial proceeding.

(j) Shall assemble and make available, to the party states and to the public, information concerning low-level waste management and disposal needs, technologies, and problems.

Records.

(k) Shall keep current and annual inventories of all generators by name and quantity of low-level waste generated within the region, based upon information provided by the party states. Inventory information shall include both volume in cubic feet and total curie content of the low-level waste and all available information on chemical composition and toxicity of such wastes.

(1) Shall keep an inventory of all regional facilities and specialized facilities, including, but not necessarily restricted to, information on their size, capacity and location, as well as specific wastes capable of being managed, and the projected useful life of each regional facility.

(m) Shall make and publish an annual report to the governors of the signatory party states and to the public detailing its programs, operations and finances, including copies of the annual budget and the independent audit required by this compact.

(n) Notwithstanding any other provision of this compact to the contrary, may, with the unanimous approval of the Commission members of the host state(s), enter into temporary agreements with non-party states or other regional boards for the emergency disposal of low-level waste at the regional facility, if so authorized by law(s) of the host state(s), or other disposal facilities located in states that are not parties to this agreement.

(o) Shall promulgate regulations, pursuant to host state law, to specifically govern and define exactly what would constitute an emergency situation and exactly what restrictions and limitations would be placed on temporary agreements.

(p) Shall not accept any donations, grants, equipment, supplies, materials or services, conditional or otherwise, from any source, except from any Federal agency and from party states which are certified as being legal and proper under the laws of the donating party state.

(C) Budget and Operation.

(1) Fiscal Year–The Commission shall establish a fiscal year which conforms to the fiscal year of the Commonwealth of Pennsylvania.

(2) Current Expense Budget–Upon legislative enactment of this compact by two party states and each year until the regional facility becomes available, the Commission shall adopt a current expense budget for its fiscal year. The budget shall include the Commission's estimated expenses for administration. Such expenses shall be allocated to the party states according to the following formula:

Each designated initial host state will be allocated costs equal to twice the costs of the other party states, but such costs will not exceed $200,000.

Each remaining party state will be allocated a cost of one half the cost of the initial host state, but such cost will not exceed $100,000.

The party states will include the amounts allocated above in their respective budgets, subject to such review and approval as may be required by their respective budgetary processes. Such amounts shall be due and payable to the Commission in quarterly installments during the fiscal year.

(3) Annual Budget Request–For continued funding of its activities, the Commission shall submit an annual budget request to each party state for funding, based upon the percentage of the region's waste generated in each state in the region, as reported in the latest available annual inventory required under Article 2(B)(k). The percentage of waste shall be based on volume of waste or total curie content as determined by the Commission.

(4) Annual Report to Include Budget–The Commission shall prepare and include in the annual report a budget showing anticipated receipts and disbursements for the ensuing year.

(5) Annual Independent Audit–

(a) As soon as practible after the closing of the fiscal year, an audit shall be made of the financial accounts of the Commission. The audit shall be made by qualified certified public accountants selected by the Commission, who have no personal direct or indirect interest in the financial affairs of the Commission or any of its officers or employees

Reports.

The report of audit shall be prepared in accordance with accepted accounting practices and shall be filed with the chairman and such other officers s the Commission shall direct. Copies of the report shall be distributed to each Commission member and shall be made available for public distribution.

Public information.

(b) Each signatory party, by its duly authorized officers, shall be entitled to examine and audit at any time all of the books, documents, records, files and accounts and all other papers, things or property of the Commission. The representatives of the signatory parties shall have access to all books, documents, records, accounts, reports, files and all other papers, things or property belonging to or in use by the Commission and necessary to facilitate the audit; and they shall be afforded full facilities for verifying transactions with the balances or securities held by depositaries, fiscal agents and custodians.

Article III–Rights, Responsibilities, and Obligations of Party States[1]

(A) Regional Facilities.

There shall be regional facilities sufficient to dispose of the low-level waste generated within the region. Each regional facility shall be capable of disposing of such low-level waste but in the form(s) required by regulations or license conditions. Specialized facilities for particular types of low-level waste management, reduction or treatment may not be developed in any party state unless they are in accordance with the laws and regulations of such state and applicable Federal laws and regulations.

(B) Equal Access to Regional Facilities.

Public health and safety.

Each party state shall have equal access as other party states to regional facilities located within the region and accepting low-level waste, provided, however, that the host state may close the regional facility located within its borders when necessary for public health and safety. However, a host state shall send notification to the Commission in writing within three (3) days of this action and shall, within thirty (30) working days, provide in writing the reasons for the closing.

(C) Initial Host State.

Pennsylvania and party states which generate 25 percent or more of the volume or curies of low-level waste generated by Pennsylvania, based on a comparison of averages over the three years 1982 through 1984, are designated as "initial host states" and are required to develop and host low-level waste sites as regional facilities. The percentage of waste from each state shall be determined by cubic foot volume or total curie content, whichever is greater.

---

[1]   There was no "Article II" in the original slip law.

(D) Exemption From Being Initial Host State.

Party states which generate less than 25 percent of the volume or curies of low-level waste generated by Pennsylvania, based on a comparison of averages over the years 1982 through 1984, shall be exempt from initial host state responsibilities. These states shall continue to be exempt as long as they generate less that 25 percent threshold over successive 3-year periods. Once a state generates an average of 25 percent or more of the volume or curies generated by Pennsylvania over a successive 3-year period, it shall be designated as a "host state" for a 30-year period by the Commission and shall immediately initiate development of a regional facility to be operational within five years. Such host state shall be prepared to accept at its regional facility low-level waste at least equal to that generated in the state. With Commission approval, any party state may volunteer to host a regional facility. The percentage of waste from each state shall be determined by either a cubic foot volume or total curie content, whichever is greater.

(E) Useful Life of Regional Facilities.

Pennsylvania and other host states are obligated to develop regional facilities for the duration of this compact. All regional facilities shall be designed for at least a 30-year useful life. At the end of the facility's life, normal closure and maintenance procedures shall be initiated in accordance with the applicable requirements of the host state and the Federal Government. Each host state's obligation for operating regional facilities shall remain as long as the state continues to produce over a 3-year period 25 percent or more of the volume or curies of low-level waste generated by Pennsylvania.

(F) Duties of Host State.

Each host state shall:

(a) Cause a regional facility to be sited and developed on a timely basis.

Environmental protection. Public health and safety.

(b) Ensure by law, consistent with applicable state and Federal law, the protection and preservation of public health, safety and environmental quality in the siting, design, development, licensure or other regulation, operation, closure, decommissioning, long-term care and the institutional control period of the regional facility within the state. To the extent authorized by Federal law, a host state may adopt more stringent laws, rules or regulations than required by Federal law.

(c) Ensure and maintain a manifest system which documents all waste-related activities of generators, brokers, carriers and related activities of generators, brokers, carriers and operators, and establish the chain of custody of waste from its initial generation to the end of its hazardous life. Copies of all such manifests shall be submitted to the Commission on a timely basis.

(d) Ensure that charges for disposal of low-level waste at the regional facility are sufficient to fully fund the safe disposal and perpetual care of the regional facility and that charges are assessed without discrimination as to the party state of origin.

Reports.

(e) Submit an annual report to the Commission on the status of the regional facility which contains projections of the anticipated future capacity.

(f) Notify the Commission immediately if any exigency arises requiring the possible temporary or permanent closure of a regional facility within the state at a time earlier than was projected in the state's most recent annual report to the Commission.

(g) Require that the institutional control period of any disposal facility be at least as long as the hazardous life, as defined in Article 1(h), of the radioactive materials that are disposed at that facility.

(h) Prohibit the use of any shallow land burial, as defined in Article (r), and develop alternative means for treatment, storage and disposal of low-level waste.

(i) Establish by law, to the extent not prohibited by Federal law, requirements for financial responsibility, including, but not limited to:

(i) Requirements for the purchase and maintenance of adequate insurance by generators, brokers, carriers and operators of the regional facility;

(ii) Requirements for the establishment of a long-term care fund to be funded by a fee placed on generators to pay for preventative or corrective measures of low-level waste to the regional facility; and

(iii) Any further financial responsibility requirements that shall be submitted by generators, brokers, carriers and operators as deemed necessary by the host state.

(G) Duties of Party State.

Each party state:

(a) Shall appropriate its portion of the Commission's initial and annual budgets as set out in Article 2(C)(2) and (3).

(b) To the extent authorized by Federal law shall develop and enforce procedures requiring low-level waste shipments originating within its borders and destined for a regional facility to conform to volume reduction, packaging and transportation requirements and regulations as well as any other requirements specified by the regional facility. Such procedures shall include, but are not limited to:

(i) Periodic inspections of packaging and shipping practices;

(ii) Periodic inspections of low-level waste containers while in custody of carriers; and

(iii) Appropriate enforcement actions with respect to violations.

(c) To the extent authorized by Federal law, shall, after receiving notification from a host state or other person that a person in a party state has violated volume reduction, packaging, shipping or transportation requirements or regulations, take appropriate action to ensure that violations do not recur. Appropriate action shall include, but is not limited to, the requirement that a bond be posted by the violator to pay the cost of repackaging at the regional facility and the requirement that future shipments be inspected. Appropriate action may also include suspension of the violator's use of the regional facility. Should such suspension be imposed, the suspension shall remain in effect until such time as the violator has, to the satisfaction of the party state imposing such suspension, complied with the appropriate requirements or regulations upon which the suspension was based and has taken appropriate action to ensure that such violation or violations do not recur.

Records.

(d) Shall maintain a registry of all generators and quantities generated within the state.

(H) Liability.

In the event of liability arising from the operation of any regional facility and during and after closure of that facility, each party state shall share in that liability in an amount equal to that state's share of the region's low-level waste disposed of at the facility. part of a host state or any party state, then any other host or party state(s) may make any claim allowable under law for that negligence, malfeasance or neglect. If such liability arises from a particular waste shipment or shipments to, or quantity of waste or condition at, the regional facility, then any host or party state may make any claim allowable under law for such liability. The percentage of waste shall be based on volume of waste or total curie content.

(I) Failure of Party State to Fulfill Obligations.

Claims.

A party state which fails to fulfill its obligations, including timely funding of the Commission, may have its privileges under the Compact suspended or its membership in the Compact revoked by the Commission and be subject to any other legal equitable remedies available to the party states.

Article IV–Prohibited Acts and Penalties

(A) Prohibition.

It shall be unlawful for any person to dispose of low-level waste within the region except at a regional facility unless authorized by the Commission.

(B) Waste Disposed of Within Region.

After establishment of the regional facility(s), it shall be unlawful for any person to dispose of any low-level waste within the region unless the waste was generated within the region or unless authorized to do so both by the Commission and by law of the host state in which said disposal takes place. For the purpose of this compact, waste generated within the region excludes radioactive material shipped from outside the party states to a waste management facility within the region. In determining whether to grant such authorization, the factors to be considered by the Commission shall include, but not be limited to, the following:

Public health and safety.

(a) The impact on the health, safety and environmental quality of the citizens of the party states;

Environmental protection.

(b) The impact of importing waste on the available capacity and projected life of the regional facility;

(c) The availability of a regional facility appropriate for the safe disposal of the type of low-level waste involved.

(C) Waste Generated Within Region.

Any and all low-level waste generated within the region shall be disposed of at a regional facility, except for specific cases agreed upon by the Commission, with the affirmative votes by a majority of the Commission members of the host state(s) affected by the decision.

(D) Liability.

Generators, brokers and carriers of wastes, and owners and operators of sites shall be liable for their acts, omissions, conduct or relationships in accordance with all laws relating thereto. The party states shall impose a fine for any violation in an amount equal to the present and future costs associated with correcting any harm caused by the violation and shall assess punitive fines or penalties if it is deemed necessary. In addition, the host state shall bar any person who violates host state or Federal regulations from using the regional facility until that person demonstrates to the satisfaction of the host state the ability and willingness to comply with the law.

(E) Conflict of Interest.

(1) Prohibitions–

No commissioner, officer or employee shall:

(a) Be financially interested, either directly or indirectly, in a contract, sale, purchase, lease or transfer of real or personal property to which the Commission is a party.

(b) Solicit or accept money or any other thing of value in addition to the expenses paid to him by the Commission for services performed within the scope of his official duties.

(c) Offer money or anything of value for or in consideration of obtaining an appointment, promotion or privilege in his employment with the Commission.

Employment
and
unemployment.
Contracts.

(2) Forfeiture of Office or Employment–Any officer or employee who shall willfully violate any of the provisions of this section shall forfeit his office or employment.

(3) Agreement Void–Any contract or agreement knowingly made in contravention of this section is void.

(4) Criminal and Civil Sanctions–

Officers and employees of the Commission shall be subject, in addition to the provisions of this section, to such criminal and civil sanctions for misconduct in office as may be imposed by Federal law and the law of the signatory state in which such misconduct occurs.

Article V–Eligibility, Entry Into Effect, Congressional Consent, Withdrawal

(A) Eligibility.

Delaware.
Maryland
Pennsylvania.
West Virginia.

Only the States of Pennsylvania, West Virginia, Delaware and Maryland are eligible to become parties to this compact.

(B) Entry into Effect.

An eligible state may become a party state by legislative enactment of this compact or by executive order of the governor adopting this compact; provided, however, a state becoming a party state by executive order shall cease to be a party state upon adjournment of the first federal session of its legislature convened thereafter, unless the legislature shall have enacted this compact before such adjournment.

(C) Congressional Consent.

Effective dates.

This compact shall take effect when it has been enacted by the legislatures of Pennsylvania and one or more eligible states. However, Article 4 (B) and (C) shall not take effect until Congress has consented to this compact. Every fifth year after such consent has been given, Congress may withdraw consent.

(D) Withdrawal.

A party state may withdraw from the compact by repealing the enactment of this compact, but no such withdrawal shall become effective until two years after enactment of the repealing legislation. If the withdrawing state is a host state, any regional facility in that state shall remain available to receive low-level waste generated within the region until five years after the effective date of the withdrawal.

Article VI–Construction and Severability

(A) Construction.

The provisions of this compact shall be broadly construed to carry out the purposes of the compact, but the sovereign powers of a party state shall not unnecessarily be infringed.

(B) Severability.

If any part or application of this compact is held invalid, the remainder, or its application to other situations or persons, shall not be affected.

D. SOUTHWESTERN LOW-LEVEL RADIOACTIVE WASTE DISPOSAL COMPACT CONSENT ACT

Public Law 100–712   **102 Stat. 4773**

November 23, 1988

An Act

Southwestern
Low-Level
Radioactive
Waste Disposal
Compact
Consent Act.
42 USC 2021d
note.
Environmental
protection.
42 USC 2021d
note.

42 USC 2021d
note.

42 USC 2021d
note.

State listing.

To grant the consent of the Congress to the Southwestern Low-Level Radioactive Waste Disposal Compact.

Be it enacted by the Senate and House of Representatives of the United States of America in Congress assembled,

Sec. 1. Short Title

This Act may be cited as the "Southwestern Low-Level Radioactive Waste Disposal Compact Consent Act."

Sec. 2. Congressional Finding

The Congress finds that the compact set forth in section 5 is in furtherance of the Low-Level Radioactive Waste Policy Act.

Sec. 3. Conditions of Consent to Compact

The consent of the Congress to the compact set forth in section 5–

(1) shall become effective on the date of the enactment of this Act;

(2) is granted subject to the provisions of the Low-Level Radioactive Waste Policy Act; and

(3) is granted only for so long as the regional commission established in the compact complies with all of the provisions of such Act.

Sec. 4. Congressional Review

The Congress may alter, amend, or repeal this Act with respect to the compact set forth in section 5 after the expiration of the 10-year period following the date of enactment of this Act, and at such intervals thereafter as may be provided in such compact.

Sec. 5. Southwestern Low-Level Radioactive Waste Compact

In accordance with section 4(a)(2) of the Low-Level Radioactive Waste Policy Act (42 USC 2021d(a)(2)), the consent of Congress is given to the states of Arizona, California, and any eligible states, as defined in article VII of the Southwestern Low-Level Radioactive Waste Disposal Compact, to enter into such compact. Such compact is substantially as follows:

Article I–Compact Policy and Formation

The party states hereby find and declare all of the following :

(A) The United States Congress, by enacting the Low-Level Radioactive Waste Policy Act, Public Law 96-573, as amended by the Low-Level Radioactive Waste Policy Amendments Act of 1985 (42 USC section 2021b to 2021j, incl.), has encouraged the use of interstate compacts to provide for the establishment and operation of facilities for regional management of low-level radioactive waste.

(B) It is the purpose of this compact to provide the means for such a cooperative effort between or among party states to protect the citizens of the states and the states' environments.

(C) It is policy of party states to this compact to encourage the reduction of the volume of low-level radioactive waste requiring disposal within the compact region.

Public health
and safety.

(D) It is the policy of the party states that the protection of the health and safety of their citizens and the most ecological and economical management of low-level radioactive wastes can be accomplished through cooperation of the states by minimizing the amount of handling and transportation required to dispose of these wastes and by providing facilities that serve the compact region.

(E) Each party state, if an agreement state pursuant to section 2021 of Title 42 of the United States Code, or the Nuclear Regulatory Commission if not an agreement state, is responsible for the primary regulation of radioactive materials within its jurisdiction.

Article II–Definitions

As used in this compact, unless the context clearly indicates otherwise, the following definitions apply:

(A) "Commission" means the Southwestern Low-Level Radioactive Waste Commission established in article III of this compact.

(B) "Compact region" or "region" means the combined geographical area with the boundaries of the party states.

(C) "Disposal" means the permanent isolation of low-level radioactive waste pursuant to requirements established by the Nuclear Regulatory Commission and the Environmental Protection Agency under applicable laws, or by a party state if the state hosts a disposal facility.

(D) "Generate," when used in relation to low-level radioactive waste, means to produce low-level radioactive waste.

(E) "Generator" means a person whose activity, excluding the management of low-level radioactive waste, results in the production of low-level radioactive waste.

(F) "Host county" means a county, or similar political subdivision of a party state, in which a regional disposal facility is located or being developed.

(G) "Host state" means a party state in which a regional disposal facility is located or being developed. The state of California is the host state under this compact for the first thirty years from the date the California regional disposal facility commences operations.

(H) "Institutional control period" means that period of time in which the facility license is transferred to the disposal site owner in compliance with the appropriate regulations for long-term observation and maintenance following the postclosure period.

(I) "Low-level radioactive waste" means regulated radioactive material that meets all of the following requirements:

(1) The waste is not high-level radioactive waste, spent nuclear fuel, or by-product material (as defined in section 11e(2) of the Atomic Energy Act of 1954 (42 USC section 2014(e)(2)).

(2) The waste is not uranium mining or mill tailings.

(3) The waste is not any waste for which the Federal Government is responsible pursuant to subdivision (b) of section 3

of the Low-Level Radioactive Waste Policy Amendments Act of 1985 (42 USC section 2021c(b)).

(4) The waste is not an alpha emitting transuranic nuclide with a half-life greater than five years and with a concentration greater than one hundred nanocuries per gram, or plutonium-241 with a concentration greater than three thousand five hundred nanocuries per gram, or curium-242 with a concentration greater than twenty thousand nanocuries per gram.

(J) "Management" means collection, consolidation, storage, packaging, or treatment.

(K)"Major generator state" means a party state which generates 10 percent of the total amount of low-level radioactive waste produced within the compact region and disposed of at the regional disposal facility. If no party state other than California generates at least ten percent of the total amount, "Major generator state" means the party state which is second to California in the amount of waste produced within the compact region and disposed of at the regional disposal facility.

(L) "Operator" means a person who operates a regional disposal facility.

(M) "Party state" means any state that has become a party in accordance with article VII of this compact.

(N) "Person" means an individual, corporation, partnership, or other legal entity, whether public or private.

(O) "Postclosure period" means that period of time after completion of closure of a disposal facility during which the licensee shall observe, monitor, and carry out necessary maintenance and repairs at the disposal facility to assure that the disposal facility will remain stable and will not need ongoing active maintenance. This period ends with the beginning of the institutional control period.

(P) "Regional disposal facility" means a non-Federal low-level radioactive waste disposal facility established and operated under this compact.

(Q) "Site closure and stabilization" means the activities of the disposal facility operator taken at the end of the disposal facility's operating life to assure the continued protection of the public from any residual radioactive or other potential hazards present at the disposal facility.

(R) "Transporter" means a person who transports low-level radioactive waste.

(S) "Uranium mine and mill tailings" means waste resulting from mining and processing of ores containing uranium.

Article III–The Commission

Establishment.

(A) There is hereby established the Southwestern Low-Level Radioactive Waste Commission.

(1) The Commission shall consist of one voting member from each party state to be appointed by the governor, confirmed by the senate of that party state, and to serve at the pleasure of the governor of each party state, and one voting member from the host county. The appointing authority of each party state shall notify the Commission in writing of the identity of the member and any alternates. An alternate may act in the member's absence.

(2) The host state shall also appoint that number of additional voting members of the Commission which is necessary for the host state's members to compose at least 51 percent of the membership on the Commission. The host state's additional members shall be appointed by the host state governor and confirmed by the host state senate.

If there is more than one host state, only the state in which is located the regional disposal facility actively accepting low-level radioactive waste pursuant to this compact may appoint these additional members.

(3) If the host county has not been selected at the time the Commission is appointed, the governor of the host state shall appoint an interim local government member, who shall be an elected representative of a local government. After a host county is selected, the interim local government member shall resign and the governor shall appoint the host county member pursuant to paragraph (4).

(4) The governor shall appoint the host county member from a list of at least seven candidates compiled by the board of supervisors of the host county.

(5) In recommending and appointing the host county member pursuant to paragraph (4), the board of supervisors and the governor shall give first consideration to recommending and appointing the members of the board of supervisors in whose district the regional disposal facility is located or being developed. If the board of supervisors of the host county does not provide a list to the governor of at least seven candidates from which to choose, the governor shall appoint a resident of the host county as the host county member.

(6) The host county member is subject to confirmation by the senate of that party and shall serve at the pleasure of the governor of the host state.

(B) The Commission is a legal entity separate and distinct from the party states and shall be so liable for its actions. Members of the Commission shall not be personally liable for actions taken in their official capacity. The liabilities of the Commission shall not be deemed liabilities of the party states.

(C) The Commission shall conduct its business affairs pursuant to the laws of the host state and disputes arising out of Commission action shall be governed by the laws of the host state. The Commission shall be located in the capital city of the host state in which the regional disposal facility is located.

Public information. Records.

(D) The Commission's records shall be subject to the host state's public records law, and the meetings of the Commission shall be open and public in accordance with the host state's open meeting law.

(E) The Commission members are public officials of the appointing state and shall be subject to the conflict of interest laws, as well as any other law, of the appointing state. The Commission members shall be compensated according to the appointing state's law.

(F) Each Commission member is entitled to one vote. A majority of the Commission constitutes a quorum. Unless otherwise provided in this capacity, a majority of the total number of votes on the Commission is necessary for the Commission to take any action.

(G) The Commission has all of the following duties and authority:

(1) The Commission shall do, pursuant to the authority granted by this compact, whatever is reasonably necessary to ensure that low-

level radioactive wastes are safely disposed of and managed within the region.

(2) The Commission shall meet at least once a year and otherwise as business requires.

(3) The Commission shall establish a compact surcharge to be imposed upon party state generators. The surcharge shall be based upon the cubic feet of low-level radioactive waste and the radioactivity of the low-level radioactive waste and shall be collected by the operator of the disposal facility.

The host state shall set, and the Commission shall impose, the surcharge after congressional approval of the compact. The amount of the surcharge shall be sufficient to establish and maintain at a reasonable level funds for all of the following purposes:

(a) The activities of the Commission and Commission staff.

(b) At the discretion of the host state, a third-party liability fund to provide compensation for injury to persons or property during the operational, closure, stabilization, and postclosure and institutional control periods of the regional disposal facility. This subparagraph does not limit the responsibility or liability of the operator, who shall comply with any federal or host state statutes or regulations regarding third-party liability claims.

(c) A local government reimbursement fund, for the purpose of reimbursing the local government entity or entities hosting the regional disposal facility for any costs or increased burdens on the local governmental entity for services, including, but not limited to, general fund expenses, the improvement and maintenance of roads and bridges, fire protection, law enforcement, monitoring by local health officials, and emergency preparation and response related to the hosting of the regional disposal facility.

(4) The surcharges imposed by the Commission for purposes of subparagraphs (b) and (c) of paragraph (3) and surcharges pursuant to paragraph (3) of subdivision (E) of article IV shall be transmitted on a monthly basis to the host state for distribution to the proper accounts.

(5) The Commission shall establish a fiscal year which conforms to the fiscal years of the party states to the extent possible.

Records.

(6) The Commission shall keep an accurate account of all receipts and disbursements. An annual audit of the books of the Commission shall be conducted by an independent certified public accountant, and the audit report shall be made a part of the annual report of the Commission.

Reports.

(7) The Commission shall prepare and include in the annual report a budget showing anticipated receipts and disbursements for the subsequent fiscal year.

Gifts and property.

(8) The Commission may accept any grants, equipment, supplies, materials, or services, conditional or otherwise, from the federal or state government. The nature, amount and condition, if any, of any donation, grant or other resources accepted pursuant to this paragraph and the identity of the donor or grantor shall be detailed in the annual report of the Commission.

However, the host state shall receive, for the uses specified in subparagraph (E) of paragraph (2) of subsection (d) of section 2021e of Title 42 of the United States Code, any payments paid from the special escrow account for which the Secretary of Energy is trustee pursuant to subparagraph (A) of paragraph (2) of subsection (d) of section 2021e of Title 42 of the United States Code.

Reports.

(9) The Commission shall submit communications to the governors to the presiding officers of the legislatures of the party states regarding the activities of the Commission, including an annual report to be submitted on or before January 15 of each year. The Commission shall include in the annual report a review of, and recommendations for, low-level radioactive waste disposal methods which are alternative technologies to the shallow land burial of low-level radioactive waste.

Public information.

(10) The Commission shall assemble and make available to the party states, and to the public, information concerning low-level radioactive waste management needs, technologies, and problems.

Records.

(11) The Commission shall keep a current inventory of all generators within the region, based upon information provided by the party states.

Records.

(12) The Commission shall keep a current inventory of all regional disposal facilities, including information on the size, capacity, location, specific low-level radioactive wastes capable of being managed, and the projected useful life of each regional disposal facility.

(13) The Commission may establish advisory committees for the purpose of advising the Commission on the disposal and management of low-level radioactive waste.

(14) The Commission may enter into contracts to carry out its duties and authority, subject to projected resources. No contract made by the Commission shall bind a party state.

(15) The Commission shall prepare contingency plans, with the cooperation and approval of the host state, for the disposal and management of low-level radioactive waste in the event that any regional disposal facility should be closed.

(16) The Commission may sue and be sued and, when authorized by a majority vote of the members, may seek to intervene in an administrative or judicial proceeding related to this compact.

(17) The Commission shall be managed by an appropriate staff, including an executive director. Notwithstanding any other provision of law, the Commission may hire or retain, or both, legal counsel.

(18) The Commission may, subject to applicable federal and state laws, recommend to the appropriate host state authority suitable land and rail transportation routes for low-level radioactive waste carriers.

(19) The Commission may enter into an agreement to import low-level radioactive waste into the region only if both of the following requirements are met–

(a) The Commission approves the importation agreement by a two-thirds vote of the Commission.

(b) The Commission and the host state assess the affected regional disposal facilities' capability to handle imported low-level radioactive wastes and any relevant environmental or economic factors, as defined by the host state's appropriate regulatory authorities.

(20) The Commission may, upon petition, allow an individual generator, a group of generators, or the host state of the compact, to export low-level radioactive wastes to a low-level radioactive waste disposal facility located outside the region. The Commission may approve the petition only by a two-thirds vote of the Commission. The permission to export low-level radioactive wastes shall be effective for that period of time and for the amount of low-level

radioactive waste, and subject to any other term or condition, which may be determined by the Commission.

(21) The Commission may approve, only by a two-thirds vote of the Commission, the exportation outside the region of material, which otherwise meets the criteria of low-level radioactive waste, if the sole purpose of the exportation is to process the material for recycling.

(22) The Commission shall, not later than 10 years before the closure of the initial or subsequent regional disposal facility, prepare a plan for the establishment of the next regional disposal facility.

Article IV–Rights, Responsibilities, and Obligations of Party States

(A) There shall be regional disposal facilities sufficient to dispose of the low-level radioactive waste generated within the region.

(B) Low-level radioactive waste generated within the region shall be disposed of at regional disposal facilities and each party state shall have access to any regional disposal facility without discrimination.

(C)(1) Upon the effective date of this compact, the State of California shall serve as the host state and shall comply with the requirements of subdivision (E) for at least 30 years from the date the regional disposal facility begins to accept low-level radioactive waste for disposal. The extension of the obligation and duration shall be at the option of the State of California.

If the State of California does not extend this obligation, the party state, other than the State of California, which is the largest major generator state shall then serve as the host state for the second regional disposal facility.

The obligation of a host state which hosts the second regional disposal facility shall also run for thirty years from the date the second regional disposal facility begins operations.

Public health and safety.

(2) The host state may close its regional disposal facility when necessary for public health or safety.

(D) The party states of this compact cannot be members of another regional low-level radioactive waste compact entered into pursuant to the Low-Level Radioactive Waste Policy Act, as amended by the Low-Level Radioactive Waste Policy Amendments Act of 1985 (42 USC sections 2021b to 2021j, incl.).

(E) A host state shall do all of the following:

(1) Cause a regional disposal facility to be developed on a timely basis.

Public health and safety.

(2) Ensure by law, consistent with any applicable federal laws, the protection and preservation of public health and safety in the siting, design, development, licensing, regulation, operation, closure, decommissioning, and long-term care of the regional disposal facilities within the state.

(3) Ensure that charges for disposal of low-level radioactive waste at the regional disposal facility are reasonably sufficient to do all of the following:

(a) Ensure the safe disposal of low-level radioactive waste and long-term care of the regional disposal facility.

(b) Pay for the cost of inspection, enforcement, and surveillance activities at the regional disposal facility.

(c) Assure that charges are assessed without discrimination as to the party state of origin.

(4) Submit an annual report to the Commission on the status of the regional disposal facility including projections of the facility's anticipated future capacity.

(5) The host state and the operator shall notify the Commission immediately upon the occurrence of any event which could cause a possible temporary or permanent closure of a regional disposal facility.

(F) Each party state is subject to the following duties and authority:

(1) To the extent authorized by federal law, each party shall develop and enforce procedures requiring low-level radioactive waste shipments originating within its borders and destined for a regional disposal facility to conform to packaging and transportation requirements and regulations. These procedures shall include, but are not limited to, all of the following requirements:

(a) Periodic inspections of packaging and shipping practices.

(b) Periodic inspections of low-level radioactive waste containers while in the custody of transporters.

(c) Appropriate enforcement actions with respects to violations.

(2) A party state may impose a surcharge on the low-level radioactive waste generators within the state to pay for activities required by paragraph (1).

(3) To the extent authorized by federal law, each party state shall, after receiving notification from a host state that a person in a party state has violated packaging, shipping, or transportation requirements or regulations, take appropriate actions to ensure that these violations do not continue. Appropriate actions may include, but are not limited to, requiring that a bond be posted by the violator to pay the cost of repackaging at the regional disposal facility and prohibit future shipments to the regional disposal facility.

(4) Each party state shall maintain a registry of all generators within the state that may have low-level radioactive waste to be disposed of at a regional disposal facility, including, but not limited to, the amount of low-level radioactive waste and the class of low-level radioactive waste generated by each generator.

(5) Each party state shall encourage generators within its borders to minimize the volume of low-level radioactive waste requiring disposal.

(6) Each party state may rely on the good faith performance of the other party states to perform those acts which are required by this compact to provide regional disposal facilities, including the use of the regional disposal facilities in a manner consistent with this compact.

(7) Each party state shall provide the Commission with any data and information necessary for the implementation of the Commission's responsibilities, including taking those actions necessary to obtain this data or information.

(8) Each party state shall agree that only low-level radioactive waste generated within the jurisdiction of the party states shall be disposed of in the regional disposal facility, except as provided in paragraph (19) of subdivision (G) of article III.

(9) Each party state shall agree that if there is any injury to persons or property resulting from the operation of a regional disposal facility, the damages resulting from the injury may be paid from the third-party liability fund pursuant to subparagraph (b) of paragraph

(3) of subdivision (G) of article III, only to the extent that the damages exceed the limits of liability insurance carried by the operator. No party state, by joining this compact, assumes any liability resulting from the siting, operation, maintenance, long-term care, or other activity relating to a regional facility, and no party state shall be liable for any harm or damage resulting from a regional facility not located within the state.

Article V–Approval of Regional Facilities

A regional disposal facility shall be approved by the host state in accordance with its laws. This compact does not confer any authority on the Commission regarding the siting, design, development, licensing, or other regulation, or the operation, closure, decommissioning, or long-term care of, any regional disposal facility within a party state.

Article VI–Prohibited Acts and Penalties

(A) No person shall dispose of low-level radioactive waste within the region unless the disposal is at a regional disposal facility, except as otherwise provided in paragraphs (20) and (21) of subdivision (G) of article III.

(B) No person shall dispose of or manage any low-level radioactive waste within the region unless the low-level radioactive waste was generated within the region, except as provided in paragraphs (19), (20), and (21) of subdivision (G) of article III.

(C) Violations of this section shall be reported to the appropriate law enforcement agency within the party state's jurisdiction.

(D) Violations of this section may result in prohibiting the violator from disposing of low-level radioactive waste in the regional disposal facility, as determined by the Commission or the host state.

Article VII–Eligibility, Entry into Effect, Congressional Consent, Withdrawal, Exclusion

(A) The States of Arizona, North Dakota, South Dakota, and California are eligible to become parties to this compact. Any other state may be made eligible by a majority vote of the Commission and ratification by the legislatures of all of the party states by statute, and upon compliance with those terms and conditions for eligibility which the host state may establish. The host state may establish all terms and conditions for the entry of any state, other than the states named in this subparagraph, as a member of this compact.

(B) Upon compliance with the other provisions of this compact, an eligible state may become a party state by legislative enactment of this compact or by executive order of the governor of the state adopting this compact. A state becoming a party state by executive order shall cease to be a party state upon adjournment of the first general session of its legislature convened after the executive order is issued, unless before the adjournment the legislature enacts this compact.

(C) A party state, other than the host state, may withdraw from the compact by repealing the enactment of this compact, but this withdrawal shall not become effective until two years after the effective date of the repealing legislation. If a party state which is a major generator of low-level radioactive waste voluntarily withdraws from the compact pursuant to this subdivision, that state shall make arrangements for the disposal of the other party states' low-level radioactive waste for a time period equal the period of time it was a member of this compact.

If the host state withdraws from the compact, the withdrawal shall not become effective until five years after the effective date of the repealing legislation.

(D) A party state may be excluded from this compact by a two-thirds vote of the Commission members, acting in a meeting, if the state to be excluded has failed to carry out any obligations required by compact.

Effective date.

(E) This compact shall take effect upon the enactment by statute by the legislatures of the State of California and at least 1 other eligible state and upon the consent of Congress and shall remain in effect until otherwise provided by federal law. This compact is subject to review by Congress and the withdrawal of the consent of Congress every 5 years after its effective date, pursuant to federal law.

Article VIII–Construction and Severability

(A) The provisions of this compact shall be broadly construed to carry out the purposes of the compact, but the sovereign powers of a party state shall not be infringed unnecessarily.

(B) This compact does not affect any judicial proceeding pending on the effective date of this compact.

(C) If any provision of this compact or the application thereof to any person or circumstances is held invalid, that invalidity shall not affect other provision or applications of the compact which can be given effect without the invalid provision or application, and to this end the provisions of this compact are severable.

(D) Nothing in this compact diminishes or otherwise impairs the jurisdiction, authority, or discretion of either of the following:

(1) The Nuclear Regulatory Commission pursuant to the Atomic Energy Act of 1954, as amended (42 USC section 2011 *et seq.*).

(2) An agreement state under section 274 of the Atomic Energy Act of 1954, as amended (42 USC section 2021).

(E) Nothing in this compact confers any new authority on the states or Commission to do any of the following:

(1) Regulate the packaging or transportation of low-level radioactive waste in a manner inconsistent with the regulations of the Nuclear Regulatory Commission or the United States Department of Transportation.

(2) Regulate health, safety, or environmental hazards from source, by-product, or special nuclear material.

(3) Inspect the activities of licensees of the agreement states or of the Nuclear Regulatory Commission.

E. TEXAS LOW-LEVEL RADIOACTIVE WASTE DISPOSAL COMPACT CONSENT ACT

Public Law 105–236   **112 Stat. 1542**

September 20, 1998

An Act

To grant the consent of the Congress to the Texas Low-Level Radioactive Waste Disposal Compact.

Be it enacted by the Senate and House of Representatives of the United States of America in Congress assembled,

42 USC 2021d
note.

Sec. 1. Short Title

This Act may be cited as the "Texas Low-Level Radioactive Waste Disposal Compact Consent Act."

42 USC 2021d
note.

Sec. 2. Congressional Finding

The Congress finds that the compact set forth in section 5 is in furtherance of the Low-Level Radioactive Waste Policy Act (42 USC 2021b et seq.).

42 USC 2021d
note.
Effective date.

Sec. 3. Conditions of Consent to Compact

The Consent of the Congress to the compact set forth in section 5–

(1) shall become effect on the date of the enactment of this Act;

(2) is granted subject to the provisions of the Low-Level Radioactive Waste Policy Act (42 USC 2021b et seq.); and

(3) is granted only for so long as the regional commission established in the compact complies with all of the provisions of such Act.

42 USC 2021d
note.

Sec. 4. Congressional Review

The Congress may alter, amend, or repeal this Act with respect to the compact set forth in section 5 after the expiration of the 10-year period following the date of the enactment of this Act, and at such intervals thereafter as may be provided in such compact.

42 USC 2021d
note.
Maine.
Vermont.

Sec. 5. Texas Low-Level Radioactive Waste Compact

(a) CONSENT OF CONGRESS.–In accordance with section 4(a)(2) of the Low-Level Radioactive Waste Policy Act (42 USC 2021d(a)(2)), the consent of Congress is given to the States of Texas, Maine, and Vermont to enter into the compact set forth in subsection (b).

Article I. Policy and Purpose

SEC. 1.01. The party states recognize a responsibility for each state to seek to manage low-level radioactive waste generated within its boundaries, pursuant to the Low-Level Radioactive Waste Policy Act, as amended by the Low-Level Radioactive Waste Policy Amendments Act of 1985 (42 USC 2021b-2021j). They also recognize that the United States Congress, by enacting the Act, has authorized and encouraged states to enter into compacts for the efficient management and disposal of low-level radioactive water. It is the policy of the party states to cooperate in the protection of the health, safety, and welfare of their citizens and the environment and to provide for and encourage the economical management and disposal of low-level radioactive waste. It is the purpose of this compact to provide the framework for such a cooperative effort; to promote the health, safety, and welfare of the citizens and the environment of the party states; to limit the number of facilities needed to effectively, efficiently, and economically manage low-level radioactive waste and to encourage the reduction of the generation thereof; and to distribute the costs, benefits, and obligations among the party states; all in accordance with the terms of this compact.

Article II. Definitions

SEC. 2.01. As used in this compact, unless the context clearly indicates otherwise, the following definitions apply:

(1) "Act" means the Low-Level Radioactive Waste Policy Act, as amended by the Low-Level Radioactive Waste Policy Amendments Act of 1985 (42 USC 2021b-2021j).

(2) "Commission" means the Texas Low-Level Radioactive Waste Disposal compact Commission established in Article III of this compact.

(3) "Compact facility" or "facility" means any site, location, structure, or property located in and provided by the host state for the purpose of management or disposal of low-level radioactive waste for which the party states are responsible.

(4) "Disposal" means the permanent isolation of low-level radioactive waste pursuant to requirements established by the United States Nuclear Regulatory Commission and the United States Environmental Protection Agency under applicable laws, or by the host state.

(5) "Generate" when used in relation to low-level radioactive waste, means to produce low-level radioactive waste.

(6) "Generator" means a person who produces or processes low-level radioactive waste in the course of its activities, excluding persons who arrange for the collection, transportation, management, treatment, storage, or disposal of waste generated outside the party states, unless approved by the commission.

(7) "Host county" means a county in the host state in which a disposal facility is located or is being developed.

(8) "Host state" means a party state in which a compact facility is located or is being developed. The State of Texas is the host state under this compact.

(9) "Institutional control period" means that period of time following closure of the facility and transfer of the facility license from the operator to the custodial agency in compliance with the appropriate regulations for long-term observation and maintenance.

(10) "Low-Level radioactive waste" has the same meaning as that term is defined in section 2(9) of the Act (42 USC 2021b(9)), or in the host state statute so long as the waste is not incompatible with management and disposal at the compact facility.

(11) "Management" means collection, consolidation, storage, packaging, or treatment.

(12) "Operator" means a person who operates a disposal facility.

(13) "Party state" means any state that has become a party in accordance with Article VII of this compact. Texas, Maine, and Vermont are initial party states under this compact.

(14) "Person" means an individual, corporation, partnership or other legal entity, whether public or private.

(15) "Transporter" means a person who transports low-level radioactive waste.

Article III. The Commission

Establishment.

SEC. 3.01. There is hereby established the Texas Low-Level Radioactive Waste disposal Compact Commission. The commission shall consist of one voting member from each party state except that the host state shall be entitled to six voting members. Commission members shall be appointed by the party state governors, as provided by the laws of each party state. Each party state may provide alternates for each appointed member.

SEC. 3.02. A quorum of the commission consists of a majority of the members. Except as otherwise provided in this compact, an official act of the commission must receive the affirmative vote of a majority of its members.

SEC. 3.03. The commission is a legal entity separate and distinct from the party states and has governmental immunity to the same extent as an entity created under the authority of Article XVI, section 59, of the Texas Constitution. Members of the commission shall not be personally liable for actions taken in their official capacity. The liabilities of the commission shall not be deemed liabilities of the party states.

SEC. 3.04. The Commission shall:

(1) Compensate its members according to the host state's law.

Records.

(2) Conduct its business, hold meetings, and maintain public records pursuant to laws of the host state, except that notice of public meetings shall be given in the non-host party states in accordance with their respective statutes.

(3) Be located in the capital city of the host state.

(4) Meet at least once a year and upon the call of the chair, or any member. The governor of the host state shall appoint a chair and vice-chair.

Records.

(5) Keep an accurate account of all receipts and disbursements. An annual audit of the books of the commission shall be conducted by an independent certified public accountant, and the audit report shall be made a part of the annual report of the commission.

(6) Approve a budget each year and establish a fiscal year that conforms to the fiscal year of the host state.

(7) Prepare, adopt, and implement continency plans for the disposal and management of low-level radioactive waste in the event that the compact facility should be closed. Any plan which requires the host state to store or otherwise manage the low-level radioactive waste from all the party states must be approved by at least four host state members of the commission. The commission, in a contingency plan or otherwise, may not require a non-host party state to store low-level radioactive waste generated outside of the state.

Reports.
Deadlines.

(8) Submit communications to the governors and to the presiding officers of the legislatures of the party states regarding the activities of the commission, including an annual report to be submitted on or before January 31 of each year.

Public information.

(9) Assembly and make available to the party states, and to the public, information concerning low-level radioactive waste management needs, technologies, and problems.

Records.

Deadlines.

(10) Keep a current inventory of all generators within the party states, based upon information provided by the party states.

(11) By no later than 180 days after all members of the commission are appointed under section 3.01 of this article, establish by rule the total volume of low-level radioactive waste that the host state will dispose of in the compact facility in the years 1995-2045, including decommissioning waste. The shipments of low-level radioactive waste from all non-host party states shall not exceed 20 percent of the volume estimated to be disposed of by the host state during the 50-year period. When averaged over such 50-year period, the total of all shipments from non-host party states shall not exceed 20,000 cubic feet a year. The commission shall coordinate the volumes, timing, and frequency of shipments from generators in the non-host party states in order to assure that over the life of this agreement shipments from the non-host party states do not exceed 20 percent of the volume projected by the commission under this paragraph.

SEC 3.05. The commission may:

(1) Employ staff necessary to carry out its duties and functions. The commission is authorized to use to the extent practicable the services of existing employees of the party states. Compensation shall be as determined by the commission.

(2) Accept any grants, equipment, supplies, materials, or services, conditional or otherwise, from the federal or state government. The nature, amount and condition , if any, of any donation, grant or other resources accepted pursuant to this paragraph and the identity of the donor or grantor shall be detailed in the annual report of the commission.

(3) Enter into contracts to carry out its duties and authority, subject to projected resources. No contract made by the commission shall bind a party state.

(4) Adopt, by a majority vote, bylaws and rules necessary to carry out the terms of this compact. Any rules promulgated by the commission shall be adopted in accordance with the Administrative Procedure and Texas Register Act (Article 6252-13a, Vernon's Texas Civil Statutes).

(5) Sue and be sued and, when authorized by a majority vote of the members, seek to intervene in administrative or judicial proceedings related to this compact.

(6) Enter into an agreement with any person, state, regional body, or group of states for the importation of low-level radioactive waste into the compact for management or disposal, provided that the agreement receives a majority vote of the commission. The commission may adopt such conditions and restrictions in the agreement as it deems advisable.

(7) Upon petition, allow an individual generator, a group of generators, or the host state of the compact, to export low-level waste to a low-level radioactive waste disposal facility located outside the party states. The commission may approve the petition only by a majority vote of its members. The permission to export low-level radioactive waste shall be effective for that period of time and for the specified amount of low-level radioactive waste, and subject to any other term or condition, as is determined by the commission.

(8) Monitor the exportation outside of the party states of material, which otherwise meets the criteria of low-level radioactive waste,

where the sole purpose of the exportation is to manage or process the material for recycling or waste reduction and return it to the party states for disposal in the compact facility.

SEC. 3.06. Jurisdiction and venue of any action contesting any action of the commission shall be in the United States District Court in the district where the commission maintains its office.

Article IV. Rights, Responsibilities, and Obligations of Party

SEC. 4.01. The host state shall develop and have full administrative control over the development, management and operation of a facility for the disposal of low-level radioactive waste generated within the party states. The host state shall be entitled to unlimited use of the facility over its operating life. Use of the facility by the non-host party states for disposal of low-level radioactive waste, including such waste resulting from decommissioning of any nuclear electric generation facilities located in the party states, is limited to the volume requirements of section 3.04(11) of Article III.

SEC. 4.02. Low-level radioactive waste generated within the party states shall be disposed of only at the compact facility, except as provided in section 3.05(7) of Article III.

SEC. 4.03. The initial states of this compact cannot be members of another low-level radioactive waste compact entered into pursuant to the Act.

SEC. 4.04. The host state shall do the following

(1) Cause a facility to be developed in a timely manner and operated and maintained through the institutional control period.

(2) Ensure, consistent with any applicable federal and host state laws, the protection and preservation of the environment and the public health and safety in the siting, design, development, licensing, regulation, operation, closure, decommissioning, and long-term care of the disposal facilities within the host state.

Notification.

(3) Close the facility when reasonably necessary to protect the public health and safety of its citizens or to protect its natural resources from harm. However, the host state shall notify the commission of the closure within three days of its action and shall, within 30 working days of its action, provide a written explanation to the commission of the closure, and implement any adopted contingency plan.

(4) Establish reasonable fees for disposal at the facility of low-level radioactive waste generated in the party states based on disposal fee criteria set out in sections 402,272 and 402,273, Texas Health and Safety Code. The same fees shall be charged for the disposal of low-level radioactive waste that was generated in the host state and in the non-host party states. Fees shall also be sufficient to reasonably support the activities of the Commission.

Reports.

(5) Submit an annual report to the commission on the status of the facility, including projections of the facility's anticipated future capacity, and on the related funds.

Notification.

(6) Notify the Commission immediately upon the occurrence of any event which could cause a possible temporary or permanent closure of the facility and identify all reasonable options for the disposal of low-level radioactive waste at alternate compact facilities or, by arrangement and commission vote, at non-compact facilities.

(7) Promptly notify the other party states of any legal action involving the facility.

(8) Identify and regulate, in accordance with federal and host state law, the means and routes of transportation of low-level radioactive waste in the host state.

SEC. 4.05. Each party state shall do the following:

(1) Develop and enforce procedures requiring low-level radioactive waste shipments originating within its borders and destined for the facility to conform to packaging, processing, and waste form specifications of the host state.

(2) Maintain a registry of all generators within the state that may have low-level radioactive waste to be disposed of at a facility, including, but not limited to, the amount of low-level radioactive waste and the class of low-level radioactive waste generated by each generator.

(3) Develop and enforce procedures requiring generators within its borders to minimize the volume of low-level radioactive waste requiring disposal. Nothing in this compact shall prohibit the storage, treatment, or management of waste by a generator.

(4) Provide the commission with any data and information necessary for the implementation of the commission's responsibilities, including taking those actions necessary to obtain this data or information.

(5) Pay for community assistance projects designated by the host county in an amount for each non-host party state equal to 10 percent of the payment provided for in Article V for each such state. One-half of the payment shall be due and payable to the host county on the first day of the month following ratification of this compact agreement by Congress and one-half of the payment shall be due and payable on the first day of the month following the approval of a facility operating license by the host state's regulatory body.

(6) Provide financial support for the commission's activities prior to the date of facility operation and subsequent to the date of congressional ratification of this compact under section 7.07 of Article VII. Each party state will be responsible for annual payments equaling its pro-rata share of the commission's expenses, incurred for administrative, legal, and other purposes of the commission.

(7) If agreed by all parties to a dispute, submit the dispute to arbitration or other alternate dispute resolution process. If arbitration is agreed upon, the governor of each party state shall appoint an arbitrator. If the number of party states is an even number, the arbitrators so chosen shall appoint an additional arbitrator. The determination of a majority of the arbitrators shall be binding on the party states. Arbitration proceedings shall be conducted in accordance with the provisions of 9 USC sections 1 to 16. If all parties to a dispute do not agree to arbitration or alternate dispute resolution process, the United States District Court in the district where the commission maintains its office shall have original jurisdiction over any action between or among parties to this compact.

(8) Provide on a regular basis to the commission and host state–

(A) an accounting of waste shipped and proposed be shipped to the compact facility, by volume and curies;

(B) proposed transportation methods and routes; and

(C) proposed shipment schedules.

(9) Seek to join in any legal action by or against the host state to prevent nonparty states or generators from disposing of low-level radioactive waste at the facility.

SEC. 4.06. Each party state shall act in good faith and may rely on the good faith performance of the other party states regarding requirements of this compact.

Article V. Party State Contributions

SEC. 5.01. Each party state, except the host state, shall contribute a total of $25 million to the host state. Payments shall be deposited in the host state treasure to the credit of the low-level waste fund in the following manner except as otherwise provided. Not later than the 60th day after the date of congressional ratification of this compact, each state shall pay to the host state $12.5 million. Not later than the 60th day after the date of the opening of the compact facility, each non-host party state shall pay to the host state an additional $12.5 million.

Deadline.

SEC. 5.02. As an alternative, the host state and the non-host states may provide for payments in the same total amount as stated above to be made to meet the principal and interest expense associated with the bond indebtedness or other form of indebtedness issued by the appropriate agency of the host state for purposes associated with the development, operation, and post-closure monitoring of the compact facility. In the event the member states proceed in this manner, the payment schedule shall be determined in accordance with the schedule of debt repayment. This schedule shall replace the payment schedule described in section 5.01 of this article.

Article VI. Prohibited Acts and Penalties

SEC. 6.01. No person shall dispose of low-level radioactive waste generated within the party states unless the disposal is at the compact facility, except as otherwise provided in section 3.05(7) of Article III..

SEC. 6.02. No person shall manage or dispose of any low-level radioactive waste within the party states unless the low-level radioactive waste was generated within the party states, except as provided in section 3.05(6) of Article III. Nothing herein shall be construed to prohibit the storage or management of low-level radioactive waste by a generator, nor its disposal pursuant to 10 CFR Part 20.302.

SEC. 6.03. Violations of this article may result in prohibiting the violator from disposing of low-level radioactive waste in the compact facility, or in the imposition of penalty surcharges on shipments to the facility, as determined by the commission.

Article VII. Eligibility, Entry into Effect; Congressional Consent; Withdrawal; Exclusion

SEC. 7.01. The states of Texas, Maine, and Vermont are party states to this compact. Any other state may be made eligible for party status by a majority vote of the commission and ratification by the legislature of the host state, subject to fulfillment of the rights of the initial non-host party states under section 3.04(11) of Article III and section 4.01 of Article IV, and upon compliance with those terms and conditions for eligibility that the host state may establish. The host state may establish all terms and conditions for the entry of any state, other than the states names in this section, as a member of this compact; provided, however, the specific provisions of this compact, except for those pertaining to the composition of the commission and those pertaining to section 7.09 of this article, may not be changed except upon ratification by the legislatures of the party states.

SEC. 7.02. Upon compliance with the other provisions of this compact, a state made eligible under section 7.01 of this article may become a party state by legislative enactment of this compact or by executive order of the governor of the state adopting this compact. A state becoming a party state by executive order shall cease to be a party state upon adjournment of the first general session of its legislature convened after the executive order is issued, unless before the adjournment, the legislature enacts this compact.

SEC. 7.03. Any party state may withdraw from this compact by repealing enactment of this compact subject to the provisions herein. In the event the host state allows an additional state or additional states to join the compact, the host state's legislature, without the consent of the non-host party states, shall have the right to modify the composition of the commission so that the host state shall have a voting majority on the commission, provided, however, that any modification maintains the right of each initial party state to retain one voting member on the commission.

Effective Date. SEC. 7.04. If the host state withdraws from the compact, the withdrawal shall not become effective until five years after enactment of the repealing legislation and the non-host party states may continue to use the facility during that time. The financial obligation of the non-host party states under Article V shall cease immediately upon enactment of the repealing legislation. If the host state withdraws from the compact or abandons plans to operate a facility prior to the date of any non-host party state payment under sections 4.05(5) and (6) of Article IV or Article V, the non-host party states are relieved of any obligations to make the contributions. This section sets out the exclusive remedies for the non-host party states if the host state withdraws from the compact or is unable to develop and operate a compact facility.

Effective date. SEC. 7.05. A party state, other than the host state, may withdraw from the compact by repealing the enactment of this compact, but this withdrawal shall not become effective until two years after the effective date of the repealing legislation. During this two-year period the party state will continue to have access to the facility. The withdrawing party shall remain liable for any payments under sections 4.05(5) and (6) of Article IV that were due during the two-year period, and shall not be entitled to any refund of payments previously made.

SEC. 7.06. Any party state that substantially fails to comply with the terms of the compact or to fulfill its obligations hereunder may have its membership in the compact revoked by a seven-eighths vote of the commission following notice that a hearing will be scheduled not less than six months from the date of the notice. In all other respects, revocation proceedings undertaken by the commission will be subject to the Administrative Procedure and Texas Register Act (Article 6252-13a, Vernon's Texas Civil Statutes), except that a party state may appeal the Effective date. commission's revocation decision to the United States District Court in accordance with section 3.06 of Article III. Revocation shall take effect one year from date such party state receives written notice from the commission of a final action.

Written notice of revocation shall be transmitted immediately following the vote of the commission, by the chair, to the governor of the affected party state, all other governors of party states, and to the United States Congress.

Effective date.

SEC. 7.07. This compact shall take effect following its enactment under the laws of the host state and any other party state and thereafter upon the consent of the United States Congress and shall remain in effect until otherwise provided by federal law. If Texas and either Maine or Vermont ratify this compact, the compact shall be in full force and effect as to Texas and the other ratifying state, and this compact shall be interpreted as follows:

(1) Texas and the other ratifying state are the initial party states.

(2) The commission shall consist of two voting members from the other ratifying state and six from Texas.

(3) Each party state is responsible for its pro-rata share of the commission's expenses.

SEC. 7.08. This compact is subject to review by the United States Congress and the withdrawal of the consent of Congress every five years after its date, pursuant to federal law.

SEC. 7.09. The host state legislature, with the approval of the governor, shall have the right and authority, without the consent of the non-host party states, to modify the provisions contained in section 3.04(11) of Article III to comply with section 402.219(c)(1), Texas Health & Safety Code, as long as the modification does not impair the rights of the initial non-hose party states.

Article VIII. Construction and Severability

SEC. 8.01. The provisions of this compact shall be broadly construed to carry out the purposes of the compact, but the sovereign powers of a party shall not be infringed upon unnecessarily.

SEC. 8.02. This compact does not affect any judicial proceeding pending on the effective date of this compact.

SEC. 8.03. No party state acquires any liability, by joining this compact, resulting from the siting, operation, maintenance, long-term care or any other activity relating to the compact facility. No non-host party state shall be liable for any harm or damage from the siting, operation, maintenance, or long-term care relating to the compact facility. Except as otherwise expressly provided in this compact, nothing in this compact shall be construed to alter the incidence of liability of any kind for any act or failure to act. Generators, transporters, owners and operators of the facility shall be liable for their acts, omissions, conduct or relationships in accordance with applicable law. By entering. into this compact and securing the ratification by Congress of its terms, no party state acquires a potential liability under section 5(d)(2)(C) of the Act (42 USC section 202le(d)(2)(C)) that did not exist prior to entering into this compact.

SEC. 8.04. If a party state withdraws from the compact pursuant to section 7.03 of Article VII or has its membership in this compact revoked pursuant to section 7.06 of Article VII, the withdrawal or revocation shall not affect any liability already incurred by or chargeable to the affected state under section 8.03 of this article.

SEC. 8.05. The provisions of this compact shall be severable and if any phrase, clause, sentence, or provision of this compact is declared by a court of competent jurisdiction to be contrary to the constitution of any participating state or of the United States or the applicability thereof to any government, agency, person or circumstances is held invalid, the validity of the remainder of this compact and the applicability thereof to any government, agency, person, or circumstance shall not be affected thereby to the extent the remainder can in all fairness be given effect. If any provision of this compact shall be held contrary to the constitution of any state participating therein, the compact shall remain in full force and effect as to the state affected as to all severable matters.

SEC. 8.06. Nothing in this compact diminishes or otherwise impairs the jurisdiction, authority, or discretion of either of the following:

(1) The United States Nuclear Regulatory Commission pursuant to the Atomic Energy Act of 1954, as amended (42 USC section 2011 *et. seq.*).

(2) An agreement state under section 274 of the Atomic Energy Act of 1954, as amended (42 USC 2021).

SEC. 8.07. Nothing in this compact confers any new authority on the states or commission to do any of the following:

(1) Regulate the packaging or transportation of low-level radioactive waste in a manner inconsistent with the regulations of the United States Nuclear Regulatory Commission or the United States Department of Transportation.

(2) Regulate health, safety, or environmental hazards from source, by-product, or special nuclear material.

(3) Inspect the activities of licensees of the agreement states or of the United States Nuclear Regulatory Commission.

6. High-Level Radioactive Waste

6. High-Level Radioactive Waste
Contents

A. NUCLEAR WASTE POLICY ACT OF 1982, AS AMENDED

Public Law 97–425   **96 Stat. 2201**

January 7, 1983

Sec. 1. Short Title and Table of Contents

This Act may be cited as the "Nuclear Waste Policy Act of 1982." (TOC not duplicated here.)

42 USC 10101.   **Sec. 2. Definitions**

For purposes of this Act:

(1) The term "Administrator" means the Administrator of the Environmental Protection Agency.

(2) The term "affected Indian tribe" means any Indian tribe–

(A) within whose reservation boundaries a monitored retrievable storage facility, test and evaluation facility, or a repository for high-level radioactive waste or spent fuel is proposed to be located;

(B) whose federally defined possessory or usage rights to other lands outside of the reservation's boundaries arising out of congressionally ratified treaties may be substantially and adversely affected by the locating of such a facility: Provided, That the Secretary of the Interior finds, upon the petition of the appropriate governmental officials of the tribe, that such effects are both substantial and adverse to the tribe;

(3) the term "atomic energy defense activity" means any activity of the Secretary performed in whole or in part in carrying out any of the following functions:

(A) naval reactors development;

(B) weapons activities including defense inertial confinement fusion;

(C) verification and control technology;

(D) defense nuclear materials production;

(E) defense nuclear waste and materials by-products management;

(F) defense nuclear materials security and safeguards and security investigations; and

(G) defense research and development.

(4) The term "candidate site" means an area, within a geologic and hydrologic system, that is recommended by the Secretary under section 112 for site characterization, approved by the President under section 112 for site characterization, or undergoing site characterization under section 113.

(5) The term "civilian nuclear activity" means any atomic energy activity other than an atomic energy defense activity.

(6) The term "civilian nuclear power reactor" means a civilian nuclear power plant required to be licensed under section 103 or 104b. of the Atomic Energy Act of 1954 (42 USC 2133, 2134(b)).

(7) The term "Commission" means the Nuclear Regulatory Commission.

(8) The term "Department" means the Department of Energy.

(9) The term "disposal" means the emplacement in a repository of high-level radioactive waste, spent nuclear fuel, or other highly radioactive material with no foreseeable intent of recovery, whether or not such emplacement permits the recovery of such waste.

(10) The terms "disposal package" and "package" mean the primary container that holds, and is in contact with, solidified high-level radioactive waste, spent nuclear fuel, or other radioactive materials, and any overpacks that are emplaced at a repository.

(11) The term "engineered barriers" means manmade components of a disposal system designed to prevent the release of radionuclides into the geologic medium involved. Such term includes the high-level radioactive waste form, high-level radioactive waste canisters, and other materials placed over and around such canisters.

(12) The term "high-level radioactive waste" means–

(A) the highly radioactive material resulting from the reprocessing of spent nuclear fuel, including liquid waste produced directly in reprocessing and any solid material derived from such liquid waste that contains fission products in sufficient concentrations; and

(B) other highly radioactive material that the Commission, consistent with existing law, determines by rule requires permanent isolation.

(13) The term "Federal agency" means any Executive agency, as defined in section 105 of title 5, United States Code.

(14) The term "Governor" means the chief executive officer of a State.

(15) The term "Indian tribe" means any Indian tribe, band, nation, or other organized group or community of Indians recognized as eligible for the services provided to Indians by the Secretary of the Interior because of their status as Indians, including any Alaska Native village, as defined in section 3(c) of the Alaska Native Claims Settlement Act (43 USC 1602(c)).

(16) The term "low-level radioactive waste" means radioactive material that–

(A) is not high-level radioactive waste, spent nuclear fuel, transuranic waste, or by-product material as defined in section 11e(2) of the Atomic Energy Act of 1954 (42 USC 2014(e)(2)); and

(B) the Commission, consistent with existing law, classifies as low level radioactive waste.

(17) The term "Office" means the Office of Civilian Radioactive Waste Management established in section 305.

(18) The term "repository" means any system licensed by the Commission that is intended to be used for, or may be used for, the permanent deep geologic disposal of high-level radioactive waste and spent nuclear fuel, whether or not, such system is designed to permit the recovery, for a limited period during initial operation, of any materials placed in such system. Such term includes both surface and subsurface areas at which high-level radioactive waste and spent nuclear fuel handling activities are conducted.

(19) The term "reservation" means–

(A) any Indian reservation or dependent Indian community referred to in clause 9a) or (b) of section 1151 of title 18, United States Code; or

(B) any land selected by an Alaska Native village or regional corporation under the provisions of the Alaska Native Claims Settlement Act (43 USC 1601 *et seq*.).

(20) The term "Secretary" means the Secretary of Energy.

(21) The term "site characterization" means–

(A) siting research activities with respect to a test and evaluation facility at a candidate site; and

(B) activities, whether in the laboratory or in the field, undertaken to establish the geologic condition and the ranges of the parameters of a candidate site relevant to the location of a repository, including borings, surface excavations, excavations of exploratory shafts, limited subsurface lateral excavations and borings, and in situ testing needed to evaluate the suitability of a candidate site for the location of a repository, but not including preliminary borings and geophysical testing needed to assess whether site characterization should be undertaken.

(22) The term "siting research" means activities, including borings, surface excavations, shaft excavations, subsurface lateral excavations and borings, and in situ testing, to determine the suitability of a site for a test and evaluation facility.

(23) The term "spent nuclear fuel" means fuel that has been withdrawn from a nuclear reactor following irradiation, the constituent elements of which have not been separated by reprocessing.

(24) The term "State" means each of the several States, the District of Columbia, the Commonwealth of Puerto Rico, the Virgin Islands, Guam, American Samoa, the Northern Mariana Islands, the Trust Territory of the Pacific Islands, and any other territory or possession of the United States.

(25) The term "storage" means retention of high-level radioactive waste, spent nuclear fuel, or transuranic waste with the intent to recover such waste or fuel for subsequent use, processing, or disposal.

(26) The term "Storage Fund" means the Interim Storage Fund established in section 137(c).

(27) The term "test and evaluation facility" means an at-depth, prototypic, underground cavity with subsurface lateral excavations extending from a central shaft that is used for research and development purposes, including the development of data and experience for the safe handling and disposal of solidified high-level radioactive waste, transuranic waste, or spent nuclear fuel.

(28) The term "unit of general local government" means any borough, city, county, parish, town, township, village, or other general purpose political subdivision of a State.

(29) The term "Waste Fund" means the Nuclear Waste Fund established in section 302(c).

(30) The term "Yucca Mountain site" means the candidate site in the State of Nevada recommended by the Secretary to the President under section 112(b)(1)(B) on May 27, 1986.

(31) The term "affected unit of local government" means the unit of local government with jurisdiction over the site of a repository or a monitored retrievable storage facility. Such term may, at the discretion of the Secretary, include units of local government that are contiguous with such unit.

(32) The term "Negotiator" means the Nuclear Waste Negotiator.

(33) As used in title IV, the term "Office" means the Office of the Nuclear Waste Negotiator established under title IV of this Act.

(34) The term "monitored retrievable storage facility" means the storage facility described in section 141(b)(1).[1]

42 USC 10102.
Sec. 3. Separability
If any provision of this Act, or the application of such provision to any person or circumstance, is held invalid, the remainder of this Act, or the application of such provisions to persons or circumstances other than those as to which it is held invalid, shall not be affected thereby.

42 USC 10103.
Sec. 4. Territories and Possessions
Nothing in this Act shall be deemed to repeal, modify, or amend the provisions of section 605 of the Act of March 12, 1980.

42 USC 10104.
Sec. 5. Ocean Disposal
Nothing in this Act shall be deemed to affect the Marine Protection, Research, and Sanctuaries Act of 1972.

42 USC 10105.
Sec. 6. Limitation on Spending Authority
The authority under this Act to incur indebtedness, or enter into contracts, obligating amounts to be expended by the Federal Government shall be effective for any fiscal year only to such extent or in such amounts as are provided in advance by appropriation Acts.

42 USC 10106.
Sec. 7. Protection of Classified National Security Information
Nothing in this Act shall require the release or disclosure to any person or to the Commission of any classified national security information.

42 USC 10107.
Sec. 8. Applicability
(a) ATOMIC ENERGY DEFENSE ACTIVITIES–Subject to the provisions of subsection (c), the provisions of this Act shall not apply with respect to any atomic energy defense activity or to any facility used in conjunction with any such activity.

(b) EVALUATION BY PRESIDENT–(1) Not later than 2 years after the date of the enactment of this Act, the President shall evaluate the use of disposal capacity at one or more repositories to be developed under subtitle A of title I for the disposal of high-level radioactive waste resulting from atomic energy defense activities. Such evaluation shall take into consideration factors relating to cost efficiency, health and safety, regulation, transportation, public acceptability, and national security.

(2) Unless the President finds, after conducting the evaluation required in paragraph (1), that the development of a repository for the disposal of high-level radioactive waste resulting from atomic energy defense activities only is required, taking into account all of the factors describe in such subsection, the Secretary shall proceed promptly with arrangement for the use of one or more of the repositories to be developed under subtitle A of title I for the disposal of such waste. Such arrangements shall include the allocation of costs of developing, constructing, and operating this repository of repositories. The costs resulting from permanent disposal of high-level radioactive waste from atomic energy defense activities shall be paid by the Federal Government, into the especial account established under section 302.

(3) Any repository for the disposal of high-level radioactive waste resulting from atomic energy defense activities only shall (A) be subject to licensing under section 202 of the Energy Reorganization

---

[1]   Subsections (30) through (34) were added by P.L. 100–203, § 5002, 101 Stat. 1330 (1987).

Act of 1973 (42 USC 5842); and (B) comply with all requirements of the Commission for the siting, development, construction, and operation of a repository.

(c) APPLICABILITY TO CERTAIN REPOSITORIES–The provisions of this Act shall apply with respect to any repository not used exclusively for the disposal of high-level radioactive waste or spent nuclear fuel resulting from atomic energy defense activities, research and development activities of the Secretary, or both.

42 USC 10108.   **Sec. 9. Applicability**

TRANSPORTATION–NOTHING in this Act shall be construed to affect Federal, State, or local laws pertaining to the transportation of spent nuclear fuel or high-level radioactive waste.

Title I–Disposal and Storage of High-Level Radioactive Waste, Spent Nuclear Fuel, and Low-Level Radioactive Waste

42 USC 10121.   **Sec. 101. State and Affected Indian Tribe Participation in Development of Proposed Repositories for Defense Waste**

(a) NOTIFICATION TO STATES AND AFFECTED INDIAN TRIBES–Notwithstanding the provisions of section 8, upon any decision by the Secretary or the President to develop a repository for the disposal of high-level radioactive waste or spent nuclear fuel resulting exclusively from atomic energy defense activities, research and development activities of the Secretary, or both, and before proceeding with any site-specific investigations with respect to such repository, the Secretary shall notify the Governor and legislature of the State in which such repository is proposed to be located, or the governing body of the affected Indian tribe on whose reservation such repository is proposed to be located, as the case may be, of such decision.

(b) PARTICIPATION OF STATES AND AFFECTED INDIAN TRIBES–Following the receipt of any notification under subsection (a), the State or Indian tribe involved shall be entitled, with respect to the proposed repository involved, to rights of participation and consultation identical to those provided in sections 115 through 118, except that any financial assistance authorized to be provided to such State or affected Indian tribe under section 116(c) or 118(b) shall be made from amounts appropriated to the Secretary for purposes of carrying out this section.

Subtitle A–Repositories for Disposal of High-Level Radioactive Waste and Spent Nuclear Fuel

42 USC 10131.   **Sec. 111. Findings and Purposes**

(a) FINDINGS–THE Congress finds that–

(1) radioactive waste creates potential risks and requires safe and environmentally acceptable methods of disposal;

(2) a national problem has been created by the accumulation of (A) spent nuclear fuel from nuclear reactors; and (B) radioactive waste from (i) reprocessing of spent nuclear fuel; (ii) activities related to medical research, diagnosis, and treatment; and (iii) other sources;

(3) Federal efforts during the past 30 years to devise a permanent solution to the problems of civilian radioactive waste disposal have not been adequate;

(4) while the Federal Government has the responsibility to provide for the permanent disposal of high-level radioactive waste and such spent nuclear fuel as may be disposed of in order to protect the public health and safety and the environment, the costs of such

disposal should be the responsibility of the generators and owners of such waste and spent fuel;

(5) the generators and owners of high-level radioactive waste and spent nuclear fuel have the primary responsibility to provide for, and the responsibility to pay the costs of, the interim storage of such waste and spent fuel until such waste and spent fuel is accepted by the Secretary of Energy in accordance with the provisions of this Act;

(6) State and public participation in the planning and development of repositories is essential in order to promote public confidence in the safety of disposal of such waste and spent fuel; and

(7) high-level radioactive waste and spent nuclear fuel have become major subjects of public concern, and appropriate precautions may be taken to ensure that such waste and spent fuel do not adversely affect the public health and safety and the environment for this or future generations.

(b) PURPOSES–The purposes of this subtitle are–

(1) to establish a schedule for the siting, construction, and operation of repositories that will provide a reasonable assurance that the public and the environment will be adequately protected from the hazards posed by high-level radioactive waste and such spent nuclear fuel as may be disposed of in a repository;

(2) to establish the Federal responsibility, and a definite Federal policy, for the disposal of such waste and spent fuel;

(3) to define the relationship between the Federal Government and the State government with respect to the disposal of such waste and spent fuel; and

(4) to establish a Nuclear Waste Fund, composed of payments made by the generators and owners of such waste and spent fuel, that will ensure that the costs of carrying out activities relating to the disposal of such waste and spent fuel will be borne by the persons responsible for generating such waste and spent fuel.

42 USC 10132.

Sec. 112. Recommendation of Candidate Sites for Site Characterization

(a) GUIDELINES–Not later than 180 days after the date of the enactment of this Act, the Secretary, following consultation with the Council on Environmental Quality, the Administrator of the Environmental Protection Agency, the Director of the United States Geological Survey, and interested Governors, and the concurrence of the Commission shall issue general guidelines for the recommendation of sites for repositories. Such guidelines shall specify detailed geologic considerations that shall be primary criteria for the selection of sites in various geologic media. Such guidelines shall specify factors that qualify or disqualify any site from development as a repository, including factors pertaining to the location of valuable natural resources, hydrology, geophysics, seismic activity, and atomic energy defense activities, proximity to water supplies, proximity to populations, the effect upon the rights of users of water, and proximity to components of the National Park System, the National Wildlife Refuge System, the National Wild and Scenic Rivers System, the National Wilderness Preservation System, or National Forest Lands. Such guidelines shall take into consideration the proximity to sites where high-level radioactive waste and spent nuclear fuel is generated or temporarily stored and the transportation and safety factors involved in moving such waste to a repository. Such guidelines shall specify population factors that will disqualify any site from development as a repository if any surface facility of such repository would be located (1) in a highly populated area; or (2) adjacent to an area 1 mile by 1 mile having a population of not less than 1,000 individuals. Such guidelines also shall require the Secretary to consider the cost and impact of transporting to the repository site the solidified high-level radioactive waste and spent fuel to be disposed of in the repository and the advantages of regional distribution in the siting of repositories. Such guidelines shall require the Secretary to consider the various geologic media in which sites for repositories may be located and, to the extent practicable, to recommend sites in different geologic media. The Secretary shall use guidelines established under this subsection in considering candidate sites for recommendation under subsection (b). The Secretary may revise such guidelines from time to time, consistent with the provisions of this subsection.

(b)   RECOMMENDATION   BY   SECRETARY   TO   THE PRESIDENT–(1)(A) Following the issuance of guidelines under subsection (a) and consultation with the Governors of affected States, the Secretary shall nominate at least 5 sites that he determines suitable for site characterization for selection of the first repository site.

Recommendation date.

(B) Subsequent to such nomination, the Secretary shall recommend to the President 3 of the nominated sites not later than January 1, 1985 for characterization as candidate sites.

(C) Such recommendations under subparagraph (B) shall be consistent with the provisions of section 305.

Environmental assessment.

(D) Each nomination of a site under this subsection shall be accompanied by an environmental assessment, which shall include a detail statement of the basis for such recommendation and of the probable impacts of the site characterization activities planned for such site, and a discussion of alternative activities relating to site characterization that may be undertaken to avoid such impacts. Such environmental assessment shall include–

(i) an evaluation by the Secretary as to whether such site is suitable for site characterization under the guidelines established under subsection (a);

(ii) an evaluation by the Secretary as to whether such site is suitable for development as a repository under each such guideline that does not require site characterization as a prerequisite for application of such guidelines;

(iii) an evaluation by the Secretary of the effects of the site characterization activities at such site on the public health and safety and the environment;

(iv) a reasonable comparative evaluation by the Secretary of such site with other sites and locations that have been considered:

(v) a description of the decision process by which such site was recommended; and

(vi) an assessment of the regional and local impacts of locating the proposed repository at such site.

(E)(i) The issuance of any environmental assessment under this paragraph shall be considered to be a final agency action subject to judicial review in accordance with the provisions of chapter 7 of title 5, United States Code, and section 119. Such judicial review shall be limited to the sufficiency of such environmental assessment with respect to the items described in clauses (i) through (vi) of subparagraph (D).

(F) Each environmental assessment prepared under this paragraph shall be made available to the public.

(G) Before nominating a site, the Secretary shall notify the Governor and legislature of the State in which such site is located, or the governing body of the affected Indian tribe where such site is located, as the case may be, of such nomination and the basis for such nomination.

(2) Before nominating any site the Secretary shall hold public hearings in the vicinity of such site to inform the residents of the area in which such site is located of the proposed nomination of such site and to receive their comments. At such hearings, the Secretary shall also solicit and receive any recommendations of such residents with respect to issues that should be addressed in the environmental assessment described in paragraph (1) and the site characterization plan described in section 113(b)(l).

(3) In evaluating the sites nominated under this section prior to any decision to recommend a site as a candidate site, the Secretary shall use available geophysical, geologic, geochemical and hydrologic, and other information and shall not conduct any preliminary borings or excavations at a site unless (i) such preliminary boring or excavation activities were in progress upon the date of enactment of this Act or (ii) the Secretary certifies that such available information from other sources, in the absence of preliminary borings or excavations, will not be adequate to satisfy applicable requirements of this Act or any other law: *Provided,* That preliminary borings or excavations under this section shall not exceed a diameter of 6 inches.

(c)   PRESIDENTIAL   REVIEW   OF   RECOMMENDED CANDIDATE SITES–

Decision transmittal or notification.

(1) The President shall review each candidate site recommendation made by the Secretary under subsection (b). Not later than 60 days after the submission by the Secretary of a recommendation of a candidate site, the President, in his discretion, may either approve or disapprove such candidate site, and shall transmit any such decision to the Secretary and to either the Governor and legislature of the State in which such candidate site is located, or the governing body of the affected Indian tribe where such candidate site is located, as the case may be. If, during such 60-day period, the President fails to approve or disapprove such candidate site, or fails to invoke his authority under paragraph (2) to delay his decision, such candidate site shall be considered to be approved, and the Secretary shall notify such Governor and legislature, or governing body of the affected Indian tribe, of the approval of such candidate site by reason of the inaction of the President.

(2) The President may delay for not more than 6 months his decision under paragraph (1) to approve or disapprove a candidate site, upon determining that the information provided with the recommendation of the Secretary is insufficient to permit a decision within the 60-day period referred to in paragraph (1). The President may invoke his authority under this paragraph by submitting written notice to the Congress, within such 60-day period of his intent to invoke such authority. If the President invokes such authority, but fails to approve or disapprove the candidate site involved by the end of such 6-month period, such candidate site shall be considered to be approved, and the Secretary shall notify such Governor and legislature, or governing body of the affected Indian tribe of the approval of such candidate site by reason of the inaction of the President.

(d) PRELIMINARY ACTIVITIES–Except as otherwise provided in this section , each activity of the President or the Secretary under this section shall be considered to be a preliminary decision making activity. No such activity shall require the preparation of an environmental impact statement under section 102(2)(C) of the National Environmental Policy Act of 1969 (42 USC 4332(2)(C)), or to require any environmental review under subparagraph (E) or (F) of section 102(2) of such Act.[2]

42 USC 10133.

Sec. 113. Site Characterization

(a) IN GENERAL–The Secretary shall carry out, in accordance with the provisions of this section, appropriate site characterization activities at the Yucca Mountain site. The Secretary shall consider fully the comments received under subsection (b)(2) and section 112(b)(2) and shall, to the maximum extent practicable and in consultation with the Governor of the State of Nevada conduct site characterization activities in a manner that minimizes any significant adverse environmental impacts identified in such comments or in the environmental assessment submitted under subsection (b)(1).

Plan submittal, review and comment.

(b) COMMISSION AND STATES–(1) Before proceeding to sink shafts at the Yucca Mountain site, the Secretary shall submit for such candidate site to the Commission and to the Governor or legislature of the State of Nevada for their review and comment–

(A) a general plan for site characterization activities to be conducted at such candidate site, which plan shall include–

(i) a description of such candidate site;

(ii) a description of such site characterization activities, including the following: the extent of planned excavations, plans for any onsite testing with radioactive or nonradioactive material, plan for any investigation activities that may affect the capabilities of such candidate site to isolate high-level radioactive waste and spent nuclear fuel, and plans to control any adverse, safety-related impacts from such site characterization activities;

(iii) plan for the decontamination and decommissioning of such candidate site, and for the mitigation of any significant adverse environmental impacts caused by the site characterization activities if it is determined unsuitable for application for a construction authorization for a repository;

(iv) criteria to be used to determine the suitability of such candidate site for the location of a repository, developed pursuant to section 112(a); and

(v) any other information required by the Commission;

(B) a description of the possible form or packaging for the high-level radioactive waste and spent nuclear fuel to be emplaced in such repository, a description, to the extent practicable, of the relationship between such waste form or packaging and the geologic medium of such site, and a description of the activities being conducted by the Secretary with respect to such possible waste form or packaging or such relationship; and

(C) a conceptual repository design that takes into account likely site-specific requirements.

---

[2]   Amended by P.L. 100–203, § 5011, 101 Stat. 1330 (1987); P.L. 102–154, Title I, 105 Stat. 1000 (1991).

(2) Before proceeding to sink shafts at the Yucca Mountain site, the Secretary shall (A) make available to the public the site characterization plan described in paragraph (1); and (B) hold public hearings in the vicinity of such candidate site to inform the residents of the area in which such candidate site is located of such plan, and to receive their comments.

(3) During the conduct of site characterization activities at the Yucca Mountain site, the Secretary shall report not less than once every 6 months to the Commission and to the Governor and legislature of the State of Nevada on the nature and extent of such activities and the information developed from such activities.

(c) RESTRICTIONS–(1) The Secretary may conduct at the Yucca Mountain site only such site characterization activities as the Secretary considers necessary to provide the data required for evaluation of the suitability of such site for an application to be submitted to the Commission for a construction authorization for a repository at such site, and for compliance with the National Environmental Policy Act of 1969 (42 USC 4321 *et seq.*).

(2) In conducting site characterization activities–

(A) the Secretary may not use any radioactive material at a site unless the Commission concurs that such use is necessary to provide data for the preparation of the required environmental reports and an application for a construction authorization for a repository at such site; and

(B) if any radioactive material is used at a site–

(i) the Secretary shall use the minimum quantity necessary to determine the suitability of such sites for a repository, but in no event more than the curie equivalent of 10 metric tons of spent nuclear fuel; and

(ii) such radioactive material shall be fully retrievable.

(3) If the Secretary at any time determines the Yucca Mountain site to be unsuitable for development as a repository, the Secretary shall–

(A) terminate all site characterization activities at such site;

(B) notify the Congress, the Governor and legislature of Nevada of such termination and the reasons for such termination;

(C) remove any high-level radioactive waste, spent nuclear fuel, or other radioactive materials at or in such site as promptly as practicable;

(D) take reasonable and necessary steps to reclaim the site and to mitigate any significant adverse environmental impacts caused by site characterization activities at such site;

(E) suspend all future benefits payments under subtitle F with respect to such site; and

(F) report to Congress not later than 6 months after such determination the Secretary's recommendations for further action to assure the safe, permanent disposal of spent nuclear fuel and high-level radioactive waste, including the need for new legislative authority.

(d) PRELIMINARY ACTIVITIES–Each activity of the Secretary under this section that is in compliance with the provisions of subsection (c) shall be considered a preliminary decision making activity. No such activity shall require the preparation of an environmental impact statement under section 102(2)(C) of the National Environmental Policy Act of 1969 (42 USC 4332(2)(C)), or to require any environmental review under subparagraph (E) or (F) of section 102(2) of such Act.[3]

42 USC 10134.

Notification of decision.

Public availability.

Sec. 114. Site Approval and Construction Authorization

(a) HEARINGS AND PRESIDENTIAL RECOMMENDATION–The Secretary shall hold public hearings in the vicinity of the Yucca Mountain site for the purposes of informing the residents of the area of such consideration and receiving their comments regarding the possible recommendation of such site. If, upon completion of such hearings and completion of site characterization activities at the Yucca Mountain site under section 113, the Secretary decides to recommend approval of such site to the President, the Secretary shall notify the Governor and legislature of the State of Nevada of such decision. No sooner than the expiration of the 30-day period following such notification, the Secretary shall submit to the President a recommendation that the President approve such site for the development of a repository. Any such recommendation by the Secretary shall be based on the record of information developed by the Secretary under section 113 and this section, including the information described in subparagraph (A) through subparagraph (G).

Together with any recommendation of a site under this paragraph, the Secretary shall make available to the public, and submit to the President, a comprehensive statement of the basis of such recommendation, including the following:

(A) a description of the proposed repository, including preliminary engineering specifications for the facility;

(B) a description of the waste form or packaging proposed for use at such repository, and an explanation of the relationship between such waste form or packaging and the geologic medium of such site;

(C) a discussion of data, obtained in site characterization activities, relating to the safety of such site;

(D) a final environmental impact statement prepared for the Yucca Mountain site pursuant to subsection (f) and the National Environmental Policy Act of 1969 (42 USC 4321 *et seq.*), together with comments made concerning such environmental impact statement by the Secretary of the Interior, the Council on Environmental Quality, the Administrator, and the Commission, except that the Secretary shall not be required in any such environmental impact statement to consider the need for a repository, the alternatives to geological disposal, or alternative sites to the Yucca Mountain site;

(E) preliminary comments of the Commission concerning the extent to which the at-depth site characterization analysis and the waste form proposal for such site seem to be sufficient for inclusion in any application to be submitted by the Secretary for licensing of such site as a repository;

(F) the views and comments of the Governor and legislature of any State, or the governing body of any affected Indian tribe, as determined by the Secretary, together with the response of the Secretary to such views;

---

[3]   Amended by P.L. 100–203, § 5011, 101 Stat. 1330 (1987).

(G) such other information as the Secretary considers appropriate; and

(H) any impact report submitted under section 116(c)(2)(B) by the State of Nevada.

(2)(A) If, after recommendation by the Secretary, the President considers the Yucca Mountain site qualified for application for a construction authorization for a repository, the President shall submit a recommendation of such site to Congress.

(B) The President shall submit with such recommendation a copy of the statement for such site prepared by the Secretary under paragraph (1).

(3)(A) The President may not recommend the approval of Yucca Mountain site unless the Secretary has recommended to the President under paragraph (1) approval of such site and has submitted to the President a statement for such site as required under such paragraph.

(B) No recommendation of a site by the President under this subsection shall require the preparation of an environmental impact statement under section 102(2)(C) of the National Environmental Policy Act of 1969 (42 USC 4332(2)(C), or to require any environmental review under subparagraph (E) or (F) of section 102(2) of such Act.

(b) SUBMISSION OF APPLICATION–If the President recommends to the Congress the Yucca Mountain site under subsection (a) and the site designation is permitted to take effect under section 115, the Secretary shall submit to the Commission an application for a construction authorization for a repository at such site not later than 90 days after the date on which the recommendation of the site designation is effective under such section and shall provide to the Governor and legislature of the State of Nevada a copy of such application.

(c) STATUS REPORT ON APPLICATION–Not later than 1 year after the date on which an application for a construction authorization is submitted under subsection (b), and annually thereafter until the date on which such authorization is granted, the Commission shall submit a report to the Congress describing the proceeding undertaken through the date of such report with regard to such application, including a description of–

(1) any major unresolved safety issues, and the explanation of the Secretary with respect to design and operation plans for resolving such issues;

(2) any matters of contention regarding such application; and

(3) any Commission actions regarding the granting of denial of such authorization.

Construction
authorization
applications.

(d) COMMISSION ACTION–The Commission shall consider an application for a construction authorization for all or part of a repository in accordance with the laws applicable to such applications, except that the Commission shall issue a final decision approving or disapproving the issuance of a construction authorization not later that the expiration of 3 years after the date of the submission of such application, except that the Commission may extend such deadlines by not more than 12 months if, not less than 30 days before such deadlines, the Commission complies with the reporting requirements established in subsection (e)(2). The Commission decision approving the first such application shall prohibit the emplacement in the first repository of a quantity of spent fuel containing in excess of 70,000 metric tons of heavy metal or a quantity of solidified high-level radioactive waste resulting from the reprocessing of such a quantity of spent fuel until such time as a second repository is in operation. In the event that a monitored retrievable storage facility, approved pursuant to subtitle C of this Act, shall be located, or is planned to be located, within 50 miles of the first repository, then the Commission decision approving the first such application shall prohibit the emplacement of a quality of spent fuel containing in excess of 70,000 metric tons of heavy metal or a quantity of solidified high-level radioactive waste resulting from the reprocessing of spent fuel in both the repository and monitored retrievable storage facility until such time as a second repository is in operation.

(e) PROTECT DECISION SCHEDULE–(1) The Secretary shall prepare and update, as appropriate, in cooperation with all affected Federal agencies, a project decision schedule that portrays the optimum way to attain the operation of the repository within the time periods specified in this subtitle. Such schedule shall include a description of objectives and a sequence of deadlines for all Federal agencies required to take action, including an identification of the activities in which a delay in the start, or completion, of such activities will cause a delay in beginning repository operation.

(2) Any Federal agency that determines that it cannot comply with any deadline in the project decision schedule, or fails to so comply, shall submit to the Secretary and to the Congress a written report explaining the reason for its failure or expected failure to meet such deadlines, the reason why such agency could not reach an agreement with the Secretary, the estimated time for completion of the activity or activities involved, the associated effect on its other deadlines in the project decision schedule, and any recommendations it may have or actions it intends to take regarding any improvements in its operation or organization, or changes to its statutory directives or authority, so that it will be able to mitigate the delay involved. The Secretary, within 30 days after receiving any such report, shall file with the Congress his response to such report, including the reasons why the Secretary could not amend the project decision schedule to accommodate the Federal agency involved.

(f) ENVIRONMENTAL IMPACT STATEMENT–

(1) Any recommendation made by the Secretary under this section shall be considered a major Federal action significantly affecting the quality of the human environment for purposes of the National Environmental Policy Act of 1969 (42 USC 4321 *et seq.*). A final environmental impact statement prepared by the Secretary under such Act shall accompany any recommendation to the President to approve a site for a repository.

(2) With respect to the requirements imposed by the National Environmental Policy Act of 1969 (42 USC 4321 *et seq.*), compliance with the procedures and requirements of this Act shall be deemed adequate consideration of the need for a repository, the time of the initial availability of a repository, and all alternatives to the isolation of high-level radioactive waste and spent nuclear fuel in a repository.

(3) For purposes of complying with the requirements of the National Environmental Policy Act of 1969 (42 USC 4321 *et seq.*) and this section, the Secretary need not consider alternative sites to the Yucca Mountain site for the repository to be developed under this subtitle.

(4) Any environmental impact statement prepared in connection with a repository proposed to be constructed by the Secretary under this subtitle shall, to the ex tent practicable, be adopted by the Commission in connection with the issuance by the Commission of a

construction authorization and license for such repository. To the extent such statement is adopted by the Commission, such adoption shall be deemed to also satisfy the responsibilities of the Commission under the National Environmental Policy Act of 1969 (42 USC 4321 *et seq.*) and no further consideration shall be required, except that nothing in this subsection shall affect any independent responsibilities of the Commission to protect the public health under the Atomic Energy Act of 1954 (42 USC 2011 *et seq.*).

(5) Nothing in this Act shall be construed to amend or otherwise detract from the licensing requirements of the Nuclear Regulatory Commission established in title II of the Energy Reorganization Act of 1974 (42 USC 5841 *et seq.*).

(6) In any such statement prepared with respect to the repository to be constructed under this subtitle, the Nuclear Regulatory Commission need not consider the need for a repository, the time of initial availability of a repository, alternate sites to the Yucca Mountain site, or nongeologic alternatives to such site.[4]

42 USC 10135.

Sec. 115. Review of Repository Site Selection

(a) DEFINITION–For purposes of this section, the term "resolution of repository siting approval" means a joint resolution of the Congress, the matter after the resolving clause of which is as follows: That there hereby is approved the site at . for a repository, with respect to which a notice of disapproval was submitted by ___ on ___. The first blank space in such resolution shall be filled with the name of the geographic location of the proposed site of the repository to which such resolution pertains; the second blank space in such resolution shall be filled with the designation of the State Governor and legislature or Indian tribe governing body submitting the notice of disapproval to which such resolution pertains; and the last blank space in such resolution shall be filled with the date of such submission.

(b) STATE OR INDIAN TRIBE PETITIONS–The designation of a site as suitable for application for a construction authorization for a repository shall be effective at the end of the 60-day period beginning on the date that the President recommend such site to the Congress under section 114, unless the Government and legislature of the State in which such site is located, or the governing body of an Indian tribe on whose reservation such site is located, as the case may be, has submitted to the Congress a notice of disapproval under section 116 or 118 notice of disapproval has been submitted, the designation of such site shall not be effective except as provided under subsection (c).

Notice of disapproval, submittal to Congress.

(c) CONGRESSIONAL REVIEW OF PETITIONS–If any notice of disapproval of a repository site designation has been submitted to the Congress under section 116 or 118 after a recommendation for approval of such site is made by the President under section 114, such site shall be disapproved unless, during the first period of 90 calendar days of continuous session of the Congress after the date of the receipt by the Congress of such notice of disapproval, the Congress passes a resolution of repository siting approval in accordance with this subsection approving such site, and such resolution thereafter becomes law.

(d) PROCEDURES APPLICABLE TO THE SENATE–(1) The provisions of this subsection are enacted by the Congress–

---

[4]   Amended by P.L. 100–203, § 5011, 101 Stat. 1330 (1987).

(A) as an exercise of the rulemaking power of the Senate, and as such they are deemed a part of the rules of the Senate, but applicable only with respect to the procedure to be followed in the Senate in the case of resolutions of repository siting approval, and such provisions supersede other rules of the Senate only to the extent that they are inconsistent with such other rules; and

(B) with full recognition of the constitutional right of the Senate to change the rules (so far as relating to the procedure of the Senate) at any time, in the same, manner and to the same extent as in the case of any other rule of the Senate.

Introduction of resolution.

(2)(A) Not later than the first day of session following the day on which any notice of disapproval of a repository site selection is submitted to the Congress under section 116 or 118, a resolution of repository siting approval shall be introduced (by request) in the Senate by the chairman of the committee to which such notice of disapproval is referred, or by a Member of Members of the Senate designated by such chairman.

Committee recommendations.

(B) Upon introduction, a resolution of repository siting approval shall be referred to the appropriate committee or committees of the Senate by the President of the Senate, and all such resolutions with respect to the same repository site shall be referred to the same committee or committees. Upon the expiration of 60 calendar days of continuous session after the introduction of the first resolution of repository siting approval with respect to any site, each committee to which such resolution was referred shall make its recommendations to the Senate.

Discharge of committee.

(3) If any committee to which is referred a resolution of siting approval introduced under paragraph (2)(A), or, in the absence of such a resolution, any other resolution of siting approval introduced with respect to the site involved, has not reported such resolution at the end of 60 days of continuous session of Congress after introduction of such resolution, such committee shall be deemed to be discharged from further consideration of such resolution, and such resolution shall be placed on the appropriate calendar of the Senate.

(4)(A) When each committee to which a resolution of siting approval has been referred has reported, or has been deemed to be discharged from further consideration of, a resolution described in paragraph (3), it shall at any time thereafter be in order (even though a previous motion to the same effect has been disagreed to) for any Member of the Senate to move to proceed to the consideration of such resolution. Such motion shall be highly privilege and shall not be debatable. Such motion shall not be subject to amendment, to a motion to postpone, or to a motion to proceed to the consideration of other business. A motion to reconsider the vote by which such motion is agreed to or disagreed to shall not be in order. If a motion to proceed to the consideration of such resolution is agreed to, such resolution shall remain the unfurnished business of the Senate until disposed of.

Debate.

(B) Debate on a resolution of siting approval, and on all debatable motions and appeals in connection with such resolution, shall be limited to not more than 10 hours, which shall be divided equally between Members favoring and Members opposing such resolution. A motion further to limit debate shall be in order and shall not be debatable. Such motion shall not be subject to amendment, to a motion to postpone, or to a motion to proceed to the consideration of other business, and a motion to recommit such resolution shall not be in order. A motion to reconsider the vote by which such resolution is agreed to or disagreed to shall not be in order.

(C) Immediately following the conclusion of the debate on a resolution of siting approval, and a single quorum call at the conclusion of such debate if requested in accordance with rules of the Senate, the vote on final approval of such resolution shall occur.

Appeals.

(D) Appeals from the decisions of the Chair relating to the application of the rules of the Senate to the procedure relating to a resolution of siting approval shall be decided without debate.

(5) If the Senate receives from the House a resolution of repository siting approval with respect to any site, then the following procedure shall apply:

(A) The resolution of the House with respect to such site shall not be referred to a committee.

(B) With respect to the resolution of the Senate with respect to such site–

(i) the procedure with respect to that or other resolutions of the Senate with respect to such site shall be the same as if no resolution from the House with respect to such site had been received; but

(ii) on any vote on final passage of a resolution of the Senate with respect to such site, a resolution from the House with respect to such site where the text is identical shall be automatically substituted for the resolution of the Senate.

(e) PROCEDURES APPLICABLE TO THE HOUSE OF REPRESENTATIVES–

(1) The provisions of this section are enacted by the Congress–

(A) as an exercise of the rulemaking power of the House of Representatives, and as such they are deemed a part of the rules of the House, but applicable only with respect to the procedure to be followed in the House in the case of resolutions of repository siting approval, and such provisions supersede other rules of the House only to the extent that they are inconsistent with such other rules; and

(B) with full recognition of the constitutional right of the House to change the rules (so far as relating to the procedure of the House) at any time, in the same manner and to the same extent as in the case of any other rule of the House.

(2) Resolutions of repository siting approval shall upon introduction be immediately referred by the Speaker of the House to the appropriate committee or committees of the House. Any such resolution received from the Senate shall be held at the Speaker's table.

Discharge of committee.

(3) Upon the expiration of 60 days of continuous session after the introduction of the first resolution of repository siting approval with respect to any site, each committee to which such resolution was referred shall be discharged from further consideration of such resolution, and such resolution shall be referred to the appropriate calendar, unless such resolution or an identical resolution was previously reported by each committee to which it was referred.

Resolution, consideration and debate.

(4) It shall be in order for the Speaker to recognize a Member favoring a resolution to call up a resolution of repository siting approval after it has been on the appropriate calendar for 5 legislative days. When any such resolution is called up, the House shall proceed

to its immediate consideration and the Speaker shall recognize the Member calling up such resolution and a Member opposed to such resolution for 2 hours of debate in the House, to be equally divided and controlled by such Members. When such time has expired, the previous question shall be considered as ordered on the resolution to adoption without intervening motion. No amendment to any such resolution shall be in order, nor shall it be in order to move to reconsider the vote by which such resolution is agreed to or disagreed to.

(5) If the House receives from the Senate a resolution of repository siting approval with respect to any site, then the following procedure shall apply:

(A) The resolution of the Senate with respect to such site shall not be referred to a committee.

(B) With respect to the resolution of the House with respect to such site–

(i) the procedure with respect to that or other resolutions of the House with respect to such site shall be the same as if no resolution from the Senate with respect to such site had been received; but

(ii) on any vote on final passage of a resolution of the House with respect to such site, a resolution from the Senate with respect to such site where the text is identical shall be automatically substituted for the resolution of the House.

(f) COMPUTATION OF DAYS–For purposes of this section–

(1) continuity of session of Congress is broken only by an adjournment sine die; and

(2) the days on which either House is not in session because of an adjournment of more than 3 days to a day certain are excluded in the computation of the 90-day period referred to in subsection (c) and the 60-day period referred to in subsections (d) and (e).

(g) INFORMATION PROVIDED TO CONGRESS–In considering any notice of disapproval submitted to the Congress under section 116 or 118, the Congress may obtain any comments of the Commission with respect to such notice of disapproval. The provision of such comments by the Commission shall not be construed as binding the Commission with respect to any licensing or authorization action concerning the repository involved.

Sec. 116. Participation of States

42 USC 10136.

(a) NOTIFICATION OF STATES AND AFFECTED TRIBES–The Secretary shall identify the States with one or more potentially acceptable sites for a repository within 90 days after the date of enactment of this Act. Within 90 days of such identification, the Secretary shall notify the Governor, the State legislature, and the tribal council of any affected Indian tribe in any State of the potentially acceptable sites within such State. For the purposes of this title, the term "potentially acceptable site" means any site at which, after geologic studies and field mapping but before detailed geologic data gathering, the Department undertakes preliminary drilling and geophysical testing for the definition of site location.

Potentially acceptable site.

(b) STATE PARTICIPATION IN REPOSITORY SITING DECISIONS–(1) Unless otherwise provided by State law, the Governor or legislature of each State shall have authority to submit a notice of disapproval to the Congress under paragraph (2). In any case in which State law provides for submission of any such notice of disapproval by any other person or entity, any reference in this subtitle to the Governor or legislature of such State shall be considered to refer instead to such other person or entity.

(2) Upon the submission by the President to the Congress of a recommendation of a site for a repository, the Governor or legislature of the State in which such site is located may disapprove the site designation and submit to the Congress a notice of disapproval. Such Governor or legislature may submit such a notice of disapproval to the Congress not later than the 60 days after the date that the President recommends such site to the Congress under section 114. A notice of disapproval shall be considered to be submitted to the Congress on the date of the transmittal of such notice of disapproval to the Speaker of the House and the President pro tempore of the Senate. Such notice of disapproval shall be accompanied by a statement of reasons explaining why such Governor or legislature disapproved the recommended repository site involved.

(3) The authority of the Governor or legislature of each State under this subsection shall not be applicable with respect to any site located on a reservation.

(c) FINANCIAL ASSISTANCE–(1)(A) The Secretary shall make grants to the State of Nevada and any affected unit of local government for the purpose of participating in activities required by this section and section 117 or authorized by written agreement entered into pursuant to section 117(c). Any salary or travel expense that would ordinarily be incurred by such State or affected unit of local government, may not be considered eligible for funding under this paragraph.

(B) The Secretary shall make grants to the State of Nevada and any affected unit of local government for purposes of enabling such State or affected unit of local government–

(i) to review activities taken under this subtitle with respect to the Yucca Mountain site for purposes of determining any potential economic, social, public health and safety, and environmental impacts of a repository on such State, or affected unit of local government and its residents;

(ii) to develop a request for impact assistance under paragraph (2);

(iii) to engage in any monitoring, test, or evaluation activities with respect to site characterization programs with regard to such site;

(iv) to provide information to Nevada residents regarding any activities of such State , the Secretary, or the Commission with respect to such site; and

(v) to request information from, and make comments and recommendations to, the Secretary regarding any activities taken under this subtitle with respect to such site.

(C) Any salary or travel expense that would ordinarily be incurred by the State of Nevada or any affected unit of local government may not be considered eligible for funding under this paragraph.

(2)(A)(i) The Secretary shall provide financial and technical assistance to the State of Nevada, and any affected unit of local government requesting such assistance.

(ii) Such assistance shall be designed to mitigate the impact on such State or affected unit of local government of the development of such repository and the characterization of such site.

(iii) Such assistance to such State or affected unit of local government of such State shall commence upon the initiation of site characterization activities.

(B) The State of Nevada and any affected unit of local government may request assistance under this subsection by preparing and submitting to the Secretary a report on the economic, social, public health and safety, and environmental impacts that are likely to result from site characterization activities at the Yucca Mountain site. Such report shall be submitted to the Secretary after the Secretary has submitted to the State a general plan for site characterization activities under section 113(b).

(C) As soon as practicable after the Secretary has submitted such site characterization plan, the Secretary shall seek to enter into a binding agreement with the State of Nevada setting forth–

(i) the amount of assistance to be provided under this subsection to such State or affected unit of local government; and

(ii) the procedures to be followed in providing such assistance.

(3)(A) In addition to financial assistance provided under paragraphs (1) and (2), the Secretary shall grant to the State of Nevada and any affected unit of local government an amount each fiscal year equal to the amount such State or affected unit of local government, respectively, would receive if authorized to tax site characterization activities at such site, and the development and operation of such repository, as such State or affected unit of local government taxes the non-Federal real property and industrial activities occurring within such State or affected unit of local government.

(B) Such grants shall continue until such time as all such activities, development, and operation are terminated at each such site.

(4)(A) The State of Nevada or any affected unit of local government may not receive any grant under paragraph (1) after the expiration of the 1 year period following–

(i) the date on which the Secretary notifies the Governor and legislature of the State of Nevada of the termination of site characterization activities at the site in such State;

(ii) the date on which the Yucca Mountain site is disapproved under section 115; or

(iii) the date on which the Commission disapproves an application for a construction authorization for a repository at such site; whichever occurs first.

(B) The State of Nevada or any affected unit of local government may not receive any further assistance under paragraph (2) with respect to a site if repository construction activities or site characterization activities at such site are terminated by the Secretary or if such activities are permanently enjoined by any court.

(C) At the end of the 2-year period beginning on the effective date of any license to receive and possess for a repository in a State, no Federal funds, shall be made available to such State or affected unit of local government under paragraph (1) or (2), except for–

(i) such funds as may be necessary to support activities related to any other repository located in, or proposed to be located in, such State, and for which a license to receive and possess has not been in effect for more than 1 year;

(ii) such funds as may be necessary to support State activities pursuant to agreements or contracts for impact assistance entered into, under paragraph (2), by such State with the Secretary during such 2-year period; and

(iii) such funds as may be provided under an agreement entered into under title IV.

(5) Financial assistance authorized in this subsection shall be made out of amounts held in the Waste Fund.

(6) No State, other than the State of Nevada, may receive financial assistance under this subsection after the date of the enactment of the Nuclear Waste Policy Amendments Act 1987.[5]

(d) ADDITIONAL NOTIFICATION AND CONSULTATION–Whenever the Secretary is required under any provision of this Act to notify or consult with the governing body of an affected Indian tribe where a site is located, the Secretary shall also notify or consult with, as the case may be, the Governor of the State in which such reservation is located.

42 USC 10137.

Sec. 117. Consultation with States and Affected Indian Tribes

(a) PROVISION OF INFORMATION–(1) The Secretary, the Commission, and other agencies involved in the construction, operation, or regulation of any aspect of a repository in a State shall provide to the Governor and legislature of such State, and to the governing body of any affected Indian tribe, timely and complete information regarding determinations or plans made with respect to the site characterization siting, development, design, licensing, construction, operation, regulation, or decommissioning of such repository.

Information request, response.

(2) Upon written request for such information by the Governor or legislature of such State, or by the governing body of any affected Indian tribe, as the case may be, the Secretary shall provide a written response to such request within 30 days of the receipt of such request. Such response shall provide the information requested or, in the alternative, the reasons why the information cannot be so provided. If the Secretary fails to so respond within such 30 days, the Governor or legislature of such State, or the governing body of any affected Indian tribe, as the case may be, may transmit a formal written objection to such failure to respond to the President. If the President or Secretary fails to respond to such written request within 30 days of the receipt by the President of such formal written objection, the Secretary shall immediately suspend all activities in such State authorized by this subtitle, and shall not renew such activities until the Governor or legislature of such State, or the governing body of any affected Indian tribe, as the case may be, has received the written response to such written request required by this subsection.

(b) CONSULTATION AND COOPERATION–In performing any study of an area within a State for the purpose of determining the suitability of such area for a repository pursuant to section 112(c), and in subsequently developing and loading any repository within such State, the Secretary shall consult and cooperate with the Governor and

---

[5]   Amended by P.L. 100–203, § 5032, 101 Stat. 1330 (1987).

legislature of such State and the governing body of any affected Indian tribe in an effort to resolve the concerns of such State and any affected Indian tribe regarding the public health and safety, environmental, and economic impacts of any such repository. In carrying out his duties under this subtitle, the Secretary shall take such concerns into account to the maximum extent feasible and as specified in written agreements entered into under subsection (c).

Report to Congress.

(c) WRITTEN AGREEMENT–Not later than 60 days after (1) the approval of a site for site characterization for such a repository under section 112(c), or (2) the written request of the State or Indian tribe in any affected State notified under section 116(a) to the Secretary, whichever, first occurs, the Secretary shall seek to enter into a binding written agreement, and shall begin negotiations, with such State and, where appropriate, to enter into a separate binding agreement with the governing body of any affected Indian tribe, setting forth (but not limited to) the procedures under which the requirements of subsections (a) and (b), and the provisions of such written agreement, shall be carried out. Any such written agreement shall not affect the authority of the Commission under existing law. Each such written agreement shall, to the maximum extent feasible, to completed no later than 6 months after such notification.[6]

Such written agreement shall specify procedures–

(1) by which such State or governing body of an affected Indian tribe, as the case may be, may study, determine, comment on, and make recommendations with regard to the possible public health and safety, environmental, social, and economic impacts of any such repository;

(2) by which the Secretary shall consider and respond to comments and recommendations made by such State or governing body of an affected Indian tribe, including the period in which the Secretary shall so respond;

(3) by which the Secretary and such State or governing body of an affected Indian tribe may review or modify the agreement periodically;

(4) by which such State or governing body of an affected Indian tribe is to submit an impact report and request for impact assistance under section 116(c) or section 118(b), as the case may be;

(5) by which the Secretary shall assist such State, and the units of general local government in the vicinity of the repository site, in resolving the offsite concerns of such State and units of general local government, including, but not limited to, questions of State liability arising from accidents, necessary road upgrading and access to the site, ongoing emergency preparedness and emergency response, monitoring of transportation of high-level radioactive waste and spent nuclear fuel through such State, conduct of baseline health studies of

---

[6]   Amended by P.L. 104–66, § 1051(i), 109 Stat. 716 (1995), deleting the following from subsection (c):

> If such written agreement is not completed within such period, the Secretary shall report to the Congress in writing within 30 days on the status of negotiation to develop such agreement and the reasons why such agreement has not been completed. Prior to submission of such report to the Congress, the Secretary shall transmit such report to the Governor of such State or the governing body of such affected Indian tribe, as the case may be, for their review and comments. Such comments shall be included in such report prior to submission to the Congress.

inhabitants in neighboring communities near the repository site and reasonable periodic monitoring thereafter, and monitoring of the repository site upon any decommissioning and decontamination;

(6) by which the Secretary shall consult and cooperate with such State on a regular, ongoing basis and provide for an orderly process and timely schedule for State review and evaluation, including identification in the agreement of key events, milestones, and decision points in the activities of the Secretary at the potential repository site;

State notification. Transportation of radioactive waste and spent nuclear fuel. Monitoring and testing.

(7) by which the Secretary shall notify such State prior to the transportation of any high-level radioactive waste and spent nuclear fuel into such State for disposal at the repository site;

(8) by which such State may conduct reasonable independent monitoring and testing of activities on the repository site, except that such monitoring and testing shall not unreasonably interfere with or delay onsite activities;

(9) for sharing, in accordance with applicable law, of all technical and licensing information, the utilization of available expertise, the facilitating of permit procedures, joint project review, and the formation of joint surveillance and monitoring arrangements to carry out applicable Federal and State laws;

(10) for public notification of the procedures specified under the preceding paragraphs; and

(11) for resolving objections of a State and affected Indian tribes at any stage of the planning, siting, development, construction, operation, or closure of such a facility within such State through negotiation, arbitration, or other appropriate mechanisms.

(d)[7] ON-SITE REPRESENTATIVE–The Secretary shall offer to any State, Indian tribe or unit of local government within whose jurisdiction a site for a repository or monitored retrievable storage facility is located under this title an opportunity to designate a representative to conduct on-site oversight activities at such site. Reasonable expenses of such representatives shall be paid out of the Waste Fund.

42 USC 10138. Notice of disapproval, submittal to Congress.

Sec. 118. Participation of Indian Tribes

(a) PARTICIPATION OF INDIAN TRIBES IN REPOSITORY SITING DECISIONS–Upon the submission by the President to the Congress of a recommendation of a site for a repository located on the reservation of an affected Indian tribe, the governing body of such Indian tribe may disapprove the site designation and submit to the Congress a notice of disapproval. The governing body of such Indian tribe may submit such a notice of disapproval to the Congress not later than the 60 days after the date that the President recommends such site to the Congress under section 114. A notice of disapproval shall be considered to be submitted to the Congress on the date of the transmittal of such notice of disapproval to the Speaker of the House and the President pro tempore of the Senate. Such notice of disapproval shall be accompanied by a statement of reasons explaining why the governing body of such Indian tribe disapproved the recommended repository site involved.

Grants.

(b) FINANCIAL ASSISTANCE–(1) The Secretary shall make grants to each affected tribe notified under section 116(a) for the purpose of participating in activities required by section 117 or authorized by written agreement entered into pursuant to section 117(c). Any salary or travel

---

[7]   Added by P.L. 100–203, § 5011, 101 Stat. 1330 (1987).

expense that would ordinarily be incurred by such tribe, may not be considered eligible for funding under this paragraph.

(2) (A) The Secretary shall make grants to each affected Indian tribe where a candidate site for a repository is approved under section 112(c). Such grants may be made to each such Indian tribe only for purposes of enabling such Indian tribe–

(i) to review activities taken under this subtitle with respect to such site for purposes of determining any potential economic, social, public health and safety, and environmental impacts of such repository on the reservation and its residents;

(ii) to develop a request for impact assistance under paragraph (2);

(iii) to engage in any monitoring, testing, or evaluation activities with respect to site characterization programs with regard to such site;

(iv) to provide information to the residents of its reservation regarding any activities of such Indian tribe, the Secretary, or the Commission with respect to such site; and

(v) to request information from, and make comments and recommendations to, the Secretary regarding any activities taken under this subtitle with respect to such site.

(B) The amount of funds provided to any affected Indian tribe under this paragraph in any fiscal year may not exceed 100 percent of the costs incurred by such Indian tribe with respect to the activities described in clauses (i) through (v) of subparagraph (A). Any salary or travel expense that would ordinarily be incurred by such Indian tribe may not be considered eligible for funding under this paragraph.

(3) (A) The Secretary shall provide financial and technical assistance to any affected Indian tribe requesting such assistance and where there is a site with respect to which the Commission has authorized construction of a repository. Such assistance shall be designed to mitigate the impact on such Indian tribe of the development of such repository. Such assistance to such Indian tribe shall commence within 6 months following the granting by the Commission of a construction authorization for such repository and following the initiation of construction activities at such site.

Report
submittal.

(B) Any affected Indian tribe desiring assistance under this paragraph shall prepare and submit to the Secretary a report on any economic, social, public health and safety, and environmental impacts that are likely as a result of the development of a repository at a site on the reservation of such Indian tribe. Such report shall be submitted to the Secretary following the completion of site characterization activities at such site and before the recommendation of such site to the President by the Secretary for application for a construction authorization for a repository. As soon as practicable following the granting of a construction authorization for such repository, the Secretary shall seek to enter into a binding agreement with the Indian tribe involved setting forth the amount of assistance to be provided to such Indian tribe under this paragraph and the procedures to be followed in providing such assistance.

(4) The Secretary shall grant to each affected Indian tribe where a site for a repository is approved under section 112(c) an amount each fiscal year equal to the amount such Indian tribe would receive were it authorized to tax site characterization activities at such site, and the

development and operation of such repository, as such Indian tribe taxes the other commercial activities occurring on such reservation. Such grants shall continue until such time as all such activities, development, and operation are terminated at such site.

Grants, limitation.

(5) An affected Indian tribe may not receive any grant under paragraph (1) after the expiration of the 1-year period following–

(i) the date on which the Secretary notifies such Indian tribe of the termination of site characterization activities at the candidate site involved on the reservation of such Indian tribe;

(ii) the date on which such site is disapproved under section 115;

(iii) the date on which the Commission disapproves an application for a construction authorization for a repository at such site;

(iv) the date of the enactment of the Nuclear Waste Policy Amendments Acts of 1987;[8] whichever occurs first, unless there is another candidate site on the reservation of such Indian tribe that is approved under section 112(c) and with respect to which the actions described in clauses (i), (ii), and (iii) have not been taken.

(B) An affected Indian tribe may not receive any further assistance under paragraph (2) with respect to a site if repository construction activities at such site are terminated by the Secretary or if such activities are permanently enjoined by any court.

Funding.

(C) At the end of the 2-year period beginning on the effective date of any license to receive and possess for a repository at a site on the reservation of an affected Indian tribe, no Federal funds shall be made available under paragraph (1) or (2) to such Indian tribe, except for–

(i) such funds as may be necessary to support activities of such Indian tribe related to any other repository where a license to receive and possess has not been in effect for more than 1 year; and

(ii) such funds as may be necessary to support activities of such Indian tribe pursuant to agreements or contracts for impact assistance entered into, under paragraph (2), by such Indian tribe with the Secretary during such 2-year period.

(6) Financial assistance authorized in this subsection shall be made out of amounts held in the Nuclear Waste Fund established in section 302.

42 USC 10139.

Sec. 119. Judicial Review of Agency Actions

(a) JURISDICTION OF UNITED STATES COURTS OF APPEALS–

(1) Except for review in the Supreme Court of the United States courts of appeals shall have original and exclusive jurisdiction over any civil action–

(A) for review of any final decision or action of the Secretary, the President , or the Commission under this subtitle;

(B) alleging the failure of the Secretary, the President, or the Commission to make any decision, or take any action, required under this subtitle;

---

[8] Amended by P.L. 100–203, § 5033, 101 Stat. 1330 (1987).

(C) challenging the constitutionality of any decision made, or action taken, under any provision of this subtitle;

(D) for review of any environmental impact statement prepared pursuant to the National Environmental Policy Act of 1969 (42 USC 4321 *et seq.*) with respect to any action under this subtitle, or as required under section 135(c) (1), or alleging a failure to prepare such statement with respect to any such action;

(E) for review of any environmental assessment prepared under section 112(b) (1) or 135(c)(2); or

(F) for review of any research and development activity under title II.

(2) The venue of any proceeding under this section shall be in the judicial circuit in which the petitioner involved resided or has its principle office, or in the United States Court of Appeals for the District of Columbia.

(c) Deadline For Commencing Action–A civil action for judicial review described under subsection (a)(1) may be brought not later than the 180th day after the date of the decision or action or failure to act involved, as the case may be, except that if a party shows that he did not know of the decision or action complained of (or of the failure to act), and that a reasonable person acting under the circumstances would not have known, such party may bring a civil action not later than the 180th day after the date such party acquired actual or constructive knowledge of such decision, action, or failure to act.

Sec. 120. Expedited Authorizations

42 USC 10140.

(a) ISSUANCE OF AUTHORIZATION–(1) To the extent that the taking of any action related to the site characterization of a site or the construction or initial operation of a repository under this subtitle requires a certificate, right-of-way, permit, lease, or other authorization from a Federal agency or officer, such agency or officer shall issue or grant any such authorization at the earliest practicable date, to the extent permitted by the applicable provisions of law administered by such agency or officer. All actions of a Federal agency or officer with respect to consideration of applications or requests for the issuance or grant of any such authorization shall be expedited, and any such application or request shall take precedence over any similar applications or requests not related to such repositories.

(2) The provisions of paragraph (1) shall not apply to any certificate, right-of-way, permit, lease, or other authorization issued or granted by, or requested from, the Commission.

(b) Terms of Authorizations.–Any authorization issued or granted pursuant to subsection (a) shall include such terms and conditions as may be required by law, and may include terms and conditions permitted by law.

Sec. 121. Certain Standards and Criteria

42 USC 10141.

(a) ENVIRONMENTAL PROTECTION AGENCY STANDARDS– Not later than 1 year after the date of the enactment of this Act, the Administrator, pursuant to authority under other provisions of law, shall, by rule, promulgate generally applicable standards for protection of the general environment from offsite releases from radioactive material in repositories.

(b) Commission Requirements and Criteria–(1) (A) Not later than January 1, 1984, the Commission, pursuant to authority under other provisions of law, shall, by rule promulgate technical requirements and criteria that it will apply, under the Atomic Energy Act of 1954 (42 USC

2011 *et seq.*) and the Energy Reorganization Act of 1974 (42 USC 5801 *et seq.*), in approving or disapproving.–

(i) applications for authorization to construct repositories;

(ii) applications for licenses to receive and possess spent nuclear fuel and high-level radioactive waste in such repositories; and

(iii) applications for authorization for closure and decommissioning of such repositories.

(B) Such criteria shall provide for the use of a system of multiple barriers in the design of the repository and shall include such restrictions on the retrievability of the solidified high-level radioactive waste and spent fuel emplaced in the repository as the Commission deems appropriate.

(C) Such requirements and criteria shall not be inconsistent with any comparable standards promulgated by the Administrator under subsection (a).

(2) For purposes of this Act, nothing in this section shall be constructed to prohibit the Commission from promulgating requirements and criteria under paragraph (1) before the Administrator promulgates standards under subsection (a). If the Administrator promulgates standards under subsection (a) after requirements and criteria are promulgated by the Commission under paragraph (1), such requirements and criteria shall be revised by the Commission if necessary to comply with paragraph (1) (C).

(c) Environmental Impact Statements–The promulgation of standards or criteria in accordance with the provisions of this section shall not require the preparation of an environmental impact statement under section 102(2)(C) of the National Environmental Policy Act of 1969 (42 USC 4332(2)(C)), or to require any environmental review under subparagraph (E) or (F) of section 102(2) of such Act.

42 USC 10142.   **Sec. 122. Disposal of Spent Nuclear Fuel**

Notwithstanding any other provision of this subtitle, any repository constructed on a site approved under this subtitle shall be designed and constructed to permit the retrieval of any spent nuclear fuel placed in such repository, during an appropriate period of operation of the facility, for any reason pertaining to the public health and safety, or the environment, or for the purpose of permitting the recovery of the economically valuable contents of such spent fuel. The Secretary shall specify the appropriate period of retrievability with respect to any repository at the time of design of such repository, and such aspect of such repository shall be subject to approval or disapproval by the Commission as part of the construction authorization process under subsections(b) through (d) of section 114.

42 USC 10143.   **Sec. 123. Title to Material**

Delivery, and acceptance by the Secretary, of any high-level radioactive waste or spent nuclear fuel for a repository constructed under this subtitle shall constitute a transfer to the Secretary of title to such waste or spent fuel.

42 USC 10144.   **Sec. 124. Consideration of Effect of Acquisition of Water Rights**

The Secretary shall give full consideration to whether the development, construction, and operation of a repository may require any purchase or other acquisition of water rights that will have a significant adverse effect on the present or future development of the area in which such repository is located. The Secretary shall mitigate any such adverse effects to the maximum extent practicable.

42 USC 10145.

Sec. 125. Termination of Certain Provisions

Sections 119 and 120 shall cease to have effect at such time as a repository developed under this subtitle is licensed to receive and possess high-level radioactive waste and spent nuclear fuel.

Subtitle B–Interim Storage Program

42 USC 10151.

Sec. 131. Findings and Purposes

(a) FINDINGS–The congress finds that–

(1) the persons owning and operating civilian nuclear power reactors have the primary responsibility for providing interim storage of spent nuclear fuel from such reactors by maximizing, to the extent practical, the effective use of existing storage facilities at the site of each civilian nuclear power reactor, and by adding new onsite storage capacity in a timely manner where practical;

(2) the Federal Government has the responsibility to encourage and expedite the effective use of existing storage facilities and the addition of needed new storage capacity at the site of each civilian nuclear power reactor; and

(3) the Federal Government has the responsibility to provide, in accordance with the provisions of this subtitle, not more than 1,900 metric tons of capacity for interim storage of spent nuclear fuel for civilian nuclear power reactors that cannot reasonably provide adequate storage capacity at the sites of such reactors when needed to assure the continued, orderly operation of such reactors.

(b) Purposes.–The purposes of this subtitle are–

(1) to provide for the utilization of available spent nuclear fuel pools at the site of each civilian nuclear power reactor to the extent practical and the addition of new spent nuclear fuel storage capacity where practical at the site of such reactor; and

(2) to provide, in accordance with the provisions of this subtitle, for the establishment of a federally owned and operated system for the interim storage of spent nuclear fuel at one or more facilities owned by the Federal Government with not more than 1,900 metric tons of capacity to prevent disruptions in the orderly operation of any civilian nuclear power reactor that cannot reasonably provide adequate spent nuclear fuel storage capacity at the site of such reactor when needed.

42 USC 10152.

Sec. 132. Available Capacity for Interim Storage of Spent Nuclear Fuel

The Secretary, the Commission, and other authorized Federal officials shall each take such actions as such official considers necessary to encourage and expedite the effective use of available storage and necessary additional storage, at the site of each civilian nuclear power reactor consistent with–

(1) the protection of the public health and safety, and the environment;

(2) economic considerations;

(3) continued operation of such reactor;

(4) any applicable provisions of law; and

(5) the views of the population surrounding such reactor.

42 USC 10153.
Licensing
procedures.

Sec. 133. Interim at Reactor Storage

The Commission shall, by rule, establish procedures for the licensing of any technology approved by the Commission under section 219(a)[9] for use at the site of any civilian nuclear power reactor. The establishment of such procedures shall not preclude the licensing, under any applicable procedures or rules of the Commission in effect prior to such establishments, of any technology for the storage of civilian spent nuclear fuel at the site of any civilian nuclear power reactor.

42 USC 10154.

Sec. 134. Licensing of Facility Expansions and Transshipments

a) ORAL ARGUMENT–In any Commission hearing under section 189 of the Atomic Energy Act of 1954 (42 USC 2239) on an application for a license, or for an amendment to an existing license, filed after the date of the enactment of this Act, to expand the spent nuclear fuel storage capacity at the site of a civilian nuclear power reactor, through the use of high--density fuel storage racks, fuel rod compaction, the transshipment of spent nuclear fuel to another civilian nuclear power reactor within the same utility system, the construction of additional spent nuclear fuel pool capacity or dry storage capacity, or by other means, the Commission shall, at the request of any party, provide an opportunity for oral argument with respect to any matter which the Commission determines to be in controversy among the parties. The oral arguments shall preceded by such discovery procedures as the rules of the Commission shall provide The Commission shall require each party, including the Commission staff, to submit in written form, at the time of the oral argument, a summary of the facts, data, and arguments upon which such party proposes to rely that are known at such time to such party. Only facts and data in the form of sworn testimony or written submission may be relied upon by the parties during oral arguments. of the material that may be submitted by the parties during oral arguments, the Commission shall only consider those facts and data that are submitted in the form of sworn testimony or written submission.

Summary
submittal of
facts, data and
arguments.

(b) ADJUDICATORY HEARING–(1) At the conclusion of any oral argument under subsection (a), the Commission shall designate any disputed questions of fact, together with any remaining questions of law, for resolution in an adjudicatory hearing only if it determines that–

(A) there is a genuine and substantial dispute of fact which can only be resolved with sufficient accuracy by the introduction of evidence in an adjudicatory hearing; and

(B) the decision of the Commission is likely to depend in whole or in part on the resolution of such dispute.

(2) In making a determination under this subsection, the Commission–

(A) shall designate in writing the specific facts that are in genuine and substantial dispute, the reason why the decision of the agency is likely to depend on the resolution of such facts, and the reason why an adjudicatory hearing is likely to resolve the dispute; and

(B) shall not consider–

(i) any issue relating to the design, construction, or operation of any civilian nuclear power reactor already licensed to operate at such site, or any civilian nuclear power reactor for which a construction permit has been granted at

---

[9]   This should be referring to 218(a). This is a mistake made in the original statute.

such site, unless the Commission determines that any such issue substantially affects the design, construction, or operation of the facility or activity for which such license application, authorization, or amendment is being considered; or

(ii) any siting or design issue fully considered and decided by the Commission in connection with the issuance of a construction permit or operating license for a civilian nuclear power reactor at such site, unless (I) such issue results from any revision of siting or design criteria by the Commission following such decision; and (II) the Commission determines that such issue substantially affects the design, construction, or operation of the facility or activity for which such license application, authorization, or amendment is being considered.

(3) The provisions of paragraph (2)(B) shall apply only with respect to licenses, authorizations, or amendments to licenses or authorizations, applied for under the Atomic Energy Act of 1954 (42 USC 2011 *et seq.*) before December 31, 2005.

(4) The provisions of this section shall not apply to the first application for a license or license amendment received by the Commission to expand onsite spent fuel storage capacity by the use of a new technology not previously approved for use at any nuclear power plant by the Commission.

(c) Judicial Review.–No court shall hold unlawful or set aside a decision of the Commission in any proceeding described in subsection (a) because of a failure by the Commission to use a particular procedure pursuant to this section unless–

(1) an objection to the procedure used was presented to the Commission in a timely fashion or there are extraordinary circumstances that excuse the failure to present a timely objection; and

(2) the court finds that such failure has precluded a fair consideration and informed resolution of a significant issue of the proceeding taken as a whole.

42 USC 10155.

Sec. 135. Storage of Spent Nuclear Fuel

(a) STORAGE CAPACITY–(1) Subject to section 8, the Secretary shall provide, in accordance with paragraph (5), not more than 1,900 metric tons of capacity for the storage of spent nuclear fuel from civilian nuclear power reactors. Such storage capacity shall be provided through any one or more of the following methods, used in any combination determined by the Secretary to be appropriate:

(A) use of available capacity at one or more facilities owned by the Federal Government on the date of the enactment of this Act, including the modification and expansion of any such facilities, if the Commission determines that such use will adequately protect the public health and safety, except that such use shall not–

(i) render such facilities subject to licensing under the Atomic Energy Act of 1954 (42 USC 2011 *et seq.*) or the Energy Reorganization Act of 1974 (42 USC 5801 *et. seq.*); or

(ii) except as provided in subsection (c) require the preparation of an environmental impact statement under section 102(2)(C) of the National Environmental Policy Act of 1969 (42 USC 4332(2)(C)), such facility is already being used, or has previously been used, for such storage or for any similar purpose.

(B) acquisition of any modular or mobile spent nuclear fuel storage equipment, including spent nuclear fuel storage casks, and provision of such equipment, to any person generating or holding title to spent nuclear fuel, at the site of any civilian nuclear power reactor operated by such person or at any site owned by the Federal Government on the date of enactment of this Act;

(C) construction of storage capacity at any site of a civilian nuclear power reactor.

(2) Storage capacity authorized by paragraph (1) shall not be provided at any Federal or non-Federal site within which there is a candidate site for a repository. The restriction in the preceding sentence shall only apply until such time as the Secretary decides that such candidate site is no longer a candidate site under consideration for development as a repository.

(3) In selecting methods of providing storage capacity under paragraph (1), the Secretary shall consider the timeliness of the availability of each such method and shall seek to minimize the transportation of spent nuclear fuel, the public health and safety impacts, and the costs of providing such storage capacity.

(4) In providing storage capacity through any method described in paragraph (1), the Secretary shall comply with any applicable requirements for licensing or authorization of such method, except as provided in paragraph (1)(A)(i).

(5) The Secretary shall ensure that storage capacity is made available under paragraph (1) when needed, as determined on the basis of the storage needs specified in contracts entered into under section 136(a), and shall accept upon request any spent nuclear fuel as covered under such contracts.

Facility.

(6) For purposes of paragraph (1)(A), the term "facility" means any building of structure.

(b) CONTRACTS–(1) Subject to the capacity limitation established in subsections (a)(1) and (d), the Secretary shall offer to enter into, and may enter into contracts under section 136(a) with any person generating or owning spent nuclear fuel for purposes of providing storage capacity for such spent fuel under this section only if the Commission determines that–

(A) adequate storage capacity to ensure the continued orderly operation of the civilian nuclear power reactor at which such spent nuclear fuel is generated cannot reasonably be provided by the person owning and operating such reactor at such site, or at the site, of any other civilian nuclear power reactor operated by such person, and such capacity cannot be made available in a timely manner through any method described in subparagraph (B); and

(B) such person is diligently pursuing licensed alternatives to the use of Federal storage capacity for the storage of spent nuclear fuel expected to be generated by such person in the future, including-

(i) expansion of storage facilities at the site of any civilian nuclear power reactor operated by such person;

(ii) construction of new or additional storage facilities at the site of any civilian nuclear power reactor operated by such person;

(iii) acquisition of modular or mobile spent nuclear fuel storage equipment, including spent nuclear fuel storage casks, for use at the site of any civilian nuclear power reactor operated by such person; and

(iv) transshipment to another civilian nuclear power reactor owned by such person.

(2) In making the determination described in paragraph (1)(A), the Commission shall ensure maintenance of a full core reserve storage capability at the site of the civilian nuclear power reactor involved unless the Commission determines that maintenance of such capability is not necessary for the continued orderly operation of such reactor.

(3) The Commission shall complete the determinations required in paragraph (1) with respect to any request for storage capacity not later than 6 months after receipt of such request by the Commission.

(c) ENVIRONMENTAL REVIEW–(1) The provision of 300 or more metric tons of storage capacity at any one Federal site under subsection (a)(1)(A) shall be considered to be a major Federal action requiring preparation of an environmental impact statement under section 102(2)(C) of the National Environmental Policy Act of 1969 (42 USC 4332(2)(C)).

Public availability.

(2) (A) The Secretary shall prepare, and make available to the public, an environmental assessment of the probable impacts of any provision of less than 300 metric tons of storage capacity at any one Federal site under subsection (a)(1)(A) that requires the modification or expansion of any facility at the site, and a discussion of alternative activities that may be undertaken to avoid such impacts. Such environmental assessment shall include–

(i) an estimate of the amount of storage capacity to be made available at such site;

(ii) an evaluation as to whether the facilities to be used at such site are suitable for the provision of such storage capacity;

(iii) a description of activities planned by the Secretary with respect to the modification or expansion of the facilities to be used at such site;

(iv) an evaluation of the effects of the provision of such storage capacity at such site on the public health and safety, and the environment;

(v) a reasonable comparative evaluation of current information with respect to such site and facilities and other sites and facilities available for the provision of such storage capacity;

(vi) a description of any other sites and facilities that have been considered by the Secretary for the provision of such storage capacity; and

(vii) an assessment of the regional and local impacts of providing such storage capacity at such site, including the impacts on transportation.

5 USC 701 et. seq. Judicial review.

(B) The issuance of any environmental assessment under this paragraph shall be considered to be final agency action subject to judicial review in accordance with the provisions of chapter 7 of title 5, United States Code. Such judicial review shall be limited to the sufficiency of such assessment with respect to the items described in clauses (i) through (vii) of subparagraph (A).

(3) Judicial review of any environmental impact statement or environmental assessment prepared pursuant to this subsection shall be conducted in accordance with the provisions of section 119.

(d) REVIEW OF SITES AND STATE PARTICIPATION–(1) In carrying out the provisions of this subtitle with regard to any interim storage of spent fuel from civilian nuclear power reactors which the Secretary is authorized by section 135 to provide, the Secretary shall, as soon as practicable, notify, in writing, the Governor and the State legislature of any State and the Tribal Council of any affected Indian tribe in such State in which is located a potentially acceptable site or facility for such interim storage of spent fuel of his intention to investigate that site or facility.

Investigation.

(2) During the course of investigation of such site or facility, the Secretary shall keep the Governor, State legislature, and affected Tribal Council currently informed of the progress of the work, and results of the investigation. At the time of selection by the Secretary of any site or existing facility, but prior to undertaking any site-specific work or alterations, the Secretary shall promptly notify the Governor, the legislature, and any affected Tribal Council in writing of such selection and subject to the provisions of paragraph (6) of this subsection, shall promptly enter into negotiations with such State and affected Tribal Council to establish a cooperative agreement under which such State and Council shall have the right to participate in a process of consultation and cooperation, based on public health and safety and environmental concerns, in all stages of the planning, development, modification, expansion, operation, and closure of storage capacity at a site or facility within such State for the interim storage of spent fuel from civilian nuclear power reactors. Public participation in the negotiation of such an agreement shall be provided for and encouraged by the Secretary, the State, and the affected Tribal Council. The Secretary in cooperation with the State and Indian tribes, shall develop and publish minimum guideline for public participation in such negotiations, but the adequacy of such guidelines or any failure to comply with such guidelines shall not be a basis for judicial review.

Guidelines.

Cooperative agreement.

(3) The cooperative agreement shall include, but need not be limited to, the sharing in accordance with applicable law of all technical and licensing information, the utilization of available expertise, the facilitating of permitting procedures, joint project review, and the formulation of joint surveillance and monitoring arrangements to carry out applicable Federal and State laws. The cooperative agreement also shall include a detailed plan or schedule of milestones, decision points and opportunities for State or eligible Tribal Council review and objection. Such cooperative agreement shall provide procedures for negotiating and resolving objections of the State and affected Tribal Council in any stage of planning, development, modification, expansion, operation, or closure of storage capacity at a site or facility within such State. The terms of any cooperative agreement shall not affect the authority of the Nuclear Regulatory Commission under existing law.

Process of consultation and cooperation.

(4) For the purpose of this subsection, "process of consultation and cooperation" means a methodology by which the Secretary (A) keeps the State and eligible Tribal Council fully and currently informed about the aspects of the project related to any potential impact on the public health and safety and environment; (B) solicits, receives, and evaluates concerns and objections of such State and Council with regard to such aspects of the project on an ongoing basis; and (C) works diligently and cooperatively to resolve, through arbitration or other appropriate mechanisms, such concerns and

objections. The process of consultation and cooperation shall not include the grant of a right to any State or Tribal Council to exercise an absolute veto of any aspect of the planning, development, modification, expansion, or operation of the project.

Report to Congress.

(5) The Secretary and the State and affected Tribal Council shall seek to conclude the agreement required by paragraph (2) as soon as practicable, but not later than 180 days following the date of notification of the selection under paragraph (2).The Secretary shall periodically report to the Congress thereafter on the status of the agreements approved under paragraph (3). Any report to the Congress on the status of negotiations of such agreement by the Secretary shall be accompanied by comments solicited by the Secretary from the State and eligible Tribal Council.

(6) (A) Upon deciding to provide an aggregate of 300 or more metric tons of storage capacity under subsection (a)(1) at any one site, the Secretary shall notify the Governor and legislature of the State where such site is located, or the governing body of the Indian tribe in whose reservation such site is located, as the case may be, of such decision. During the 60-day period following receipt of notification by the Secretary of his decision to provide an aggregate of 300 or more metric tons of storage capacity at any one site, the Governor or legislature of the State in which such site is located, or the governing body of the affected Indian tribe where such site is located, as the case may be, may disapprove the provision of 300 or more metric tons of storage capacity at the site involved and submit to the Congress a notice of such disapproval. A notice of disapproval shall be considered to be submitted to the Congress on the date of the transmittal of such notice of disapproval to the Speaker of the House and

Notice of disapproval, submittal to Congress.

(B) Unless otherwise provided by State law, the Governor or legislature of each State shall have authority to submit a notice of disapproval to the Congress under subparagraph (A). In any case in which State law provides for submission of any such notice of disapproval by any other person or entity, any reference in this subtitle to the Governor or legislature of such State shall be considered to refer instead to such other person or entity.

(C) The authority of the Governor and legislature of each State under this paragraph shall not be applicable with respect to any site located on a reservation.

(D) If any notice of disapproval is submitted to the Congress under subparagraph (A), the proposed provision of 300 or more metric tons of storage capacity at the site involved shall be disapproved unless, during the first period of 90 calendar days of continuous session of the Congress following the date of the receipt by the Congress of such notice of disapproval, the Congress passes a resolution approving such proposed provision of storage capacity in accordance with the procedures established in this paragraph and subsections (d) through (f) of section 115 and such resolution thereafter becomes law. For purposes of this paragraph, the term "resolution" means a joint resolution of either House of the Congress, the matter after the resolving clause of which is as follows: That there hereby is approved the provision of 300 or more metric tons of spent nuclear fuel storage capacity at the site located at _____, with respect to which a notice of disapproval was submitted by _____ on _____. The first blank space in such resolution

Resolution.

shall be filled with the geographic location of the site involved; the second blank space in such resolution shall be filled with the designation of the State Governor and legislature or affected Indian tribe governing body submitting the notice of disapproval involved; and the last blank space in such resolution shall be filled with the date of submission of such notice of disapproval.

(E) For purposes of the consideration of any resolution described in subparagraph (D), each reference in subsections (d) and (e) of section 115 to a resolution of repository siting approval shall be considered to refer to the resolution described in such subparagraph.

Affected Tribal Council.

(7) As used in this section, the term "affected Tribal Council" means the governing body of any Indian tribe within whose reservation boundaries there is located a potentially acceptable site for interim storage capacity of spent nuclear fuel from civilian nuclear power reactors, or within whose boundaries a site for such capacity is selected by the Secretary, or whose federally defined possessory or usage rights to other lands outside of the reservation's boundaries arising out of congressionally ratified treaties, as determined by the Secretary of the Interior pursuant to a petition filed with him by the appropriate governmental officials of such tribe, may be substantially and adversely affected by the establishment of any such storage capacity.

(e) LIMITATIONS–Any spent nuclear fuel stored under this section shall be removed from the storage site or facility involved as soon as practicable, but in any event not later than 3 years following the date on which a repository or monitored retrievable storage facility developed under this Act is available for disposal of such spent nuclear fuel.

(f) REPORT.–The Secretary shall annually prepare and submit to the Congress a report on any plans of the Secretary for providing storage capacity under this section. Such report shall include a description of the specific manner of providing such storage selected by the Secretary, if any. The Secretary shall prepare and submit the first such report not later than 1 year after the date of the enactment of this Act.

5 USC 553.

(g) CRITERIA FOR DETERMINING ADEQUACY OF AVAILABLE STORAGE CAPACITY–Not later than 90 days after the date of the enactment of this Act, the Commission pursuant to section 553 of the Administrative Procedures Act, shall propose, by rule, procedures and criteria for making the determination required by subsection (b) that a person owning and operating a civilian nuclear power reactor cannot reasonably provide adequate spent nuclear fuel storage capacity at the civilian nuclear power reactor site when needed to ensure the continued orderly operation of such reactor. Such criteria shall ensure the maintenance of a full core reserve storage capability at the site of such reactor unless the Commission determines that maintenance of such capability is not necessary for the continued orderly operation of such reactor. Such criteria shall identify the feasibility of reasonably providing such adequate spent nuclear fuel storage capacity, taking into account economic, technical, regulatory, and public health and safety factors, through the use of high-density fuel storage racks, fuel rod compaction, transshipment of spent nuclear fuel to another civilian nuclear power reactor within the same utility system, construction of additional spent nuclear fuel pool capacity, or such other technologies as may be approved by the Commission.

(h) APPLICATION–Notwithstanding any other provision of law, nothing in this Act shall be construed to encourage, authorize, or require the private or Federal use, purchase, lease, or other acquisition of any storage facility located away from the site of any civilian nuclear power reactor and not owned by the Federal Government on the date of the enactment of this Act.

(i) COORDINATION WITH RESEARCH AND DEVELOPMENT PROGRAM–To the extent available, and consistent with the provisions of this section, the Secretary shall provide spent nuclear fuel for the research and development program authorized in section 217 from spent nuclear fuel received by the Secretary for storage under this section. Such spent nuclear fuel shall not be subject to the provisions of subsection (e).

Sec. 136. Interim Storage Fund

(a) CONTRACTS–

(1) During the period following the date of the enactment of this Act, but not later than January 1, 1990, the Secretary is authorized to enter into contracts with persons who generate or own spent nuclear fuel resulting from civilian nuclear activities for the storage of such spent nuclear fuel in any storage capacity provided under this subtitle: *Provided, however,* That the Secretary shall not enter into contracts for spent nuclear fuel in amounts in excess of the available storage capacity specified in section 135(a). Those contracts shall provide that the Federal Government will (1) take title at the civilian nuclear power reactor site, to such amounts of spent nuclear fuel from the civilian nuclear power reactor as the Commission determines cannot be stored onsite, (2) transport the spent nuclear fuel to a federally owned and operated interim away-from-reactor storage facility, and (3) store such fuel in the facility pending further processing, storage, or disposal. Each such contract shall (A) provide for payment to the Secretary of fees determined in accordance with the provisions of this section; and (B) specify the amount of storage capacity to be provided for the person involved.

(2) The Secretary shall undertake a study and, not later than 180 days after the date of the enactment of this Act, submit to the Congress a report, establishing payment charges that shall be calculated on an annual basis, commencing on or before January 1, 1984. Such payment charges and the calculation thereof shall be published in the Federal Register, and shall become effective not less than 30 days after publication. Each payment charge published in the Federal Register under this paragraph shall remain effective for a period of 12 months from the effective date as the charge for the cost of the interim storage of any spent nuclear fuel. The report of the Secretary shall specify the method and manner of collection (including the rates and manner of payment) and any legislative recommendations determined by the Secretary to be appropriate.

(3) Fees for storage under this subtitle shall be established on a nondiscriminatory basis. The fees to be paid by each person entering into a contract with the Secretary under this subsection shall be based upon an estimate of the pro rata costs of storage and related activities under this subtitle with respect to such person, including the acquisition, construction, operation, and maintenance of any facilities under this subtitle.

(4) The Secretary shall establish in writing criteria setting forth the terms and conditions under which such storage services shall be made available.

(5) Except as provided in section 137, nothing in this or any other Act requires the Secretary, in carrying out the responsibilities of this

42 USC 10156.

Study; report to Congress.

Publication in *Federal Register.*

Fees.

section, to obtain a license or permit to possess or own spent nuclear fuel.

(b) LIMITATION–No spent nuclear fuel generated or owned by any department of the United States referred to in section 101 or 102 of title 5, United States Code, may be stored by the Secretary in any storage capacity provided under this subtitle unless such department transfers to the Secretary, for deposit in the Interim Storage Fund, amounts equivalent to the fees that would be paid to the Secretary under the contracts referred to in this section if such spent nuclear fuel were generated by any other person.

(c) ESTABLISHMENT OF INTERIM STORAGE FUND–There hereby is established in the Treasury of the United States a separate fund, to be known as the Interim Storage Fund. The Storage Fund shall consist of–

(1) all receipts, proceeds, and recoveries realized by the Secretary under subsections (a), (b), and (e), which shall be deposited in the Storage Fund immediately upon their realization;

(2) any appropriations made by the Congress to the Storage Fund; and

(3) any unexpended balances available on the date of the enactment of this Act for functions or activities necessary or incident to the interim storage of civilian spent nuclear fuel, which shall automatically be transferred to the Storage Fund on such date.

(d) USE OF STORAGE FUND–The Secretary may make expenditures from the Storage Fund, subject to subsection (e), for any purpose necessary or appropriate to the conduct of the functions and activities of the Secretary, or the provision or anticipated provision of services, under this subtitle, including–

(1) the identification, development, licensing, construction, operation, decommissioning, and post-decommissioning maintenance and monitoring of any interim storage facility provided under this subtitle;

(2) the administrative cost of the interim storage program;

(3) the costs associated with acquisition, design, modification, replacement, operation, and construction of facilities at an interim storage site, consistent with the restrictions in section 135;

(4) the cost of transportation of spent nuclear fuel; and

(5) impact assistance as described in subsection (e).

Payments.   (e) IMPACT ASSISTANCE–(1) Beginning the first fiscal year which commences after the date of the enactment of this Act, the Secretary shall make annual impact assistance payments to a State or appropriate unit of local government, or both, in order to mitigate social or economic impacts occasioned by the establishment and subsequent operation of any interim storage capacity within the jurisdictional boundaries of such government or governments and authorized under this subtitle: *Provided, however,* That such impact assistance payments shall not exceed (A) ten percentum of the costs incurred in paragraphs (1) and (2), or (B) $15 per kilogram of spent fuel, whichever is less:

(2) Payments made available to States and units of local government pursuant to this section shall be–

(A) allocated in a fair and equitable manner with a priority to those States or units of local government suffering the most severe impacts; and

(B) utilized by States or units of local governments only for (i) planning, (ii) construction and maintenance of public services, (iii) provision of public services related to the providing of such interim storage authorized under this title, and (iv) compensation for loss of taxable property equivalent to that if the storage had been provided under private ownership.

Regulations.

(3) Such payments shall be subject to such terms and conditions as the Secretary determines necessary to ensure that the purposes of this subsection shall be achieved. The Secretary shall issue such regulations as may be necessary to carry out the provisions of this subsection.

(4) Payments under this subsection shall be made available solely from the fees determined under subsection (a).

(5) The Secretary is authorized to consult with States and appropriate units of local government in advance of commencement of establishment of storage capacity authorized under this subtitle in an effort to determine the level of the payment such government would be eligible to receive pursuant to this subsection.

Unit of local government.

(6) As used in this subsection, the term "unit of local government" means a county, parish, township, municipality, and shall include a borough existing in the State of Alaska on the date of the enactment of this subsection, and any other unit of government below the State level which is a unit of general government as determined by the Secretary.

Report to Congress.

(f) ADMINISTRATION OF STORAGE FUND–(1) The Secretary of the Treasury shall hold the Storage Fund and, after consultation with the Secretary, annually report to the Congress on the financial condition and operations of the Storage Fund during the preceding fiscal year.

Budget submittal.

(2) The Secretary shall submit the budget of the Storage Fund to the Office of Management and Budget triennially along with the budget of the Department of Energy submitted at such time in accordance with chapter 11 of title 31, United States Code. The budget of the Storage Fund shall consist of estimates made by the Secretary of expenditures from the Storage Fund and other relevant financial matters for the succeeding 3 fiscal years, and shall be included in the Budget of the United States Government. The Secretary may make expenditures from the Storage Fund, subject to appropriations which shall remain available until expended. Appropriations shall be subject to triennial authorization.

(3) If the Secretary determines that the Storage Fund contains at any time amounts in excess of current needs, the Secretary may request the Secretary of the Treasury to invest such amounts, or any portion of such amounts as the Secretary determines to be appropriate, in obligations of the United States–

(A) having maturities determined by the Secretary of the Treasury to be appropriate to the needs of the Storage Fund; and

(B) bearing interest at rates determined to be appropriate by the Secretary of the Treasury, taking into consideration the current average market yield on outstanding marketable obligations of the United States with remaining periods to maturity comparable to the maturities of such investments, except that the interest rate on such investments shall not exceed the average interest rate applicable to existing borrowings.

(4) Receipts, proceeds, and recoveries realized by the Secretary under this section, and expenditures of amounts from the Storage Fund, shall be exempt from annual apportionment under the provisions of subchapter II of chapter 15 of title 31, United States Code.

(5) If at any time the moneys available in the Storage Fund are insufficient to enable the Secretary to discharge his responsibilities under this subtitle, the Secretary shall issue to the Secretary of the Treasury obligations in such forms and denominations, bearing such maturities, and subject to such terms and conditions as may be agreed to by the Secretary and the Secretary of the Treasury. The total of such obligations shall not exceed amounts provided in appropriation Acts. Redemption of such obligations shall be made by the Secretary from moneys available in the Storage Fund. Such obligations shall bear interest at a rate determined by the Secretary of the Treasury, which shall be not less than a rate determined by taking into consideration the average market yield on outstanding marketable obligations of the United States of comparable maturities during the month preceding the issuance of the obligations under this paragraph. The Secretary of the Treasury shall purchase any issued obligations, and for such purpose the Secretary of the Treasury is authorized to use as a public debt transaction the proceeds from the sale of any securities issued under chapter 31 of title 31, United States Code, and the purposes for which securities may be issued under such Act are extended to include any purchase of such obligations. The Secretary of the Treasury may at any time sell any of the obligations acquired by him under this paragraph. All redemptions, purchases, and sales by the Secretary of the Treasury of obligations under this paragraph shall be treated as public debt transactions of the United States.

Interest payments.

(6) Any appropriations made available to the Storage Fund for any purpose described in subsection (d) shall be repaid into the general fund of the Treasury, together with interest from the date of availability of the appropriations until the date of repayment. Such interest shall be paid on the cumulative amount of appropriations available to the Storage Fund, less the average undisbursed cash balance in the Storage Fund account during the fiscal year involved. The rate of such interest shall be determined by the Secretary of the Treasury taking into consideration the average market yield during the month preceding each fiscal year on outstanding marketable obligations of the United States of comparable maturity. Interest payments may be deferred with the approval of the Secretary of the Treasury, but any interest payments so deferred shall themselves bear interest.

Deferral.

42 USC 10157.

Sec. 137. Transportation

(a) TRANSPORTATION–(1) Transportation of spent nuclear fuel under section 136(a) shall be subject to licensing and regulation by the Commission and by the Secretary of Transportation as provided for transportation of commercial spent nuclear fuel under existing law.

(2) The Secretary, in providing for the transportation of spent nuclear fuel under this Act, shall utilize by contract private industry to the fullest extent possible in each aspect of such transportation. The Secretary shall use direct Federal services for such transportation only upon a determination of the Secretary of Transportation, in consultation with the Secretary, that private industry is unable or unwilling to provide such transportation services at reasonable cost.

Subtitle C–Monitored Retrievable Storage

42 USC 10161.   **Sec. 141. Monitored Retrievable Storage**

(a) FINDINGS–The Congress finds that–

(1) long-term storage of high-level radioactive waste or spent nuclear fuel in monitored retrievable storage facilities is an option for providing safe and reliable management of such waste or spent fuel;

(2) the executive branch and the Congress should proceed as expeditiously as possible to consider fully a proposal for construction of one or more monitored retrievable storage facilities to provide such long-term storage;

(3) the Federal Government has the responsibility to ensure that site-specific designs for such facilities are available as provided in this section;

(4) the generators and owners of the high-level radioactive waste and spent nuclear fuel to be stored in such facilities have the responsibility to pay the costs of the long-term storage of such waste and spent fuel; and

(5) disposal of high-level radioactive waste and spent nuclear fuel in a repository developed under this Act should proceed regardless of any construction of a monitored retrievable storage facility pursuant to this section.

(b) SUBMISSION OF PROPOSAL BY SECRETARY–(1) On or before June 1, 1985, the Secretary shall complete a detailed study of the need for and feasibility of, and shall submit to the Congress a proposal for, the construction of one or more monitored retrievable storage facilities for high-level radioactive waste and spent nuclear fuel. Each such facility shall be designed–

(A) to accommodate spent nuclear fuel and high-level radioactive waste resulting from civilian nuclear activities;

(B) to permit continuous monitoring, management, and maintenance of such spent fuel and waste for the foreseeable future;

(C) to provide for the ready retrieval of such spent fuel and waste for further processing or disposal; and

(D) to safely store such spent fuel and waste as long as may be necessary by maintaining such facility through appropriate means, including any required replacement of such facility.

(2) Such proposal shall include–

(A) the establishment of a Federal program for the siting, development, construction, and operation of facilities capable of safely storing high-level radioactive waste and spent nuclear fuel, which facilities are to be licensed by the Commission;

(B) a plan for the funding of the construction and operation of such facilities, which plan shall provide that the costs of such activities shall be borne by the generators and owners of the high-level radioactive waste and spent nuclear fuel to be stored in such facilities;

(C) site-specific designs, specifications, and cost estimates sufficient to (i) solicit bids for the construction of the first such facility; (ii) support congressional authorization of the construction of such facility; and (iii) enable completion and operation of such facility as soon as practicable following congressional authorization of such facility; and

(D) a plan for integrating facilities constructed pursuant to this section with other storage and disposal facilities authorized in this Act.

Consultations.

(3) In formulating such proposal, the Secretary shall consult with the Commission and the Administrator, and shall submit their comments on such proposal to the Congress at the time such proposal is submitted.

(4) The proposal shall include, for the first such facility, at least 3 alternative sites and at least 5 alternative combinations of such proposed sites and facility designs consistent with the criteria of paragraph (b)(1). The Secretary shall recommend the combination among the alternatives that the Secretary deems preferable. The environmental assessment under subsection (c) shall include a full analysis of the relative advantages and disadvantages of all 5 such alternative combinations of proposed sites and proposed facility designs.

Environmental assessment.

(c) ENVIRONMENTAL IMPACT STATEMENTS–(1) Preparation and submission to the Congress of the proposal required in this section shall not require the preparation of an environmental impact statement under section 102(2)(C) of the National Environmental Policy Act of 1969 (42 USC 4332(2)(C)). The Secretary shall prepare, in accordance with regulations issued by the Secretary implementing such Act, an environmental assessment with respect to such proposal. Such environmental assessment shall be based upon available information regarding alternative technologies for the storage of spent nuclear fuel and high-level radioactive waste. The Secretary shall submit such environmental assessment to the Congress at the time such proposal is submitted.

Submittal to Congress.

(2) If the Congress by law, after review of the proposal submitted by the Secretary under subsection (b), specifically authorizes construction of a monitored retrievable storage facility, the requirements of the National Environmental Policy Act of 1969 (42 USC 4321 et seq.) shall apply with respect to construction of such facility, except that any environmental impact statement prepared with respect to such facility shall not be required to consider the need for such facility or any alternative to the design criteria for such facility set forth in subsection (b) (1).

(d) LICENSING–Any facility authorized pursuant to this section shall be subject to licensing under section 202(3)) of the Energy Reorganization Act of 1974   (42 USC 5842(3). In reviewing the application filed by the Secretary for licensing of the first such facility, the Commission may not consider the need for such facility or any alternative to the design criteria for such facility set forth in subsection (b) (1).

(e) CLARIFICATION–Nothing   in   this   section   limits   the consideration of alternative facility designs consistent with the criteria of paragraph (b)(1) in any environmental impact statement, or in any licensing procedure of the Commission, with respect to any monitored, retrievable facility authorized pursuant to this section.

Payments.

(f) IMPACT ASSISTANCE–(1) Upon receipt by the Secretary of congressional authorization to construct a facility described in subsection (b), the Secretary shall commence making annual impact aid payments to appropriate units of general local government in order to migrate any social or economic impacts resulting from the construction and subsequent operation of any such facility within the jurisdictional boundaries of any such unit.

(2) payments made available to units of general local government under this subsection shall be–

(A) allocated in a fair and equitable manner, with priority given to units of general local government determined by the Secretary to be most severely affected; and

(B) utilized by units of general local government only for planning, construction, maintenance, and provision of public services related to the siting of such facility.

Regulations.

(3) Such payments shall be subject to such terms and conditions as the Secretary determines are necessary to ensure achievement of the purposes of this subsection. The Secretary shall issue such regulations as may be necessary to carry out the provisions of this subsection.

(4) Such payments shall be made available entirely from funds held in the Nuclear Waste Fund established in section 302 (c) and shall be available only to the extent provided in advance in appropriation Acts.

Consultations.

(5) The Secretary may consult with appropriate units of general local government in advance of commencement of construction of any such facility in an effort to determine the level of payments each such unit is eligible to receive under this subsection.

(g) LIMITATION–No monitored retrievable storage facility development pursuant to this section may be constructed in any State in which there is located any site approved for site characterization under section 112. The restriction in the preceding sentence shall only apply until such time as the Secretary decides that such candidate site is no longer a candidate site under consideration for development as a repository. Such restriction shall continue to apply to any site selected for construction as a repository.

(h) PARTICIPATION OF STATES AND INDIAN TRIBES–Any facility authorized pursuant to this section shall be subject to the provisions of sections 115, 116(a), 116(b), 116(d), 117, and 118. For purposes of carrying out the provisions of this subsection, any reference in sections 115 through 118 to a repository shall be considered to refer to a monitored retrievable storage facility.

42 USC 10162.

Sec. 142. Authorization of Monitored Retrievable Storage

(a) NULLIFICATION OF OAK RIDGE SITTING PROPOSAL–The proposal the Secretary (EC-1022, 100th Congress) to locate a monitored retrievable storage facility at a site on the Clinch River in the Roane County portion of Oak Ridge, Tennessee, with alternative sites on the Oak Ridge Reservation of the Department of Energy and on the former site of a proposed nuclear power plant in Hartsville, Tennessee, is annulled and revoked. In carrying out the provisions of sections 144 and 145, the Secretary shall make no presumption or preference to such sites by reason of their previous selection.

(b) Authorization.–The Secretary is authorized to site, construct, and operate one monitored retrievable storage facility subject to the conditions described in sections 143 through 149.[10]

42 USC 10163.

Sec. 143. Monitored Retrievable Storage Commission

(a) ESTABLISHMENT–(1) (A) There is established a Monitored Retrievable Storage Review Commission (hereinafter in this section

---

[10]   Sections 142 through 149 added by P.L. 100–203, Title V, § 5021, 101 Stat. 1330–232 (1987).

referred to as the "MRS Commission"), that shall consist of 3 members who shall be appointed by and serve at the pleasure of the President pro tempore of the Senate and the Speaker of the House of Representatives.

(B)[75] Members of the MRS Commission shall be appointed not later than 30 days after the date of the enactment of the Nuclear Waste Policy Amendments Act of 1987 from among persons who as a result of training, experience and attainments are exceptionally well qualified to evaluate the need for a monitored retrievable storage facility as a part of the Nation's nuclear waste management system.

Reports.

(C) The MRS Commission shall prepare a report on the need for a monitored retrievable storage facility as a part of a national nuclear waste management system that achieves the purposes of this Act. In preparing the report under this subparagraph, the MRS Commission shall–

(i) review the status and adequacy of the Secretary's evaluation of the systems advantages and disadvantages of bringing such a facility into the national nuclear waste disposal system;

(ii) obtain comment and available data on monitored retrievable storage from affected parties, including States containing potentially acceptable sites;

(iii) evaluate the utility of a monitored retrievable storage facility from a technical perspective; and

(iv) make a recommendation to Congress as to whether such a facility should be included in the national nuclear waste management system in order to achieve the purposes of this Act, including meeting needs for packaging and handling of spent nuclear fuel, improving the flexibility of the repository development schedule, and providing temporary storage of spent nuclear fuel accepted for disposal.

(2) In preparing the report and making its recommendation under paragraph (1) the MRS Commission shall compare such a facility to the alternative of at-reactor storage of spent nuclear fuel prior to disposal of such fuel in a repository under this Act. Such comparison shall take into consideration the impact on–

(A) repository design and construction;

(B) waste package design, fabrication and standardization;

(C) waste preparation;

(D) waste transportation systems;

(E) the reliability of the national system for the disposal of radioactive waste;

(F) the ability of the Secretary to fulfill contractual commitments of the Department under this Act to accept spent nuclear fuel for disposal; and

(G) economic factors, including the impact on the costs likely to be imposed on ratepayers of the Nation's electric utilities for temporary at-reactor storage of spent nuclear fuel prior to final disposal in a repository, as well as the costs likely to be imposed on ratepayers of the Nation's electric utilities in building and operating such a facility.

Reports.

(3)[11] The report under this subsection, together with the recommendation of the MRS Commission, shall be transmitted to Congress on November 1, 1989.

(4) (A) (i) Each member of the MRS Commission shall be paid at the rate provided for level III of the Executive Schedule for each day (including travel time) such member is engaged in the work of the MRS Commission, and shall receive travel expenses, including per diem in lieu of subsistence in the same manner as is permitted under sections 5702 and 5703 of title 5, United States Code.

(ii) The MRS Commission may appoint and fix compensation, not to exceed the rate of basic pay payable for GS-18 of the General Schedule, for such staff as may be necessary to carry out its functions.

(B) (i) The MRS Commission may hold hearings, sit and act at such times and places, take such testimony and receive such evidence as the MRS Commission considers appropriate. Any member of the MRS Commission may administer oaths or affirmations to witnesses appearing before the MRS Commission.

(ii) The MRS Commission may request any Executive agency, including the Department, to furnish such assistance or information, including records, data, files, or documents, as the Commission considers necessary to carry out its functions. Unless prohibited by law, such agency shall promptly furnish such assistance or information.

(iii) To the extent permitted by law, the Administrator of the General Services Administration shall, upon request of the MRS Commission, provide the MRS Commission with necessary administrative services, facilities, and support on a reimbursable basis.

(iv) The MRS Commission may procure temporary and intermittent services from experts and consultants to the same extent as is authorized by section 3109(b) of title 5, United States Code, at rates and under such rules as the MRS Commission considers reasonable.

(C) The MRS Commission shall cease to exist 60 days after the submission to Congress of the report required under this subsection.

42 USC 10164.

Sec. 144. Survey

After the MRS Commission submits its report to the Congress under section 143, the Secretary may conduct a survey and evaluation of potentially suitable sites for a monitored retrievable storage facility. In conducting such survey and evaluation, the Secretary shall consider the extent to which siting a monitored retrievable storage facility at each site surveyed would–

(1) enhance the reliability and flexibility of the system for the disposal of spent nuclear fuel and high-level radioactive waste established under this Act;

(2) minimize the impacts of transportation and handling of such fuel and waste;

(3) provide for public confidence in the ability of such system to safely dispose of the fuel and waste;

---

[11]   Added by P.L. 100–203, Title V, § 5021, 101 Stat. 1330–232 (1987); P.L. 100–507, § 2, 102 Stat. 2541 (1988), extended the report deadline from June 1, 1989 to November 1, 1989.

(4) impose minimal adverse effects on the local community and the local environment;

(5) provide a high probability that the facility will meet applicable environmental, health, and safety requirements in a timely fashion;

(6) provide such other benefits to the system for the disposal of spent nuclear fuel and high-level radioactive waste as the Secretary deems appropriate; and

(7) unduly burden a State in which significant volumes of high-level radioactive waste resulting from atomic energy defense activities are stored.

42 USC 10165.   **Sec. 145. Site Selection**

(a) GENERAL–The Secretary may select the site evaluated under section 144 that the Secretary determines on the basis of available information to be the most suitable for a monitored retrievable storage facility that is an integral part of the system for the disposal of spent nuclear fuel and high-level radioactive waste established under this Act.

(b) LIMITATION–The Secretary may not select a site under subsection (a) until the Secretary recommends to the President the approval of a site for development as a repository under section 114(a).

(c) SITE SPECIFIC ACTIVITIES–The Secretary may conduct such site specific activities at each site surveyed under section 144 as he determines may be necessary to support an application to the Commission for a license to construct a monitored retrievable storage facility at such site.

(d) ENVIRONMENTAL ASSESSMENT–Site specific activities and selection of a site under this section shall not require the preparation of an environmental impact statement under section 102(2)(C) of the National Environmental Policy Act of 1969 (42 USC 4332(2)(C)). The Secretary shall prepare an environmental assessment with respect to such selection in accordance with regulations issued by the Secretary implementing such Act. Such environmental assessment shall be based upon available information regarding alternative technologies for the storage of spent nuclear fuel and high-level radioactive waste. The Secretary shall submit such environmental assessment to the Congress at the time such site is selected.

(e) NOTIFICATION BEFORE SELECTION–(1) At least 6 months before selecting a site under subsection (a), the Secretary shall notify the Governor and legislature of the State in which such site is located, or the governing body of the affected Indian tribe where such site is located, as the case may be, of such potential selection and the basis for such selection.

(2) Before selecting any site under subsection (a), the Secretary shall hold at least one public hearing in the vicinity of such site to solicit any recommendations of interested parties with respect to issues raised by the selection of such site.

(f) NOTIFICATION OF SELECTION–The Secretary shall promptly notify Congress and the appropriate State or Indian tribe of the selection under subsection (a).

(g) LIMITATION–No monitored retrievable storage facility authorized pursuant to section 142 (b) may be constructed in the State of Nevada.

42 USC 10166.   **Sec. 146. Notice of Disapproval**

(a) IN GENERAL–The selection of a site under section 145 shall be effective at the end of the period of 60 calendar days beginning on the date of notification under such subsection, unless the governing body of the Indian tribe on whose reservation such site is located, or, if the site is not on a reservation, the Governor and the legislature of the State in which the site is located, has submitted to Congress a notice of disapproval with respect to such site. If any such notice of disapproval has been submitted under this subsection, the selection of the site under section 145 shall not be effective except as provided under section 115(c).

(b) REFERENCES.–For purposes of carrying out the provisions of this subsection, references in section 115(c) to a repository shall be considered to refer to a monitored retrievable storage facility and references to a notice of disapproval of a repository site designation under section 116(b) or 118(a) shall be considered to refer to a notice of disapproval under this section.

42 USC 10167.

Sec. 147. Benefits Agreement

Once selection of a site for a monitored retrievable storage facility is made by the Secretary under section 145, the Indian tribe on whose reservation the site is located, or, in the case that the site is not located on a reservation, the State in which the site is located, shall be eligible to enter into a benefits agreement with the Secretary under section 170.

42 USC 10168.

Sec. 148. Construction Authorization

(a) ENVIRONMENTAL IMPACT STATEMENT–(1) Once the selection of a site is effective under section 146, the requirements of the National Environmental Policy Act of 1969 (42 USC 4321 *et seq.*) shall apply with respect to construction of a monitored retrievable storage facility, except that any environmental impact statement prepared with respect to such facility shall not be required to consider the need for such facility or any alternative to the design criteria for such facility set forth in section 141 (b) (1).

(2) Nothing in this section shall be construed to limit the consideration of alternative facility designs consistent with the criteria described in section 141(b)(1) in any environmental impact statement, or in any licensing procedure of the Commission, with respect to any monitored retrievable storage facility authorized under section 142(b).

(b) APPLICATION FOR CONSTRUCTION LICENSE–Once the selection of a site for a monitored retrievable storage facility is effective under section 146, the Secretary may submit an application to the Commission for a license to construct such a facility as part of an integrated nuclear waste management system and in accordance with the provisions of this section and applicable agreements under this Act affecting such facility.

(c) LICENSING–Any monitored retrievable storage facility authorized pursuant to section 142(b) shall be subject to licensing under section 202(3) of the Energy Reorganization Act of 1974 (42 USC 5842(3)). In reviewing the application filed by the Secretary for licensing of such facility, the Commission may not consider the need for such facility or any alternative to the design criteria for such facility set forth in section 141(b)(1).

(d) LICENSING CONDITIONS–Any license issued by the Commission for a monitored retrievable storage facility under this section shall provide that–

(1) construction of such facility may not begin until the Commission has issued a license for the construction of a repository under section 115(d);

(2) construction of such facility or acceptance of spent nuclear fuel or high-level radioactive waste shall be prohibited during such

time as the repository license is revoked by the Commission or construction of the repository ceases;

(3) the quantity of spent nuclear fuel or high-level radioactive waste at the site of such facility at any one time may not exceed 10,000 metric tons of heavy metal until a repository under this Act first accepts spent nuclear fuel or solidified high-level radioactive waste; and

(4) the quantity of spent nuclear fuel or high-level radioactive waste at the site of the facility at any one time may not exceed 15,000 metric tons of heavy metal.

42 USC 10169.   **Sec. 149. Financial Assistance**

The provisions of section 116(c) or 118(b) with respect to grants, technical assistance, and other financial assistance shall apply to the State, to affected Indian tribes and to affected units of local government in the case of a monitored retrievable storage facility in the same manner as for a repository.

Subtitle D–Low-Level Radioactive Waste

42 USC 10171.   **Sec. 151. Financial Arrangements for Low-Level Radioactive Waste Site Closure**

(a) FINANCIAL ARRANGEMENTS–(1) The Commission shall establish by rule, regulation, or order, after public notice, and in accordance with section 181 of the Atomic Energy Act of 1954 (42 USC 2231), such standards and instructions as the Commission may deem necessary or desirable to ensure in the case of each license for the disposal of low-level radioactive waste that an adequate bond, surety, or other financial arrangement (as determined by the Commission) will be provided by a licensee to permit completion of all requirements established by the Commission for the decontamination, decommissioning, site closure, and reclamation of sites, structures, and equipment used in conjunction with such low-level radioactive waste. Such financial arrangements shall be provided and approved by the Commission, or, in the case of sites within the boundaries of any agreement State under section 274 of the Atomic Energy Act of 1954 (42 USC 2021), by the appropriate State or State entity, prior to issuance of licenses for low-level radioactive waste disposal or, in the case of licenses in effect on the date of the enactment of this Act, prior to termination of such licenses.

(2) If the Commission determines that any long-term maintenance or monitoring, or both, will be necessary at a site described in paragraph (1), the Commission shall ensure before termination of the license involved that the licensee has made available such bonding, surety, or other financial arrangements as may be necessary to ensure that any necessary long-term maintenance or monitoring needed for such site will be carried out by the person having title and custody for such site following license termination.

(b) TITLE AND CUSTODY–(1) The Secretary shall have authority to assume title and custody of low-level radioactive waste and the land on which such waste is disposed of, upon request of the owner of such waste and land and following termination of the license issued by the Commission for such disposal, if the Commission determines that–

(A) the requirements of the Commission for site closure, decommissioning, and decontamination have been met by the licensee involved and that such licensee is in compliance with the provisions of subsection (a);

(B) such title and custody will be transferred to the Secretary without cost to the Federal Government; and

(C) Federal ownership and management of such site is necessary or desirable in order to protect the public health and safety, and the environment.

(2) If the Secretary assumes title and custody of any such waste and land under this subsection, the Secretary shall maintain such waste and land in a manner that will protect the public health and safety, and the environment.

(c) SPECIAL SITES–If the low-level radioactive waste involved is the result of a licensed activity to recover zirconium, hafnium, and rare earths from source material, the Secretary, upon request of the owner of the site involved, shall assume title and custody of such waste and the land on which it is disposed when such site has been decontaminated and stabilized in accordance with the requirements established by the Commission and when such owner has made adequate financial arrangements approved by the Commission for the long-term maintenance and monitoring of such site.

Subtitle E–Redirection of the Nuclear Waste Program[12]

42 USC 10172.

Sec. 160. Selection of Yucca Mountain Site

(a) IN GENERAL–(1) The Secretary shall provide for an orderly phase-out of site specific activities at all candidate sites other than the Yucca Mountain site.

(2) The Secretary shall terminate all site specific activities (other than reclamation activities) at all candidate sites, other than the Yucca Mountain site, within 90 days after the date of enactment of the Nuclear Waste Policy Amendments Act of 1987.

(b) Effective on the date of the enactment of the Nuclear Waste Policy Amendments Act of 1987, the State of Nevada shall be eligible to enter into a benefits agreement with the Secretary under section 170.

42 USC 10172a.

Sec. 161. Siting a Second Repository[13]

(a) CONGRESSIONAL ACTION REQUIRED–The Secretary may not conduct site-specific activities with respect to a second repository unless Congress has specifically authorized and appropriated funds for such activities.

(b) REPORT–The Secretary shall report to the President and to Congress on or after January 1, 2007, but not later than January 1, 2010, on the need for a second repository.

(c) TERMINATION OF GRANITE RESEARCH–Not later than 6 months after the date of the enactment of the Nuclear Waste Policy Amendments Act of 1987, the Secretary shall phase out in an orderly manner funding for all research programs in existence on such date of enactment designed to evaluate the suitability of crystalline rock as a potential repository host medium.

(d) ADDITIONAL SITING CRITERIA–In the event that the Secretary at any time after such date of enactment considers any sites in crystalline rock for characterization or selection as a repository, the Secretary shall consider (as a supplement to the siting guidelines under section 112) such potentially disqualifying factors as–

[12]   Added by P.L. 100–203, § 5011, 101 Stat. 1330 (1987).
[13]   Added by P.L. 100–203, § 5012, 101 Stat. 1330 (1987).

(1) seasonal increases in population;

(2) proximity to public drinking water supplies, including those of metropolitan areas; and

(3) the impact that characterization or siting decisions would have on lands owned or placed in trust by the United States for Indian tribes.

Subtitle F–Benefits[14]

42 USC 10173.

Sec. 170. Benefits Agreements

(a) IN GENERAL–(1) The Secretary may enter into a benefits agreement with the State of Nevada concerning a repository or with a State or an Indian tribe concerning a monitored retrievable storage facility for the acceptance of high-level radioactive waste or spent nuclear fuel in that State or on the reservation of that tribe, as appropriate.

(2) The State or Indian tribe may enter into such an agreement only if the State Attorney General or the appropriate governing authority of the Indian tribe or the Secretary of the Interior, in the absence of an appropriate governing authority, as appropriate, certifies to the satisfaction of the Secretary that the laws of the State or Indian tribe provide adequate authority for that entity to enter into the benefits agreement.

(3) Any benefits agreement with a State under this section shall be negotiated in consultation with affected units of local government in such State.

(4) Benefits and payments under this subtitle may be made available only in accordance with a benefits agreement under this section.

(b) AMENDMENT–A benefits agreement entered into under subsection (a) may be amended only by the mutual consent of the parties to the agreement and terminated only in accordance with section 173.

(c) AGREEMENT WITH NEVADA–The Secretary shall offer to enter into a benefits agreement with the Governor of Nevada. Any benefits agreement with a State under this subsection shall be negotiated in consultation with any affected units of local government in such State.

(d) MONITORED RETRIEVABLE STORAGE–The Secretary shall offer to enter into a benefits agreement relating to a monitored retrievable storage facility with the governing body of the Indian tribe on whose reservation the site for such facility is located, or, if the site is not located on a reservation, with the Governor of the State in which the site is located and in consultation with affected units of local government in such State.

(e) LIMITATION–Only one benefits agreement for a repository and only one benefits agreement for a monitored retrievable storage facility may be in effect at any one time.

(f) JUDICIAL REVIEW–Decisions of the Secretary under this section are not subject to judicial review.[15]

---

14   Amended by P.L. 100–203, § 5031, 101 Stat. 1330 (1987), by adding Subtitle F.
15   As added by P.L. 100–202, § 101(d) [Title III, § 300], 101 Stat. 1329–121, December 22, 1987, and amended by P.L. 100–203, Title V, § 5031, 101 Stat. 1330–237 (1987).

42 USC 10173a.   **Sec. 171. Content of Agreements**

(a) IN GENERAL–(1) In addition to the benefits to which a State, an affected unit of local government or Indian tribe is entitled under title I, the Secretary shall make payments to a State or Indian tribe that is a party to a benefits agreement under section 170 in accordance with the following schedule:

BENEFITS SCHEDULE

(amounts in $ millions)

Event	MRS	Repository
(A) Annual payments prior to first spent fuel receipt	5	10
(B) Upon first spent fuel receipt	10	20
(C) Annual payments after the first spent fuel receipt until closure of the facility	10	20

(2) For purposes of this section, the term–

(A) "MRS" means a monitored retrievable storage facility,

(B) "spent fuel" means high-level radioactive waste or spent nuclear fuel, and

(C) "first spent fuel receipt" does not include receipt of spent fuel or high-level radioactive waste for purposes of testing or operational demonstration.

(3) Annual payments prior to first spent fuel receipt under paragraph (1)(A) shall be made on the date of execution of the benefits agreement and thereafter on the anniversary date of such execution. Annual payments after the first spent fuel receipt until closure of the facility under paragraph (1)(C) shall be made on the anniversary date of such first spent fuel receipt.

(4) If the first spent fuel payment under paragraph (1)(B) is made within six months after the last annual payment prior to the receipt of spent fuel under paragraph (1)(A), such first spent fuel payment under paragraph (1)(B) shall be reduced by an amount equal to one-twelfth of such annual payment under paragraph (1)(A) for each full month less than six that has not elapsed since the last annual payment under paragraph (1)(A).

(5) Notwithstanding paragraph (1), (2), or (3), no payment under this section may be made before January 1, 1989, and any payment due under this title before January 1, 1989, shall be made on or after such date.

(6) Except as provided in paragraph (7), the Secretary may not restrict the purposes for which the payments under this section may be used.

(7) (A) Any State receiving a payment under this section shall transfer an amount equal to not less than one-third of the amount of such payment to affected units of local government of such State.

(B) A plan for this transfer and appropriate allocation of such portion among such governments shall be included in the benefits agreement under section 170 covering such payments.

(C) In the event of a dispute concerning such plan, the Secretary shall resolve such dispute, consistent with this Act and applicable State law.

(b) CONTENTS–A benefits agreement under section 170 shall provide that–

(1) a Review Panel be established in accordance with section 172;

(2) the State or Indian tribe that is party to such agreement waive its rights under title I to disapprove the recommendation of a site for a repository;

(3) the parties to the agreement shall share with one another information relevant to the licensing process for the repository of monitored retrievable storage facility, as it becomes available;

(4) the State or Indian tribe that is party to such agreement participate in the design of the repository or monitored retrievable storage facility and in the preparation of documents required under law or regulation governing the effects of the facility on the public health and safety; and

(5) the State or Indian tribe waive its rights, if any, to impact assistance under sections 116(c)(1)(B)(ii), 116(c)(2), 118(b)(2)(A)(ii), and 118(b)(3).

(c) The Secretary shall make payments to the States or affected Indian tribes under a benefits agreement under this section from the Waste Fund. The signature of the Secretary on a valid benefits agreement under section 170 shall constitute a commitment by the United States to make payments in accordance with such agreement.

Sec. 172. Review Panel

42 USC 10173b.

(a) IN GENERAL–The Review Panel required to be established by section 171(b)(1) of this Act shall consist of a Chairman selected by the Secretary in consultation with the Governor of the State or governing body of the Indian tribe, as appropriate, that is party to such agreement and six other members as follows:

(1) two members selected by the Governor of such State or governing body of such Indian tribe;

(2) two members selected by units of local government affected by the repository or monitored retrievable storage facility;

(3) one member to represent persons making payments into the Waste Fund, to be selected by the Secretary; and

(4) one member to represent other public interests, to be selected by the Secretary.

(b) TERMS–

(1) The members of the Review Panel shall serve for terms of four years each.

(2) Members of the Review Panel who are not full-time employees of the Federal Government, shall receive a per diem compensation for each day spent conducting work of the Review Panel, including their necessary travel or other expenses while engaged in the work of the Review Panel.

(3) Expenses of the Panel shall be paid by the Secretary from the Waste Fund.

(c) DUTIES–The Review Panel shall–

(1) advise the Secretary on matters relating to the proposed repository or monitored retrievable storage facility, including issues relating to design, construction, operation, and decommissioning of the facility;

(2) evaluate performance of the repository or monitored retrievable storage facility, as it considers appropriate;

(3) recommend corrective actions to the Secretary;

(4) assist in the presentation of State or affected Indian tribe and local perspectives to the Secretary; and

(5) participate in the planning for and the review of preoperational data on environmental, demographic, and socioeconomic conditions of the site and the local community.

(d) INFORMATION–The Secretary shall promptly make available promptly any information in the Secretary's possession requested by the Panel or its Chairman.

(e) FEDERAL ADVISORY COMMITTEE ACT–The requirements of the Federal Advisory Committee Act shall not apply to a Review Panel established under this title.

42 USC 10173c.

Sec. 173. Termination

(a) IN GENERAL–The Secretary may terminate a benefits agreement under this title if–

(1) the site under consideration is disqualified for its failure to comply with guidelines and technical requirements established by the Secretary in accordance with this Act; or

(2) the Secretary determines that the Commission cannot license the facility within a reasonable time.

(b) TERMINATION BY STATE OR INDIAN TRIBE–A State or Indian tribe may terminate a benefits agreement under this title only if the Secretary disqualifies the site under consideration for its failure to comply with technical requirements established by the Secretary in accordance with this Act or the Secretary determines that the Commission cannot license the facility within a reasonable time.

(c) DECISIONS OF THE SECRETARY–Decisions of the Secretary under this section shall be in writing, shall be available to Congress and the public, and are not subject to judicial review.

Subtitle G–Other Benefits[16]

42 USC 10174.

Sec. 174. Consideration in Siting Facilities

The Secretary, in siting Federal research projects, shall give special consideration to proposals from States where a repository is located.

42 USC 10174a.

Sec. 175. Report

(a) IN GENERAL–Within one year of the date of the enactment of the Nuclear Waste Policy Amendments Act of 1987, the Secretary shall report to Congress on the potential impacts of locating a repository at the Yucca Mountain site, including the recommendations of the Secretary for mitigation of such impacts and a statement of which impacts should be dealt with by the Federal Government, which should be dealt with by the State with State resources, including the benefits payments under section 171, and which should be a joint Federal-State responsibility. The report under this subsection shall include the analysis of the Secretary of the authorities available to mitigate these impacts and the appropriate sources of funds for such mitigation.

(b) IMPACTS TO BE CONSIDERED–Potential impacts to be addressed in the report under this subsection (a) shall include impacts on–

(1) education, including facilities and personnel for elementary and secondary schools, community colleges, vocational and technical schools and universities;

(2) public health, including the facilities and personnel for treatment and distribution of water, the treatment of sewage, the control of pests and the disposal of solid waste;

---

[16]   Amended by P.L. 100–203, § 5031, 101 Stat. 1330 (1987), by adding Subtitle G.

(3) law enforcement, including facilities and personnel for the courts, police and sheriff's departments, district attorneys and public defenders and prisons;

(4) fire protection, including personnel, the construction of fire stations, and the acquisition of equipment;

(5) medical care, including emergency services and hospitals;

(6) cultural and recreational needs, including facilities and personnel for libraries and museums and the acquisition and expansion of parks;

(7) distribution of public lands to allow for the timely expansion of existing, or creation of new, communities and the construction of necessary residential and commercial facilities;

(8) vocational training and employment services;

(9) social services, including public assistance programs, vocational and physical rehabilitation programs, mental health services, and programs relating to the abuse of alcohol and controlled substances;

(10) transportation, including any roads, terminals, airports, bridges, or railways associated with the facility and the repair and maintenance of roads, terminals, airports, bridges, or railways damaged as a result of the construction, operation, and closure of the facility;

(11) equipment and training for State and local personnel in the management of accidents involving high-level radioactive waste;

(12) availability of energy;

(13) tourism and economic development, including the potential loss of revenue and future economic growth; and

(14) other needs of the State and local governments that would not have arisen but for the characterization of the site and the construction, operation, and eventual closure of the repository facility.

Subtitle H–Transportation[17]

42 USC 10175.

Sec. 180. Transportation

(a) No spent nuclear fuel or high-level radioactive waste may be transported by or for the Secretary under subtitle A or under subtitle C except in packages that have been certified for such purpose by the Commission.

(b) The Secretary shall abide by regulations of the Commission regarding advance notification of State and local governments prior to transportation of spent nuclear fuel or high-level radioactive waste under subtitle A or under subtitle C.

(c) The Secretary shall provide technical assistance and funds to States for training for public safety officials of appropriate units of local government and Indian tribes through whose jurisdiction the Secretary plans to transport spent nuclear fuel or high-level radioactive waste under subtitle A or under subtitle C. Training shall cover procedures required for safe routine transportation of these materials, as well as procedures for dealing with emergency response situations. The Waste Fund shall be the source of funds for work carried out under this subsection.

---

[17]   Amended by P.L. 100–203, § 5061, 101 Stat. 1330 (1987), by adding Subtitle H.

Title II–Research, Development, and Demonstration Regarding Disposal of High-Level Radioactive Waste and Spent Nuclear Fuel

42 USC 10191.

Sec. 211. Purpose

It is the purpose of this title–

(1) to provide direction to the Secretary with respect to the disposal of high-level radioactive waste and spent nuclear fuel;

(2) to authorize the Secretary, pursuant to this title–

(A) to provide for the construction, operation, and maintenance of a deep geologic test and evaluation facility; and

(B) to provide for a focused and integrated high-level radioactive waste and spent nuclear fuel research and development program, including the development of a test and evaluation facility to carry out research and provide an integrated demonstration of the technology for deep geologic disposal of high-level radioactive waste, and the development of the facilities to demonstrate dry storage of spent nuclear fuel; and

(3) to provide for an improved cooperative role between the Federal Government and States, affected Indian tribes, and units of general local government in the siting of a test and evaluation facility.

42 USC 10192.

Sec. 212. Applicability

The provisions of this title are subject to section 8 and shall not apply to facilities that are used for the disposal of high-level radioactive waste, low-level radioactive waste, transuranic waste, or spent nuclear fuel resulting from atomic energy defense activities.

42 USC 10193.

Sec. 213. Identification of Sites

(a) GUIDELINES–Not later than 6 months after the date of the enactment of this Act and notwithstanding the failure of other agencies to promulgate standards pursuant to applicable law, the Secretary, in consultation with the Commission, the Director of the United States Geological Survey, the Administrator, the Council on Environmental Quality, and such other Federal agencies as the Secretary considers appropriate, is authorized to issue, pursuant to section 553 of title 5, United States Code, general guidelines for the selection of a site for a test and evaluation facility. Under such guidelines the Secretary shall specify factors that qualify or disqualify a site for development as a test and evaluation facility, including factors pertaining to the location of valuable natural resources, hydrogeophysics, seismic activity, and atomic energy defense activities, proximity to water supplies, proximity to populations, the effect upon the rights of users of water, and proximity to components of the National Park System, the National Wildlife Refuge System, the National Wild and Scenic Rivers System, the National Wilderness Preservation System, or National Forest Lands. Such guidelines shall require the Secretary to consider the various geologic media in which the site for a test and evaluation facility may be located and, to the extent practicable, to identify sites in different geologic media. The Secretary shall use guidelines established under this subsection in considering and selecting sites under this title.

(b) SITE IDENTIFICATION BY THE SECRETARY–(1) Not later than 1 year after the date of the enactment of this Act, and following promulgation of guidelines under subsection (a), the Secretary is authorized to identify 3 or more sites, at least 2 of which shall be in different geologic media in the continental United States, and at least 1 of which shall be in media other than salt. Subject to Commission requirements, the Secretary shall give preference to sites for the test and evaluation facility in media possessing geochemical characteristics that retard aqueous transport of radionuclides in order to provide a greater possible protection of public health and safety as operating experience is gained at the test and operation facility, and with the exception of the primary areas under review by the Secretary on the date of the enactment of this Act for the location of a test and evaluation facility or repository, all sites identified under this subsection shall be more than 15 statute miles from towns having a population of greater than 1000 persons as determined by the most recent census unless such sites contain high-level radioactive waste prior to identification under this title. Each identification of a site shall be supported by an environmental assessment, which shall include a detailed statement of the basis for such identification and of the probable impacts of the siting research activities planned for such site, and a discussible impact of the siting research activities planned for such site, and a discussion of alternative activities relating to siting research that may be undertaken to avoid such impacts. Such environmental assessment shall include–

Environmental assessment.

(A) an evaluation by the Secretary as to whether such site is suitable for siting research under the guidelines established under subsection (a);

(B) an evaluation by the Secretary of the effects of the siting research activities at such site on the public health and safety and the environment;

(C) a reasonable comparative evaluation by the Secretary of such site with other sites and locations that have been considered;

(D) a description of the decision process by which such site was recommended; and

(E) an assessment of the regional and local impacts of locating the proposed test and evaluation facility at such site.

(2) When the Secretary identifies a site, the Secretary shall as soon as possible notify the Governor of the State in which such site is located, or the governing body of the affected Indian tribe where such site is located, of such identification and the basis of such identification. Additional sites for the location of the test and evaluation facility authorized in section 302(d) may be identified after such 1 year period, following the same procedure as if such sites had been identified within such period.[18]

42 USC 10194.

Sec. 214. Siting Research and Related Activities

(a) IN GENERAL–Not later than 30 months after the date on which the Secretary completes the identification of sites under section 213, the Secretary is authorized to complete sufficient evaluation of 3 sites to select a site for expanded siting research activities and for other activities under section 218. The Secretary is authorized to conduct such preconstruction activities relative to such site selection for the test and evaluation facility as he deems appropriate. Additional sites for the location of the test and evaluation facility authorized in section 302(d) may be evaluated after such 30-month period, following the same procedures as if such sites were to be evaluated within such period.

(b) Public Meetings and Environmental Assessment–Not later than 6 months after the date on which the Secretary completes the identification

[18]   Amended by P.L. 102–285, § 10(a), 106 Stat. 171 (1992), which renamed the Geological Survey and provided that on and after May 18, 1992, it shall be known as the United States Geological Survey. An earlier statute, P.L. 102–154, Title I, 105 Stat. 1000 (1991), had provided for the identical change of name effective on and after November 13, 1991.

of sites under section 213, and before beginning siting research activities, the Secretary shall hold at least 1 public meeting in the vicinity of each site to inform the residents of the area of the activities to be conducted at such site and to receive their views.

(c) Restrictions–Except as provided in section 218 with respect to a test and evaluation facility, in conducting siting research activities pursuant to subsection (a)–

(1) the Secretary shall use the minimum quantity of high-level radioactive waste or other radioactive materials, if any, necessary to achieve the test or research objectives;

(2) the Secretary shall ensure that any radioactive material used or placed on a site shall be fully retrievable; and

(3) upon termination of siting research activities at a site for any reason, the Secretary shall remove any radioactive material at or in the site as promptly as practicable.

(d) Title To Material–The Secretary may take title, in the name of the Federal Government, to the high-level radioactive waste spent nuclear fuel, or other radioactive material emplaced in a test and evaluation facility. If the Secretary takes title to any such material, the Secretary shall enter into the appropriate financial arrangements described in subsection (a) or (b) of section 302 for the disposal of such material.

42 USC 10195.   **Sec. 215. Test and Evaluation Facility Siting Review and Reports**

(a) CONSULTATION AND COOPERATION–The Governor of a State, or the governing body of an affected Indian tribe, notified of a site identification under section 213 shall have the right to participate in a process of consultation and cooperation as soon as the site involved has been identified pursuant to such section and throughout the life of the test and evaluation facility. For purposes of this section, the term "process of consultation and cooperation" means a methodology–

Process of consultation and cooperation.

(1) by which the Secretary–

(A) keeps the Governor or governing body involved fully and currently informed about any potential economic or public health and safety impacts in all stages of the siting, development, construction, and operation of a test and evaluation facility;

(B) solicits, receives, and evaluates concerns and objections of such Governor or governing body with regard to such test and evaluation facility on an ongoing basis; and

(C) works diligently and cooperatively to resolve such concerns and objections; and

(2) by which the State or affected Indian tribe involved can exercise reasonable independent monitoring and testing of onsite activities related to all stages of the siting, development, construction and operation of the test and evaluation facility, except that any such monitoring and testing shall not unreasonably interfere with onsite activities.

(b) WRITTEN AGREEMENTS–The Secretary shall enter into written agreements with the Governor of the State in which an identified site is located or with the governing body of any affected Indian tribe where an identified site is located in order to expedite the consultation and cooperation process. Any such written agreement shall specify–

(1) procedures by which such Governor or governing body may study, determine, comment on, and make recommendations with regard to the possible health, safety, and economic impacts of the test and evaluation facility;

(2) procedures by which the Secretary shall consider and respond to comments and recommendations made by such Governor or

governing body, including the period in which the Secretary shall so respond;

(3) the documents the Department is to submit to such Governor or governing body, the timing for such submissions, the timing for such Governor or governing body to identify public health and safety concerns and the process to be followed to try to eliminate those concerns;

(4) procedures by which the Secretary and either such Governor or governing body may review or modify the agreement periodically; and

(5) procedures for public notification of the procedures specified under subparagraphs (A) through (D).

(c) LIMITATION–Except as specifically provided in this section, nothing in this title is intended to grant any State or affected Indian tribe any authority with respect to the siting, development, or loading of the test and evaluation facility.

42 USC 10196.   **Sec. 216. Federal Agency Actions**

(a) COOPERATION AND COORDINATION–Federal agencies shall assist the Secretary by cooperating and coordinating with the Secretary in the preparation of any necessary reports under this title and the mission plan under section 301.

(b) ENVIRONMENTAL REVIEW–(1) No action of the Secretary or any other Federal agency required by this title or section 301 with respect to a test and evaluation facility to be taken prior to the initiation of onsite construction of a test and evaluation facility shall require the preparation of an environmental impact statement under section 102(2)(C) of the Environmental Policy Act of 1969 (42 USC 4332(2)(C)), or to require the preparation of environmental reports, except as otherwise specifically provided for in this title.

(2) The Secretary and the heads of all other Federal agencies shall, to the maximum extent possible, avoid duplication of efforts in the preparation of reports under the National Environmental Policy Act of 1969 (42 USC 4321 *et seq.*).

42 USC 10197.   **Sec. 217. Research and Development on Disposal of High-Level Radioactive Waste**

(a) PURPOSE–Not later than 64 months after the date of the enactment of this Act, the Secretary is authorized to, to the extent practicable, begin at a site evaluated under section 214, as part of and as an extension of siting research activities of such site under such section, the mining and construction of a test and evaluation facility. Prior to the mining and construction of such facility, the Secretary shall prepare an environmental assessment. the purpose of such facility shall be–

Environmental assessment.   (1) to supplement and focus the repository site characterization process;

(2) to provide the conditions under which known technological components can be integrated to demonstrate a functioning repository-like system;

(3) to provide a means of identifying, evaluating, and resolving potential repository licensing issues that could not be resolved during the siting research program conducted under section 212;

(4) to validate, under actual conditions, the scientific models used in the design of a repository;

(5) to refine the design and engineering of repository components and systems and to confirm the predicted behavior of such components and systems;

(6) to supplement the siting data, the generic and specific geological characteristics developed under section 214 relating to isolating disposal materials in the physical environment of a repository;

(7) to evaluate the design concepts for packaging, handling, and emplacement of high-level radioactive waste and spent nuclear fuel at the design rate; and

(8) to establish operating capability without exposing workers to excessive radiation.

(b) DESIGN–The Secretary shall design each test and evaluation facility–

(1) to be capable of receiving not more than 100 full-sized canisters of solidified high-level radioactive waste (which canisters shall not exceed an aggregate weight of 100 metric tons), except that spent nuclear fuel may be used instead of such waste if such waste cannot be obtained under reasonable conditions;

(2) to permit full retrieval of solidified high-level radioactive waste, or other radioactive material used by the Secretary for testing, upon completion of the technology demonstration activities; and

(3) based upon the principle that the high-level radioactive waste, spent nuclear fuel, or other radioactive material involved shall be isolated from the biosphere in such a way that the initial isolation is provided by engineered barriers functioning as a system with the geologic environment.

Testing.

(c) OPERATION–(1) Not later than 88 months after the date of the enactment of this Act, the Secretary shall begin an in situ testing program at the test and evaluation facility in accordance with the mission plan developed under section 301, for purposes of–

(A) conducting in situ tests of bore hole sealing, geologic media fracture sealing, and room closure to establish the techniques and performance for isolation of high-level radioactive waste, spent nuclear fuel, or other radioactive materials from the biosphere;

(B) conducting in situ tests with radioactive sources and materials to evaluate and improve reliable models for radionuclide mitigation, absorption, and containment within the engineered barriers and geologic media involved, if the Secretary finds there is reasonable assurance that such radioactive sources and materials will not threaten the use of such site as a repository;

(C) conducting in situ tests to evaluate and improve models for ground water or brine flow through fractured geologic media;

(D) conducting in situ tests under conditions representing the real time and the accelerated time behavior of the engineered barriers within the geologic environment involved;

(E) conducting in situ tests to evaluate the effects of heat and pressure on the geologic media involved on the hydrology of the surrounding area and on the integrity of the disposal packages;

(F) conducting in situ tests under both normal and abnormal repository conditions to establish safe design limits for disposal packages and to determine the effects of the gross release of radionuclides into surroundings, and the effects of various credible failure modes, including–

(i) seismic events leading to the coupling of aquifers through the test and evaluation facility;

(ii) thermal pulses significantly greater than the maximum calculated; and

(iii) human intrusion creating a direct pathway to the biosphere; and

(G) conducting such other research and development activities as the Secretary considers appropriate, including such activities necessary, to obtain the use of high-level radioactive waste, spent nuclear fuel, or other radioactive materials (such as any highly radioactive material from the Three Mile Island nuclear power plant or from the West Valley Demonstration Project) for test and evaluation purposes, if such other activities are reasonably necessary to support the repository program and if there is reasonable assurance that the radioactive sources involved will not threaten the use of such site as a repository.

(2) The in situ testing authorized in this subsection shall be designed to ensure that the suitability of the site involved for licensing by the Commission as a repository will not be adversely affected.

(d) USE OF EXISTING DEPARTMENT FACILITIES–During the conducting of siting research activities under section 214 and for such period thereafter as the Secretary considers appropriate, the Secretary shall use Department facilities owned by the Federal Government on the date of the enactment of this Act for the conducting of generically applicable test regarding packaging, handling, and emplacement technology for solidified high-level radioactive waste and spent nuclear fuel from civilian nuclear activities.

(e) ENGINEERED BARRIERS–The system of engineered barriers and selected geology used in a test and evaluation facility shall have a design life at least as long as that which the Commission requires by regulations issued under this Act, or under the Atomic Energy Act of 1954 (42 USC 2011 *et seq.*), for repositories.

(f) ROLE OF COMMISSION–(1)(A) Not later than 1 year after the date of the enactment of this Act, the Secretary and the Commission shall reach a written understanding establishing the procedures for review, consultation, and coordination in the planning, construction, and operation of the test and evaluation facility under this section. Such understanding shall establish a schedule, consistent with the deadlines set forth in this subtitle, for submission by the Secretary of, and review by the Commission of and necessary action on–

(i) the mission plan prepared under section 301; and

(ii) such reports and other information as the Commission may reasonably require to evaluate any health and safety impacts of the test and evaluation facility.

(B) Such understanding shall also establish the conditions under which the Commission may have access to the test and evaluation facility for the purpose of assessing any public health and safety concerns that it may have. No shafts may be excavated for the test and evaluation until the Secretary and the Commission enter into such understanding.

(2) Subject to section 305, the test and evaluation facility, and the facilities authorized in section 217, shall be constructed and operated as research, development, and demonstration facilities, and shall not be subject to licensing under section 202 of the Energy Reorganization Act of 1974 (42 USC 5842).

(3)(A) The Commission shall carry out a continuing analysis of the activities undertaken under this section to evaluate the adequacy of the consideration of public health and safety issues.

(B) The Commission shall report to the President, the Secretary, and the Congress as the Commission considers appropriate with respect to the conduct of activities under this section.

(g) ENVIRONMENTAL REVIEW–The Secretary shall prepare an environmental impact statement under section 102(2)(C) of the National Environmental Policy Act of 1969 (42 USC 4332)(2)(C) prior to conducting tests with radioactive materials at the test and evaluation facility. Such environmental impact statement shall incorporate, to the extent practicable, the environmental assessment prepared under section 217(a). Nothing in this subsection may be construed to limit siting research activities conducted under section 214. This subsection shall apply only to activities performed exclusively for a test and evaluation facility.

(h) LIMITATIONS–(1) If the test and evaluation facility is not located at the site of a repository, the Secretary shall obtain the concurrence of the Commission with respect to the decontamination and decommissioning of such facility.

(2) If the test and evaluation facility is not located at a candidate site or repository site, the Secretary shall conduct only the portion of the in situ testing program required in subsection (c) determined by the Secretary to be useful in carrying out the purposes of this act.

Terminations.

(3) The operation of the test and evaluation facility shall terminate not later than–

(A) 5 years after the date on which the initial repository begins operation; or

(B) at such time as the Secretary determines that the continued operation of a test and evaluation facility is not necessary for research, development, and demonstration purposes;

whichever occurs sooner.

(4) Notwithstanding any other provisions of this subsection, as soon as practicable following any determination by the Secretary, with the concurrence of the Commission, that the test and evaluation facility is unsuitable for continued operation, the Secretary shall take such actions as are necessary to remove from such site any radioactive material placed on such site as a result of testing and evaluation activities conducted under this section. Such requirement may be waived if the Secretary, with the concurrence of the Commission, finds that short-term testing and evaluation activities using radioactive material will not endanger the public health and safety.

42 USC 10198.

Sec. 218. Research and Development on Spent Nuclear Fuel

(a) DEMONSTRATION AND COOPERATIVE PROGRAMS–The Secretary shall establish a demonstration program, in cooperation with the private sector, for the dry storage of spent nuclear fuel at civilian nuclear power reactor sites, with the objective of establishing one or more technologies that the Commission may, by rule, approve for use at the sites of civilian nuclear power reactors without, to the maximum extent practicable, the need for additional site-specific approvals by the Commission. Not later than 1 year after the date of the enactment of this Act, the Secretary shall select at least 1, but not more than 3, sites evaluated under section 214 at such power reactors. In selecting such site or sites, the Secretary shall give preference to civilian nuclear power reactors that will soon have a shortage of interim storage capacity for spent nuclear fuel. Subject to reaching agreement as provided in subsection (b), the Secretary shall undertake activities to assist such power reactors with demonstration projects at such sites, which may use one of the following types of alternate storage technologies; spent nuclear fuel storage casks, caissons, or silos. The Secretary shall also undertake a cooperative program with civilian nuclear power reactors to encourage the development of the technology for spent nuclear fuel rod consolidation in existing power reactor water storage basins.

(b) COOPERATIVE AGREEMENTS–To carry out the programs described in subsection (a), the Secretary shall enter into a cooperative agreement with each utility involved that specifies, at a minimum, that–

(1) such utility shall select the alternate storage technique to be used, make the land and spent nuclear fuel available for the dry storage demonstration, submit and provide site-specific documentation for a license application to the Commission, obtain a license relating to the facility involved, construct such facility, operate such facility after licensing, pay the costs required to construct such facility, and pay all costs associated with the operation and maintenance of such facility;

(2) the Secretary shall provide, on a cost-sharing basis, consultative and technical assistance, including design support and generic licensing documentation, to assist such utility in obtaining the construction authorization and appropriate license from the Commission; and

(3) the Secretary shall provide generic research and development of alternative spent nuclear fuel storage techniques to enhance utility-provided, at-reactor storage capabilities, if authorized in any other provision of this act or in any other provision of law.

(c) DRY STORAGE RESEARCH AND DEVELOPMENT–(1) The consultative and technical assistance referred to in subsection (b)(2) may include, but shall not be limited to, the establishment of a research and development program for the dry storage of not more than 300 metric tons of spent nuclear fuel at facilities owned by the Federal Government on the date of the enactment of this Act. The purpose of such program shall be to collect necessary data to assist the utilities involved in the licensing process.

(2) To the extent available, and consistent with the provisions of section 135, the Secretary shall provide spent nuclear fuel for the research and development program authorized in this subsection from spent nuclear fuel received by the Secretary for storage under section 135. Such spent nuclear fuel shall not be subject to the provisions of section 135(e).

(d) FUNDING–The total contribution from the Secretary from Federal funds and the use of Federal facilities or services shall not exceed 25 percent of the total costs of the demonstration program authorized in subsection (a), as estimated by the Secretary. All remaining costs of such program shall be paid by the utilities involved or shall be provided by the Secretary from the Interim Storage Fund established in section 136.

(e) RELATION TO SPENT NUCLEAR FUEL STORAGE PROGRAM–The spent nuclear fuel storage program authorized in section 135 shall not be construed to authorize the use of research development or demonstration facilities owned by the Department unless–

Report to
congressional
committees.

(1) a period of 30 calendar days (not including any day in which either House of Congress is not in session because of adjournment of more than 3 calendar days to a day certain) has passed after the Secretary has transmitted to the Committee on Science, Space, and Technology[19] of the House of Representatives and the Committee on Energy and Natural Resources of the Senate a written report containing a full and complete statement concerning (A) the facility involved; (B) any necessary modifications; (C) the cost thereof; and (D) the impact on the authorized research and development program; or

(2) each such committee, before the expiration of such period, has transmitted to the Secretary a written notice to the effect that such committee has no objection to the proposed use of such facility.

42 USC 10199.

Sec. 219. Payments to States and Indian Tribes

(a) PAYMENTS–Subject to subsection (b), the Secretary shall make payments to each State or affected Indian tribe that has entered into an agreement pursuant to section 215. The Secretary shall pay an amount equal to 100 percent of the expenses incurred by such State or Indian tribe in engaging in any monitoring, testing, evaluation, or other consultation and cooperation activity under section 215 with respect to any site. The amount paid by the Secretary under this paragraph shall not exceed $3,000,000 per year from the date on which the site involved was identified to the date on which the decontamination and decommission of the facility is complete pursuant to section 217(h). Any such payment may only be made to a State in which a potential site for a test and evaluation facility has been identified under section 213, or to an affected Indian tribe where the potential site has been identified under such section.

(b) LIMITATION–The Secretary shall make any payment to a State under subsection (a) only if such State agrees to provide, to each unit of general local government within the jurisdictional boundaries of which the potential site or effectively selected site involved is located, at least one-tenth of the payments made by the Secretary to such State under such subsection. A State or affected Indian tribe receiving any payment under subsection (a) shall otherwise have discretion to use such payment for whatever purpose it deems necessary, including the State or tribal activities pursuant to agreements entered into in accordance with section 215. Annual payments shall be prorated on a 365-day basis to the specified dates.

42 USC 10200.

Sec. 220. Study of Research and Development Needs for Monitored Retrievable Storage Proposal

Report to
Congress.

Not later than 6 months after the date of the enactment of this Act, the Secretary shall submit to the Congress a report describing the research and development activities the Secretary considers necessary to develop the proposal required in section 141(b) with respect to a monitored retrievable storage facility.

42 USC 10201.

Sec. 221. Judicial Review

Judicial review of research and development activities under this shall be in accordance with the provisions of section 119.

---

[19]   Amended by P.L. 103–437, § 15(c)(10), 108 Stat. 4592 (1994).

42 USC 10202.

Sec. 222. Research on Alternatives for the Permanent Disposal of High-Level Radioactive Waste

Research on Alternatives for the Permanent Disposal of High-Level Radioactive Waste–The Secretary shall continue and accelerate a program of research, development, and investigation of alternative means and technologies for the permanent disposal of high-level radioactive waste from civilian nuclear activities and Federal research and development activities except that funding shall be made from amounts appropriated to the Secretary for purposes of carrying out this section. Such program shall include examinations of various waste disposal options.

42 USC 10203.

Sec. 223. Technical Assistance to Non-Nuclear Weapon States in the Field of Spent Fuel Storage and Disposal

(a) It shall be the policy of the United States to cooperate with and provide technical assistance to non-nuclear weapon states in the field of spent fuel storage and disposal.

Joint notice, publication in *Federal Register*.

(b)(1) Within 90 days of enactment of this Act, the Secretary and the Commission shall publish a joint notice in the Federal Register stating that the United States is prepared to cooperate with and provide technical assistance to non-nuclear weapon states in the fields of at-reactor spent fuel storage; away-from-reactor spent fuel storage; monitored, retrievable spent fuel storage; geologic disposal of spent fuel; and the health, safety, and environmental regulation of such activities. The notice shall summarize the resources that can be made available for international cooperation and assistance in these fields through existing programs of the Department and the Commission, including the availability of: (i) data from past or ongoing research and development projects; (ii) consultations with expert Department or Commission personnel or contractors; and (iii) liaison with private business entities and organizations working in these fields.

Joint notice, reissuance.

(2) The joint notice described in the preceding subparagraph shall be updated and reissued annually for 5 succeeding years.

(c) Following publication of the annual joint notice referred to in paragraph (2), the Secretary of State shall inform the governments of non-nuclear weapon states and, as feasible, the organizations operating nuclear power plants in such states, that the United States is prepared to

Expressions of interest.

cooperate with and provide technical assistance to non-nuclear weapon states in the fields of spent fuel storage and disposal, as set forth in the joint notice. The Secretary of State shall also solicit expressions of interest from non-nuclear weapon state governments and non-nuclear weapon state nuclear power reactor operators concerning their participation in expanded United States cooperation and technical assistance programs in these fields. The Secretary of State shall transmit any such expressions of interest to the Department and the Commission.

(d) With his budget presentation materials for the Department and the Commission for fiscal years 1984 through 1989, the President shall include funding requests for an expanded program of cooperation and technical assistance with non-nuclear weapon states in the fields of spent fuel storage and disposal as appropriate in light of expressions of interest in such cooperation and assistance on the part of non-nuclear weapon state governments and non-nuclear weapon state nuclear power reactor operators.

Non-nuclear weapon state.

(e) For the purposes of this subsection, the term "non-nuclear weapon state" shall have the same meaning as that set forth in article IX of the Treaty on the Non-Proliferation of Nuclear Weapons (21 USC 438).

(f) Nothing in this subsection shall authorize the Department of Commission to take any action not authorized under existing law.

42 USC 10204.
Reports.

Sec. 224. Subseabed Disposal

(a) REPEALED. Public Law 104–66, title I, section 1051(d), December 21, 1995, 109 Stat. 716.

(b) OFFICE OF SUBSEABED DISPOSAL RESEARCH–

(1) There is hereby established an Office of Subseabed Disposal Research within the Office of Science of the Department of Energy. The Office shall be headed by the Director, who shall be a member of the Senior Executive Service appointed by the Director of the Office of Science, and compensated at a rate determined by applicable law.

(2) The Director of the Office of Subseabed Disposal Research shall be responsible for carrying out research, development, and demonstration activities on all aspects of subseabed disposal of high–level radioactive waste and spent nuclear fuel, subject to the general supervision of the Secretary. The Director of the Office shall be directly responsible to the Director of the Office of Science, and the first such Director shall be appointed within 30 days of December 22, 1987.

(3) In carrying out his responsibilities under this chapter, the Secretary may make grants to, or enter into contracts with, the Subseabed Consortium described in subsection (d) of this section, and other persons.

(4)(A) Within 60 days of December 22, 1987, the Secretary shall establish a university–based Subseabed Consortium involving leading oceanographic universities and institutions, national laboratories, and other organizations to investigate the technical and institutional feasibility of subseabeded disposal.

(B) The Subseabeded Consortium shall develop a research plan and budget to achieve the following objectives by 1995:

(i) demonstrate the capacity to identify and characterize potential subseabeded disposal sites;

(ii) develop conceptual designs for a Subseabeded disposal system, including estimated costs and institutional requirements; and

(iii) identify and assess the potential impacts of Subseabeded disposal on the human and marine environment.

(C) In 1990, and again in 1995, the Subseabeded Consortium shall report to Congress on the progress being made in achieving the objectives of paragraph (2).

(5) REPEALED. Public Law 104–66, title I, section 1051(d), 109 Stat. 716, December 21, 1995.[20]

Title III–Other Provisions Relating to Radioactive Waste

42 USC 10221.

Sec. 301. Mission Plan

(a) CONTENTS OF MISSION PLAN–The Secretary shall prepare a comprehensive report, to be known as the mission plan, which shall provide an informational basis sufficient to permit informed decisions to be made in carrying out the repository program and the research,

---

[20]   P.L. 97–425, Title II, § 224, as added by P.L. 100–202, § 101(d), Title III, 101 Stat. 1329–104, 1329–12 (1987); P.L. 100–203, Title V, § 5063, 101 Stat. 1330–253 (1987); Amended by P.L. 104–66, Title I, § 1051(d), 109 Stat. 716 (1995); P.L. 10–245, Title III, § 309(b)(2)(E), 112 Stat. 1853 (1998).

development, and demonstration programs required under this Act. The mission plan shall include–

(1) an identification of the primary scientific, engineering, and technical information, including any necessary demonstration of engineering or systems integration, with respect to the siting and construction of a test and evaluation facility and repositories;

(2) an identification of any information described in paragraph (1) that is not available because of any unresolved scientific, engineering, or technical questions, or undemonstrated engineering or systems integration, a schedule including specific major milestones for the research, development, and technology demonstration program required under this Act and any additional activities to be undertaken to provide such information, a schedule for the activities necessary to achieve important programmatic milestones, and an estimate of the costs required to carry out such research, development and demonstration programs;

(3) an evaluation of financial, political, legal, or institutional problems that may impede the implementation of this Act, the plans of the Secretary to resolve such problems, and recommendations for any necessary legislation to resolve such problems;

(4) any comments of the Secretary with respect to the purpose and program of the test and evaluation facility;

(5) a discussion of the significant results of research and development programs conducted and the implications for each of the different geologic media under consideration for the siting of repositories, and, on the basis of such information, a comparison of the advantages and disadvantages associated with the use of such media for repository sites;

(6) the guidelines issued under section 112(a);

(7) a description of known sites at which site characterization activities should be undertaken, a description of such siting characterization activities, including the extent of planned excavations, plans for onsite testing with radioactive or nonradioactive material, plans for any investigations activities which may affect the capability of any such site to isolate high-level radioactive waste or spent nuclear fuel, plans to control any adverse, safety-related impacts from such site characterization activities, and plans for the decontamination and decommissioning of such site if it is determined unsuitable for licensing as a repository;

(8) an identification of the process for solidifying high-level radioactive waste or packaging spent nuclear fuel, including a summary and analysis of the data to support the selection of the solidification process and packaging techniques, an analysis of the requirements for the number of solidification packaging facilities needed, a description of the state of the art for the materials proposed to be used in packaging such waste or spent fuel and the availability of such materials including impacts on strategic supplies and any requirements for new or reactivated facilities to produce any such materials needed, and a description of a plan, and the schedule for implementing such plan, for an aggressive research and development program to provide when needed a high-integrity disposal package at a reasonable price;

(9) an estimate of (A) the total repository capacity required to safely accommodate the disposal of all high-level radioactive waste and spent nuclear fuel expected to be generated through December 31, 2020, in the event that no commercial reprocessing of spent

nuclear fuel occurs, as well as the repository capacity that will be required if such reprocessing does occur; (B) the number and type of repositories required to be constructed to provide such disposal capacity; (C) a schedule for the construction of such repositories; and (D) an estimate of the period during which each repository listed in such schedule will be accepting high-level radioactive waste or spent nuclear fuel for disposal;

(10) an estimate, on an annual basis, of the costs required (A) to construct and operate the repositories anticipated to be needed under paragraph (9) based on each of the assumptions referred to in such paragraph; (B) to construct and operate a test and evaluation facility, or any other facilities, other than repositories described in subparagraph (A), determined to be necessary; and (C) to carry out any other activities under this Act; and

(11) an identification of the possible adverse economic and other impacts to the State or Indian tribe involved that may arise from the development of a test and evaluation facility or repository at a site.

(b) Submission of Mission Plan.–(1) Not later than 15 months after the date of the enactment of this Act, the Secretary shall submit a draft mission plan to the States, the affected Indian tribes, the Commission, and other Government agencies as the Secretary deems appropriate for their comments.

Public inspection and agency comments. Publication in *Federal Register.*

(2) In preparing any comments on the mission plan, such agencies shall specify with precision any objections that they may have. Upon submission of the mission plan to such agencies, the Secretary shall publish a notice in the Federal Register of the submission of the mission plan and of its availability for public inspection, and, upon receipt of any comments of such agencies respecting the mission plan, the Secretary shall publish a notice in the Federal Register of the receipt of comments and of the availability of the comments for public inspection. If the Secretary does not revise the mission plan to meet objections specified in such comments, the Secretary shall publish in the Federal Register a detailed statement for not so revising the mission plan.

Plan submittal to congressional committees.

(3) The Secretary, after reviewing any other comments made by such agencies and revising the mission plan to the extent that the Secretary may consider to be appropriate, shall submit the mission plan to the appropriate committees of the Congress not later than 17 months after the date of the enactment of this Act. The mission plan shall be used by the Secretary at the end of the first period of 30 calendar days (not including any day on which either House of Congress is not in session because of adjournment of more than 3 calendar days to a day certain) following receipt of the mission plan by the Congress.

42 USC 10222.

Sec. 302. Nuclear Waste Fund

(a) CONTRACTS–(1) In the performance of his functions under this Act, the Secretary is authorized to enter into contracts with any person who generates or holds title to high-level radioactive waste, or spent nuclear fuel, of domestic origin for the acceptance of title, subsequent transportation, and disposal of such waste or spent fuel. Such

Fees.

contracts shall provide for payment to the Secretary of fees pursuant to paragraphs (2) and (3) sufficient to offset expenditures in subsection (d).

(2) For electricity generated by a civilian nuclear power reactor and sold on or after the date 90 days after the date of enactment of

Fees.

this Act, the fee under paragraph (1) shall be equal to 1.0 mil per kilowatt-hour.

(3) For spent nuclear fuel, or solidified high-level radioactive waste derived from spent nuclear fuel, which fuel was used to generate electricity in a civilian nuclear power reactor prior to the application of the fee under paragraph (2) to such reactor, the Secretary shall, not later than 90 days after the date of enactment of this Act, establish a 1 time fee per kilogram of heavy metal in spent nuclear fuel, or in solidified high-level radioactive waste. Such fee shall be in amount equivalent to an average charge of 1.0 mil per kilowatt-hour for electricity generated by such spent nuclear fuel, or such solidified high-level waste derived therefrom, to be collected from any person delivering such spent nuclear fuel or high-level waste, pursuant to section 123, to the Federal Government. Such fee shall be paid to the Treasury of the United States and shall be deposited in the separate fund established by subsection (c)126(b). In paying such a fee, the person delivering spent fuel, or solidified high-level radioactive wastes derived therefrom, to the Federal Government shall have no further financial obligation to the Federal Government for the long-term storage and permanent disposal of such spent fuel, or the solidified high-level radioactive waste derived therefrom.

Collection and payment procedures. Review.

42 USC 6421. Transmittal to Congress.

(4) Not later than 180 days after the date of enactment of this Act, the Secretary shall establish procedures for the collection and payment of the fees established by paragraph (2) and paragraph (3). The Secretary shall annually review the amount of the fees established by paragraphs (2) and (3) above to evaluate whether collection of the fee will provide sufficient revenues to offset the costs as defined in subsection (d) herein. In the event the Secretary determines that either insufficient or excess revenues are being collected, in order to recover the costs incurred by the Federal Government that are specified in subsection (d), the Secretary shall propose an adjustment to the fee to insure full cost recovery. The Secretary shall immediately transmit this proposal for such an adjustment to Congress. The adjusted fee proposed by the Secretary shall be effective after a period of 90 days of continuous session have elapsed following the receipt of such transmittal unless during such 90-day period either House of Congress adopts a resolution disapproving the Secretary's proposed adjustment in accordance with the procedures set forth for congressional review of an energy action under section 551 of the Energy Policy and Conservation Act.

(5) Contracts entered into under this section shall provide that–

(A) following commencement of operation of a repository, the Secretary shall take title to the high-level radioactive waste or spent nuclear fuel involved as expeditiously as practicable upon the request of the generator or owner of such waste or spent fuel; and

(B) in return for the payment of fees established by this section, the Secretary, beginning not later than January 31, 1998, will dispose of the high-level radioactive waste or spent nuclear fuel involved as provided in this subtitle.

Disposal services, terms and conditions.

(6) The Secretary shall establish in writing criteria setting forth the terms and conditions under which such disposal services shall be made available.

License
renewal or
issuance.

(b) ADVANCE CONTRACTING REQUIREMENT–

(1) (A) The Commission shall not issue or renew a license to any person to use a utilization or production facility under the authority of section 103 or 104 of the Atomic Energy Act of 1954 (42 USC 2133, 2134) unless–

(i) such person has entered into a contract with the Secretary under this section; or

(ii) the Secretary affirms in writing that such person is actively and in good faith negotiating with the Secretary for a contract under this section.

(B) The Commission, as it deems necessary or appropriate, may require as a precondition to the issuance or renewal of a license under section 103 or 104 of the Atomic Energy Act of 1954 (42 USC 2133, 2134) that the applicant for such license shall have entered into an agreement with the Secretary for the disposal of high-level radioactive waste and spent nuclear fuel that may result from the use of such license.

(2) Except as provided in paragraph (1), no spent nuclear fuel or high-level radioactive waste generated or owned by any person (other than a department of the United States referred to in section 101 or 102 of title 5, United States Code) may be disposed of by the Secretary in any repository constructed under this Act unless the generator or owner of such spent fuel or waste has entered into a contract with the Secretary under this section by not later than–

(A) June 30, 1983; or

(B) the date on which such generator or owner commences generation of, or takes title to, such spent fuel or waste; whichever occurs later.

(3) The rights and duties of a party to a contract entered into under this section may be assignable with transfer of title to the spent nuclear fuel or high-level radioactive waste involved.

Disposal of
radioactive
waste or spent
nuclear fuel.

(4) No high-level radioactive waste or spent nuclear fuel generated or owned by any department of the United States referred to in section 101 or 102 of title 5, United States Code, may be disposed of by the Secretary in any repository constructed under this Act unless such department transfers to the Secretary, for deposit in the Nuclear Waste Fund, amounts equivalent to the fees that would be paid to the Secretary under the contracts referred to in this section if such waste or spent fuel were generated by any other person.

(c) ESTABLISHMENT OF NUCLEAR WASTE FUND–There hereby is established in the Treasury of the United States a separate fund, to be known as the Nuclear Waste Fund. The Waste Fund shall consist of–

(1) all receipts, proceeds, and recoveries realized by the Secretary under subsections (a), (b), and (e), which shall be deposited in the Waste Fund immediately upon their realization;

(2) any appropriations made by the Congress to the Waste Fund; and

(3) any unexpended balances available on the date of the enactment of this Act for functions or activities necessary or incident to the disposal of civilian high-level radioactive waste or civilian spent nuclear fuel, which shall automatically be transferred to the Waste Fund on such date.

(d) USE OF WASTE FUND–The Secretary may make expenditures from the Waste Fund, subject to subsection (e), only for purposes of radioactive waste disposal activities under Titles I and II, including–

(1) the identification, development, licensing, construction, operation, decommissioning, and post-decommissioning maintenance and monitoring of any repository, monitored retrievable storage facility or test evaluation facility constructed under this Act;

(2) the conducting of nongeneric research, development, and demonstration activities under this Act;

(3) the administrative cost of the radioactive waste disposal program;

(4) any costs that may be incurred by the Secretary in connection with the transportation, treating, or packaging of spent nuclear fuel or high-level radioactive waste to be disposed of in a repository, to be stored in a monitored, retrievable storage site or to be used in a test and evaluation facility;

(5) the costs associated with acquisition, design, modification, replacement, operation and construction of facilities at a repository site, a monitored, retrievable storage site or a test and evaluation facility site and necessary or incident to such repository, monitored, retrievable storage facility or test and evaluation facility; and

(6) the provision of assistance to States, units of general local government, and Indian tribes under section 116, 118 and 219.

No amount may be expended by the Secretary under this subtitle for the construction or expansion of any facility unless such construction or expansion is expressly authorized by this or subsequent legislation. The Secretary hereby is authorized to construct one repository and one test and evaluation facility.

Report to Congress.

(e) ADMINISTRATION OF WASTE FUND–(1) The Secretary of the Treasury shall hold the Waste Fund and, after consultation with the Secretary, annually report to the Congress on the financial condition and operations of the Waste Fund during the preceding fiscal year.

Budget submittal.

(2) The Secretary shall submit the budget of the Waste Fund to the Office of Management and Budget triennially along with the budget of the Department of Energy submitted at such time in accordance with chapter II of title 31, United States Code. The budget of the Waste Fund shall consist of the estimates made by the Secretary of expenditures from the Waste Fund and other relevant financial matters for the succeeding 3 fiscal years, and shall be included in the Budget of the United States Government. The Secretary may make expenditures from the Waste Fund, subject to appropriations which shall remain available until expended. Appropriations shall be subject to triennial authorization.

(3) If the Secretary determines that the Waste Fund contains at any time amounts in excess of current needs, the Secretary may request the Secretary of the Treasury to invest such amounts, or any portion of such amounts as the Secretary determines to be appropriate, in obligations of the United States–

(A) having maturities determined by the Secretary of the Treasury to be appropriate to the needs of the Waste Fund; and

(B) bearing interest at rates determined to be appropriate by the Secretary of the Treasury, taking into consideration the current average market yield on outstanding marketable obligations of the United States with remaining periods to maturity comparable to the maturities of such investments, except that the interest rate on such investments shall not exceed the average interest rate applicable to existing borrowings.

(4) Receipts, proceeds, and recoveries realized by the Secretary under this section, and expenditures of amounts from the Waste Fund,

shall be exempt from annual apportionment under the provisions of subchapter II of chapter 15 of title 31, United States Code.

(5) If at any time the moneys available in the Waste Fund are insufficient to enable the Secretary to discharge his responsibilities under this subtitle, the Secretary shall issue to the Secretary of the Treasury obligations in such forms and denominations, bearing such maturities, and subject to such terms and conditions as may be agreed to by the Secretary and the Secretary of the Treasury. A total of such obligations shall not exceed amounts provided in appropriation Acts. Redemption of such obligations shall be made by the Secretary from moneys available in the Waste Fund. Such obligations shall bear interest at a rate determined by the Secretary of the Treasury, which shall be not less than a rate determined by taking into consideration the average market yield on outstanding marketable obligations of the United States of comparable maturities during the month preceding the issuance of the obligations under this paragraph. The Secretary of the Treasury shall purchase any issued obligations, and for such purpose the Secretary of the Treasury is authorized to use as a public debt transactions the proceeds from the sale of any securities issued under chapter 31 of title 31, United States Code, and the purposes for which securities may be issued under such Act are extended to include any purchase of such obligations. The Secretary of the Treasury may at any time sell any of the obligations acquired by him under this paragraph. All redemptions, purchases, and sales by the Secretary of the Treasury of obligations under this paragraph shall be treated as public debt transactions of the United States.

Interest
payments.

(6) Any appropriations made available to the Waste Fund for any purpose described in subsection (d) shall be repaid into the general fund of the Treasury, together with interest from the date of availability of the appropriations until the date of repayment. Such interest shall be paid on the cumulative amount of appropriations available to the Waste Fund, less the average undisbursed cash balance in the Waste Fund account during the fiscal year involved.

Deferral.

The rate of such interest shall be determined by the Secretary of Treasury taking into consideration the average market yield during the month preceding each fiscal year on outstanding marketable obligations of the United States of comparable maturity. Interest payments may be deferred with the approval of the Secretary of the Treasury, but any interest payments so deferred shall themselves bear interest.

42 USC 10223.
Study.

Sec. 303. Alternative Means of Financing

The Secretary shall undertake a study with respect to alternative approaches to managing the construction and operation of all civilian radioactive waste management facilities, including the feasibility of establishing a private corporation for such purposes. In conducting such study, the Secretary shall consult with the Director of the Office of Management and Budget, the Chairman of the Commission, and such other Federal agency representatives as may be appropriate. Such study

Report to
Congress.

shall be completed, and a report containing the results of such study shall be submitted to the Congress, within 1 year after the date of the enactment of this Act.

42 USC 10224.

Sec. 304. Office of Civilian Radioactive Waste Management

(a) ESTABLISHMENT–There hereby is established within the Department of Energy an Office of Civilian Radioactive Waste Management. The Office shall be headed by a Director, who shall be appointed by the President, by and with the advice and consent of the

Senate, and who shall be compensated at the rate payable for level IV of the Executive Schedule under section 5315 of title 5, United States Code.

(b) FUNCTIONS OF DIRECTOR–The Director of the Office shall be responsible for carrying out the functions of the Secretary under this Act, subject to the general supervision of the Secretary. The Director of the Office shall be directly responsible to the Secretary.

(c) ANNUAL REPORT TO CONGRESS–The Director of the Office shall annually prepare and submit to the Congress a comprehensive report on the activities and expenditures of the Office.

(d) AUDIT BY GAO–If requested by either House of the Congress (or any committee thereof) or if considered necessary by the Comptroller General, the Government Accountability Office shall conduct an audit of the Office, in accord with such regulations as the Comptroller General may prescribe. The Comptroller General shall have access to such books, records, accounts, and other materials of the Office as the Comptroller General determines to be necessary for the preparation of such audit. The Comptroller General shall submit a report on the results of each audit conducted under this section.[21]

Report to Congress.
42 USC 10225.

Sec. 305. Location of Test and Evaluation Facility

(a) REPORT TO CONGRESS–Not later than 1 year after the date of the enactment of this Act, the Secretary shall transmit to the Congress a report setting forth whether the Secretary plans to locate the test and evaluation facility at the site of a repository.

(b) PROCEDURES–(1) If the test and evaluation facility is to be located at any candidate site or repository site (A) site selection and development of such facility shall be conducted in accordance with the procedures and requirements established in title I with respect to the site selection and development of repositories; and (B) the Secretary may not commence construction of any surface facility for such tests and evaluation facility prior to issuance by the Commission of a construction authorization for a repository at the site involved.

(2) No test and evaluation facility may be converted into a repository unless site selection and development of such facility was conducted in accordance with the procedures and requirements established in title I with respect to the site selection and development of repositories.

(3) The Secretary may not commence construction of a test and evaluation facility at a candidate site or site recommended as the location for a repository prior to the date on which the designation of such site is effective under section 115.

---

[21]   Amended by P.L. 97–425, Title III, § 304, 96 Stat. 2661 (1983); P.L. 104–66, Title I, Subtitle E, § 1052(1), 109 Stat. 719 (1995); P.L. 108–271, § 8(b), 118 Stat. 814 (2004) renaming the General Accounting Office the Government Accountability Office.

42 USC 10226.
Regulations or
guidance.

Sec. 306. Nuclear Regulatory Commission Training Authorization
NUCLEAR REGULATORY COMMISSION TRAINING AUTHORIZATION–The Nuclear Regulatory Commission is authorized and directed to promulgate regulations, or other appropriate Commission regulatory guidance, for the training and qualifications of civilian nuclear power plant operators, supervisors, technicians and other appropriate operating personnel. Such regulations or guidance shall establish simulator training requirements for applicants for civilian nuclear power plant operator licenses and for operator requalification programs; requirements governing NRC administration of requalification examinations; requirements for operating tests at civilian nuclear power plant simulators, and instructional requirements for civilian nuclear

Report to
Congress.

power plant licensee personnel training programs . Such regulations or other regulatory guidance shall be promulgated by the Commission within the 12-month period following enactment of this Act, and the Commission within the 12-month period following enactment of this Act shall submit a report to Congress setting forth the actions the Commission has taken with respect to fulfilling its obligations under this section.

Title IV–Nuclear Waste Negotiator[22]

42 USC 10241.

Sec. 401. Definition
For purposes of this title, the term "State" means each of the several States and the District of Columbia.[23]

42 USC 10242.

Sec. 402. The Office of the Nuclear Waste Negotiator
(a) ESTABLISHMENT–There is established the Office of the Nuclear Waste Negotiator that shall be an independent establishment in the executive branch.[24]

President of
U.S.

(b) THE NUCLEAR WASTE NEGOTIATOR–
(1) The Office shall be headed by a Nuclear Waste Negotiator who shall be appointed by the President, by and with the advice and consent of the Senate. The Negotiator shall hold office at the pleasure of the President, and shall be compensated at the rate provided for level III of the Executive Schedule in section 5314 of title 5, United States Code.
(2) The Negotiator shall attempt to find a state or Indian tribe willing to host a repository or monitored retrievable storage facility at a technically qualified site on reasonable terms and shall negotiate with any State or Indian tribe which expresses an interest in hosting a repository or monitored retrievable storage facility.

42 USC 10243.

Sec. 403. Duties of the Negotiator
(a) NEGOTIATIONS WITH POTENTIAL HOSTS–
(1) The Negotiator shall–
(A) seek to enter into negotiations on behalf of the United States, with–
(i) the Governor of any State in which a potential site is located; and

---

22   Added by P.L. 100–203, § 5041, 101 Stat. 1330–243 (1987).
23   Amended by P.L. 97–425, Title IV, § 401, as added by P.L. 100–202, § 101(d)
[Title III], 101 Stat. 1329–104, 1329–121 (1987), and P.L. 100–203, Title V, § 5041,
101 Stat. 1330–243 (1987); P.L. 102–486, Title VIII, § 802(b), 106 Stat. 923 (1992).
24   Amended by P.L. 100–507, § 1, 102 Stat 2541 (1988).

(ii) the governing body of any Indian tribe on whose reservation a potential site is located; and

(B) attempt to reach a proposed agreement between the United States and any such State or Indian tribe specifying the terms and conditions under which such State or tribe would agree to host a repository of monitored retrievable storage facility within such State or reservation.

(2) In any case in which State law authorizes any person or entity other than the Governor to negotiate a proposed agreement under this section on behalf of the State, any reference in this title to the Governor shall be considered to refer instead to such other person or entity.

(b)   CONSULTATION   WITH   AFFECTED   STATES, SUBDIVISIONS OF STATES, AND TRIBES–In addition to entering into negotiations under subsection (a), the Negotiator shall consult with any State, affected unit of local government, or any Indian tribe that the Negotiator determines may be affected by the siting of a repository or monitored retrievable storage facility and may include in any proposed agreement such terms and conditions relating to the interest of such States, affected units of local government, or Indian tribes as the Negotiator determines to be reasonable and appropriate.

(c) CONSULTATION WITH OTHER FEDERAL AGENCIES–The Negotiator may solicit and consider the comments of the Secretary, the Nuclear Regulatory Commission, or any other Federal agency on the suitability of any potential site for site characterization. Nothing in this subsection shall be construed to require the Secretary, the Nuclear Regulatory Commission, or any other Federal agency to make a finding that any such site is suitable for site characterization.

(d) PROPOSED AGREEMENT–(1) The Negotiator shall submit to the Congress any proposed agreement between the United States and a State or Indian tribe negotiated under subsection (a) and an environmental assessment prepared under section 404(a) for the site concerned.

(2) Any such proposed agreement shall contain such terms and conditions (including such financial and institutional arrangements) as the Negotiator and the host State or Indian tribe determine to be reasonable and appropriate and shall contain such provisions as are necessary to preserve any right to participation or compensation of such State, affected unit of local government, or Indian tribe under sections 116(c), 117, and 118(b).

(3) (A) No proposed agreement entered into under this section shall have legal effect unless enacted into Federal law.

(B) A State or Indian tribe shall enter into an agreement under this section in accordance with the laws of such State or tribe. Nothing in this section may be construed to prohibit the disapproval of a proposed agreement between a State and the United States under this section by a referendum or an act of the legislature of such State.

(4) Notwithstanding any proposed agreement under this section, the Secretary may construct a repository or monitored retrievable storage facility at a site agreed to under this title only if authorized by the Nuclear Regulatory Commission in accordance with the Atomic Energy Act of 1954 (42 USC 2012 *et seq.*), title II of the Energy Reorganization Act of 1982 (42 USC 5841 *et seq.*) and any other law applicable to authorization of such construction.

42 USC 10244.

Sec. 404. Environmental Assessment of Sites

(a) IN GENERAL–Upon the request of the Negotiator, the Secretary shall prepare an environmental assessment of any site that is the subject of negotiations under section 403(a).

(b) CONTENTS–(1) Each environmental assessment prepared for a repository site shall include a detailed statement of the probable impacts of characterizing such site and the construction and operation of a repository at such site.

(2) Each environmental assessment prepared for a monitored retrievable storage facility site shall include a detailed statement of the probable impacts of construction and operation of such a facility at such site.

(c) JUDICIAL REVIEW–The issuance of an environmental assessment under subsection (a) shall be considered to be a final agency action subject to judicial review in accordance with the provisions of chapter 7 of title 5, United States Code, section 119.

(d) PUBLIC HEARINGS–(1) In preparing an environmental assessment for any repository or monitored retrievable storage facility site, the Secretary shall hold public hearings in the vicinity of such site to inform the residents of the area in which such site is located that such site is being considered and to receive their comments.

(2) At such hearings, the Secretary shall solicit and receive any recommendations of such residents with respect to issues that should be addressed in the environmental assessment required under subsection (a) and the site characterization plan described in section 113(b)(1).

(e) PUBLIC AVAILABILITY–Each environmental assessment prepared under subsection (a) shall be made available to the public.

(f) EVALUATION OF SITES–(1) In preparing an environmental assessment under subsection (a), the Secretary shall use available geophysical, geologic, geochemical and hydrologic, and other information and shall not conduct any preliminary borings or excavations at any site that is the subject of such assessment unless–

(A) such preliminary boring or excavation activities were in progress on or before the date of the enactment of the Nuclear Waste Policy Amendments Act of 1987; or

(B) the Secretary certifies that, in the absence of preliminary borings or excavations, adequate information will not be available to satisfy the requirements of this Act or any other law.

(2) No preliminary boring or excavation conducted under this section shall exceed a diameter of 40 inches.

42 USC 10245.

Sec. 405. Site Characterization; Licensing

(a) SITE CHARACTERIZATION–Upon enactment of legislation to implement an agreement to site a repository negotiated under section 403(a), the Secretary shall conduct appropriate site characterization activities for the site that is the subject of such agreement subject to the conditions and terms of such agreement. Any such site characterization activities shall be conducted in accordance with section 113, except that references in such section to the Yucca Mountain site and the State of Nevada shall be deemed to refer to the site that is the subject of the agreement and the State of Indian tribe entering into the agreement.

(b) LICENSING–(1) Upon completion of site characterization activities carried out under subsection (a), the Secretary shall submit to the Nuclear Regulatory Commission an application for construction authorization for a repository at such site.

(2) The Nuclear Regulatory Commission shall consider an application for a construction authorization for a repository or monitored retrievable storage facility in accordance with the laws applicable to such applications, except that the Nuclear Regulatory Commission shall issue a final decision approving or disapproving the issuance of a construction authorization not later than 3 years after the date of the submission of such application.

42 USC 10246.

Sec. 406. Monitored Retrievable Storage

(a) CONSTRUCTION AND OPERATION–Upon enactment of legislation to implement an agreement negotiated under section 403(a) to site a monitored retrievable storage facility, the Secretary shall construct and operate such facility as part of an integrated nuclear waste management system in accordance with the terms and conditions of such agreement.

Grants.

(b) FINANCIAL ASSISTANCE–The Secretary may make grants to any State, Indian tribe, or affected unit of local government to assess the feasibility of siting a monitored retrievable storage facility under this section at a site under the jurisdiction of such State, tribe, or affected unit of local government.

42 USC 10247.

Sec. 407. Environmental Impact Statement

(a) IN GENERAL–Issuance of a construction authorization for a repository or monitored retrievable storage facility under section 405(b) shall be considered a major Federal action significantly affecting the quality of the human environment for purposes of the National Environmental Policy Act of 1969 (42 USC 4321 *et seq.*)

(b) PREPARATION–A final environmental impact statement shall be prepared by the Secretary under such Act and shall accompany any application to the Nuclear Regulatory Commission for a construction authorization.

(c) ADOPTION–(1) Any such environmental impact statement shall, to the extent practicable, be adopted by the Nuclear Regulatory Commission, in accordance with section 1506.3 of title 40, Code of Federal Regulations, in connection with the issuance by the Nuclear Regulatory Commission of a construction authorization and license for such repository or monitored retrievable storage facility.

(2) (A) In any such statement prepared with respect to a repository to be constructed under this title at the Yucca Mountain site, the Nuclear Regulatory Commission need not consider the need for a repository, the time of initial availability of a repository, alternate sites to the Yucca Mountain site, or nongeologic alternatives to such site.

(B) In any such statement prepared with respect to a repository to be constructed under this title at a site other than the Yucca Mountain site, the Nuclear Regulatory Commission need not consider the need for a repository, the time of initial availability of a repository, or nongeologic alternatives to such site but shall consider the Yucca Mountain site as alternative to such site in the preparation of such statement.

42 USC 10248.

Sec. 408. Administrative Powers of the Negotiator

In carrying out his functions under this title, the Negotiator may–

(1) appoint such officers and employees as he determines to be necessary and prescribe their duties;

(2) obtain services as authorized by section 3109 of title 5, United States Code, at rates not to exceed the rate prescribed for grade GS-18 of the General Schedule by section 5332 of title 5, United States Code;

(3) promulgate such rules and regulations as may be necessary to carry out such functions;

(4) utilize the services, personnel, and facilities of other Federal agencies (subject to the consent of the head of any such agency);

Contracts.

(5) for purposes of performing administrative functions under this title, and to the extent funds are appropriated, enter into and perform such contracts, leases, cooperative agreements, or other transactions as may be necessary and on such terms as the Negotiator determines to be appropriate, with any agency or instrumentality of the United States, or with any public or private person or entity;

(6) accept voluntary and uncompensated services, notwithstanding the provisions of section 1342 of title 31, United States Code;

(7) adopt an official seal, which shall be judicially noticed;

(8) use the United States mails in the same manner and under the same conditions as other departments and agencies of the United States;

(9) hold such hearings as are necessary to determine the views of interested parties and the general public; and

(10) appoint advisory committees under the Federal Advisory Committee Act (5 USC App.)

42 USC 10249.

Sec. 409. Cooperation of Other Departments and Agencies

Each department, agency, and instrumentality of the United States, including any independent agency, may furnish the Negotiator such information as he determines to be necessary to carry out his functions under this title.

42 USC 10250.

Sec. 410. Termination of the Office

The Office shall cease to exist not later than 30 days after the date 7 years after the date of the enactment of the Nuclear Waste Policy Amendments Act of 1987.[25]

42 USC 10251.

Sec. 411. Authorization of Appropriations

Notwithstanding subsection (d) of section 302, and subject to subsection (e) of such section, there are authorized to be appropriated for expenditures from amounts in the Waste Fund established in subsection (c) of such section, such sums as may be necessary to carry out the provisions of this title.

Title V–Nuclear Waste Technical Review Board[26]

2 USC 10261.

Sec. 501. Definitions

As used in this title:

(1) The term "Chairman" means the Chairman of the Nuclear Waste Technical Review Board.

(2) The term "Board" means the Nuclear Waste Technical Review Board established under section 502.

2 USC 10262.

Sec. 502. Nuclear Waste Technical Review Board

(a) ESTABLISHMENT–There is established a Nuclear Waste Technical Review Board that shall be an independent establishment within the executive branch.

President of U.S.

(b) MEMBERS–(1) The Board shall consist of 11 members who shall be appointed by the President not later than 90 days after the date of the enactment of the Nuclear Waste Policy Amendments Act of 1987 from

---

[25]   Amended by P.L. 102–486, 106 Stat. 2923 (1992).
[26]   Added by P.L. 100–203, § 5051, 101 Stat. 1330–248 (1987).

among persons nominated by the National Academy of Sciences in accordance with paragraph (3).

(2) The President shall designate a member of the Board to serve as chairman.

(3) (A) The National Academy of Sciences shall, not later than 90 days after the date of the enactment of the Nuclear Waste Policy Amendments Act of 1987, nominate not less than 22 persons for appointment to the Board from among persons who meet the qualifications described in subparagraph (C).

(B) The National Academy of Sciences shall nominate not less than 2 persons to fill any vacancy on the Board from among persons who meet the qualifications described in subparagraph (C).

(C) (i) Each person nominated for appointment to the Board shall be–

(I) eminent in a field of science or engineering, including environmental sciences; and

(II) selected solely on the basis of established records of distinguished service.

(ii) The membership of the Board shall be representative of the broad range of scientific and engineering disciplines related to activities under this title.

(iii) No person shall be nominated for appointment to the Board who is an employee of–

(I) the Department of Energy;

(II) a national laboratory under contract with the Department of Energy; or

(III) an entity performing high-level radioactive waste or spent nuclear fuel activities under contract with the Department of Energy.

(4) Any vacancy on the Board shall be filled by the nomination and appointment process described in paragraphs (1) and (3).

(5) Members of the Board shall be appointed for terms of 4 years, each such term to commence 120 days after the date of enactment of the Nuclear Waste Policy Amendments Act of 1987, except that of the 11 members first appointed to the Board, 5 shall serve for 2 years and 6 shall serve for 4 years, to be designated by the President at the time of appointment.

42 USC 10263.   **Sec. 503. Functions**

The Board shall evaluate the technical and scientific validity of activities undertaken by the Secretary after the date of the enactment of the Nuclear Waste Policy Amendments Act of 1987, including–

(1) site characterization activities; and

(2) activities relating to the packaging or transportation of high-level radioactive waste or spent nuclear fuel.

42 USC 10264.   **Sec. 504. Investigatory Powers**

(a) HEARINGS–Upon request of the Chairman or a majority of the members of the Board, the Board may hold such hearings, sit and act at such times and places, take such testimony, and receive such evidence, as the Board considers appropriate. Any member of the Board may administer oaths or affirmations to witnesses appearing before the Board.

(b) PRODUCTION OF DOCUMENTS–(1) Upon the request of the Chairman or a majority of the members of the Board, and subject to existing law, the Secretary (or any contractor of the Secretary) shall provide the Board with such records, files, papers, data, or information as may be necessary to respond to any inquiry of the Board under this title.

(2) Subject to existing law, information obtainable under paragraph (1) shall not be limited to final work products of the Secretary, but shall include drafts of such products and documentation of work in progress.

42 USC 10265.   ### Sec. 505. Compensation of Members

(a) IN GENERAL–Each member of the Board shall be paid at the rate of pay payable for level III of the Executive Schedule for each day (including travel time) such member is engaged in the work of the Board.

(b) TRAVEL EXPENSES–Each member of the Board may receive travel expenses, including per diem in lieu of subsistence, in the same manner as is permitted under sections 5702 and 5703 of title 5, United States Code.

42 USC 10266.   ### Sec. 506. Staff

(a) CLERICAL STAFF–(1) Subject to paragraph (2), the Chairman may appoint and fix the compensation of such clerical staff as may be necessary to discharge the responsibilities of the Board.

(2) Clerical staff shall be appointed subject to the provisions of title 5, United States Code, governing appointments in the competitive service, and shall be paid in accordance with the provisions of chapter 51 and subchapter III of chapter 53 of such title relating to classification and General Schedule pay rates.

(b) PROFESSIONAL STAFF.–(1) Subject to paragraphs (2) and (3), the Chairman may appoint and fix the compensation of such professional staff as may be necessary to discharge the responsibilities of the Board.

(2) Not more than 10 professional staff members may be appointed under this subsection.

(3) Professional staff members may be appointed without regard to the provisions of title 5, United States Code, governing appointments in the competitive service, and may be paid without regard to the provisions of chapter 51 and subchapter III of chapter 53 of such title relating to classification and General Schedule pay rates, except that no individual so appointed may receive pay in excess of the annual rate of basic pay payable of GS-18 of the General Schedule.

42 USC 10267.   ### Sec. 507. Support Services

(a) GENERAL SERVICES–To the extent permitted by law and requested by the Chairman, the Administrator of General Services shall provide the Board with necessary administrative services facilities, and support on a reimbursable basis.

(b) ACCOUNTING, RESEARCH, AND TECHNOLOGY ASSESSMENT SERVICES–The Comptroller General, the Librarian of Congress, and the Director of the Office of Technology Assessment shall, to the extent permitted by law and subject to the availability of funds, provide the Board with such facilities, support, funds and services, including staff, as may be necessary for the effective performance of the functions of the Board.

(c) ADDITIONAL SUPPORT–Upon the request of the Chairman, the Board may secure directly from the head of any department or agency of the United States information necessary to enable it to carry out this title.

(d) MAILS–The Board may use the United States mails in the same manner and under the same conditions as other departments and agencies of the United States.

(e) EXPERTS AND CONSULTANTS–Subject to such rules as may be prescribed by the Board, the Chairman may procure temporary and intermittent services under section 3109(b) of title 5 of the United States Code, but at rates for individuals not to exceed the daily equivalent of the maximum annual rate of basic pay payable for GS-18 of the General Schedule.

42 USC 10268.

Sec. 508. Report

The Board shall report not less than 2 times per year to Congress and the Secretary its findings, conclusions, and recommendations. The first such report shall be submitted not later than 12 months after the date of the enactment of the Nuclear Waste Policy Amendments Act of 1987.

42 USC 10269.

Sec. 509. Authorization of Appropriations

Notwithstanding subsection (d) of section 302, and subject to subsection (e) of such section, there are authorized to be appropriated for expenditures from amounts in the Waste Fund established in subsection (c) of such section such sums as may be necessary to carry out the provisions of this title.

42 USC 10270.

Sec. 510. Termination of the Board

The Board shall cease to exist not later than 1 year after the date on which the Secretary begins disposal of high-level radioactive waste or spent nuclear fuel in a repository.

B. PERTINENT PROVISIONS OF THE ENERGY POLICY ACT OF 1992—(ENVIRONMENTAL PROTECTION STANDARDS)

Title VIII–High-Level Radioactive Waste[1]

42 USC 10141
note.

Sec. 801. Nuclear Waste Disposal

(a) ENVIRONMENTAL PROTECTION AGENCY STANDARDS–

(1) PROMULGATION–Notwithstanding the provisions of section 121(a) of the Nuclear Waste Policy Act of 1982 (42 USC 10141(a)), section 161b. of the Atomic Energy Act of 1954 (42 USC 2201(b)), and any other authority of the Administrator of the Environmental Protection Agency to set generally applicable standards for the Yucca Mountain site, the Administrator shall, based upon and consistent with the findings and recommendations of the National Academy of Sciences, promulgate, by rule, public health and safety standards for protection of the public from releases from radioactive materials stored or disposed of in the repository at the Yucca Mountain site. Such standards shall prescribe the maximum annual effective dose equivalent to individual members of the public from releases to the accessible environment from radioactive materials stored or disposed of in the repository. The standards shall be promulgated not later than one year after the Administrator receives the findings and recommendations of the National Academy of Sciences under paragraph (2) and shall be the only such standards applicable to the Yucca Mountain site.

(2) STUDY BY NATIONAL ACADEMY OF SCIENCES– Within 90 days after the date of the enactment of this Act, the Administrator shall contract with the National Academy of Sciences to conduct a study to provide, by not later than December 31, 1993, findings and recommendations on reasonable standards for protection of the public health and safety, including –

(A) whether a health-based standard based upon doses to individual members of the public from releases to the accessible environment (as that term is defined in the regulations contained in subpart B of Part 191 of title 40, Code of Federal Regulations, as in effect on November 18, 1985) will provide a reasonable standard for protection of the health and safety of the general public.

(B) whether it is reasonable to assume that a system for post-closure oversight of the repository can be developed, based upon active institutional controls, that will prevent an unreasonable risk of breaching the repository's engineered or geologic barriers 'r increasing the exposure of individual members of the public to radiation beyond allowable limits; and

(C) whether it is possible to make scientifically supportable predictions of the probability that the repository's engineered or geologic barriers will be breached as a result of human intrusion over a period of 10,000 years.

(3) APPLICABILITY–The provisions of this section shall apply to the Yucca Mountain site, rather than any other authority of the

---

[1]   This Act consists of P.L. 102–486, 106 Stat. 2921 (1992), and generally appears in Title 42, United States Code.

Administrator to set generally applicable standards for radiation protection.

(b) NUCLEAR REGULATORY COMMISSION REQUIREMENTS AND CRITERIA.–

(1) MODIFICATIONS–Not later than 1 year after the Administrator promulgates standards under subsection (a0, the Nuclear Regulatory Commission shall, by rule, modify its technical requirements and criteria under section 121(b) of the nuclear Waste Policy Act of 1982 942 USC 10141(b)), as necessary, to be consistent with the Administrator's standards promulgated under subsection (a).

(2) REQUIRED ASSUMPTIONS–The Commission's requirements and criteria shall assume, to the extent consistent with the findings and recommendations of the National Academy of Sciences, that, following repository closure, the inclusion of engineered barriers and the Secretary's post-closure, oversight of the Yucca Mountain site, in accordance with the subsection (c), shall be sufficient to–

(A) prevent any activity at the site that poses an unreasonable risk of breaching the repository's engineered or geologic barriers; and

(B) prevent any increase in the exposure of individual members of the public to radiation beyond allowable limits.

(c) POST-CLOSURE OVERSIGHT–Following repository closure, the Secretary of Energy shall continue to oversee the Yucca Mountain site to prevent any activity at the site that poses an unreasonable risk of–

(1) breaching the repository's engineered or geologic barriers; or

(2) increasing the exposure of individual members of the public to radiation beyond allowable limits.

42 USC 10101 note.

Sec. 803. Nuclear Waste Management Plan

(a) PREPARATION AND SUBMISSION OF REPORT–The Secretary of Energy, in consultation with the Nuclear Regulatory Commission and the Environmental Protection Agency, shall prepare and submit to the Congress a report on whether current programs and plans for management of nuclear waste as mandated by the Nuclear Waste Policy Act of 1982 (42 USC 10101 et seq.) are adequate for management of any additional volumes or categories of nuclear waste that might be generated by any new nuclear power plants that might be constructed and licensed after the date of the enactment of this Act. The Secretary shall prepare the report for submission to the President and the Congress within 1 year after the date of the enactment of this Act. The report shall examine any new relevant issues related to management of spent nuclear fuel and high-level radioactive waste that might be raised by the addition of new nuclear-generated electric capacity, including anticipated increased volumes of spent nuclear fuel or high-level radioactive waste, any need for additional interim storage capacity prior to final disposal, transportation of additional volumes of waste, and any need for additional repositories for deep geologic disposal.

(b) OPPORTUNITY FOR PUBLIC COMMENT–In preparation of the report required under subsection (a), the Secretary of Energy shall offer members of the public an opportunity to provide information and comment and shall solicit the views of the Nuclear Regulatory Commission, the Environmental Protection Agency, and other interested parties.

(c) AUTHORIZATION OF APPROPRIATIONS–There are authorized to be appropriated such sums as may be necessary to carry out this section.

C. PERTINENT PROVISIONS OF THE RONALD W. REAGAN NATIONAL DEFENSE AUTHORIZATION ACT FOR FISCAL YEAR 2005—(WASTE INCIDENTAL TO REPROCESSING)

50 USC 2601
note.

Sec. 3116. Defense Site Acceleration Completion[1]

(a) IN GENERAL.–Notwithstanding the provisions of the Nuclear Waste Policy Act of 1982, the requirements of section 202 of the Energy Reorganization Act of 1974, and other laws that define classes of radioactive waste, with respect to material stored at a Department of Energy site at which activities are regulated by a covered State pursuant to approved closure plans or permits issued by the State, the term "high-level radioactive waste" does not include radioactive waste resulting from the reprocessing of spent nuclear fuel that the Secretary of Energy (in this section referred to as the "Secretary"), in consultation with the Nuclear Regulatory Commission (in this section referred to as the "Commission"), determines–

(1) does not require permanent isolation in a deep geologic repository for spent fuel or high-level radioactive waste;

(2) has had highly radioactive radionuclides removed to the maximum extent practical; and

(3)(A) does not exceed concentration limits for Class C low- level waste as set out in section 61.55 of title 10, Code of Federal Regulations, and will be disposed of–

(i) in compliance with the performance objectives set out in subpart C of part 61 of title 10, Code of Federal Regulations; and

(ii) pursuant to a State-approved closure plan or State-issued permit, authority for the approval or issuance of which is conferred on the State outside of this section; or

(B) exceeds concentration limits for Class C low-level waste as set out in section 61.55 of title 10, Code of Federal Regulations, but will be disposed of–

(i) in compliance with the performance objectives set out in subpart C of part 61 of title 10, Code of Federal Regulations;

(ii) pursuant to a State-approved closure plan or State-issued permit, authority for the approval or issuance of which is conferred on the State outside of this section; and

(iii) pursuant to plans developed by the Secretary in consultation with the Commission.

(b)   MONITORING   BY   NUCLEAR   REGULATORY COMMISSION.–

(1) The Commission shall, in coordination with the covered State, monitor disposal actions taken by the Department of Energy pursuant to subparagraphs (A) and (B) of subsection (a)(3) for the purpose of assessing compliance with the performance objectives set out in subpart C of part 61 of title 10, Code of Federal Regulations.

(2) If the Commission considers any disposal actions taken by the Department of Energy pursuant to those subparagraphs to be not in compliance with those performance objectives, the Commission shall, as soon as practicable after discovery of the noncompliant conditions, inform the Department of Energy, the covered State, and the following congressional committees:

---

[1]   See slip law at P.L. 108–375, Division C, Title XXXI, § 3116, 118 Stat. 2162 (2004).

(A) The Committee on Armed Services, the Committee on Energy and Commerce, and the Committee on Appropriations of the House of Representatives.

(B) The Committee on Armed Services, the Committee on Energy and Natural Resources, the Committee on Environment and Public Works, and the Committee on Appropriations of the Senate.

(3) For fiscal year 2005, the Secretary shall, from amounts available for defense site acceleration completion, reimburse the Commission for all expenses, including salaries, that the Commission incurs as a result of performance under subsection (a) and this subsection for fiscal year 2005. The Department of Energy and the Commission may enter into an interagency agreement that specifies the method of reimbursement. Amounts received by the Commission for performance under subsection (a) and this subsection may be retained and used for salaries and expenses associated with those activities, notwithstanding section 3302 of title 31, United States Code, and shall remain available until expended.

(4) For fiscal years after 2005, the Commission shall include in the budget justification materials submitted to Congress in support of the Commission budget for that fiscal year (as submitted with the budget of the President under section 1105(a) of title 31, United States Code) the amounts required, not offset by revenues, for performance under subsection (a) and this subsection.

(c) INAPPLICABILITY TO CERTAIN MATERIALS.– Subsection (a) shall not apply to any material otherwise covered by that subsection that is transported from the covered State.

(d) COVERED STATES.–For purposes of this section, the following States are covered States:

(1) The State of South Carolina.

(2) The State of Idaho.

(e) CONSTRUCTION.–

(1) Nothing in this section shall impair, alter, or modify the full implementation of any Federal Facility Agreement and Consent Order or other applicable consent decree for a Department of Energy site.

(2) Nothing in this section establishes any precedent or is binding on the State of Washington, the State of Oregon, or any other State not covered by subsection (d) for the management, storage, treatment, and disposition of radioactive and hazardous materials.

(3) Nothing in this section amends the definition of "transuranic waste" or regulations for repository disposal of transuranic waste pursuant to the Waste Isolation Pilot Plant Land Withdrawal Act or part 191 of title 40, Code of Federal Regulations.

(4) Nothing in this section shall be construed to affect in any way the obligations of the Department of Energy to comply with section 4306A of the Atomic Energy Defense Act (50 U.S.C. 2567).

(5) Nothing in this section amends the West Valley Demonstration Act (42 U.S.C. 2121a note).

(f) JUDICIAL REVIEW.–Judicial review shall be available in accordance with chapter 7 of title 5, United States Code, for the following:

(1) Any determination made by the Secretary or any other agency action taken by the Secretary pursuant to this section.

(2) Any failure of the Commission to carry out its responsibilities under subsection (b).

D. PERTINENT PROVISIONS OF THE ENERGY POLICY ACT OF 2005—(GREATER-THAN-CLASS C WASTE)

Sec. 631. Safe Disposal of Greater-Than-Class C Radioactive Waste[1]

Notification.

(a) RESPONSIBILITY FOR ACTIVITIES TO PROVIDE STORAGE FACILITY.–The Secretary shall provide to Congress official notification of the final designation of an entity within the Department to have the responsibility of completing activities needed to provide a facility for safely disposing of all greater-than-Class C low-level radioactive waste.

(b) REPORTS AND PLANS.–

(1) REPORT ON PERMANENT DISPOSAL FACILITY.–

(A) PLAN REGARDING COST AND SCHEDULE FOR COMPLETION OF EIS AND ROD.–Not later than 1 year after the date of enactment of this Act, the Secretary, in consultation with Congress, shall submit to Congress a report containing an estimate of the cost and a proposed schedule to complete an environmental impact statement and record of decision for a permanent disposal facility for greater-than-Class C radioactive waste.

(B) ANALYSIS OF ALTERNATIVES.–Before the Secretary makes a final decision on the disposal alternative or alternatives to be implemented, the Secretary shall–

(i) submit to Congress a report that describes all alternatives under consideration, including all information required in the comprehensive report making recommendations for ensuring the safe disposal of all greater-than-Class C low-level radioactive waste that was submitted by the Secretary to Congress in February 1987; and

(ii) await action by Congress.

(2) SHORT-TERM PLAN FOR RECOVERY AND STORAGE.–

Deadline.

(A) IN GENERAL.–Not later than 180 days after the date of enactment of this Act, the Secretary shall submit to Congress a plan to ensure the continued recovery and storage of greater-than-Class C low-level radioactive sealed sources that pose a security threat until a permanent disposal facility is available.

(B) CONTENTS.–The plan shall address estimated cost, resource, and facility needs.

---

[1]   Added by P.L. 109–58, 119 Stat. 788 (2005).

7. Uranium Mill Tailings

7. Uranium Mill Tailings
Contents

A. URANIUM MILL TAILINGS RADIATION CONTROL
ACT OF 1978, AS AMENDED

Public Law 95–604   **92 Stat. 3021**

November 8, 1978

An Act

Sec. 1. Short Title and Table of Contents

This Act may be cited as the "Uranium Mill Tailings Radiation Control Act of 1978." (TOC not duplicated here.)

42 USC 7901.

Sec. 2. Findings and Purposes

(a) The Congress finds that uranium mill tailings located at active and inactive mill operations may pose a potential and significant radiation health hazard to the public, and that the protection of the public health, safety, and welfare and the regulation of interstate commerce require that every reasonable effort be made to provide for the stabilization, disposal, and control in a safe and environmentally sound manner of such tailings in order to prevent or minimize radon diffusion into the environment and to prevent or minimize other environmental hazards from such tailings.

(b) The purposes of this Act are to provide–

(1) in cooperation with the interested States, Indian tribes, and the persons who own or control inactive mill tailings sites, a program of assessment and remedial action at such sites, including, where appropriate, the reprocessing of tailings to extract residual uranium and other mineral values where practicable, in order to stabilize and control such tailings in a safe and environmentally sound manner and to minimize or eliminate radiation health hazards to the public, and

(2) a program to regulate mill tailings during uranium or thorium ore processing at active mill operations and after termination of such operations in order to stabilize and control such tailings in a safe and environmentally sound manner and to minimize or eliminate radiation health hazards to the public.

Title I–Remedial Action Program

42 USC 7911.

Sec. 101. Definitions

For purposes of this title–

(1) The term "Secretary" means the Secretary of Energy.

(2) The term "Commission" means the Nuclear Regulatory Commission.

(3) The term "Administrator" means the Administrator of the Environmental Protection Agency.

(4) The term "Indian tribe" means any tribe, band, clan, group, pueblo, or community of Indians recognized as eligible for services provided by the Secretary of the Interior to Indians.

(5) The term "person" means any individual association, partnership, corporation, firm, joint venture, trust, government entity, and any other entity, except that such term does not include any Indian or Indian tribe.

(6) The term "processing site" means–

(A) any site, including the mill, containing residual radioactive materials at which all or substantially all of the uranium was produced for sale to any Federal agency prior to January 1, 1971 under a contract with any Federal agency, except in the case of a site at or near Slick Rock, Colorado, unless–

(i) such site was owned or controlled as of January 1, 1978, or is thereafter owned or controlled by any Federal agency, or

(ii) a license (issued by the Commission or its predecessor agency under the Atomic Energy Act of 1954 or by a State as permitted under section 274 of such Act) for the production at such site of any uranium or thorium product derived from ores is in effect on January 1, 1978, or is issued or renewed after such date; and

(B) any other real property or improvement thereon which–

(i) is in the vicinity of such site, and

(ii) is determined by the Secretary, in consultation with the Commission, to be contaminated with residual radioactive materials derived from such site.

42 USC 2011 note.
42 USC 2021.

Any ownership or control of an area by a Federal agency which is acquired pursuant to a cooperative agreement under this title shall not be treated as ownership or control by such agency for purposes of subparagraph (A)(i). A license for the production of any uranium product from residual radioactive materials shall not be treated as a license for production from ores within the meaning of subparagraph (A)(ii) if such production is in accordance with section 108(b).

(7) The term "residual radioactive material" means–

(A) waste (which the Secretary determines to be radioactive) in the form of tailings resulting from the processing of ores for the extraction of uranium and other valuable constituents of the ores; and

(B) other waste (which the Secretary determines to be radioactive) at a processing site which relate to such processing, including any residual stock of unprocessed ores or low-grade materials.

(8) The term "tailings" means the remaining portion of a metal-bearing ore after some or all of such metal, such as uranium, has been extracted.

(9) The term "Federal agency" includes any executive agency as defined in section 105 of title 5 of the United States Code.

(10) The term "United States" means the 48 contiguous States and Alaska, Hawaii, Puerto Rico, the District of Columbia, and the territories and possessions of the United States.

42 USC 7912.

Sec. 102. Designation of Processing Sites

(a)(1) As soon as practicable, but no later than one year after enactment of this Act, the Secretary shall designate processing sites at or near the following locations:

Salt Lake City, Utah
Green River, Utah
Mexican Hat, Utah
Durango, Colorado
Grand Junction, Rifle, Colorado (two sites)
Gunnison, Colorado
Naturita, Colorado
Maybell, Colorado
Slick Rock, Colorado (two sites)
Shiprock, New Mexico
Ambrosia Lake, New Mexico

Riverton, Wyoming
Converse County, Wyoming
Lakeview, Oregon
Falls City, Texas
Tuba City, Arizona
Monument Valley, Arizona
Lowman, Idaho
Canonsburg, Pennsylvania

Remedial
action.

Subject to the provisions of this title, the Secretary shall complete remedial action at the above listed sites before his authority terminates under this title. The Secretary shall within one year of the date of enactment of this Act also designate all other processing sites within the United States which he determines requires remedial action to carry out the purposes of this title. In making such designation, the Secretary shall consult with the Administrator, the Commission, and the affected States, and in the case of Indian lands, the appropriate Indian tribe and the Secretary of the Interior.

(2) As part of his designation under this subsection, the Secretary, in consultation with the Commission, shall determine the boundaries of each such site.

(3) No site or structure with respect to which remedial action is authorized under Public Law 92-314 in Grand Junction, Colorado, may be designated by the Secretary as a processing site under this section.

Health hazard
assessment.

(b) Within one year from the date of the enactment of this Act, the Secretary shall assess the potential health hazard to the public from the residual radioactive materials at designated processing sites. Based upon such assessment, the secretary shall, within such one year period, establish priorities for carrying out remedial action at each such site. In establishing such priorities, the Secretary shall rely primarily on the advice of the Administrator.

Notification.

(c) Within thirty days after making designations of processing sites and establishing the priorities for such sites under this section, the Secretary shall notify the Governor of each affected State, and where appropriate, the Indian tribes and the Secretary of the Interior.

(d) The designations made, and priorities established, by the Secretary under this section shall be final and not be subject to judicial review.

(e)(1) The designation of processing sites within one year after enactment under this section shall include, to the maximum extent practicable, the areas referred to in section 101(6)(B).

(2) Notwithstanding the one year limitation contained in this section, the Secretary may, after such one year period, include any areas described in section 101(6)(B) as part of a processing site designated under this section if he determines such inclusion to be appropriate to carry out the purposes of this title.

42 USC 7911.

(3)[1] The Secretary shall designate as a processing site within the meaning of section 101(6) any real property, or improvements thereon, in Edgemont, South Dakota, that–

(A) is in the vicinity of the Tennessee Valley Authority uranium mill site at Edgemont (but not including such site), and

---

[1]   Added by P.L. 97–415, 96 Stat. 2079 (1983).

(B) is determined by the Secretary to be contaminated with residual radioactive materials.

(f)(1) DESIGNATION. Notwithstanding any other provision of law, the Moab uranium milling site (referred to in this subsection as the "Moab site") located approximately three miles northwest of Moab, Utah, and identified in the Final Environmental Impact Statement issued by the Nuclear Regulatory Commission in March 1996 in conjunction with Source Materials License No. SUA-917, is designated as a processing site.

(2) APPLICABILITY. This title applies to the Moab site in the same manner and to the same extent as to other processing sites designated under subsection (a), except that–

(A) sections 103, 104(b), 107(a), 112(a), and 115(a) of this title shall not apply; and

(B) a reference in this title to the date of the enactment of this Act shall be treated as a reference to the date of the enactment of this subsection [enacted October 30, 2000].

(3) REMEDIATION. Subject to the availability of appropriations for this purpose, the Secretary shall conduct remediation at the Moab site in a safe and environmentally sound manner that takes into consideration the remedial action plan prepared pursuant to section 3405(i) of the Strom Thurmond National Defense Authorization Act for Fiscal Year 1999 (10 USC 7420 note; Public Law 105-261, including–

(A) ground water restoration; and

(B) the removal, to a site in the State of Utah, for permanent disposition and any necessary stabilization, of residual radioactive material and other contaminated material from the Moab site and the floodplain of the Colorado River.[2]

42 USC 7917.

In making the designation under this paragraph, the Secretary shall consult with the Administrator, the Commission and the State of South Dakota. The provisions of this title shall apply to the site so designated in the same manner and to the same extent as to the sites designated under subsection (a) except that, in applying such provisions to such site, any reference in this title to the date of enactment of this Act shall be treated as a reference to the date of the enactment of this paragraph and in determining the State share under section 107 of the costs of remedial action, there shall be credited to the State, expenditures made by the State prior to the date of the enactment of this paragraph which the Secretary determines would have been made by the State or the United States in carrying out the requirements of this title.

42 USC 7913.

Sec. 103. State Cooperative Agreements

(a) After notifying a State of the designation referred to in section 102 of this title, the Secretary subject to section 113, is authorized to enter into cooperative agreement with such State to perform remedial actions at each designated processing site in such State (other than a site location on Indian lands referred to in section 105). The Secretary shall, to the greatest extent practicable, enter into such agreements and carry out such remedial actions in accordance with the priorities established by him under section 102. The Secretary shall commence preparations for cooperative agreements with respect to each designated processing site as promptly as practicable following the designation of each site.

---

[2]   Amended by P.L. 106–398, § 1, 114 Stat. 1654 (2000).

Terms and
Conditions.

(b) Each cooperative agreement under this section shall contain such terms and conditions as the Secretary deems appropriate and consistent with the purposes of this Act, including, but not limited to, a limitation on the use of Federal assistance to those costs which are directly required to complete the remedial action selected pursuant to section 108.

Written
consent.

(c)(1) Except where the State is required to acquire the processing site as provided in subsection (a) of section 104, each cooperative agreement with a State under section 103 shall provide that the State shall obtain, in a form prescribed by the Secretary, written consent from any person holding any record interest in the designated processing site for the Secretary or any person designated by him to perform remedial action at such site.

Waiver.

(2) Such written consent shall include a waiver by each such person on behalf of himself, herself, his heirs, successors, and assigns–

(A) releasing the United States of any liability or claims thereof by such person, his heirs, successors, and assigns concerning such remedial action, and

(B) holding the United States harmless against any claim by such person on behalf of himself, his heirs, successors, or assigns arising out of the performance of any remedial action.

(d) Each cooperative agreement under this section shall require the State to assure that the Secretary, the Commission, and the Administrator and their authorized representatives have a permanent right of entry at any time to inspect the processing site and the site provided pursuant to section 104(b)(1) in furtherance of the provisions of this title and to carry out such agreement and enforce this Act and any rules prescribed under this Act. Such right of entry under this section or section 106 into an area described in section 101(6)(B) shall terminate on completion of the remedial action, as determined by the Secretary.

(e) Each agreement under this section shall take effect only upon the concurrence of the Commission with the terms and conditions thereof.

(f) The Secretary may, in any cooperative agreement enter into this section or section 105, provide for reimbursement of the actual costs, as determined by the Secretary, of any remedial action performed with respect to so much of a designated processing site as is described in section 101(6)(B). Such reimbursement shall be made only to a property owner of record at the time such remedial action was undertaken and only with respect to costs incurred by such property owner. No such reimbursement may be made unless–

(1) such remedial action was completed prior to enactment of this Act, and unless the application for such reimburse was filed by such owner within one year after a agreement under this section or section 105 is approved by the Secretary and the Commission, and

(2) the Secretary is satisfied that such action adequately achieves the purposes of this Act with respect to the site concerned and is consistent with the standards established by the Administrator pursuant to section 275(a) of the Atomic Energy Act of 1954.

42 USC 7914.

Sec. 104. Acquisition and Disposition of Lands and Materials

(a) Each cooperative agreement under section 103 shall require the State where determined appropriate by the Secretary with the concurrence of the Commission, to acquire any designated processing site, including where appropriate any interest therein. In determining whether to require the State to acquire a designated processing site or interest therein, consideration shall be given to the prevention of windfall profits.

Residual
radioactive
material.
removal.

(b)(1) If the Secretary with the concurrence of the Commission determines that removal of residual radioactive material from a processing site is appropriate, the cooperative agreement shall provide that the State shall acquire land (including, where appropriate, any interest therein) to be used as a site for the permanent disposition and stabilization of such residual radioactive materials in a safe and environmentally sound manner.

(2) Acquisition by the State shall not be required under this subsection if a site located on land controlled by the Secretary or made available by the Secretary of the Interior pursuant to section 106(a)(2) is designated by the Secretary, with the concurrence of the Commission, of such disposition and stabilization.

(c) No State shall be required under subsection (a) or (b) to acquire any real property or improvement outside the boundaries of–

(1) that portion of the processing site which is described in section 101(6)(A), and

(2) the site used for disposition of the residual radioactive materials.

(d) In the case of each processing site designated under this title other than a site designated on Indian land, the State shall take such action as may be necessary, and pursuant to regulations of the Secretary under this subsection, to assure that any person who purchases such a processing site after the removal of radioactive materials from such site shall be notified in any appropriate manner prior to such purchase, of the nature and extent of residual radioactive materials removed from site, including notice of the date when such action took place, and the condition of such

Notification.

site after such action. If the State is the owner if such site, the State shall so notify any prospective purchaser before entering into a contract, option or other arrangement to sell or otherwise dispose of such site. The Secretary shall issue appropriate rules and regulations to require notice in the local land records of the residual radioactive materials which were located at any processing site and notice of the nature and extent of

Rules and
regulations.

residual radioactive materials removed from the site, including notice of the date when such action took place. For purposes of this subsection, the term "site" does not include any property described in section 101(6)(B) of this title which is in a State which the Secretary has certified has a program which would achieve the purposes of this subsection.[3]

(e)(1) The terms and conditions of any cooperative agreement with a State under section 103 shall provide that in the case of any lands or interests therein acquired by the State pursuant to subsection (a), the State, with the concurrence of the Secretary and the Commission, may–

(A) sell such lands and interests,

(B) permanently retain such land and interests in lands (or donate such lands and interests therein to another governmental entity within such State) for permanent use by such State or entity solely for park, recreational, or other public purposes, or

(C) transfer such lands and interest to the United Sates as provided in subsection (f).

No lands may be sold under subparagraph (A) without the consent of the Secretary and the Commission. No site may be sold under subparagraph (A) or retained under subparagraph (B) if such site is used for the disposition of residual radioactive materials.

---

[3]   Amended by P.L. 104–259, § 4(a), 110 Stat. 3174 (1996).

(2) Before offering for sale any lands and interests therein which comprise a processing site, the State shall offer to sell such lands and interests at their fair market value to the person from whom the State acquired them.

(f)(1) Each agreement under section 103 shall provide that title to–

(A) the residual radioactive materials subject to the agreement, and

(B) any lands and interests therein which have been acquired by the State, under subsection (a) or (b), for the disposition of such materials, shall be transferred by the State to the Secretary when the Secretary (with the concurrence of the Commission) determines that remedial action is completed in accordance with the requirements imposed pursuant to this title. No payment shall be made in connection with the transfer of such property from fund appropriated for purposes of this Act other than payments for any administrative and legal costs incurred in carrying out such transfer.

(2) Custody of any property transferred to the United States under this subsection shall be assumed by the Secretary or such Federal agency as the President may designate. Notwithstanding any other provision of law, upon completion of the remedial action program authorized by this title, such property and minerals shall be maintained pursuant to a license issued by the Commission in such manner as will protect the public health, safety, and the environment. The Commission may, pursuant to such license or by rule or order, require the Secretary or other Federal agency having custody of such property and minerals to undertake such monitoring, maintenance, and emergency measures necessary to protect public health and safety and other actions as the Commission deems necessary to comply with the standards of section 275(a) of the Atomic Energy Act of 1954. The Secretary or such other Federal agency is authorized to carry out maintenance, monitoring and emergency measures under this subsection, but shall take no other action pursuant to such license, rule or order with respect to such property and minerals unless expressly authorized by Congress after the date of enactment of this Act. The United States shall not transfer title to property or interest therein acquired under this subsection to any person or State, except as provided in subsection (h).

(g) Each agreement under section 103 which permits any sale described in subsection (e)(1)(A) shall provide for the prompt reimbursement to the Secretary form the proceeds of such sale. Such reimbursement shall be in an amount equal to the lesser of–

(1) that portion of the fair market value of the lands or interests therein which bears the same ratio to such fair market value as the Federal share of the costs of acquisition by the State to such lands or interest therein bears to the total cost of such acquisition, or

(2) the total amount paid by the Secretary with respect to such acquisition.

Fair market value.

The fair market value of such lands or interest shall be determined by the Secretary as of the date of the sale by the State. Any amounts received by the Secretary under this title shall be deposited in the Treasury of the United States as miscellaneous receipts.

(h) No provision of any agreement under section 103 shall prohibit the Secretary of the Interior, with the concurrence of the Secretary of Energy and the Commission, from disposing of any subsurface mineral rights by sale or lease (in accordance with laws of the United States applicable to the sale, lease, or other disposal of such rights) which are associated with land on which residual radioactive materials are disposed and which are transferred to the United States as required under this section if the Secretary of the Interior takes such action as the Commission deems necessary pursuant to a license issued by the Commission to assure that the residual radioactive materials will not be disturbed by reason of any activity carried on following such disposition. If any such materials are disturbed by any such activity, the Secretary of the Interior shall insure, prior to the disposition of the minerals, that such materials will be restored to a safe and environmentally sound condition as determined by the Commission, and that the costs of such restoration will be borne by the person acquiring such rights from the Secretary of the Interior or from his successor or assign.

42 USC 7915.

Sec. 105. Indian Tribe Cooperative Agreements

(a) After notifying the Indian tribe of the designation pursuant to section 102 of this title, the Secretary, in consultation with the Secretary of the Interior, is authorized to enter into a cooperative agreement, subject to section 113, with any Indian tribe to perform remedial action at a designated processing site located on land of such Indian tribe. The Secretary shall, to the greatest extent practicable, enter into such agreements and carry out such remedial actions in accordance with the priorities established by him under section 102. In performing any remedial action under this section and in carrying out any continued monitoring or maintenance respecting residual radioactive materials associated with any site subject to a cooperative agreement under this section, the Secretary shall make full use of any qualified members of Indian tribes resident in the vicinity of any such site. Each such agreement shall contain such terms and conditions as the Secretary deems appropriate and consistent with the purpose of this Act. Such terms and conditions shall require the following:

Terms and conditions.

(1) The Indian tribe and any person holding any interest in such land shall execute a waiver (A) releasing the United States of any liability or claim thereof by such tribe or person concerning such remedial action and (B) holding the United States harmless against any claim arising out of the performance of any such remedial action.

(2) The remedial action shall be selected and performed in accordance with section 108 by the Secretary or such person as he may designate.

(3) The Secretary, the Commission, and the Administrator and their authorized representatives shall have a permanent right of entry at any time to inspect such processing site in furtherance of the provisions of this title, to carry out such agreement, and to enforce any rules prescribed under this Act.

Each agreement under this section shall take effect only upon concurrence of the Commission with the terms and conditions thereof.

(b) When the Secretary with the concurrence of the Commission determines removal of residual radioactive materials from a processing site on land described on subsection (a) to be appropriate, he shall provide, consistent with other applicable provisions of law, a site or sites for the permanent disposition and stabilization in a safe and environmentally sound manner of such residual radioactive materials. Such materials shall be transferred to the Secretary (without payment therefor by the Secretary) and permanently retained and maintained by the Secretary under the conditions established in a license issued by the commission, subject to section 104(f)(2) and (h).

42 USC 7916.

Uranium Mill
Tailings
Remedial Action
Amendments
Act of 1988.
42 USC 7901
note.
Public lands.
State listing.

Sec. 106. Acquisition of Lands by Secretary

Where necessary or appropriate in order to consolidate in a safe and environmentally sound manner the location of residual radioactive materials which are removed from processing sites under cooperative agreements under this title or where otherwise necessary for the permanent disposition and stabilization of such materials in such manner–

(1) the Secretary may acquire land and interest in land for such purposes by purchase, donation, or under any other authority of law or

(2) the Secretary of the Interior may transfer permanently to the Secretary to carry out the purposes of this Act, public lands under the jurisdiction of the Bureau of Land Management in the vicinity of processing sites in the following counties:

(A) Apache County in the State of Arizona;

(B) Mesa, Gunnison, Moffat, Montrose, Garfield, and San Miguel Counties in the State of Colorado;

(C) Boise County in the State of Idaho;

(D) Billings and Bowman Counties in the State of North Dakota;

(E) Grand and San Juan Counties in the State of Utah;

(F) Converse and Frement Counties in the State of Wyoming; and

(G) Any other county in the vicinity of a processing site, if no site in the county in which a processing site is located is suitable.

Any permanent transfer of lands under the jurisdiction of the Bureau of Land Management by the Secretary of the Interior to the Secretary shall not take place until the Secretary complies with the requirements of the National Environmental Policy Act (42 USC 4321 et seq.) with respect to the selection of a site for the permanent disposition and stabilization of residual radioactive materials. Section 204 of the Federal Land Policy and Management Act (43 USC 1714) shall not apply to this transfer of jurisdiction. Prior to acquisition of land under paragraph (1) or (2) of this subsection in any State, the Secretary shall consult with the Governor of such State. No lands may be acquired under such paragraph (1) or (2) in any State in which there is no (1) processing site designated under this title or (2) active uranium mill operation, unless the Secretary has obtained the consent of the Governor of such State. No lands controlled by any Federal agency may be transferred to the Secretary to carry out the purposes of this Act without the Concurrence of the chief administrative officer of such agency.[4]

42 USC 7917.

Sec. 107. Financial Assistance

(a) In the case of any designated processing site for which an agreement is executed with any State for remedial action at such site, the Secretary shall pay 90 per centum of the actual cost of such remedial action, including the actual costs of acquiring such site (and any interest therein) or any disposition site (and any interest therein pursuant to section 103 of this title, and the State shall pay the remainder of such costs from non-Federal funds. The Secretary shall not pay the administrative costs incurred by any State to develop, prepare, and carry out any cooperative agreement executed with such State under this title, except the proportionate share of the administrative costs associated with

---

[4]   Amended by P.L. 100–616, 102 Stat. 3192 (1988).

the acquisition of lands and interests therein acquired by the State pursuant to this title.

(b) In the case of any designated processing site located on Indian lands, the Secretary shall pay the entire cost of such remedial action.

42 USC 7918.

Sec. 108. Remedial Action

(a)(1) The Secretary or such person as he may designate shall select and perform remedial actions at designated processing sites and disposal sites in accordance with the general standards prescribed by the Administrator pursuant to section 275a. of the Atomic Energy Act of 1954. The State shall participate fully in the selection and performance of a remedial action for which it pays part of the cost. Such remedial action shall be selected and performed with the concurrence of the Commission and in consultation, as appropriate, with the Indian tribe and the Secretary of the Interior. Residual radioactive material from a processing site designated under this title may be disposed of at a facility licensed under title II under the administrative and technical requirements of such title. Disposal of such material at such a site in accordance with such requirements shall be considered to have been done in accordance with the administrative and technical requirements of this title.[5]

(2) The Secretary shall use technology in performing such remedial action as will insure compliance with the general standards promulgated by the Administrator under section 275a. of the Atomic Energy Act of 1954 and will assure the safe and environmentally sound stabilization of residual radioactive materials, consistent with existing law.[6]

(3) Notwithstanding paragraph (1) and (2) of this subsection, after October 31, 1982, if the Administrator has not promulgated standards under section 275a. of the Atomic Energy Act of 1954 in final form by such date, remedial action taken by the Secretary under this title shall comply with standards proposed by the Administrator under such section 275a. until such time as the Administrator promulgates the standards in final form.[7]

Evaluation.

(b) Prior to undertaking any remedial action at a designated site pursuant to this title, the Secretary shall request expressions of interest from private parties regarding the remilling of the residual radioactive materials at the site and, upon receipt of any expression of interest, the Secretary shall evaluate among other things the mineral concentration of the residual radioactive materials at each designated processing site to determine whether, as a part if any remedial action program, recovery of such minerals is practicable. The Secretary, with the concurrence of the Commission, may permit the recovery of such minerals, under such terms and conditions as he may prescribe to carry out the purposes of this title. No such recovery shall be permitted unless such recovery is consistent with remedial action. Any person permitted by the Secretary to recover such mineral shall pay to the Secretary a share of the net profits derived from such recovery, as determined by the Secretary. Such share shall not exceed the total amount paid by the Secretary for carrying out remedial action at such designated site. After payment of such share to the United States under this subsection, such person shall pay to the State in which the residual radioactive materials are located a share of the net profits derived from such recovery, as determined by the Secretary. The

---

[5]   Amended by P.L. 104–259, § 4(a), 110 Stat. 3174 (1996).
[6]   Amended by P.L. 97–415, § 18, 96 Stat. 2067 (1983).
[7]   Amended by P.L. 97–415, § 18, 96 Stat. 2067 (1983).

42 USC 2021.

person recovering such minerals shall bear all cost of such recovery. Any person carrying out mineral recovery activities under this paragraph shall be required to obtain any necessary license under the Atomic Energy Act of 1954 or under State law as permitted under section 274 of such Act.

42 USC 7919.

Sec. 109. Rules

The Secretary may prescribe such rules consistent with the purposes of this Act as he deems appropriate pursuant to title V of the Department of Energy Organization Act.

42 USC 7920.

Sec. 110. Enforcement

(a)(1) Any person who violates any provision of this title or any cooperative agreement entered into pursuant to this title or any rule prescribed under this Act concerning any designated processing site, disposition site, or remedial action shall be subject to an assessment by the Secretary of a civil penalty of not more than $1,000 per day per violation. Such assessment shall be made by order after notice and an opportunity for a public hearing, pursuant to section 554 of title 5, United States Code.

Notice, hearing opportunity.

(2) Any person against whom a penalty is assessed under this section may, within sixty calendar days after the date of the order of the Secretary assessing such penalty, institute an action in the United States court of appeals for the appropriate judicial circuit for judicial review of such order in accordance with chapter 7 of title 5, United States Code. The court shall have jurisdiction to enter a judgment affirming, modifying, or setting aside in whole or in part, the order of the Secretary, or the court may remand the proceeding to the Secretary for such further action as the court may direct.

5 USC 500 *et seq.*
Jurisdiction.

(3) If any person fails to pay an assessment of a civil penalty after it has become a final and unappealable order, the Secretary shall institute an action to recover the amount of such penalty in any appropriate district court of the United States. In such action, the validity and appropriateness of such final assessment order or judgment shall not be subject to review. Section 402(d) of the Department of Energy Organization Act shall not apply with respect to the functions of the Secretary under this section.

42 USC 7172.

(4) No civil penalty may be assessed against the United States or any State or political subdivision of a State or any official or employee of the foregoing.

(5) Nothing in this section shall prevent the Secretary from enforcing any provision of this title or any cooperative agreement or any such rule by injunction or other equitable remedy.

42 USC 2011 note.

(b) Subsection (a) shall not apply to any license requirement under the Atomic Energy Act of 1954. Such licensing requirements shall be forced by the Commission as provided in such Act.

42 USC 7921.

Sec. 111. Public Participation

In carrying out the provisions of this title, including the designation of processing sites, establishing priorities for such sites, the selection of remedial actions, and the execution of cooperative agreements, the Secretary, the Administrator, and the Commission shall encourage public participation and, where appropriate, the Secretary shall hold public hearings relative to such matters in the State where processing sites and disposal sites are located.

42 USC 7922.

Water.

Sec. 112. Termination: Authorization

(a)(1) The authority of the Secretary to perform remedial action under this title shall terminate on September 30, 1998, except that–

(A) the authority of the Secretary to perform groundwater restoration activities under this subchapter is without limitation, and

(B) the Secretary may continue operation of the disposal site in Mesa County, Colorado (known as the Cheney disposal cell) for receiving and disposing of residual radioactive material from processing sites and of byproduct material from property in the vicinity of the uranium milling site located in Monticello, Utah, until the Cheney disposal cell has been filled to the capacity for which is was designed, or September 30, 2023, whichever comes first.

(2) For purposes of this subsection, the term 'byproduct material' has the meaning given that term in section 11e.(2) of the Atomic Energy Act of 1954 (42 U.S.C. 2014(e)(2)).

(b) The amounts authorized to be appropriated to carry out the purposes of this subchapter by the Secretary, the Administrator, the Commission, and the Secretary of the Interior shall not exceed such amounts as are established in annual authorization Acts for fiscal year 1979 and each fiscal year thereafter applicable to the Department of Energy. Any sums appropriated for the purposes of this title shall be available until expended.[8]

42 USC 7923.

Sec. 113. Limitation

The authority under this title to enter into contracts or other obligations requiring the United States to make outlays may be exercised only to the extent provided in advance in annual authorization and appropriation Acts.

42 USC 7924.

Sec. 114. Reports to Congress

(a) Beginning on January 1, 1980, and each year thereafter until January 1, 1986, the Secretary shall submit a report to the Congress with respect to the status of the actions required to be taken by the Secretary, the Commission, the Secretary of the Interior, the Administrator, and the States and Indian tribes under this Act and any amendments to other laws made by this Act. Each report shall–

(1) include data on the actual and estimated costs of the program authorized by this title;

(2) described the extent of participation by the States and Indian tribe in this program;

(3) evaluate the effectiveness of remedial actions, and describe any problems associated with the performance of such actions; and

(4) contain such other information as may be appropriate.

Such report shall be prepared in consultation with the Commission, the Secretary of the Interior, and the Administrator and shall contain their separate views, comments, and recommendations, if any. The Commission shall submit to the Secretary and Congress such portion of the report under this subsection as relates to the authorities of the Commission under title II of this Act.

(b) Not later than July 1, 1979, the Secretary shall provide a report to the Congress which identifies all sites located on public or acquired lands

---

[8]   Amended by P.L. 100–616, § 3, 102 Stat. 3192 (1988); P.L. 102–486, Title X, Subtitle C, § 1031, 106 Stat. 2951 (1992); P.L. 104–259, § 2, 110 Stat. 3173 (1996).

of the United States containing residual radioactive materials and other radioactive materials and other radioactive waste (other than waste resulting from the production of electric energy) and specifies which Federal agency has jurisdiction over such sites. The report shall include the identity of property and other structures in the vicinity of such site that are contaminated or may be contaminated by such materials and actions planned or taken to remove such materials. The report shall describe in what manner such sites are adequately stabilized and otherwise controlled to prevent radon diffusion from such sites into the environment and other environmental harm. If any site is not so stabilized or controlled, the report shall describe the remedial actions planned for such site and the time frame for performing such actions. In preparing the reports under this section, the Secretary shall avoid duplication of previous or ongoing studies and shall utilize all information available from other departments and agencies of the United States respecting the subject matter of such report. Such agencies shall cooperate with the Secretary in the preparation of such report and furnish such information as available to them and necessary for such reports.

(c) Not later than January 1, 1980, the Administrator, in consultation with the Commission, shall provide a report to the Congress which identifies the location and potential health, safety, and environmental hazards of uranium mine wastes together with recommendations, if any, for a program to eliminate these hazards.

(d) Copies of the reports required by this section to be submitted to the Congress shall be separately submitted to the Committees on Interior and Insular Affairs and on Energy and Commerce of the House of Representatives and the Committee on Energy and Natural Resources of the Senate.

(e) The Commission, in cooperation with the Secretary, shall ensure that any relevant information, other than trade secrets and other proprietary information otherwise exempted from mandatory disclosure under any other provision of law, obtained from the conduct of each of the remedial actions authorized by this title and the subsequent perpetual care of those residual radioactive materials is documented systematically, and made publicly available conveniently for use.

Sec. 115. Active Operations: Liability for Remedial Action

(a) No amount may be expended under this title with respect to any site licensed by the Commission under the Atomic Energy Act of 1954 or by a State as permitted under section 274 of such Act at which production of any uranium product from ores (other than from residual radioactive materials) takes place. This subsection does not prohibit the disposal of residual radioactive material from a processing site under this subchapter at a site licensed under title II or the expenditure of funds under this subchapter for such disposal.[9]

(b) In the case of each processing site designated under this title, the Attorney General shall conduct a study to determine the identity and legal responsibility which any person (other than the United States, a State, or Indian tribe) who owned or operated or controlled (as determined by the Attorney General) such site before the date of the enactment of this Act may have under any law or rule of law for reclamation or other remedial action with respect to such site. The Attorney General shall publish the results of such study, and provide

Cooperation.

42 USC 7925.

42 USC 2011 note.

42 USC 2021. Study.

---

[9]   Amended by P.L. 104–259, § 4(a), 110 Stat. 3174 (1996).

copies thereof to the Congress, as promptly as practicable following the date of the enactment of this Act. The Attorney General, based on such study, shall, to the extent he deems it appropriate and in the public interest, take such action under any provision of law in effect when uranium was produced at such site to require payment by such person of all or any part of the costs incurred by the United States for such remedial action for which he determines such person is liable.

Title II–Uranium Mill Tailings Licensing and Regulation

Sec. 201. Definition

Section 11e. of the Atomic Energy Act of 1954, is amended to read as follows:

42 USC 2014.
Byproduct
material.

e. The term "byproduct material" means (1) any radioactive material (except special nuclear material) yielded in or made radioactive by exposure to the radiation incident to the process of producing or utilizing special nuclear material, and (2) the tailings or wastes produced by the extraction or concentration of uranium or thorium from any ore processed primarily for its source material content.

Sec. 202. Custody of Disposal Site

(a) Chapter 8 of the Atomic Energy Act of 1954, is amended by adding the following new section at the end thereof:

42 USC 2113.

Sec. 83. Ownership and Custody of Certain By-product Material and Disposal Sites.–

42 USC 2002.
42 USC 2014.
42 USC 2111.

a. Any license issued or renewed after the effective date of this section under section 62 or section 81 for any activity which results in the production of any byproduct materials, as defined in section 11e.(2), shall contain such terms and conditions as the commission determines to be necessary to assure that, prior to termination of such license–

(1) the licensee will comply with decontamination, decommissioning, and reclamation standards prescribed by the Commission for sites (A) at which ores were processed primarily for their source material content and (B) at which such byproduct material is deposited, and

42 USC 2014.

(2) ownership of any byproduct material, as defined in section 11e.(2), which resulted from such licensed activity shall be transferred to (A) the United States or (B) in the State in which such activity occurred if such State exercises the option under subsection b. (1) to acquire land used for the disposal of byproduct material.

Any license in effect on the date of the enactment of this section shall either contain such terms and conditions on renewal thereof after the effective date of this section, or comply with paragraphs (1) and (2) upon the termination of such license, whichever first occurs.

Rule,
regulation, or
order.

(b)(1)(A) The Commission shall require by rule, regulation, or order that prior to the termination of any license which is issued after the effective date of this section, title to the land, including any interests therein (other than land owned by the United States or by a State) which is used for the disposal of any byproduct material, as defined by section 11e.(2), pursuant to such license shall be transferred to–

(A) the United States, or

(B) the State in which such land is located, at the option of such State.

(2) Unless the Commission determines prior to such termination that transfer of title to such land and such byproduct material is not necessary or desirable to protect the public health, safety, or welfare or to minimize or eliminate danger to life or property. Such

determination shall be made in accordance with section 181 of this Act. Notwithstanding any other provision of law or any such determination, such property and materials shall be maintained pursuant to a license issued by the Commission pursuant to section 84(b) in such manner as will protect the public health, safety, and the environment.

(B) If the Commission determines by order that use of the surface or subsurface estates, or both, of the land transferred to the United States or to a State under subparagraph (A) would not endanger the public health, safety, welfare, or environment, the Commission, pursuant to such regulations as it may prescribe, shall permit the use of the surface or subsurface estates, or both, of such land in a manner consistent with the provisions of this section. If the Commission permits such use of such land, it shall provide the person who transferred such land with the right of first refusal with respect to such use of such land.

(2) If the transfer to the United States of title to such by-product material and such land is required under this section, the Secretary of Energy or any Federal agency designated by the President shall, follow the Commission's determination of compliance under subsection c., assume title and custody of such byproduct material and land transferred as provided in this subsection. Such Secretary or Federal agency shall maintain such material and land in such manner as will protect the public health and safety and the environment. Such custody may be transferred to another officer or instrumentality of the United States only upon approval of the President.

(3) If transfer to a State of title to such byproduct material is required in accordance with this subsection, such State shall, following the Commission's determination of compliance under subsection d., assume title and custody of such byproduct material and land transferred as provided in this subsection. Such State shall maintain such material and land in such manner as will protect the public health, safety, and the environment.

42 USC 2092.
(4) In the case of any such license under section 62, which was in effect on the effective date of this section, the Commission may require, before the termination of such license, such transfer of land and interests therein (as described in paragraph (1) of this subsection) to the United States or a State in which such land is located, at the option of such State, as may be necessary to protect the public health, wealth, and the environment from any effects associated with such byproduct material. In exercising the authority of this paragraph, the Commission shall take into consideration the status of the ownership of such land and interests therein and the ability of the licensee to transfer title and custody thereof to the United States or a State.

(5) The Commission may, pursuant to a license, or by rule or order, require the Secretary or other Federal agency or State having custody of such property and materials to undertake such monitoring, maintenance, and emergency measures as are necessary to protect the public health and safety and such other actions as the Commission deems necessary to comply with the standards promulgated pursuant to section 84 of this Act. The Secretary or such other Federal agency is authorized to carry out maintenance, monitoring, and emergency measures, but shall take no other action pursuant to such license, rule or order, with respect to such property and materials unless expressly authorized by Congress after the date of enactment of this Act.

42 USC 2014.

(6) The transfer of title to land or byproduct materials, as defined in section 11e.(2), to a State or the United States pursuant to this subsection shall not relieve any licensee of liability for any fraudulent or negligent acts done prior to such transfer.

(7) Material and land transferred to the United States or a State in accordance with this subsection shall be transferred without cost to the United States or a State (other than administrative and legal costs incurred in carrying out such transfer). Subject to the provisions of paragraph (1)(B) of this subsection, the United States or a State shall not transfer title to material or property acquired under this subsection to any person, unless such transfer is in the same manner as provided under section 104(h) of the Uranium Mill Tailings Radiation Control Act of 1978.

(8) The provisions of this subsection respecting transfer of title and custody to land shall not apply in the case of lands held in trust by the United States for any Indian tribe or lands owned by such Indian tribe subject to a restriction against alienation imposed by the United States. In the case of such lands which are used for the disposal of byproduct material, as defined in section 11e.(2), the license shall be required to enter into such arrangements with the Commission as may be appropriate to assure the long-term maintenance and monitoring of such lands by the United States.

c. Upon termination on any license to which this section applies, the Commission shall determine whether or not the licensee has complied with all applicable standards and requirements under such license.

(b) this section shall be effective three years after the enactment of this Act.

42 USC 2113 note.
Effective date.

(c) The table of contents for chapter 8 of the Atomic Energy Act of 1954, is amended by inserting the following new item after the item relating to section 82:   Sec. 83. Ownership and custody of certain byproduct material and disposal sites.

Sec. 203. Authority to Establish Certain Requirements

Section 161 of the Atomic Energy Act of 1954, is amended, by
42 USC 2201.
42 USC 2231.
adding the following new subsection at the end thereof:

x. Establish by rule, regulation, or order, after public notice, and in accordance with the requirements of section 181 of this Act, such standards and instructions as the Commission may deem necessary or desirable to ensure–

42 USC 2014.

(1) that an adequate bond, surety, or other financial arrangement (as determined by the Commission) will be provided before termination of any license for byproduct material as defined in section 11e.(2), by a licensee to permit the completion of all requirements established by the Commission for the decontamination, decommissioning, and reclamation of sites, structures, and equipment used in conjunction with byproduct material as so defined, and

(2) that–

(A) in the case of any such license issued or renewed after the date of the enactment of this subsection, the need for long term maintenance and monitoring of such sites, structures and equipment after termination of such license will be minimized and, to the maximum extent practicable, eliminated; and

(B) in the case of each license for such material (whether in effect on the date of the enactment of this section or issued or renewed thereafter), if the Commission determines that any such long-term maintenance and monitoring is necessary, the licensee, before termination of any license for byproduct material as defined in section 11e.(2), will make available such bonding, surety, or other financial arrangements as may be necessary to assure such long-term maintenance and monitoring.

Such standards and instructions promulgated by the Commission, pursuant to this subsection shall take into account, as determined by the Commission, so as to avoid unnecessary duplication and expense, performance bonds or other financial arrangements which are required by other Federal agencies or State agencies and/or other local governing bodies for such decommissioning, decontamination, and reclamation and long-term maintenance and monitoring except that nothing in this paragraph shall be construed to require that the Commission accept such bonds or arrangements if the Commission determines that such bonds or arrangements are not adequate to carry out subparagraphs (1) and (2) of this subsection.

Sec. 204. Cooperation with States

42 USC 2021.

(a) Section 274b. of the Atomic Energy Act of 1954, is amended by adding "as defined in section 11e.(1)" after the words "byproduct materials" in paragraph (1) by renumbering paragraphs (2) and (3) as paragraph (3) and (4); and by inserting the following new paragraph immediately after paragraph(1):

42 USC 2021.

(2) byproduct materials as defined in section 11e.(2);

(b) Section 274d.(2) of such Act is amended by inserting the following before the word "compatible": "in accordance with the requirements of subsection o. and in all other respects."

Agreement.

(c) Section 274n. of such Act is amended by adding the following new sentence at the end thereof: "As used in this section, the term "agreement" includes any amendment to any agreement."

(d) Section 274j. of such Act is amended–

(1) by inserting "all or part of" after "suspend";

(2) by inserting "(1)" after "finds that"; and

Review.

(3) by adding at the end before the period the following: or (2) the State has not complied with one or more of the requirements of this section. The Commission shall periodically review such agreements and actions taken by the States under the agreements to ensure compliance with the provisions of this section.

(e)(1) Section 274 of such Act is amended by adding the following new subsection at the end thereof:

o. In the licensing and regulation of byproduct material, as defined in section 11e.(2) of this Act, or of any activity which results in the production of byproduct material as so defined under an agreement entered into pursuant to subsection b., a State shall require–

(1) compliance with the requirements of subsection b. of section 83 (respecting ownership of byproduct material and land), and

(2) compliance with standards which shall be adopted by the State for the protection of the public health, safety, and the environment from hazards associated with such material which are equivalent, to the extent practicable, or more stringent than, standards adopted and enforced by the Commission for the same purpose, including requirements and standards promulgated by the Commission and the Administrator of the Environmental Protection Agency pursuant to section 83, 84, and 275, and

(3) procedures which–

(A) in the case of licenses, provide procedures under State law which include–

(i) an opportunity, after public notice, for written comments and a public hearing, with a transcript,

(ii) an opportunity for cross examination, and

(iii) a written determination which is based upon findings included in such determination and upon the evidence presented during the public comment period and which is subject to judicial review;

(B) in the case of rulemaking, provide an opportunity for public participation through written comments or a public hearing and provide for judicial review of the rule;

(C) require for each license which has a significant impact on the human environment a written analysis (which shall be available to the public before the commencement of any such proceedings) of the impact of such licenses, including any activities conducted pursuant thereto, on the environment, which analysis shall include–

(i) an assessment of the radiological and nonradiological impacts to the public health of the activities to be conducted pursuant to such license;

(ii) an assessment of any impact on any waterway and groundwater resulting from such activities;

(iii) consideration of alternatives, including alternative sites and engineering methods, to the activities to be conducted pursuant to such license; and

(iv) consideration of the long-term impacts, including decommissioning, decontamination, and reclamation impacts, associated with activities to be conducted pursuant to such license, including the management of any byproduct material, as defined by section 11e.(2); and

(D) prohibit any major construction activities with respect to such material prior to complying with the provisions of subparagraph (C).

If any State under such agreement imposes upon any licensee any requirement for the payment of funds to such State for the reclamation or long-term maintenance and monitoring of such material and if transfer to the United States of such material is required in accordance with section 83b. of this Act, such agreement shall be amended by the Commission to provide that such State shall transfer to the United States upon termination of the license issued to such licensee the total amount collected by such State from such licensee for such purpose. If such payments are required, they must be sufficient to ensure compliance with the standards established by the Commission pursuant to section 161x of this Act. No State shall be required under paragraph (3) to conduct proceedings concerning any license or regulation which would duplicate proceedings conducted by the Commission.

42 USC 2201.

(2)[10] The provisions of the amendment made by paragraph (1) of this subsection (which adds a new subsection o. to section 274 of the Atomic Energy Act of 1954) shall apply only to the maximum extent practicable during the three-year period beginning on the date of the enactment of this Act.

42 USC 2021 note.

(f) Section 274c. of such Act is amended by inserting the following new sentence after paragraph (4) thereof: The Commission shall also retain authority under any such agreement to make a determination that all applicable standards and requirements have been met prior to

42 USC 2021.
42 USC 2014.

[10] Added by P.L. 96–106, § 22(d), 93 Stat. 800 (1979).

termination of a license for byproduct material, as defined in section 11e.(2).

42 USC 2021

(g) Nothing in any amendment made by this section shall preclude any State from exercising any other authority permitted under the Atomic Energy Act of 1954 respecting any byproduct material, as defined in section 11e.(2) of the Atomic Energy Act of 1954.

42 USC 2014.
42 USC 2021.

(h)(1) During the three-year period beginning on the date of the enactment of this Act, notwithstanding any other provision of this title, any State may exercise any authority under State law (including authority exercised pursuant to an agreement entered into pursuant to section 274 of the Atomic Energy Act of 1954) respecting (A) byproduct material, as

42 USC 2014.

defined in section 11e.(2) of the Atomic Energy Act of 1954, or (B) any activity which results in the production of byproduct material as so defined, in the same manner and to the same extent as permitted before the date of the enactment of this Act, except that such State authority shall be exercised in a manner which, to the extent practicable, is consistent with the requirements of section 274o. of the Atomic Energy Act of 1954 (as added by section 204(e) of this Act). The Commission shall have the authority to ensure that such section 274o. is implemented by any such State to the extent practicable during the three-year period

42 USC 2022.

beginning on the date of the enactment of this Act. Nothing in this section shall be construed to preclude the Commission or the Administrator of the Environmental Protection Agency from taking such action under section 275 of the Atomic Energy Act of 1954 as may be necessary to implement title I of this Act.[11]

(2) An agreement entered into with any State as permitted under section 274 of the Atomic Energy Act of 1954 with respect to byproduct material as defined in section 11e.(2) of such Act, may be entered into at any time after the date of the enactment of this Act but no such agreement may take effect before the date three years after the date of the enactment of this Act.

42 USC 2014.
42 USC 2021.

(3)[12] Notwithstanding any other provision of this title, where a State assumes or has assumed, pursuant to an agreement entered into under section 274b. of the Atomic Energy Act of 1954, authority over any activity which results in the production of byproduct material, as defined in section 11e.(2) of such Act, the Commission shall not, until the end of three-year period beginning on the date of the enactment of this Act, have licensing authority over such byproduct material produced

42 USC 2021.

in any activity covered by such agreement, unless the agreement is terminated, suspended, or amended to provide for such Federal licensing. If, at the end of such three-year period, a State has not entered into such an agreement with respect to byproduct material, as defined in section 11e.(2) of the Atomic Energy Act of 1954, the Commission shall have authority over such byproduct material. *Provided, however,*[13] That, in the

---

[11]   Amended by P.L . 96–106, § 22(b), 93 Stat. 799 (1979), revised section 204(h)(1) by substituting a completely new section 204(h)(1). Before amendment, section 204(h)(1) read as follows:

(H)(1) On or before the date three years after the date of the enactment of this Act, notwithstanding any amendment made by this title, any State may exercise any authority under State law respecting byproduct material as defined in section 11e.(2) of the Atomic Energy Act of 1954, in the same manner, and to the same extent, as permitted before the enactment of this Act.

[12]   Added by P.L. 96–106, § 22(a), 93 Stat. 799 (1979).

[13]   *Provided, however…*Atomic Energy Act of 1954 added by P.L. 97–415, § 19, 96 Stat. 2067 (1983).

case of a State which has exercised any authority under State law pursuant to an agreement entered into under section 274 of the Atomic Energy Act of 1954, the State authority over such byproduct material may be terminated, and the Commission authority over such material may be exercised, only after compliance by the Commission with the same procedures as are applicable in the case of termination of agreements under section 274j. of the Atomic Energy Act of 1954.

Sec. 205. Authorities of Commission Respecting Certain Byproduct Material

(a) Chapter 8 of the Atomic Energy Act of 1954, is amended by adding the following new section at the end thereof:

42 USC 2114.

Sec. 84. Authorities of Commission Respecting Certain Byproduct Material.

42 USC 2114.

a. The Commission shall insure that the management of any byproduct material, as defined in section 11e.(2), is carried out in such manner as–

(1) the Commission deems appropriate to protect the public health and safety and the environment from radiological and nonradiological hazards associated with the processing and with the possession and transfer of such material,

(2) conforms with applicable general standards promulgated by the Administrator of the Environmental Protection Agency under section 275, and

42 USC 6091.

(3) conforms to general requirements established by the Commission, with the concurrence of the Administrator, which are, to the maximum extent practicable, at least comparable to requirements applicable to the possession, transfer, and disposal of similar hazardous material regulated by the Administrator under the Solid Waste Disposal Act, as amended.

Rule, regulation, or order.
42 USC 2111.

b. In carrying out its authority under this section, the Commission is authorized to–

(1) by rule, regulation, or order require persons, officers, or instrumentalities exempted from licensing under section 81 of this Act to conduct monitoring, perform remedial work, and to comply with such other measures as it may deem necessary or desirable to protect health or to minimize danger to life or property, and in connection with the disposal or storage of such byproduct material; and

(2) make such studies and inspections and to conduct such monitoring as may be necessary.

Civil penalty.

Any violation by any person other than the United States or any officer or employee of the United States or a State of any rule, regulation, or order or licensing provision, of the Commission established under this section or section 83 shall be subject to a civil penalty in the same manner and in the same amount as violations subject to a civil penalty under section 234.

42 USC 2282.

Nothing in this section affects any authority of the Commission under any other provision of this Act.

42 USC 2111.
42 USC 2112.

(b) The first sentence of section 81 of the Atomic Energy Act of 1954, is amended to read as follows: No person may transfer or receive in interstate commerce, manufacture, produce, transfer, acquire, own, possess, import, or export any byproduct material, except to the extent authorized by this section, section 82 or section 84.

Supra.

(c) The table of content for such chapter 8 is amended by inserting the following new item after the item relating to section 83:

Sec. 84. Authorities of Commission respecting certain byproduct material.

Sec. 206. Authority of Environmental Protection Agency Respecting Certain Byproduct Material

(a) Chapter 19 of the Atomic Energy Act of 1954, is amended by inserting after section 274 the following new section:

42 USC 2022.

Sec. 275. Health and Environmental Standards For Uranium Mill Tailings.–

Rule.

a. As soon as practicable, but not later than one year after the date of enactment of this section, the Administrator of the Environmental Protection Agency (hereinafter referred to in this section as the "Administrator") shall, by rule, promulgate standards of general application (including standards applicable to licenses under section 104(h) of the Uranium Mill Tailings Radiation Control Act of 1978) for the protection of the public health, safety, and the environment from radiological and nonradiological hazards associated with residual radioactive materials (as defined in section 101 of the Uranium Mill Tailings Radiation Control Act of 1978) located at inactive uranium mill tailings sites and depository sites for such materials selected by the Secretary of Energy, pursuant to title I of the Uranium Mill Tailing Radiation Control Act of 1978. Standards promulgated pursuant to this subsection shall, to the maximum extent practicable, be consistent with the requirements of the Solid Waste Disposal Act, as amended. The Administrator may periodically revise any standard promulgated pursuant to this subsection.

42 USC 6901 note.

42 USC 2014.
Rule.

b. (1) As soon as practicable, but not later than eighteen months after the enactment of this section, the Administrator shall, by rule, promulgate standards of general application for the protection of the public health, safety, and the environment from radiological and nonradiological hazards associated with the processing and with the possession, transfer, and disposal of byproduct material, as defined in section 11e.(2) of this Act, at sites at which ores are processed primarily for their source material content or which are used for the disposal of such byproduct material.

(2) Such generally applicable standards promulgated pursuant to this subsection for nonradiological hazards shall provide for the protection of human health and the environment consistent with the standards required under subtitle C of the Solid Waste Disposal Act, as amended, which are applicable to such hazards: *Provided, however,* That no permit issued by the Administrator is required under this Act or the Solid Waste Disposal Act, as amended, for the processing, possession, transfer, or disposal of byproduct material, as defined in section 11e.(2) of this Act. The Administrator may periodically revise any standard promulgated pursuant to this subsection. Within three years after such revision of any such standard, the Commission and any State permitted to exercise authority under section 274b.(2) shall apply such revised standard in the case of any license for byproduct material as defined in section 11e.(2) or any revision thereof.

42 USC 2021.

Notice, hearing opportunity. Publication in *Federal Register*.

c. (1) Before the promulgation of any rule pursuant to this section, the Administrator shall publish, the proposed rule in the Federal Register, together with a statement of the research, analysis, and other available information in support of such proposed rule, and provide a period of public comment of at least thirty days for written comments thereon and an opportunity, after such comment period and after public notice, for any interested person to present oral data, views, and arguments at a public hearing. There shall be a transcript of any such hearing. The Administrator shall consult with the Commission and the Secretary of Energy before promulgation of any such rule.

Consultation.

Judicial Review.

(2) Judicial review of any rule promulgated under this section may be obtained by any interested person only upon such person filing a petition for review within sixty days after such promulgation in the United States court of appeals for the Federal judicial circuit in which such person resides or has his principal place of business. A copy of the petition shall be forthwith transmitted by the clerk of court to the Administrator. The Administrator thereupon shall file in the court the written submissions to, and transcript of, the written or oral proceedings on which such rule was based as provided in section 2112 of title 28, United States Code. The court shall have jurisdiction to review the rule in accordance with chapter 7 of title 5, United States Code, and to grant appropriate relief as provided in such chapter. The judgment of the court affirming, modifying, or setting aside, in whole or in part, any such rule shall be final, subject to judicial review by the Supreme Court of the United States upon certiorari or certification as provided in section 1254 of title 28, United States Code.

5 USC 701 *et seq.*

(3) Any rule promulgated under this section shall not take effect earlier than sixty calendar days after such promulgation.

42 USC 2021.

d. Implementation and enforcement of the standards promulgated pursuant to subsection b. of this section shall be the responsibility of the Commission in the conduct of its licensing activities under this Act. States exercising authority pursuant to section 274b.(2) of this Act shall implement and enforce such standards in accordance with subsection o. of such section.

33 USC 1251 note.
42 USC 2014.
42 USC 7401 note.
42 USC 2018 *et seq.*

e. Nothing in this Act applicable to byproduct material, as defined in section 11e.(92) of this Act, shall affect the authority of the Administrator under the Clean Air Act of 1970, as amended, or the Federal Water Pollution Control Act, as amended.

(b) The table of contents for chapter 19 of the Atomic Energy Act is amended by inserting the following new item after the item relating to section 274:

Sec. 275. Health and Environmental Standards for Uranium Tailings.

Sec. 207. Authorization of Appropriation for Grants

There is hereby authorized to be appropriated for fiscal year 1980 to the Nuclear Regulatory Commission not to exceed $500,000 to be used for making grants to States which have entered into agreements with the Commission under section 274 of the Atomic Energy Act of 1954, to aid in the development of State regulatory programs under such section which implement the provisions of this Act.

42 USC 2014 note.

Sec. 208. Effective Date

Except as otherwise provided in this title the amendments made by this title shall take effect on the date of the enactment of this Act.

42 USC 2011
note.

42 USC 2113
note.

Sec. 209. Consolidation of Licenses and Procedures

The Regulatory Commission shall consolidate, to the maximum extent practicable, licenses and licensing procedures under amendments made by this title with licenses and licensing procedures under other authorities contained in the Atomic Energy Act of 1954.

Title III–Study and Designation of Two Mill Tailings Sites in New Mexico

42 USC 7941.

Sec. 301. Study

The Commission, in consultation with the Attorney General and the Attorney General of the State of New Mexico, shall conduct a study to determine the extent and adequacy of the authority of the Commission and the State of New Mexico to require, under the Atomic Energy Act of 1954 (as amended by title II of this Act) or under State authority as permitted under section 274 of such Act or under other provision of law, the owners of the following active uranium mill sites to undertake appropriate action to regulate and control all residual radioactive materials at such sites to protect public health, safety, and the environment: the former Homestake- New Mexico Partners site near Milan, New Mexico, and the Anaconda carbonate process tailing site near Bluewater, New Mexico. Such study shall be completed and a report thereof submitted to the Congress and to the Secretary within one year after enactment of this Act, together with such recommendations as may be appropriate. If the Commission determines that such authority is not adequate to regulate and control such materials at such sites in the manner provided in the first sentence of this section, the Commission shall include in the report a statement of the basis for such determination. Nothing in this Act shall be construed to prevent or delay action by a State as permitted under section 274 of the Atomic Energy Act of 1954 or under any other provision of law or by the Commission to regulate such residual radioactive materials at such sites prior to completion of such study.

42 USC 2021.

Report to
Congress.

42 USC 7942.

Sec. 302. Designation by Secretary

(a) Within ninety days from the date of his receipt of the report and recommendations submitted by the Commission under section 301, notwithstanding the limitations contained in section 301, notwithstanding the limitations contained in section 101(6)(A) and in section 115(a), if the Commission determines, based on such study, that such sites cannot be regulated and controlled by the State or the Commission in the manner described in section 301, the Secretary may designate either or both of the sites referred to in section 301 as a processing site for purposes of title I. Following such designation, the Secretary may enter into cooperative agreements with New Mexico to perform remedial action pursuant to such title concerning only the residual radioactive materials at such site resulting from uranium produced for sale to a Federal agency prior to January 1, 1971, under contract with such agency. Any such designation shall be submitted by the Secretary, together with his estimate of the cost of carrying out such remedial action at the designated site, to the Committee on Interior and Insular Affairs and the Committee on Interstate and Foreign Commerce of the House of Representatives and to the Committee on Energy and Natural Resources of the Senate.

Submittal to
congressional
committees.

(b)(1) No designation under subsection (a) shall take effect before the expiration of one hundred and twenty calendar days (not including any day in which either House of Congress is not in session because of an adjournment of more than three calendar days to a day certain or an adjournment sine die) after receipt by such Committees of such designation.

(c) Except as otherwise specifically provided in subsection (a), any remedial action under title I with respect to any sites designated under this title shall be subject to the provisions of title I (including the authorization of appropriations referred to in section 112(b)).

B. PERTINENT PROVISIONS OF THE ENERGY POLICY ACT OF 1992—(REMEDIAL ACTION AND URANIUM REVITALIZATION)

Public Law 102–486   **106 Stat. 2946**

October 24, 1992

* * * *

Title X–Remedial Action and Uranium Revitalization

Subtitle A–Remedial Action at Active Processing Sites

42 USC 2296a.   **Sec. 1001. Remedial Action Program**

(a) IN GENERAL. Except as provided in subsection (b), the costs of decontamination, decommissioning, reclamation, and other remedial action at an active uranium or thorium processing site shall be borne by persons licensed under section 62 or 81 of the Atomic Energy Act of 1954 (42 U.S.C. 2091, 2111) for any activity at such site which results or has resulted in the production of byproduct material.

(b) REIMBURSEMENT–

(1) IN GENERAL–The Secretary of Energy shall, subject to paragraph (2), reimburse at least annually a licensee described in subsection (a) for such portion of the costs described in such subsection as are–

(A) determined by the Secretary to be attributable to byproduct material generated as an incident of sales to the United States; and

(B) either–

(i) incurred by such licensee not later than December 31, 2007; or

(ii) incurred by a licensee after December 31, 2007, in accordance with a plan for subsequent decontamination, decommissioning, reclamation, and other remedial action approved by the Secretary.[1]

(2) AMOUNT.–

(A) To Individual Active Site Uranium Licensees.–The amount of reimbursement paid to any licensee under paragraph (1) shall be determined by the Secretary in accordance with regulations issued pursuant to section 2296a-1 of this title and, for uranium mill tailings only, shall not exceed an amount equal to $6.25 multiplied by the dry short tons of byproduct material located on October 24, 1992 at the site of the activities of such licensee described in subsection (a) of this section, and generated as an incident of sales to the United States.[2]

---

[1]   Amended by P.L. 104–259, § 3(a), 110 Stat. 3173 (1996); P.L. 105–388, § ll(a), 112 Stat. 3484 (1998); P.L. 106–317, § 1, 114 Stat. 1277 (2000).

[2]   Amended by P.L. 104–259, § 3(a), 110 Stat. 3173 (1996); P.L. 105–388, § ll(a), 112 Stat. 3484 (1998); P.L. 106–317, § 1, 114 Stat. 1277 (2000).

(B) TO ALL ACTIVE SITE URANIUM LICENSEES–Payments made under paragraph (1) to active site uranium licensees shall not in the aggregate exceed $350,000,000.[3]

(C) TO THORIUM LICENSEES–Payments made under paragraph (1) to the licensee of the active thorium site shall not exceed $ 365,000,000, and may only be made for off-site disposal. Such payments shall not exceed the following amounts:

(i) $ 90,000,000 in fiscal year 2002.

(ii) $ 55,000,000 in fiscal year 2003.

(iii) $ 20,000,000 in fiscal year 2004.

(iv) $ 20,000,000 in fiscal year 2005.

(v) $ 20,000,000 in fiscal year 2006.

(vi) $ 20,000,000 in fiscal year 2007.[4]

(D) INFLATION ESCALATION INDEX–The amounts in subparagraphs (A), (B), and (C) of this paragraph shall be increased annually based upon an inflation index. The Secretary shall determine the appropriate index to apply.

(E) ADDITIONAL REIMBURSEMENT–

(i) DETERMINATION OF EXCESS–The Secretary shall determine as of July 31, 2005, whether the amount authorized to be appropriated pursuant to section 1003, when considered with the $6.25[5] per dry short ton limit on Reimbursement, exceeds the amount reimbursable to the licensees under subsection (b)(2).

(ii) IN THE EVENT OF EXCESS–If the Secretary determines under clause (i) that there is an excess, the Secretary may allow reimbursement in excess of $6.25 per dry short ton on a prorated basis at such sites where the costs reimbursable under subsection (b)(1) of this section exceed the $6.25 per dry short ton limitation described in paragraph (2) of such subsection.[6]

(3) BYPRODUCT LOCATION–Notwithstanding the requirement of paragraph (2)(A) that byproduct material be located at the site on the date of the enactment of this Act, byproduct material moved from the site of the Edgemont Mill to a disposal site as the result of the decontamination, decommissioning, reclamation, and other remedial action of such mill shall be eligible for reimbursement to the extent eligible under paragraph (1).

42 USC 2296a-1.   **Sec. 1002. Regulations**

Within 180 days of the date of the enactment of this Act, the Secretary shall issue regulations government reimbursement under section 1001. An active uranium or thorium processing site owner shall apply for reimbursement hereunder by submitting a request for the amount of reimbursement together with reasonable documentation in support thereof, to the Secretary. Any such request for reimbursement, supported by reasonable documentation, shall be approved by the

---

[3]   Amended by P.L. 104–259, § 3(a), 110 Stat. 3173 (1996); P.L. 105–388, § ll(a), 112 Stat. 3484 (1998); P.L. 106–317, § 1, 114 Stat. 1277 (2000).

[4]   Amended by P.L. 104–259, § 3(a), 110 Stat. 3173 (1996); P.L. 105–388, 112 Stat. 3484 (1998); P.L. 107–222, § 1(a), rewrote subparagraph (C).

[5]   Amended by P.L. 104–259, § 3(a), 110 Stat. 3173 (1996); P.L. 105–388, § ll(a), 112 Stat. 3484 (1998); P.L. 106–317, § 1, 114 Stat. 1277 (2000).

[6]   Amended by P.L. 104–259, § 3(a), 110 Stat. 3173 (1996); P.L. 105–388, § ll(a), 112 Stat. 3484 (1998); P.L. 106–317, § 1, 114 Stat. 1277 (2000).

Secretary and reimbursement therefor shall be made in a timely manner subject only to the limitations of section 1001.

42 USC 2296a-2.   **Sec. 1003. Authorization of Appropriations**

(a) IN GENERAL–There is authorized to be appropriated $715,000,000 to carry out this part. The aggregate amount authorized in the preceding sentence shall be increased annually as provided in section 2296a of this title, based upon an inflation index to be determined by the Secretary.[7]

(b) SOURCE.–Funds described in subsection (a) shall be provided from the Fund established under section 1801 of the Atomic Energy Act of 1954.

42 USC 2296a-3.   **Sec. 1004. Definitions**

For purposes of this subtitle:

(1) The term "active uranium or thorium processing site" means–

(A) any uranium or thorium processing site, including the mill containing byproduct material for which a license (issued by the Nuclear Regulatory Commission or its predecessor agency under the Atomic Energy Act of 1954, or by a State as permitted under section 274 of such Act (42 U.S.C. 2021)) for the production at such site of any uranium or thorium derived from ore–

(i) was in effect on January 1, 1978;

(ii) was issued or renewed after January 1, 1978; or

(iii) for which an application for renewal or issuance was pending on, or after January 1, 1978; and

(B) any other real property or improvement on such real property that is determined by the Secretary or by a State as permitted under section 274 of the Atomic Energy Act of 1954 (42 U.S.C. 2021) to be–

(i) in the vicinity of such site; and

(ii) contaminated with residual byproduct material;

(2) The term "byproduct material" has the meaning given such term in section 11e.(2) of the Atomic Energy Act of 1954, (42 U.S.C. 2014(e)(2)); and

(3) The term "decontamination, decommissioning, reclamation, and other remedial action" means work performed prior to or subsequent to the date of the enactment of this Act which is necessary to comply with all applicable requirements of the Uranium Mill Tailings Radiation Control Act of 1978 (42 U.S.C. 7901 et. seq.), or where appropriate, with requirements established by a State that is a party to a discontinuance agreement under section 274 of the Atomic Energy Act of 1954 (42 U.S.C. 2021).

Subtitle B–Uranium Revitalization

42 USC 2296b.   **Sec. 1011. Overfeed Program**

(a) URANIUM PURCHASES–To the maximum extent permitted by sound business practice, the Corporation shall purchase uranium in accordance with subsection (b) and overfeed it into the enrichment process to reduce the amount of power required to produce the enriched uranium ordered by enrichment services customers, taking into account costs associated with depleted tailings.

---

[7]   Amended by P.L. 104–259, § 3(b), 110 Stat. 3174 (1996); P.L. 105–388, § 11(b), 112 Stat. 3485, November 13, 1998; P.L. 107–222, § 1(b), struck out "$490,000,000" and inserted "$715,000,000."

(b) USE OF DOMESTIC URANIUM–Uranium purchased by the Corporation for purposes of this section shall be of domestic origin and purchased from domestic uranium producers to the extent permitted under the multilateral trade agreements (as defined in section 2(4) of the Uruguay Round Agreements Act and the North American Free Trade Agreement.[8]

42 USC 2296b-1.   **Sec. 1012. National Strategic Uranium Reserve**

There is hereby established the National Strategic Uranium Reserve under the direction and control of the Secretary. The Reserve shall consist of natural uranium and uranium equivalents contained in stockpiles or inventories currently held by the United States for defense purposes. Effective on the date of the enactment of this Act and for 6 years thereafter, use of the Reserve shall be restricted to military purposes and government research. Use of the Department of Energy's stockpile of enrichment tails existing on the date of the enactment of this Act shall be restricted to military purposes for 6 years thereafter.

42 USC 2296b-2.   **Sec. 1013. Sale of Remaining DOE Inventories**

The Secretary, after making the transfer required under section 1407 of the Atomic Energy Act of 954, may sell, from time to time, portions of the remaining inventories of raw or low-enriched uranium of the Department that are not necessary to national security needs, to the Corporation, at a fair market price. Sales under this section may be made only if such sales will not have a substantial adverse impact on the domestic uranium mining industry. Proceeds from sales under this subsection shall be deposited into the general fund of the United States Treasury.

42 USC 2296b-3.   **Sec. 1014. Responsibility for the Industry**

(a) CONTINUING SECRETARIAL RESPONSIBILITY–The Secretary shall have a continuing responsibility for the domestic uranium industry to encourage the use of domestic uranium. The Secretary, in fulfilling this responsibility, shall not use any supervisory authority over the Corporation. The Secretary shall report annually to the appropriate committees of Congress on action taken with respect to the domestic uranium industry, including action to promote the export of domestic uranium pursuant to subsection (b).

(b) ENCOURAGE EXPORT.–The Department, with the cooperation of the Department of Commerce, the United States Trade Representative and other governmental organization, shall encourage the export of domestic uranium. Within 180 days after the date of the enactment of this Act, the Secretary shall develop recommendations and implement government programs to promote the export of domestic uranium.

42 USC 2296b-4.   **Sec. 1015. Annual Uranium Purchase Reports**

(a) IN GENERAL–By January 1 of each year, the owner or operator of any civilian nuclear power reactor shall report to the Secretary, acting through the Administrator of the Energy Information Administration, for activities of the previous fiscal year–

(1) the country of origin and the seller of any uranium or enriched uranium purchased or imported into the United States either directly or indirectly by such owner or operator, and

(2) the country of origin and the seller of any enrichment services purchased by such owner or operator.

---

[8]   Amended by P.L. 102–486, Title X, § 1011, 106 Stat. 2948 (1992); P.L. 106–36, Title I, § 1002(g)(1), 113 Stat. 133 (1999).

(b) CONGRESSIONAL ACCESS–The information provided to the Secretary pursuant to this section shall be made available to the Congress by March 1 of each year.

42 USC 2296b-5. **Sec. 1016. Uranium Inventory Study**

Within one year after the date of the enactment of this Act, the Secretary shall submit to the Congress a study and report that includes–

(1) a comprehensive inventory of all Government owned uranium or uranium equivalents, including natural uranium, depleted tailings, low-enriched uranium, and highly enriched uranium available for conversion to commercial use;

(2) a plan for the conversion of inventories of foreign and domestic highly enriched uranium to low-enriched uranium for commercial use;

(3) an estimation of the potential need of the United States for inventories of highly enriched uranium;

(4) an analysis and summary of technological requirements and costs associated with converting highly enriched uranium to low-enriched uranium, including the construction of facilities if necessary;

(5) an estimation of potential net proceeds from the conversion and sale of highly enriched uranium;

(6) recommendations for implementing a plan to convert highly enriched uranium to low-enriched uranium; and

(7) recommendations for the future use and disposition of such inventories.

42 USC 2296b-6. **Sec. 1017. Regulatory Treatment of Uranium Purchases**

(a) ENCOURAGEMENT–The Secretary shall encourage States and utility regulatory authorities to take into consideration the achievement of the objectives and purposes of this subtitle, including the national need to avoid dependence on imports, when considering whether to allow the owner or operator of any electric power plant to recover in its rates and charges to customers any cost of purchase of domestic uranium, enriched uranium, or enrichment services from a non-affiliated seller greater than the cost of non-domestic uranium, enriched uranium or enrichment services.

(b) REPORT.–Within 1 year after the date of the enactment of this Act, and annually thereafter, the Secretary shall report to the Congress on the progress of the Secretary in encouraging actions by State regulatory authorities pursuant to subsection (a). Such report shall include detailed information on programs initiated by the Secretary to encourage appropriate State regulatory action and recommendations, if any, on further action that could be taken by the Secretary, other Federal agencies, or the Congress in order to further the purposes of this subtitle.

(c) SAVINGS PROVISION.–This section may not be construed to authorize the Secretary to take any action in violation of the multilateral trade agreements (as defined in section 2(4) of the Uruguay Round Agreements Act) or the North American Trade Agreement.[9]

42 USC 2296b-7. **Sec. 1018. Definitions**

For purposes of this subtitle:

(1) The term "Corporation" means the United States Enrichment Corporation established under section 1301 of the Atomic Energy Act of 1954, as added by this Act or its successor.[10][11]

[9] Amended by P.L. 106–36, Title I, § 1002(g)(2), 113 Stat. 133 (1999).
[10] Amended by P.L. 104–134, Title III, Ch. 1, Subch. A, § 3117(b), 110 Stat. 1321–350 (1996).

(2) The term "country of origin" means–

(A) with respect to uranium, that country where the uranium was mined;

(B) with respect to enriched uranium, that country where the uranium was mined and enriched; or

(C) with respect to enrichment services, that country where the enrichment services were performed.

(3) The term "domestic origin" refers to any uranium that has been mined in the United States including uranium recovered from uranium deposits in the United States by underground mining, open-pit mining, strip mining, in situ recovery, leaching, and ion recovery, or recovered from phosphoric acid manufactured in the United States.

(4) The term "domestic uranium producer" mans a person or entity who produces domestic uranium and who has, to the extent required by State and Federal agencies having jurisdiction, licenses and permits for the operation, decontamination, decommissioning, and reclamation of sites, structures and equipment.

(5) The term "non-affiliated" refers to a seller who does not control, and is not controlled by or under common control with the buyer.

(6) The term "overfeed" means to use uranium in the enrichment process in excess of the amount required at the transactional tails assay.

(7) The term "utility regulatory authority" means any State agency or Federal agency that has ratemaking authority with respect to the sale of electric energy by an electric utility or independent power producer. For purposes of this paragraph, the terms "electric utility", "State agency", "Federal agency", and "ratemaking authority" have the respective meanings given such terms in section 3 of the Public Utility Regulatory Policies Act of 1978.

C. NATIONAL DEFENSE AUTHORIZATION
FISCAL YEAR 2001—(PROVISIONS PERTAINING
TO REMEDIAL ACTION AT MOAB SITE)

Public Law 106–398   **114 Stat. 1654A–484**

October 30, 2000

Title XXXIV–Naval Petroleum Reserves

10 USC 7420
note.

Sec. 3403. Disposal of Oil Shale Reserve Numbered 2
(a) Transfer to Indian Tribe.--Section 3405 of the Strom Thurmond National Defense Authorization Act for Fiscal Year 1999 (10 U.S.C. 7420 note; Public Law 105-261) is amended to read as follows:

Sec. 3405. Disposal of Oil Shale Reserve Numbered 2
(i)Remedial action at Moab site.

(1)(A) The Secretary of Energy shall prepare a plan for remediation, including ground water restoration, of the Moab site in accordance with title I of the Uranium Mill Tailings Radiation Control Act of 1978 (42 U.S.C. 7911 et seq.). The Secretary of Energy shall enter into arrangements with the National Academy of Sciences to obtain the technical advice, assistance, and recommendations of the National Academy of Sciences in objectively evaluating the costs, benefits, and risks associated with various remediation alternatives, including removal or treatment of radioactive or other hazardous materials at the site, ground water restoration, and long-term management of residual contaminants. If the Secretary prepares a remediation plan that is not consistent with the recommendations of the National Academy of Sciences, the Secretary shall submit to Congress a report explaining the reasons for deviation from the National Academy of Sciences' recommendations.

(B) The remediation plan required by subparagraph (A) shall be completed not later than one year after the date of the enactment of the Floyd D. Spence National Defense Authorization Act for Fiscal Year 2001 [enacted Oct. 30, 2000], and the Secretary of Energy shall commence remedial action at the Moab site as soon as practicable after the completion of the plan.

(C) The license for the materials at the Moab site issued by the Nuclear Regulatory Commission shall terminate one year after the date of the enactment of the Floyd D. Spence National Defense Authorization Act for Fiscal Year 2001 [enacted Oct. 30, 2000], unless the Secretary of Energy determines that the license may be terminated earlier. Until the license is terminated, the Trustee, subject to the availability of funds appropriated specifically for a purpose described in clauses (i) through (iii) or made available by the Trustee from the Moab Mill Reclamation Trust, may carry out–

(i) interim measures to reduce or eliminate localized high ammonia concentrations in the Colorado River, identified by the United States Geological Survey in a report dated March 27, 2000;

(ii) activities to dewater the mill tailings at the Moab site; and

(iii) other activities related to the Moab site, subject to the authority of the Nuclear Regulatory Commission and in consultation with the Secretary of Energy.

(D) As part of the remediation plan for the Moab site required by subparagraph (A), the Secretary of Energy shall develop, in consultation with the Trustee, the Nuclear Regulatory Commission, and the State of Utah, an efficient and legal means for transferring all responsibilities and title to the Moab site and all the materials therein from the Trustee to the Department of Energy.

(2) The Secretary of Energy shall limit the amounts expended in carrying out the remedial action under paragraph (1) to–

(A) amounts specifically appropriated for the remedial action in an appropriation Act; and

(B) other amounts made available for the remedial action under this subsection.

(3)(A) The royalty payments received by the Secretary of Energy under subsection (e) shall be available to the Secretary, without further appropriation, to carry out the remedial action under paragraph (1) until such time as the Secretary determines that all costs incurred by the United States to carry out the remedial action (other than costs associated with long-term monitoring) have been paid.

(B) Upon making the determination referred to in subparagraph (A), the Secretary of Energy shall transfer all remaining royalty amounts to the general fund of the Treasury and release to the Tribe the royalty interest retained by the United States under subsection (e).

(4)(A) Funds made available to the Department of Energy for national security activities shall not be used to carry out the remedial action under paragraph (1), except that the Secretary of Energy may use such funds for program direction directly related to the remedial action.

(B) There are authorized to be appropriated to the Secretary of Energy to carry out the remedial action under paragraph (1) such sums as are necessary.

(5) If the Moab site is sold after the date on which the Secretary of Energy completes the remedial action under paragraph (1), the seller shall pay to the Secretary of Energy, for deposit in the general fund of the Treasury, the portion of the sale price that the Secretary determines resulted from the enhancement of the value of the Moab site as a result of the remedial action. The enhanced value of the Moab site shall be equal to the difference between–

(A) the fair market value of the Moab site on the date of enactment of the Floyd D. Spence National Defense Authorization Act for Fiscal Year 2001 [enacted Oct. 30, 2000], based on information available on that date; and

(B) the fair market value of the Moab site, as appraised on completion of the remedial action.

(6)[1] (A) Not later than October 1, 2019, the Secretary of Energy shall complete remediation at the Moab site and removal of the tailings to the Crescent Junction site in Utah.

(B) In the event the Secretary of Energy is unable to complete remediation at the Moab Site by October 1, 2019, the Secretary shall submit to Congress a plan setting forth the projected completion date and the estimated funding to meet the revised date. The Secretary shall submit the plan, if required, to Congress not later than October 2, 2019.

(b) URANIUM MILL TAILINGS–Section 102 of the Uranium Mill Tailings Radiation Control Act of 1978 (42 USC 7912) is amended by adding at the end the following new subsection:

"(f) DESIGNATION OF MOAB SITE AS PROCESSING SITE–

(1) DESIGNATION–Notwithstanding any other provision of law, the Moab uranium milling site (referred to in this subsection as the 'Moab site') located approximately three miles northwest of Moab, Utah, and identified in the Final Environmental Impact Statement issued by the Nuclear Regulatory Commission in March 1996 in conjunction with Source Materials License No. SUA-917, is designated as a processing site.

(2) APPLICABILITY–This title applies to the Moab site in the same manner and to the same extent as to other processing sites designated under subsection (a), except that–

(A) sections 103, 104(b), 107(a), 112(a), and 115(a) of this title shall not apply; and

(B) a reference in this title to the date of the enactment of this Act shall be treated as a reference to the date of the enactment of this subsection.

(3) REMEDIATION–Subject to the availability of appropriations for this purpose, the Secretary shall conduct remediation at the Moab site in a safe and environmentally sound manner that takes into consideration the remedial action plan prepared pursuant to section 3405(i) of the Strom Thurmond National Defense Authorization Act for Fiscal Year 1999 (10 USC 7420 note; Public Law 105-261), including–

(A) ground water restoration; and

(B) the removal, to a site in the State of Utah, for permanent disposition and any necessary stabilization, of residual radioactive material and other contaminated material from the Moab site and the floodplain of the Colorado River.".

(c) CONFORMING AMENDMENT–Section 3406 of the Strom Thurmond National Defense Authorization Act for Fiscal Year 1999 (10 USC 7420 note; Public Law 105-261) is amended by adding at the end the following new subsection:

(f) Oil Shale Reserve Numbered 2.–This section does not apply to the transfer of Oil Shale Reserve Numbered 2 under section 3405.".

---

[1]   6(A) and (B) added by P.L. 110–181, Div. C, Title XXXIV, § 3402, 122 Stat. 590 (2008).

NRC FORM 335 (12-2010) NRCMD 3.7 BIBLIOGRAPHIC DATA SHEET *(See instructions on the reverse)*	U.S. NUCLEAR REGULATORY COMMISSION	1. REPORT NUMBER (Assigned by NRC, Add Vol., Supp., Rev., and Addendum Numbers, If any.) NUREG 0980 Vol. 1, No. 10

2. TITLE AND SUBTITLE Nuclear Regulatory Legislation 112th Congress, 2nd Session	3. DATE REPORT PUBLISHED	
	MONTH	YEAR
	September	2013
	4. FIN OR GRANT NUMBER	

5. AUTHOR(S)	6. TYPE OF REPORT Compilation of Nuc Legislation
	7. PERIOD COVERED (Inclusive Dates) Through the 112th Con. (Dec '12)

8. PERFORMING ORGANIZATION - NAME AND ADDRESS (If NRC, provide Division, Office or Region, U. S. Nuclear Regulatory Commission, and mailing address; if contractor, provide name and mailing address.)
Office of the General Counsel
U.S. Nuclear Regulatory Legislation
Washington D.C. 20555-0001

9. SPONSORING ORGANIZATION - NAME AND ADDRESS (If NRC, type "Same as above", if contractor, provide NRC Division, Office or Region, U. S. Nuclear Regulatory Commission, and mailing address.)
Same as 8 above

10. SUPPLEMENTARY NOTES

11. ABSTRACT (200 words or less)
This document is a compilation of nuclear regulatory legislation and other relevant material through the 112th Congress, 2nd Session. This compilation has been prepared for use as a resource document, which the NRC intends to update at the end of every congress.

The contents of NUREG-0980 include the Commissioner Tenure; Atomic Energy Act of 1954, as Amended; Energy Reorganization Act of 1974, as Amended; Reorganization Plan of 1980 and Other Documents Pertaining to NRC Jurisdiction; Low-Level Radioactive Waste; High-Level Radioactive Waste; Uranium Mill Tailings. NUREG-0980 also includes statutes and treaties on export licensing, nuclear non-proliferation, and environmental protection.

12. KEY WORDS/DESCRIPTORS (List words or phrases that will assist researchers in locating the report.) Nuclear Regulatory Legislation; Atomic Energy Act; Energy Reorganization Act; Nuclear Waste Policy Act; NRC Authorization and Appropriations Acts; Statutes; Treaties; Agreements on Export Licensing and Non-Proliferation	13. AVAILABILITY STATEMENT unlimited
	14. SECURITY CLASSIFICATION
	(This Page) unclassified
	(This Report) unclassified
	15. NUMBER OF PAGES
	16. PRICE

NRC FORM 335 (12-2010)

NUREG-0980
Vol. 1, No. 10

Nuclear Regulatory Legislation
112th Congress; 2nd Session

September
2013

www.ingramcontent.com/pod-product-compliance
Lightning Source LLC
Chambersburg PA
CBHW080227180526
45167CB00006B/2234